Business
Statistics

Business
Statistics

Richard P. Runyon

Audrey Haber

1982

RICHARD D. IRWIN, INC. Homewood, Illinois 60430
Irwin-Dorsey Limited Georgetown, Ontario L7G 4B3

ISBN 0-256-02655-6

Library of Congress Catalog Card No. 81–81549

Printed in the United States of America

1 2 3 4 5 6 7 8 9 0 H 9 8 7 6 5 4 3 2

Dedicated to
Lois Runyon
and
Jerome Jassenoff

Preface

It is our belief that statistics should be one of the most enjoyable and mind-expanding courses in the curriculum. After all, statistics and probability relate to and impact on virtually every aspect of our daily lives, professional and personal. We are continuously assessing information (perceptual as well as numerical), summarizing it, drawing conclusions, and acting upon these conclusions in the countless decisions we make each and every day. Statistics, the discipline, attempts to bring some order, coherence, and a degree of formality to procedures we routinely use to lessen the degree of uncertainty in our knowledge base, the validity of our conclusions, and the cogency of our daily decisions and actions.

In other words, statistics is relevant in our lives. But more than that, the subject is itself inherently interesting. It deserves being presented in a lively and vibrant manner without sacrificing the mathematical and logical integrity of the subject. This we have attempted to do. How well we have succeeded is for others to judge.

Following are some of the features of this book that make it different from most other statistical texts:

1. The writing style is informal. We attempt to "talk" directly with the student. When our teaching experience shows us that students typically encounter difficulties, we attempt to act as the students' surrogate. We raise the questions that frequently crop up in class and we attempt to answer them.

2. Many chapters are introduced by a scenario, drawn from real life, that anticipates the main instructional points in the chapter. Thus, in Chapter 1, when we introduce the basic concepts of descriptive and inferential statistics, we focus on a serious contemporary problem, namely, computer fraud. Noting that computer criminals are often brilliant practitioners of their trade, we observe that the usual techniques for uncovering white-collar crimes (i.e., exhaustive auditing of the books) are frequently unable to detect the numerical skullduggery that takes place in the complex world of the floppy discs and random access. We introduce a character, Montana Rincon, who attempts to use sampling techniques as a means of uncovering the presence of computer

fraud. We return to Montana Rincon's statistical sleuthing throughout the book.

3. We have included many within-chapter and end-of-chapter exercises. We have given many of these exercises a context so that the student can appreciate the fact that statistical analyses do not take place in the abstract. Rather they are methods of dealing with real problems in the real world. Consequently, many of the data sets are obtained from the literature of business, economics, finance, and management.

4. Each key term is defined in the margin of the page in which the given concept is introduced. This technique assures that the definitions are given a degree of prominence in accordance with their importance. Moreover, reading the marginal definitions provides a useful chapter summary.

5. We regard the various tables within the text as an integral part of the instructional program. Thus, the captions are usually quite detailed, reinforcing concepts and principles introduced in the textual materials themselves.

6. The statistical tables appearing in the Appendixes were also given much thought and care. Perhaps nothing is more frustrating to a student than to analyze the problem correctly but to falter when using the decision table. This is often not the student's fault since many tables fail to indicate whether or not the values shown are one- or two-tailed, whether the critical region is equal or greater to the tabled values, or equal to or less than this value, etc. To obviate this difficulty, most tables are preceded by a brief explanation of their use, along with an illustrative example.

All in all, we have attempted an intuitive and conceptual approach to statistical analysis. A student with a satisfactory background in high school algebra should have no difficulty in mastering the statistical procedures presented in this text. An accompanying Study Guide provides additional guided practice in each of the statistical procedures appearing in the text. The study guide also augments the text by introducing additional explanatory materials and topics or, in some cases, by presenting statistical procedures that are useful in specialized applications. In addition, the answers to the exercises provided in the text are given in greater detail in the study guide. Thus the student is able to see more than the bottom line, since various intermediate steps are included.

In an undertaking of this sort, there are many others besides the authors who have actively participated. We wish to express the deepest appreciation to the following colleagues who provided invaluable comments and criticisms during the evolution of the text: (Needless to say, we take full responsibility for the final form of the manuscript.)

Paul D. Berger, Boston University; Richard Green, University of California; Ronald Koot, Pennsylvania State University; R. Burt Madden, Shippensburg State College; Donald S. Miller, Emporia State University; Vasanth Solomon, Drake University; Paul Van Ness, Rochester Institute of Technology; Charles Warnock, Colorado State University; Richard J. Westfall, Cabrillo Community College; Howard J. Williams, Wilkes College; and Jack Yurkiewicz, Hofstra University.

A special note of thanks to Teri French who labored cheerfully over our chicken scratches and managed to arrange them into a semblance of coherent language. We love you, Teri.

Thank you also, Laurie Jassenoff, for your invaluable help with some of the more difficult calculations.

Then there are our spouses—Lois Runyon and Jerome Jassenoff—and our children. They endured (or perhaps welcomed?) many long separations, both physical and psychological, as we concentrated on bringing this project to a conclusion. We are profoundly grateful for their patience and understanding during this time.

Richard P. Runyon
Audrey Haber

Contents

What Is Statistics?

1

1.1 STATISTICS AND UNCERTAINTY

Montana Rincon usually awakened in the morning with a surge of good spirits. He liked his work as a CPA, found it challenging, and enjoyed interacting with his many clients. On this particular morning, however, he recoiled at the strident summons of the alarm. He did not wish to relinquish sleep. It had been that bad a night. Unlike his family and friends, his idea of an evening of relaxation involved reading lengthy treatises on the latest developments in his field. But something went wrong last evening. Somehow, he had selected a horror story. Entitled innocently enough, "Computer Crime," the paper delved into the many ways that brilliant minds had corrupted that marvelous tool, the computer, and had put it to work for their own criminal purposes. While he found the names used to describe the crimes both fanciful and humorous, the crimes themselves made his flesh crawl. Data Diddling, the Trojan Horse, the Salami Technique, Logic Bombs, Superzapping, Scavenging—these were but a few of the many types of computer crimes. What they all had in common was a devious scheme to deprive the victims of something that was rightfully theirs.

The diabolic part was that the victims were totally unaware of the crimes against them and the crimes were all executed in but a fraction of a second. A person or a corporation could be financially wiped out in the blink of an eye. What made the whole thing absolutely terrifying was the fact that computer crimes were rarely discovered as a result of internal controls, dogged investigation, or audits of the books. Their discovery was usually an accident, plain and simple. He shuddered to think of the computer crimes that were being executed at that very moment in which the chances of detection were remote. Could it be happening to any of his clients?

He was now involved in auditing the books of a large credit card company. What if someone had used the Salami Technique to take thin slices off selected accounts? It would be simple to do, but almost impossible to catch. All told, the company handled about 5 million accounts. What if someone had built a subroutine into the computer program that randomly selected 1 percent of these accounts and added a small amount (averaging, perhaps, 10 cents) to each bill, which was then deposited into a favored account (his own)? That's 50,000 people swindled out of 10 cents each, on the average. The books would balance and the culprit would be $5,000 richer each month. Would he get caught? How many people would notice an overpayment of 10 or 15 cents? If they did, how many would complain? The postage alone would cost as much as or more than the overcharge. It simply wouldn't be worth the time and effort.

The thought was frightening. He could easily certify the books as correct while a thief drained company profits. Admittedly, not much of a drain in this case. But there is a potential for ripoffs that would make the Brink's robbery look like a fraternity prank. Again, a cold chill passed through his body. How could he avoid being an unwilling accessory in a computer crime? There was no way he could examine the computer program and all of its many ramifications. He simply wasn't sufficiently knowledgeable. There was also no way he could examine all 5 million accounts.

What about a statistical approach? The more he thought about the idea, the better he liked it. He could select, say, 1,000 accounts for a detailed analysis. He then would locate those in which there were errors. He could quickly rule out the most common errors, as for example, a key punch operator entering the wrong number. For those remaining, he would raise a number of penetrating questions. Of the thousand selected, what percentage of these were in favor of the company; what percentage in favor of the customer? What was the average amount of error?

Could he consider the thousand accounts he selected as representative of the entire 5 million accounts? If so, could he estimate the percentage of explained errors in *all* 5 million accounts? Could he estimate the total dollar volume involved in these errors? Finally, could he venture an informed guess as to whether these errors represented haphazard and honest mistakes, or did they represent a systematic effort to commit fraud?

Let's pause for a moment and reflect on what Montana Rincon has been doing. The scenario started with uncertainty. Reading the report on computer crime had made him aware of his own professional vulnerability. Following a restless night, he decided to pursue a course that would resolve some of the uncertainty. After considering various alternatives, he decided to *collect, organize, analyze, interpret, and present information.* This is, in fact, a good working definition of what *statistics* is all about. Note that he would not have raised these questions if he had all the information he required. Statistics is called into play when there are doubts and when we are exploring the unknown.

Once he decided to collect *data* as a means of answering his questions, he realized that there were some serious logistical problems. He was interested in drawing conclusions that would apply to all 5 million accounts. These accounts represented the *population* of interest. Like so many populations that interest statisticians, it was inconceivable that Montana could ever study every member of the population. There simply weren't sufficient time, personnel, and resources to do so. This is, in fact, a common problem. Except in those rare instances when a *census* is undertaken, populations are rarely knowable. Some are infinite and others are so large that they might as well be infinite. When we wish to predict the outcome of a national election, the population is all registered voters who will cast ballots in the election. It is inconceivable that we could poll all of them prior to an election. Likewise, if we are considering marketing a new product with a mass appeal, we would not attempt to survey all possible users.

What would we do? In a word, we would collect a *sample.* However, a sample is useless unless it mirrors the population to which we wish to generalize. In a sense, we want our sample to be a miniature model of the population. One way we may approach this model is to select a *random sample.* This is what Montana Rincon did. He activated a computer program that selected 1,000 accounts in such a way that any of the 5 million accounts in the population could have been selected in the sample. Incidentally, judging the population from a sample is not the exclusive domain of statisticians. We all do it at one time or another; in fact, more frequently than we may realize.

statistics

The collection, organization, analysis, interpretation, and presentation of numerical information.

data

Numbers obtained by measuring, observing, or counting real objects or events in the physical world.

population

The complete set of measurements or observations that interest the person collecting a sample.

census

The process by which we collect data on every element of the population.

sample

A subset of a population.

random sample

A subset of a population selected in such a way that each member has an equal chance of being chosen; each sample of a given size has an equal chance of being selected.

Many of us form opinions about a minority group by talking to only a few members. We judge an actor after watching a couple of his movies. We may estimate weather conditions in a particular locale by visiting there once or twice. Similarly, a few sips are usually sufficient to tell you whether the taste of a wine is satisfactory. It is not necessary to chug-a-lug the whole decanter. We return to the subject of random samples in Chapter 6.

Having decided on the method of selecting his sample, Montana next had to decide what *variables* to study. Since he was interested in assessing error, he collected the following information on each account: the debits (including interest) as of the billing date (variable 1) and the actual amount billed to each customer (variable 2). It would then be a simple matter to subtract the value of variable 2 from the value of variable 1. A difference of zero would mean no error; a positive difference would be an error in favor of the customer; and a negative difference, an error favoring the credit card company.

After Montana had collected the samples, he organized the numerical information and conducted some calculations that yielded summary statements, such as the proportion of accounts in the sample that contained errors and the average size of the error. Each of these summary numerical values represents a *statistic*. A statistic is calculated from a sample and yields summary descriptive statements about some aspect of that sample. Some commonly used statistics describe the central values in a set of measurements (e.g., the arithmetic mean and median) or the spread of these measurements (e.g., the range and the standard deviation). Descriptive statistics are traditionally symbolized by italic letters (e.g., $\overline{X}$ for mean). We explain and demonstrate these various statistics in Chapter 3.

For each descriptive statistic there is a corresponding *parameter*. Thus, there is a population mean, median, range, and standard deviation. Traditionally, parameters are symbolized by Greek letters (e.g., μ = the mean of the population).

Table 1.1 illustrates several populations, variables measured, sample statistics of interest, and corresponding population parameters.

variable

A characteristic of an individual or object that is measurable and takes on different values.

statistic

A summary numerical value calculated from a sample.

parameter

A summary numerical value calculated from a population.

Example A

An electronics firm mass-produces chips for use by the computer industry. Occasionally, the daily output of the plant (the *population*) contains so many defective components that shipping them out to customers invites the risk of losing all future sales. The problems the firm faces are as follows: The daily production of the plant involves hundreds of thousands of chips. Testing each chip is a time-consuming process. To hire and train sufficient personnel to run tests on every chip would be so expensive that the company could not remain competitive. Nevertheless, it is clear that some procedures must be undertaken to estimate the percentage of defective components in the daily output (the *parameter*). If this estimate

Illustration of several populations, variables measured, sample statistics of interest, and corresponding population parameters.   Table 1.1

Purpose	Population of interest	Variable measured	Sample selected from	Sample statistics	Parameter
To predict outcome of a national election	Registered voters	Choice of candidate	Voter registration lists	Proportion favoring each candidate	Actual proportion in the population favoring each candidate
To estimate the resistance to stress of a metallic casting	All castings made of this alloy	The breaking point (time to break under continued stress or amount of total stress applied before breaking)	Actual castings of this alloy	Mean time or mean amount of stress	The actual mean time or stress in the population
To estimate retail cost of unleaded gasoline in a state	Cost of unleaded gasoline at retail outlets in the state	Price per gallon	List of retail gasoline outlets in state	The mean or average cost per gallon	The actual mean cost per gallon in the state
To decide on the packaging of a new product	Potential purchasers of the product	Ratings of each of several purchase options	List of potential users	Average rating of each packaging option	Actual consumer ratings of each option throughout the population
To ascertain possible employee reaction to a wage freeze	Employees in a given manufacturing plant	Favor or oppose the anticipated freeze	List of all employees in the plant	Proportion favoring the freeze	Actual proportion favoring the freeze among all the employees

exceeds a certain critical value (e.g., 2 percent defective), the entire day's output must be scrapped. How would you go about estimating the percentage of defective components?

Since there is no economical way to study the characteristics of each and every chip, we randomly select a small portion of the total output (the *sample*). We then subject each chip in this sample to a standard performance test. For our purposes we'll assume that performance on this test reveals the presence or absence of a defect (the *variable*). The *data* consist of the number of defective and nondefective chips in the sample. The data are then analyzed according to certain rules to yield *descriptive statistics*. In this example, the appropriate statistic is the percentage of defectives. We then use *inferential statistics* to make estimates of the parameter, i.e., the true percentage of defectives in the population.

6

Example B

A flour mill receives numerous orders for large sacks of bleached flour from its customers. Because of the high costs of labor each sack is not individually weighed; rather, they are filled by automated equipment set to deliver approximately 250 pounds. For billing purposes, it is decided to select 50 sacks at random from the warehouse, weigh them, and then use the mean as the best estimate of the average weight per sack.

In this example, the *variable* of interest is the weights of flour sacks. The *sample* consists of the 50 randomly selected sacks. The *data* are the weights of the sacks. The *population* is the weights of all sacks actually packaged by the mill. The *parameter* is the actual mean weight of the population.

1.2 THE TWO FACES OF STATISTICS: DESCRIPTIVE AND INFERENTIAL

descriptive statistics

Procedures used to summarize information about samples in a convenient and understandable form.

When Montana manipulated the data according to certain rules to yield summary statements, he was engaging in an important function of statistical analysis: the *descriptive* function. Some of the activities involve computation, as when he calculates a statistic, such as an arithmetic mean or proportion. Others may involve the use of graphic techniques, which are visual aids to thinking about data.

It is only rare that the investigator's interest focuses on the descriptive function as such. To illustrate, Montana was interested in drawing conclusions about the population rather than simply describing the results obtained from his sample. In other words, he was interested in the inferences he could make about the proportion of errors in the population based on the proportion of errors he found in the sample. He was also interested in formulating

Figure 1.1   A flow diagram of various stages in a statistical investigation

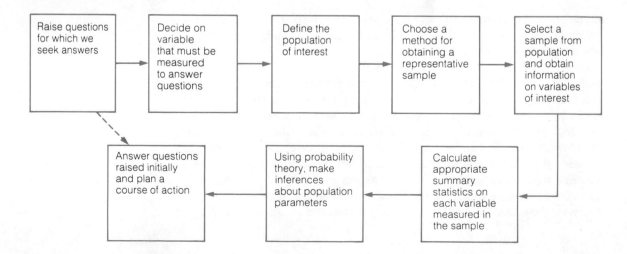

decisions concerning future possible courses of action. These activities make up the *inferential* function of statistics. Probability theory occupies a central position in inferential statistics. To illustrate, imagine that Montana found that the proportion of errors in his sample of 1,000 accounts was 2 percent. This is a descriptive statistic. He would then turn to inferential statistics to answer such questions as: What is the probability that the true proportion of errors in the population is as high as 3 percent? Or as low as 1 percent? Is the incidence of errors sufficiently large to justify a more detailed investigation of the possibility of computer fraud?

Figure 1.1 shows a flow diagram of the various stages in a study aimed at providing statistical answers to questions.

inferential statistics

Making or drawing conclusions about populations from samples and deciding upon courses of action.

EXERCISES

Exercises 1.1–1.6 are based on the following scenario:

When business executives set production goals for specific time periods, they are making explicit forecasts about the demand for their products, the general state of the economy, and the impact of competing products, among other things. Business surveys can often provide a valuable input to this process. Imagine that a regional home building materials supplier is attempting to set production goals for the following six-month period. From a mailing list of contractors and builders, it randomly selects a sample to whom a survey will be sent. Among other things, all recipients are asked to indicate the dollar volume of construction they anticipate completing during the six-month period.

1.1 The builders and contractors receiving the survey constituted:
   a. The sample.
   b. The parameter.
   c. The statistic.
   d. The population.

1.2 The actual average dollar volume of all contractors and builders during the six-month period constitutes:
   a. A statistic.
   b. A population.
   c. A sample.
   d. A parameter.

1.3 The average dollar volume of the sample of builders and contractors constitutes:
   a. A statistic.
   b. A population.
   c. A sample.
   d. A parameter.

1.4 Summary statements based on the sample constitute:
   a. Descriptive statistics.
   b. Parameters.
   c. Inferential statistics.
   d. Populations.

1.5 Estimates of the population parameters constitute:
  a. Descriptive statistics.
  b. Inferential statistics.
  c. Parameters.
  d. Populations.

1.6 All builders and contractors in the region constitute:
  a. A statistic.
  b. A population.
  c. A sample.
  d. A parameter.

1.3 STATISTICS AND BUSINESS DECISIONS

Apart from the inevitability of death and taxes, all aspects of life are fraught with varying degrees of uncertainty. No responsible corporate executive would consider making important business decisions and recommending policy changes without first consulting and digesting a wealth of statistical facts. Sometimes these facts can inform; at other times, they may mislead or deceive. Business people who are illiterate in statistics face a two-edged sword—they are deprived of a powerful tool that may provide a competitive edge and they are unable to distinguish intelligently between useful and useless statistical information.

Let us look at a few examples and see how intelligent business decision makers must use statistics to plan, control, and reduce the uncertainty involved in any business decision.

> An investment counselor calculates the rate of return on a specific kind of investment for the most recent year. He compares this with rates obtained on the same investment in other years. He also compares this rate with the current rates on other investments. Can you see how he might use these statistics to make intelligent investment decisions?

> The personnel manager uses data on the proportion of members of minorities in various job categories to determine compliance with affirmative action committee decisions.

> The head of marketing reviews statistics on income distribution in specific locales to set prices for a new product. Based on obtained data, decisions may be made to set prices differently for different areas.

> The head of personnel reviews production requirements for each month to determine such things as vacation schedules, overtime, and needs for part-time employees.

"Where am I drawing my customers from?" is a question raised by retailers that often goes unanswered. Urban Science Applications and R. L. Polk have developed a license plate survey as a means of providing an answer. Enumerators from Urban Science select a sample of license plate numbers in the parking lot of a shopping center or retail store. Polk, which has collected auto registration data since 1924, feeds the information into a computer. In turn, the computer prints out a map that plots customers' homes by census tracts or zip codes without revealing names or addresses.

Used in conjunction with demographic data, the maps can help retailers select new locations. To illustrate, General Motors is using the technique to assist dealers in selecting sites for billboard space. It has even authorized license plate surveys of the parking lots of competitive dealers in order to determine characteristics of their customers.

EXERCISES

1.7 Many magazines or periodicals collect and publish data on their readers. Information on such things as age, occupation, and income are then considered by potential advertisers. Discuss how this kind of information might be useful in making advertising decisions by certain types of businesses.

1.8 An executive of a soft-drink bottling company reviews data on different age categories so that she may predict the numbers of persons in the 13-to-24 age group one year, three years, and five years from now. Why would these data be of interest to this particular executive?

1.4 QUALITATIVE AND QUANTITATIVE VARIABLES

Variables may be classified into two broad classes—qualitative and quantitative variables. *Qualitative variables* are classified into categories according to the characteristics by which they differ rather than in terms of the degree to which they share a particular characteristic. In other words, they differ in kind, rather than by "how much."

Qualitative variables that are either-or with respect to the characteristic of interest are called *dichotomous variables*. For example, a person is either employed or unemployed; a flash cube is either defective or satisfactory; a candidate for office is either elected or defeated. Qualitative variables that involve more than two categories are referred to as *multinomial*. Examples of multinomial variables include types of business (proprietorship, partnership, or corporation), types of occupation (professional, clerical, supervisory, etc.), and types of computer language (COBOL, FORTRAN, BASIC, etc.).

The data obtained with qualitative variables are limited to counting. You have unquestionably read the results of Gallup polls in your daily newspaper. Registered voters are asked, "If the national elections were held today, would you vote for candidate A, B, or C?" The number selecting each candidate is counted. The statistic most commonly used to summarize qualitative data

qualitative variables

Variables that differ in kind rather than degree.

dichotomous variables

Qualitative variables in which there are only two categories (either-or).

multinomial variables

Qualitative variables in which there are more than two categories.

is either a proportion or its variant, a percentage. Thus, the results of a poll may be summarized as follows: Twenty-eight percent of the registered voters indicated they would vote for candidate A, 19 percent favored B, and 53 percent expressed a preference for C.

In contrast, *quantitative variables* differ in terms of *how much* individuals, objects, or events possess a given characteristic. Thus, the numbers we use to deal with quantitative variables can be added, subtracted, multiplied, and divided. Quantitative variables may themselves be described as either discrete or continuous.

1.4.1 Quantitative variables: Discrete and continuous

discrete variables

Variables that can take on values only at specific points along a scale of values.

Some variables can take on values only at specific points along a scale of values. Such variables are referred to as *discrete* or *discontinuous*. They are characterized by gaps in which no real values are found. A few examples are: the number of children in a family; the number of automobiles in stock at a dealership; the number of employees in an industry; and the inventory of spare parts in a repair shop. Typically, the data accumulated with discrete variables result from the process of counting. There may be three children in a family, 205 automobiles in stock, 132,543 employees in an industry, and 6,005 spare parts in inventory. Discrete variables often increase by whole numbers; however, many discrete variables encountered in the business world have fractional values, e.g., closing stock prices (changes of one eighth), the Dow Jones Industrial Average (changes of 0.01), or the prime interest rate (changes of 0.25 percent).

continuous variables

Variables that can take on any fractional value along a scale.

In contrast, values of a *continuous variable* may be found at *any* point along the scale of values. They are obtained by the process of measuring. The units of measurement used with continuous variables depend on the sensitivity of our measuring instrument. Thus, on a crude instrument, we may record time to the nearest hour; in the most sophisticated nuclear timing devices, accuracy may be obtained to the millionth of a second and beyond. It is theoretically possible to obtain even smaller fractional values with a more sensitive measuring instrument. Several additional examples of continuous variables are weight, height, and temperature.

Often, values of continuous variables are expressed as whole numbers, conveying the impression that they are discrete. Thus, we may report our weight as 162 pounds and our height as 68 inches. However, these decisions were ours to make. With sufficiently sensitive measuring instruments, we could just as easily have reported weight to the nearest ounce or tenth of an ounce and our height to a fractional value. We do not have this option when reporting the number of children in a family or the number of automobiles currently in stock at a new car dealership.

Incidentally, the distinction between continuous and discrete variables is not always clear-cut and unambiguous. Take currency as an example. Some will argue that a penny is to money what an atom is to matter—a basic building block. It is the fundamental unit from which larger monetary structures are erected. From this point of view, money is a discrete variable. The money we carry on our persons and with which we pay our bills is discrete. However, in the world of finance in which actual money may never change hands, money is frequently expressed in fractional terms. For example, tax rates are expressed in mills and, in the field of cost accounting, the cost per unit may be stated to the fifth decimal place. In this sense, then, money is continuous. For this reason, we shall regard money as a continuous variable.

Finally, although most discrete variables do not take on fractional values, statistics based on them may be continuous. The average number of cars in inventory at various dealers at any given time can take on any fractional value. It is, therefore, continuous. Indeed, the first authenticated case of computer fraud was based on the fact that computed interest on savings account is a continuous variable, whereas the interest is reported to the penny. Using the Salami Technique, a computer programmer rounded the interest down to the lower cent and assigned the remainder to a privileged account (his own). Since the average amount rounded is half a cent, he was able to pick up about $1,000 a month on 200,000 accounts.

Table 1.2 provides a few examples of qualitative variables and discrete and continuous quantitative variables.

| | Quantitative variables | |
Qualitative variables	Discrete	Continuous
Gender (male-female)	Children in family (0,1,2 . . .)	Height of buildings in a city
Type of manufacturer (textile, automobile, petroleum, etc.)	Number of cars sold	Weights of contents of packaged cereals
Religion	Number of students in your statistics class	Age of perishable items on a supermarket shelf
Race	Number of items in inventory	Tread wear on test automobiles
Occupation	White blood cell count (number of white cells per cubic millimeter)	EPA miles-per-gallon ratings
Color (e.g., hair color, eye color)	Crop yields (in terms of number of items, e.g., bushels of corn, bales of cotton)	Temperature of wines during aging
Absence vs. nonabsence		Distances between stops on a delivery route
		Crop yields (in terms of number of pounds or kilos)
		Late records (measured in hours, minutes, etc.)

Table 1.2
Some examples of qualitative and discrete and continuous quantitative variables

EXERCISES

1.9 Indicate the qualitative variable in each pair:
  a. Dollar sales; gender.
  b. Employment status; life span of a transistor.
  c. Color of soap packaging; weight of the soap.
  d. Brand of beer; calorie content of a 12-ounce bottle.
  e. Type of milk cow; weight of milk produced daily.

1.10 Indicate the quantitative variable in each pair:
  a. Brand of breakfast cereal; family size.
  b. Type of farming tractors, weights of farming tractors.
  c. Weight of nonferrous ores consumed in a month; types of nonferrous ores consumed in a month.
  d. Weight of grain products imported in a year; types of grain products imported.
  e. Number of building units constructed in a month; types of building units constructed in a month.

1.5   MISUSE OF STATISTICS

Perhaps no field of human endeavor has been so maligned as statistics. We have all heard such expressions as, "A politician uses statistics as a drunk uses a lamp post—more for support than illumination," or "Statistics is the last refuge of the uninformed," or "You can prove anything you want to with statistics," or the famous quote from the 19th-century British prime minister, Benjamin Disraeli, "There are lies, damned lies, and statistics." In truth, statistics don't lie but people do. Some statistical lies are told out of ignorance with no malice aforethought. Others represent bald-faced efforts to obscure the facts, to deceive the public, and to defraud the gullible. Statistical chicanery may take many forms—a misleading chart, statistical facts quoted out of context, and incorrect applications of legitimate statistical procedures, to name a few. We shall present examples of these misuses from time to time. Our goal is not to make you proficient in the various techniques of statistical larceny, but rather to improve your ability to recognize when someone else is indulging in a bit of statistical mischief.

SUMMARY

1.   In this chapter, we saw that there is a wedding between statistics and uncertainty. Since much of daily life involves uncertainty, statistics has many potential and actual uses in human endeavors. Intelligent business decision makers use statistics to plan, control, and reduce the uncertainty-involved in business decisions.

2. Statistical analysis often involves collecting data from samples that are representative of some population. Summary descriptive statistics are calculated and charts are constructed to provide summary statements about the samples. Probability theory is then invoked as a tool for estimating parameters and drawing inferences about populations from sample data.
3. The descriptive statistics are rarely the target of interest in and of themselves. More often, we are interested in the inferential function.
4. The variables we study may be classified as qualitative and quantitative variables. Qualitative variables differ in kind and quantitative variables differ in degree or "how much."
5. Quantitative variables may be either discrete or continuous.

TERMS TO REMEMBER

statistics (3)	descriptive statistics (6)
data (3)	inferential statistics (7)
population (3)	qualitative variables (9)
census (3)	dichotomous variables (9)
sample (3)	multinomial variables (9)
random sample (3)	quantitative variables (10)
variable (4)	discrete variables (10)
statistic (4)	continuous variables (10)
parameter (4)	

EXERCISES

1.11 The public service commission of a state with nuclear electric power has received heated requests that it place a moratorium on planning and constructing nuclear power plants. The chairperson of the commission requests that the Concerned Citizens against Nuclear Power undertake a questionnaire to ascertain voter reactions to this proposal. What are some of the risks in this suggestion?

1.12 For the following example, define the population of interest and specify whether a census or a sample would be more appropriate. Explain your answer.

In order to estimate the durability of a new line of luggage, the company conducts a series of tests including: dropping the luggage from varying heights onto a concrete pad and hitting it with various size sledge hammers on the top and bottom surfaces and on each of the four sides.

1.13 What type of data—discrete or continuous—would the following variables produce?
   a. The number of people who purchase new automobiles on any given day.
   b. The period of time between the installation of various pieces of equipment and the time when they must be replaced due to obsolescence.
   c. The numbers of people expressing preferences for different color cars.
   d. The amount of pressure (in terms of pounds per square inch) required to shatter a plastic molding.

1.14 On March 10, 1980, the following announcement appeared in the *Los Angeles Times:*

ADIDAS WILL OPEN ITS FIRST U.S. PLANT IN PA.

Adidas, which currently produces 280,000 pairs of shoes daily at plants around the world, said it would make 2,000 to 3,000 pairs a day in Pennsylvania by the end of the year. Eventually, production capacity is expected to reach 20,000 pairs daily.

How do you suppose Adidas executives made decisions regarding the size of the new plant? What are some of the factors involved in their forecast of product requirements and planning production capacity?

1.15 The Los Angeles *Times* reported a survey that showed that the biggest companies do not necessarily pay sales and marketing executives the highest salaries. Suppose you were looking for a position as a sales or marketing executive. How might you use the information reported in this survey?

1.16 On August 11, 1980, *The Wall Street Journal* reported:

The housing industry recovered in June, but industry economists believe the recovery might soon stall. New-home sales climbed 16 percent to an adjusted annual rate of 535,000 units from May's revised 461,000.

What are some of the variables that could affect the accuracy of the prediction? For example, how might mortgage rates affect the housing demand? Can we expect the results of the upcoming national election to have any effect on this prediction?

1.17 As the deadline approached for the negotiation of a new contract with Local 542, a poll was taken of 245 of its members. One hundred thirty-five favored a strike whereas 110 favored no labor stoppage during negotiations.
Identify each of the following as a sample, population, or data.
    a. All voting members of Local 542.
    b. The tabulation of the strike vote.
    c. The members who were polled.

1.18 A manufacturer is considering the marketing of a new product line. However, he is unsure of consumer acceptance. Outline the procedures he might undertake to estimate consumer acceptance.

1.19 There are a number of occasions when a continuous quantitative variable is divided into arbitrary categories and treated as a qualitative variable. For example, many pharmaceutical houses, when they test prospective drugs on laboratory animals, use two dosage levels of the test drug, calling one a low level and the other a high level of drug administration. A case in point is the evaluation of the dye FD&C Red No. 2 by the Food and Drug Administration. Eighty-eight animals received daily injections of the dye. Among 44 receiving low daily dosages, 4 contracted cancer. Among the remaining 44, receiving high dosages, 14 fell prey to cancer.
Identify the following:
    a. The population.
    b. The sample.
    c. Variable.
    d. Data.

1.20 The board of education in a large midwestern city voted to close one of its high schools because of financial difficulties. Afterwards, a group of parents and teenagers barricaded themselves in the high school and threatened to stay until the board reversed itself. The chairman of the board expressed dismay, saying "The letters we received from the community were running about two to one in favor of closing the school."

  a. What was the population sampled by the letters?

  b. Comment on the adequacy of the sampling method. Are there possible biases in the sample?

1.21 For each of the following, specify the population of interest and comment on any possible sources of error in the selection of the sample or collection of the data.

  a. A bank is interested in determining the average monthly checking account balances of its customers. For convenience, it selects its sample from among those accounts that show little activity over the course of the month.

  b. A committee consisting of three people, all members of a minority group, conduct a survey of the effects of a new affirmative action program within a specific industry.

  c. A union is involved with direct negotiations with management. Management presents its offer to union officials. In order to assess union members' reaction to the proposed contract, members are asked, "Will you vote to accept the wage increases management is offering, even though they are less than those we proposed?"

  d. A questionnaire, prepared to study attitudes on safety measures in a particular firm, is mailed to all employees of that firm. The questionnaire is received the day after a major accident at one of the plants.

Organizing Statistical Data

2.1 FREQUENCY TABLES, RELATIVE FREQUENCY TABLES, AND BAR CHARTS: QUALITATIVE VARIABLES

Montana Rincon had finally finished his careful audit of 1,000 different credit card accounts. He sorted his records into two broad classes—those that he found free of errors and those that contained errors. This is a qualitative distinction similar to the one we make between male and female people, right and wrong on a test item, employed and unemployed workers, or defective and satisfactory products. What he discovered, by actual count, was that 60 records contained errors and 940 were error free. Further examination of the 60 erroneous records revealed that none was the result of a clerical error. Thus, he was left with 60 accounts containing unexplained errors.

2.1.1 Frequency table

The following table summarizes these findings:

Table 2.1

Frequency table of unexplained errors found in a sample of 1,000 accounts of a credit card company. The column headed by f shows the frequency of occurrence in each category. The sum of the frequencies (Σf)* equals n, the sample size.

	Frequency (f)
Errors	60
No errors	940
n	1,000

* See the Study Guide for information about summation (Σ) notation.

2.1.2 Relative frequency table

Often, expressing frequency counts in the abstract is neither very helpful or informative. What do we know when we learn that 16 women serve as chief executive officers in a given industry, or that 23 black football players endorse products in media advertising, or that 60 errors were found in the sample of credit card accounts? More often we are interested in the rate or relative frequency with which these events occur. What proportion of women serve as chief executive officers in a given industry? What proportion of black football players obtain rewarding advertising contracts? What proportion of credit card accounts contain error? Since we are dealing with qualitative data, a more useful measure to describe these findings would be the *proportion* (or *percentage*) of errors that occurred in this sample. Thus, Montana converted his frequencies and prepared the following summary table:

proportion

With qualitative variables, relative frequency of occurrence; found by dividing the frequency in a given category by n.

percentage

Proportion multiplied by 100.

Table 2.2

Relative frequency table of unexplained errors found in a sample of 1,000 accounts of a credit card company. The proportion in each category is found by dividing the frequency in that category by n. Multiplying by 100 converts a proportion to a percentage.

	f	Proportion	Percentage
Errors	60	0.06	6
No errors	940	0.94	94

2.1.3 Bar chart

We can visually display Montana's findings in a bar chart, such as shown in Figure 2.1.

As we look at the display of the data, we might be tempted to disregard this apparently low incidence of errors. "sixty errors, 6 percent, bah! Humbug! What's the big deal?" But, suppose that this sample truly mirrors the population from which it was drawn. In other words, imagine there is a corresponding 6 percent of errors in the population of all measurements. With a total of 5 million accounts, somewhere around 300,000 could well contain errors! No more humbug, huh?

2.1.4 Putting it together: Examples

Example A

Based on preliminary tax returns, the Internal Revenue Service published the following breakdown of type of organization of industries in

Figure 2.1

Bar chart showing frequency of errors and no errors in a sample of 1,000 accounts. Note that the *height* of each bar corresponds to the frequency in that category.

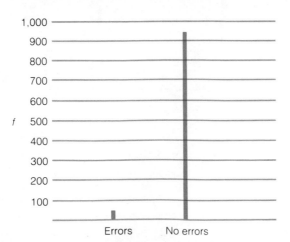

the United States: proprietorships, 11,358,000; active partnerships, 1,096,000; active corporations, 2,105,000.

Prepare *(a)* a frequency table of these data, *(b)* a relative frequency table, and *(c)* a bar chart.

Solution A

a.

Type of organization	f (000) THOUSANDS
Proprietorships	11,358
Active partnerships	1,096
Active corporations	2,105
n	14,559

b.

Type of organization	f (000)	Proportion	Percent
Proprietorships	11,358	11,358/14,559 = 0.78	78
Active partnerships	1,096	1,096/14,559 = 0.08	8
Active corporations	2,105	2,105/14,559 = 0.14	14
	14,559	1.00	100

c.

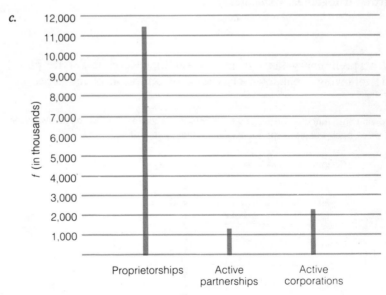

Source: Adapted from J. B. Poe, *An Introduction to the American Business Enterprise,* 4th ed. (Homewood, Ill.: Richard D. Irwin, 1980), p. 15. © 1980 by Richard D. Irwin, Inc.

Example B

A firm produces 10,000 flash-cubes every day. Since sales are adversely affected by defective cubes, it is necessary to set up quality-control procedures. Therefore, the quality-control department randomly selects 150 from the daily output, and applies the voltage necessary to flash the cube. They find six that are defective, i.e., they do not flash.

a. Why did they use only a sample rather than the entire day's output (the population)?

b. Summarize the results in a relative frequency table and a bar chart.

Solution B

a. Most of the time we study samples because the population itself is too large. In this case, however, the test to determine a defect destroys the product. Therefore, it would be economic suicide to test an entire day's output—the population.

b.

Quality	f	Proportion	Percent
Defective	6	0.04	4
Nondefective	144	0.96	96
	150	1.00	100

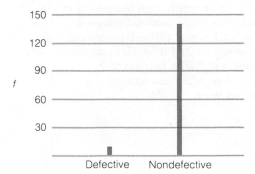

EXERCISES

2.1 The table on page 22 shows the business enterprises in the United States according to the major type of economic activity they perform operating within the three legal forms of business organizations.

Industry	Number of firms (000)		
	Proprietorships	Active partnerships	Active corporations
Agriculture, forestry, and fisheries	3,470	121	62
Mining	60	18	15
Construction	963	60	198
Manufacturing	223	31	214
Transportation, communication, and public utilities	346	17	81
Trade	2,282	195	646
Finance, insurance, and real estate	827	447	414
Services	3,153	207	473
Totals*	11,324	1,096	2,103

* Note: Because of multiple-industry listings, totals are sometimes slightly different from the sum of the individual items; however, for simplicity, we have adjusted the totals to correspond with the sums of the parts.

Source: Adapted from J. B. Poe, *An Introduction to The American Business Enterprise,* 4th ed. (Homewood, Ill.: Richard D. Irwin, 1980), p. 15. © 1980 by Richard D. Irwin, Inc.

    a. Set up three relative frequency distributions corresponding to the types of business organizations.

    b. Construct a bar chart summarizing these distributions, using relative frequency (proportion) on the vertical axis.

2.2 The Jajoba shrub has been widely hailed as a possible source of economic gain in the Southwest and as a potential substitute for sperm whale oil. It grows wild in the Sonoran Desert, and its seed yields an extremely high quantity of oil. In order to obtain some data on the survival rate of seedlings, Sherbrooke studied 219 seedlings over a four-year period. [Cited by K. F. Foder, "1980—Environmental effects of Harvesting the Wild Desert," *Desert Plants* 2, no. 2, (1980)]. He found that 27 survived after one year, 8 after two, 4 after three, and 3 after four years.

    a. Calculate the survival rates per year (use percentages).

    b. Plot a bar chart of the survival rates.

2.2 FREQUENCY DISTRIBUTIONS, RELATIVE FREQUENCY DISTRIBUTIONS, HISTOGRAMS, FREQUENCY POLYGONS: QUANTITATIVE VARIABLES

Let us return to Montana Rincon's analysis of credit-card errors. Thus far, he has only just begun to describe the data he has collected. By making the qualitative distinction (errors vs no-errors), he now knows that 60 errors exist in his sample. It is now time to take a closer look at these errors. As a first step, he lists the amount of each error and comes up with the following table:

18	22	13	8	33	11
1	4	19	18	8	7
8	11	29	5	14	17
23	15	17	21	3	10
12	20	17	17	19	35
16	7	11	14	29	22
17	28	20	26	18	9
31	17	6	19	10	25
5	36	22	15	16	12
14	9	32	7	13	13

Table 2.3
Amount of error, in cents, of the 60 accounts containing unexplained errors

Because the data in Table 2.3 are not arranged in any systematic way, they are referred to as *raw data*. Looking at the data in this form, you can see that it would be difficult to find any meaningful patterns.

raw data

Numerical observations prior to any statistical treatment.

2.2.1   An array

One way of organizing this hodgepodge of data is to arrange the values of the variable into an *array*. This involves ordering the data from low to high values (an ascending array) or from high to low (a descending array). Table 2.4 shows Rincon's data arranged into an ascending array.

array

Data that are ordered either from high to low or vice versa.

1	8	12	16	19	25
3	8	13	17	19	26
4	9	13	17	19	28
5	9	13	17	20	29
5	10	14	17	20	29
6	10	14	17	21	31
7	11	14	17	22	32
7	11	15	18	22	33
7	11	15	18	22	35
8	12	16	18	23	36

Table 2.4
An ascending array of amount of error, in cents, of 60 accounts

It is now possible to discern some features of the errors. For example, we can see that the amount of error ranges from a low of one cent to a high of 36 cents. An array is one of the simplest techniques that may be used to sort data. However, an array is not always the most informative way to organize data, particularly with large quantities of data. One of the most convenient ways of summarizing data in a meaningful manner is to group the values into a frequency distribution. This may be done directly from the raw data or from an array.

2.2.2   The frequency distribution

We have all heard the expression that we are often unable to see the forest for the trees. The implication is that we are frequently so overwhelmed

frequency distribution

When data are arranged according to their magnitude, a frequency distribution shows the frequency of occurrence in each class.

by details that larger patterns go unseen. So it is with data. Grouping data into a *frequency distribution,* while removing some of the fine detail, enables us to see features and patterns that may otherwise not be immediately apparent. Constructing a frequency distribution involves compressing the individual values into groupings of values (classes). The way in which we group the values is largely a matter of judgment. There is no single correct way to set up a frequency distribution for a given set of data. We must first decide approximately into how many equal classes we wish to group our data. As a rule of thumb, statisticians generally use anywhere from 5 to 20 classes.[1]

Table 2.5 shows Rincon's data cast in the form of a frequency distribution. For example, nine errors fell in the class from 9 to under 13 cents; whereas only three errors were equal to or greater than 33 cents but less than 37 cents.

Table 2.5

Frequency distribution of amounts of errors in 60 accounts

Class limits	f
1–under 5	3
5–under 9	9
9–under 13	9
13–under 17	10
17–under 21	14
21–under 25	5
25–under 29	3
29–under 33	4
33–under 37	3
Total	60

Note: Since we are dealing with a continuous variable, we use "under" to designate any value less than the lower limit of the next class. This assures us of no overlapping between classes.

width

The width of a class is equal to its upper boundary minus its lower boundary.

In this case, we have arbitrarily decided to divide our data into nine classes. In order to determine the *width* of each class we use the following:

$$\text{Approximate width} = \frac{\text{Highest value} - \text{lowest value}}{\text{Number of classes desired}}$$

$$= \frac{36 - 1}{9}$$

$$= 3.89$$

Since it is more convenient to work with whole numbers, we rounded off and used 4 as the width of each class. In general we try to set up the frequency distribution with equal class widths, although this is not always possible with certain kinds of economic and business data (see Section 3.2.6). Note that the classes have been set up so that there is no question as to

[1] Some statisticians prefer to use Sturgess's rule. According to this rule, $k = 1 + 3.3 \log n$, in which k is the number of classes and n is the sample size. If we used Sturgess's rule in constructing the frequency distribution in Table 2.5 we would find that, with $n = 60$, $\log 60 = 1.778$ (see Table XIII). Thus, $k = 1 + (3.3)(1.778) = 6.9$. This rounds to 7.

where each observation falls. That is, the classes do not overlap in any way—they are *mutually exclusive*. Each class is bounded by an upper and a lower *class limit*. For example, all values equal to or greater than 1 cent and less than 5 cents are included in the first class. An observation or measurement falls in a particular class if it is equal to or greater than the lower limit of that class and less than its upper limit.

In addition to identifying the class limits, it is often useful to find the *midpoint* of each class. With continuous variables, the midpoint is found by adding together two adjacent lower class limits and dividing by two. To illustrate, the midpoint of the class 17–under 21 is $(17 + 21)/2 = 19$.

A frequency distribution shows the number of observations that fall into each of the classes. Notice how features that were dimly discerned in the array become clear in the frequency distribution. The data are not uniformly spread out between the two extreme values. Rather, there are relatively few values at each extreme and a piling up of values toward the center classes. Moreover, there are no values so extreme that they cannot be conveniently accommodated by the nine classes. This is not always the case. Some variables, such as personal income, include some values so extreme (e.g., the income of a sheik collecting oil royalties) that many classes are required to incorporate all the data. Finally, by adding together the frequencies in adjacent classes, we can make such statements as: 29 errors are 17 cents or greater; only 12 are less than 9 cents.

2.2.3 The relative frequency distribution

Just as we saw earlier with qualitative variables, an extremely useful measure to describe a distribution is the proportion (or percentage) of cases that fall within each class. We call this a *relative frequency distribution*. To obtain the relative frequency for a given class, we divide the number of cases in that class by the total number of observations *(n)*.

Table 2.6 presents the same frequency distribution as Table 2.5 as well as the relative frequency for each class.

Note that the sum of the relative frequencies is 1.00 (or 100 percent if expressed as a percentage). As we shall see later, the relative frequency distri-

mutually exclusive

Two events are said to be mutually exclusive if they cannot occur simultaneously.

class limit

Upper or lower boundary of a class.

midpoint

A value halfway between the two class limits.

relative frequency distribution

When data are arranged according to their magnitude, a relative frequency distribution shows the proportion or percentage within each class.

Class limits	f	Proportion	Percent
1–under 5	3	0.05	5
5–under 9	9	0.15	15
9–under 13	9	0.15	15
13–under 17	10	0.17	17
17–under 21	14	0.23	23
21–under 25	5	0.08	8
25–under 29	3	0.05	5
29–under 33	4	0.07	7
33–under 37	3	0.05	5
	60	1.00	100

Table 2.6

Frequency and relative frequency distribution of amounts of errors in 60 accounts

bution is extremely important in inferential statistics. If the sample accurately mirrors the population from which it was drawn, the relative frequency distribution provides the probability or likelihood that a given element selected from the population will be selected from a given class. As we shall see in Chapter 4, one way of expressing probability is in terms of the relative frequency with which an event occurs. Thus, given that an account contains an error, the probability that it will be equal to or greater than 29 cents and less than 33 cents is about seven in 100.

2.2.4 The frequency histogram and relative frequency histogram

histogram

A bar chart used with quantitative variables.

Two types of graphic representation usually used with quantitative variables are the *frequency histogram* and the *relative frequency histogram*. The histogram for Rincon's error data is shown in Figure 2.2. The *X* or horizontal axis represents the classes and the *Y* or vertical axis, the frequency scale. When classes of equal widths are used, each bar spans its class limit and

Figure 2.2

Frequency histogram of data presented in Table 2.5

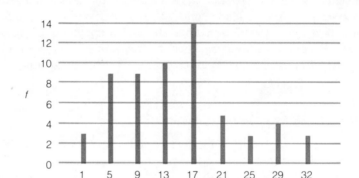

Figure 2.3

Relative frequency histogram of data presented in Table 2.6. Note that the *area* of a given bar equals the relative frequency in the corresponding class. The total area equals 1.00.

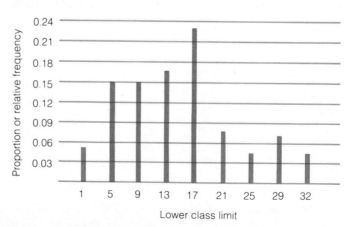

its height corresponds to the frequency of that class. Since the lower class limits are shown, we read the graph in the following manner: 3 cases had errors equal to or greater than 1 and less than 5 cents, or 14 cases had errors equal to or greater than 17 and less than 21 cents.

A relative frequency histogram (Figure 2.3) converts the Y-scale to proportions or percentages. The proportion of cases that fall into a class equals the proportion of area corresponding to that class. Thus, the total area under a relative frequency distribution is 1.00 (or 100 percent if expressed as a percentage). We return to this important point in Chapter 5 when we discuss theoretical probability distributions.

EXERCISE

The techniques we have learned thus far may readily be applied to discrete variables. The primary difference, when dealing with a discrete variable, is that no real values exist between the upper limit of one class and the lower limit of the next class. Therefore, the class limits will reflect a gap.

2.3 The table below shows the total take-offs and landings at the 24 busiest airports in 1977 in the United States and Canada.[2]

741,272	282,222
639,624	258,494
592,863	243,173
573,858	240,063
516,558	235,825
500,976	231,714
487,005	225,295
466,206	213,060
457,469	208,170
450,288	201,758
429,703	201,608
424,030	194,846

 a. Construct a frequency distribution using six classes. (Since we are dealing with a discrete variable, the upper limit of the first class will be 279,999 and the lower limit of the next class will be 280,000. Note the gap that exists between these two values.)

 b. Construct a histogram for the obtained distribution.

---

[2] Source: *The World Almanac & Book of Facts 1979*, p. 145, copyright © Newspaper Enterprise Association, New York, N.Y. 10166.

2.2.5 Putting it together: Examples

Example C: Frequency distribution, histogram, frequency polygon

Suppose you are the personnel director of a large company with thousands of employees. One part of your annual report deals with time lost per thousand work-hours due to absences and latenesses. During the course of the year you have compiled the following data:

Table 2.7
Weekly number of hours lost per thousand work-hours

Week of	Hours lost	Week of	Hours lost	Week of	Hours lost
Jan 2	23.9	May 1	16.8	Sept 4	12.8
9	22.5	8	11.3	11	11.2
16	14.3	15	19.0	18	19.1
23	18.5	22	23.4	25	9.9
30	20.5	29	12.1		
Feb 6	9.7	June 5	19.3	Oct 2	21.0
13	19.0	12	14.0	9	11.0
20	21.6	19	16.9	16	18.7
27	15.6	26	11.3	23	15.2
				30	21.3
Mar 6	17.0	July 3	11.8	Nov 6	13.5
13	20.8	10	16.5	13	17.4
20	14.9	17	20.0	20	10.2
27	15.9	24	13.3	27	19.4
		31	17.8		
Apr 3	22.1	Aug 7	15.0	Dec 4	17.3
10	10.2	14	16.3	11	12.4
17	11.5	21	18.2	18	25.3
24	16.8	28	13.6	25	26.4

How might you report these data? Certainly it would be tedious, to say the least, to report hours lost for each of the 52 weeks. Therefore, you decide to organize these data into a frequency distribution. Let us summarize how to construct a frequency distribution.

1. Decide the number of classes. Most data can be accommodated by 5 to 20 classes. However, there are no hard and fast rules. The choice usually depends on such factors as the amount of data *(n)* and the spread of the values of the variable. Where either or both is large, a larger number of classes is called for. Remember that telescoping data into classes suppresses some of the information. If too few are used, important features of the distribution may be obscured. If too many, the goal of achieving presentational economy may be defeated. In the final analysis, the decision is a compromise based on the willingness to give up some detail in order to achieve presentational economy.

2. Decide the width of the classes. A good way to approximate the size of each class is to find the difference between the largest and smallest measurement and then divide this difference by the number of classes desired. The resulting value can be rounded off to one that is convenient for the particular distribution.

3. Decide the class limits. The classes should be set up so that there is no ambiguity about where a particular measurement falls. In other words, there should be no overlap of classes.

Looking at your data (Table 2.7), you decide to condense into about seven classes.[3] Thus, the approximate width of each class would be $(26.9 - 9.5)/7 = 2.486$. For convenience, you round off and use 2.5 as the width of each class. You construct the following frequency distribution directly from the raw data.

Class limits	f
9.5–under 12.0	10
12.0–under 14.5	8
14.5–under 17.0	10
17.0–under 19.5	12
19.5–under 22.0	6
22.0–under 24.5	4
24.5–under 27.0	2
	52

Table 2.8
Frequency distribution of hours lost per thousand work-hours

In addition to presenting a frequency distribution, you decide to portray your data graphically. Thus, you set up the histogram in Figure 2.4.

An alternative graphic device, when dealing with continuous variables, is the *frequency polygon*. This device, as we shall see in the next example, is particularly useful when comparing two or more distributions.

Each point in a frequency polygon is positioned over the midpoint of the corresponding class. To illustrate, the midpoint of the first class is $(9.5 + 12)/2 = 10.75$ and a dot is placed at this point at a height equal to the class frequency (10). Similarly, the midpoint of the second class is $(12 + 14.5)/2 = 13.25$. Since a polygon is a closed figure, you must add one class at both ends of the distribution. Since the frequency is zero for these classes, the polygon is brought down to the horizontal axis at the midpoint of the two classes that are added. Thus, the midpoint of the class added at the lower end of the distribution is $(7 + 9.5)/2 = 8.25$. At the upper end, it $(27 + 29.5)/2 = 28.25$. (See Figure 2.5).

You are now in a position to summarize your findings quickly and efficiently. A glance at the obtained frequency distribution (Table 2.8)

frequency polygon

A closed geometric figure used to represent a frequency distribution.

---

[3] This is the number of intervals that would be obtained by the use of Sturgess's rule, i.e., $k = 1 + 3.3 \log 52 = 6.66$, rounded to 7.

and the visual displays (Figs. 2.4 and 2.5) show that losses of 22 hours or more (per thousand) are relatively rare. It appears that losses between 14.5 to less than 19.5 have the greatest likelihood of occurrence. In addition, the chances are quite small that more than 24.5 work-hours (per thousand) will be lost in any week during the course of a year.

Figure 2.4
Histogram for distribution of work-hours lost per thousand

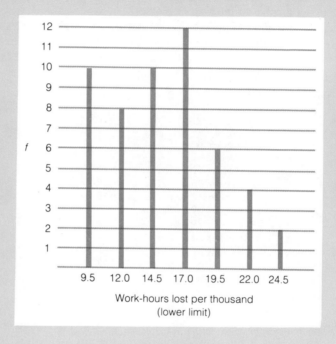

Work-hours lost per thousand
(lower limit)

Figure 2.5
Frequency polygon for distribution of work-hours lost per thousand

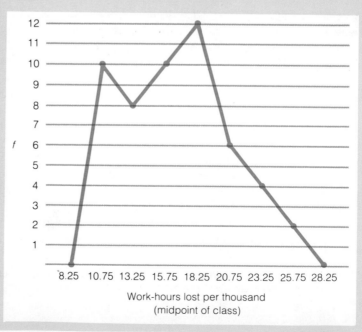

Work-hours lost per thousand
(midpoint of class)

EXERCISE

2.4 Cutting drills with diamond edges are used in many industrial processes. In one plant, records were kept of the number of hours each drill lasted prior to replacement. Following is the frequency distribution of hours prior to replacement.

a. Construct a relative frequency distribution for these data.

b. Construct a frequency polygon from the relative frequency distribution.

Hours prior to replacement	Number of drills
0–under 15	4
15–under 30	8
30–under 45	15
45–under 60	32
60–under 75	36
75–under 90	14
90–under 105	11
105–under 120	6
	126

Example D: Comparing two distributions with comparable ns: the frequency polygon

Let us suppose that there is another division of your company located in another part of the country. You meet with its personnel director and

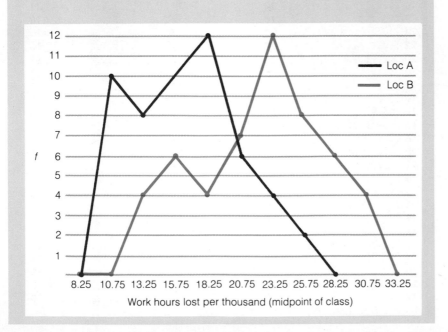

Figure 2.6

Frequency polygons for distributions of work-hours lost (per thousand) at two different locations compare work-hours lost at your location (location A) with work-hours lost at this other location (location B).

You decide to convey this information graphically. The frequency polygon is more helpful than the histogram when we wish to compare one distribution directly with another. Superimposing one histogram over another is often confusing. This is not the case with the frequency polygon. Figure 2.6 shows both polygons in the same chart.

Some interesting differences emerge as we compare these two polygons. It seems apparent that many more work-hours are being lost at location B. For example, whereas weekly losses of 24.5 hours or more are rare at location A, they appear fairly common at location B.

EXERCISES

2.5 Shown below are the EPA miles-per-gallon ratings for 59 automobiles with *manual* transmissions and 65 automobiles with *automatic* transmissions.

Manual transmission		Automatic transmissions	
EPA	MPG	EPA	MPG
27	16	12	11
16	16	12	11
19	16	11	16
19	16	11	16
19	24	12	11
16	24	11	11
19	18	15	12
24	23	15	11
22	17	12	13
20	22	12	11
20	16	11	11
19	18	10	15
19	19	10	15
19	29	13	15
16	22	13	11
16	21	11	12
23	21	10	11
21	26	13	11
16	24	13	10
22	24	12	15
21	18	11	15
19	18	10	15
17	15	13	13
19	15	13	13
21	20	12	11
16	16	11	13
15	15	11	10
17	13	10	10

Manual transmission		Automatic transmissions	
EPA	MPG	EPA	MPG
27	19	12	12
19		12	15
		12	12
		12	12
		15	

   a. Construct separate frequency distributions for manual and automatic transmissions. (To facilitate comparisons, use a width of 1.0 for each distribution.)

   b. Construct a frequency polygon for each distribution, using the same chart.

Write a brief summary of your findings. What are your chances of achieving at
least 25 MPG?

   c. Combine the data for manual and automatic transmissions into one frequency distribution, again using a width of 1.0 for each class.

   d. Construct a frequency polygon for the combined data.

Compare the frequency polygons obtained in *(b)* with the one obtained in *(d)*.
Which provides more meaningful information? Why? What are your chances of achieving at least 25 MPG?

Example E: Comparing two distributions with different *n*s: the relative frequency polygon

We are sometimes called upon to compare two distributions, each of
which is based on a different number of observations. Even though the
class designations are the same, differences in the absolute frequencies
associated with each class make direct comparisons difficult. Under these
circumstances, relative frequency distributions and *relative frequency polygons* simplify these comparisons. In Table 2.9 are shown the frequency
distributions of cotton crop yields in pounds per acre on two different

relative frequency polygon

A closed geometric figure used to represent a relative frequency distribution.

Table 2.9

Frequency distributions of yield per acre on a 500-acre and a 200-acre farm

Class limits	*f*	
	Farm A	Farm B
240–under 480	10	6
480–under 720	50	20
720–under 960	90	34
960–under 1,200	195	82
1,200–under 1,440	95	40
1,440–under 1,680	50	16
1,680–under 1,920	10	2
n	500	200

Table 2.10
Relative frequency distributions of yield per acre on a 500-acre and a 200-acre farm

Class limits	Proportion	
	Farm A	Farm B
240–under 480	0.02	0.03
480–under 720	0.10	0.10
720–under 960	0.18	0.17
960–under 1,200	0.39	0.41
1,200–under 1,440	0.19	0.20
1,440–under 1,680	0.10	0.08
1,680–under 1,920	0.02	0.01

farms. Farm A occupies 500 acres and Farm B occupies 200 acres. We wish to know whether the distributions of their yields differ markedly from one another.

A glance at the two frequency distributions does little to inform us of any similarities or differences in the yields of the two farms. However, the comparison is facilitated by use of relative frequency distributions (Table 2.10).

A class by class comparison reveals the relative frequencies of yields to be quite comparable. Figure 2.7 provides a visual display in the form of two relative frequency polygons superimposed on the same chart. At a glance we can see that the distributions of yields are highly similar.

Figure 2.7
Relative frequency polygons of yield per acre on a 500-acre and a 200-acre farm

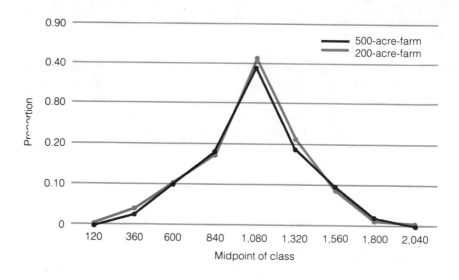

EXERCISE

2.6 An executive search firm has two divisions. The owner sets up different systems for the divisions whereby employees can earn commissions and bonuses. The following table shows commissions earned during the first week under these new systems.

| | f | |
Class limits	Division A	Division B
100–under 200	6	4
200–under 300	10	6
300–under 400	25	13
400–under 500	22	12
500–under 600	18	10
600–under 700	13	6
700–under 800	3	0
Total........................	97	51

a. Construct a relative frequency distribution for each division.
b. Construct and superimpose relative frequency polygons for each division.

2.3 THE CUMULATIVE FREQUENCY DISTRIBUTION AND OGIVE

A frequency distribution permits us to specify the number of observations that fall within each class. There are numerous occasions, however, when we wish to describe the total number of observations that fall below various points in a distribution. One widely used technique is referred to as a "less than" *cumulative frequency distribution.* Table 2.11 shows both a frequency distribution and a cumulative frequency distribution of the miles-per-gallon data on 60 different model automobiles.

We obtain the cumulative frequencies by successively adding the frequencies associated with each class as illustrated in Table 2.11. The cumulative frequency for each class denotes the number of cases that fall below the lower limit of the succeeding class. For example, 40 automobiles achieve less than 18 miles per gallon. All 60 automobiles score less than 30 mpg, and none achieves less than 9 miles per gallon.

The graphic technique used to portray the cumulative frequency distribution for these data is shown in Figure 2.8. The points on the vertical axis

cumulative frequency distribution

A distribution that shows the total number of observations that fall below various points in a distribution.

Class limits	f	Cumulative frequency *(cf)*
6–under 9	0	0
9–under 12	14	$0 + 14 = 14$
12–under 15	10	$14 + 10 = 24$
15–under 18	16	$24 + 16 = 40$
18–under 21	8	$40 + 8 = 48$
21–under 24	8	$48 + 8 = 56$
24–under 27	2	$56 + 2 = 58$
27–under 30	2	$58 + 2 = 60$
	60	

Table 2.11

Frequency and cumulative frequency distribution for miles-per-gallon scores of 60 automobiles

Figure 2.8   Ogive for cumulative frequency distribution of miles-per-gallon scores of 60 automobiles. The figure shows that 14 cars obtained less than 12 miles per gallon, 24 obtained less than 15 miles per gallon, and so forth.

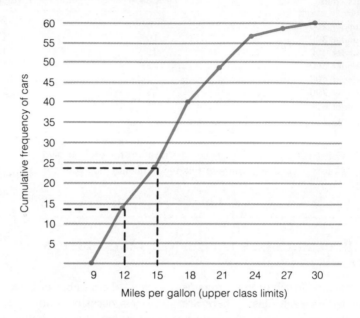

ogive

A chart of a cumulative frequency distribution.

represent the cumulative frequencies for values less than the corresponding value on the horizontal axis. This type of curve is called an *ogive*.

EXERCISES

2.7 Using the frequency distribution in Exercise 2.4:
  a. Construct a cumulative frequency distribution.
  b. Construct an ogive.

2.8 Based on the cumulative frequency distribution constructed in the preceding exercise, answer the following questions:
  a. What percentage of drills lasted less than 30 hours?
  b. What percentage lasted less than 75 hours?
  c. What percentage lasted 90 or more hours?
  d. What percentage of drills lasted less than 105 hours?

2.4 OTHER GRAPHIC DEVICES

Thus far, we have looked at graphic techniques used to portray frequency or relative frequency distributions. However, other types of charts are commonly used to display various aspects of business data. Figure 2.9 presents some examples of these other methods.

a.   The pie chart divides the total pie into 360 degrees. The proportion represented by each component is multiplied by 360°, and the resulting angle is used to allocate the size of the "slice" of each component. For example, the proportion of the federal budget obtained from individual income taxes is 0.43. The angle of this slice is, therefore, $360° \times 0.43 = 155°$ (rounded).

b.   This type of line chart accumulates the contribution of each component at all points along the horizontal axis. These appear as bands corresponding to each component. To find the contribution of a given component at a given time, find the difference between the values corresponding to the top and the bottom of its band. The dashed lines show the contribution of dividends in 1975.

c.   This line chart shows three different categories of interest rates corresponding to various time periods. Unlike *(b)*, these are not components of a single variable. Therefore, they cannot be cumulated.

Corporate profits before taxes, taxes, dividends (profits paid to stockholders), and undistributed profits (profits retained by the business enterprise) from 1947 through 1978.

Figure 2.9
Other examples of graphic displays

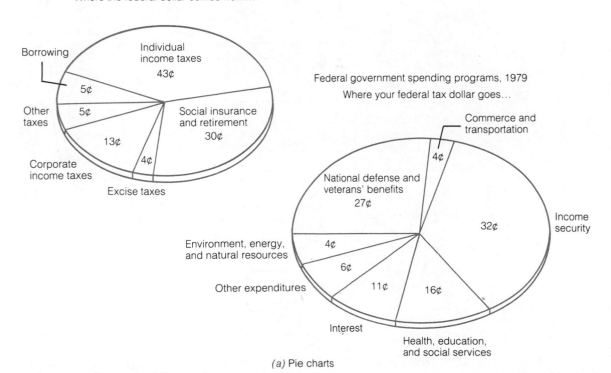

Sources of federal government revenue dollars, 1979-1980

Where the federal dollar comes from...

(a) Pie charts

Source: U.S. Department of Commerce.

Figure 2.9 *(continued)*

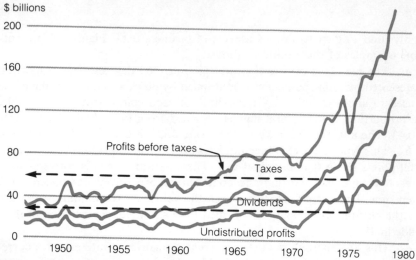

(b) Line chart showing the contribution of each of the components making up the total

Source: U.S. Department of Commerce.

(c) line chart

*At the end of the month

Source: Survey of Current Business, Jan. 1980.

Figure 2.9 *(continued)*

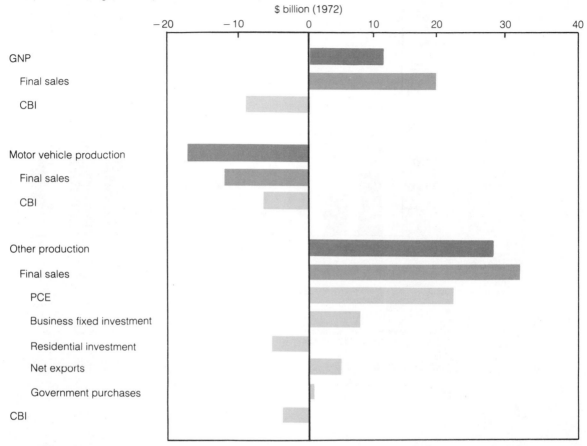

Real product: Change, fourth quarter 1978 to fourth quarter 1979

Note: PCE is personal consumption expenditures, and CBI is change in business inventories.

(*d*) Bar chart

Source: U.S. Department of Commerce, Bureau of Economic Analysis.

40

Figure 2.9 *(concluded)*

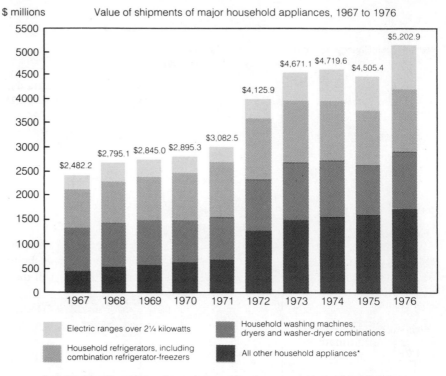

$ millions   Value of shipments of major household appliances, 1967 to 1976

Electric ranges over 2¼ kilowatts

Household washing machines, dryers and washer-dryer combinations

Household refrigerators, including combination refrigerator-freezers

All other household appliances*

*Includes domestic cooking appliances, home and farm freezers, water heaters (electric and nonelectric), dishwashing machines, food waste disposers, household trash compactors, and floor waxing and polishing machines.

(e) Bar chart showing the contribution of each of the components making up the total

Source: U.S. Department of Commerce, Bureau of the Census.

  d. This bar chart is used to show changes from one time period to another.

  e. This bar chart is used in the same way as the line chart in *(b).* It shows the contribution of each component to the total during different time periods.

EXERCISE

2.9 The tables below show the amount of consumer credit outstanding by *(a)* type of holder and *(b)* type of credit extended. For each distribution, calculate the proportion of the various parts of the total. Using these proportions, calculate the number of degrees assigned to each part and construct pie charts to represent each distribution.

Holder	Amount ($ millions)	Type	Amount ($ millions)
Commercial banks	$114,756	Automobiles	$ 88,767
Finance companies	47,147	Mobile homes	15,309
Credit unions	41,388	Home improvement	14,037
Retailers and others	30,125	Revolving	18,925
Total	$233,416	All others	96,378
			$233,416

Consumer installment credit, end of June 1978

Source: Adapted from *The World Almanac & Book of Facts, 1979*, p. 82, copyright © Newspaper Enterprise Association, New York, N.Y. 10166.

2.5 GRAPHIC VICES

Graphic devices should not be considered as substitutes for further statistical analysis, but rather as aids to thinking about data. Their purpose should be to provide a visually attractive and statistically accurate overview of data, to summarize relationships among variables, to expose trends, or to clarify fine distinctions. Unfortunately, either through ignorance or malice the effect of graphs is sometimes to confuse and mislead the unsophisticated reader. For every legitimate use of a graphic device there are numerous available techniques for perpetrating graphic larceny. Take the line graph as a case in point. The impressions conveyed by the line is intimately related to our choice of units along the vertical and horizontal axes. By the appropriate manipulation of these axes, one can make a business calamity look like a minor irritant or vice versa. Take a look at Figure 2.10. We see two different ways of depicting the drop in housing starts during a four-month period in 1979. Both convey precisely the same statistical information but few would deny that the visual impressions suggest very different trends.

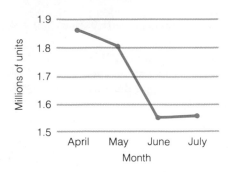
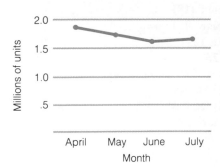

Figure 2.10
Two different line charts displaying the same information but conveying different impressions

SUMMARY

In this chapter, we looked at various ways to bring coherence to masses of raw data.

1. The frequency table is used to summarize the number of observations falling in each category of qualitative variables. Converting the frequencies to proportions or percentages expresses the data as relative frequencies.
2. A bar chart is often used to display frequency or relative frequency of qualitative data.
3. The frequency distribution is used to summarize masses of quantitative data. Constructing a frequency distribution involves organizing the variable into mutually exclusive classes and assigning each case to its appropriate class. Decisions must be made concerning the number of classes, the width of the classes, and the class limits.
4. Converting the frequencies to either percentages or proportions results in a relative frequency distribution. The total area under a relative frequency distribution is 1.00.
5. We may construct a cumulative frequency distribution by cumulating the frequency falling below specified points in a distribution. An ogive is a graphic device used to provide visual displays of cumulative frequency distributions.
6. Frequency and relative frequency distributions are often visually displayed in the form of a histogram: a bar graph of quantitative data.
7. A frequency polygon is commonly used to display frequency and relative frequency distributions of continuous variables. Frequency polygons are particularly useful when we wish to compare different distributions. When different n's are involved, relative frequency polygons expedite comparisons of different distributions.
8. Other ways of visually displaying data include pie charts and line charts.
9. Graphic devices should be thought of as aids to thinking about data. Care should be taken in designing visual displays so that erroneous impressions are not inadvertently conveyed.

TERMS TO REMEMBER

proportion (18)
percentage (18)
raw data (23)
array (23)
frequency distribution (24)
width (24)
mutually exclusive (25)
class limit (25)

midpoint (25)
relative frequency distribution (25)
histogram (26)
frequency polygon (29)
relative frequency polygon (33)
cumulative frequency distribution (35)
ogive (36)

EXERCISES

2.10 Daylaur Department store has its own credit card accounts. The credit department randomly selects 40 accounts and records the number of days within which the bill is paid:

16	9	5	8	6
10	16	4	11	4
3	19	21	16	15
24	45	11	8	19
37	59	14	72	3
22	10	6	14	11
20	9	16	6	75
21	7	15	12	10

a. Using a width of 10, set up a frequency distribution and a relative frequency distribution (using percentages).
b. Construct a histogram, using frequency to represent the points along the vertical axis.
c. Currently no finance charge is charged on overdue accounts. The credit department decides to charge interest on accounts that are more than 30 days late. What percentage of these accounts will be affected?
d. The credit department decides to terminate all accounts that are more than 60 days overdue. What percentage of these accounts will be terminated?

2.11 Shown below is the breakdown of 606 alcohol-related automobile collisions among motorists in London, Ontario, between 1968 and 1973 (as of July 1, each year).
a. Convert each frequency into a percentage of the total.
b. Draw three separate line charts showing the year-by-year changes within each age group.
c. On July 1, 1971, the legal age limit for purchasing and consuming alcohol beverages was reduced from 21 to 18. Does there appear to be any change in the distribution of alcohol-related collisions following this ruling?

	Age group		
Year	16–17	18–20	24
1968–69	17	47	5
1969–70	26	39	7
1970–71	25	48	9
1971–72	33	133	14
1972–73	23	153	27

2.12 Using the table in Exercise 2.11,
a. Convert the frequencies *within each age group* to percentages.
b. Draw a line chart showing the percentage of alcohol related collisions for each age group during the five time periods.
c. How does the summary information in this exercise differ from the information in Exercise 2.11?

2.13 In 1978, the following amounts of money were received as royalties and licensing fees from foreign affiliates of the indicated manufacturing industries:

Machinery	$1,242,080,000
Chemical and allied products	443,600,000
Primary and fabricated metals	44,360,000
Transportation equipment	44,360,000
Other	443,600,000
	$2,218,000,000

a. Express each as a percentage of the whole.

b. Construct a pie chart to show the distribution of royalties among the indicated industries.

2.14 The following data show the kilowatt-hours used per day in a one-bedroom apartment in Los Angeles.

8	6	8
8	5	4
8	7	4
8	7	4
9	8	5
8	6	4
8	5	7
9	6	6
9	6	5
8	8	8

a. Comment on each of the following as possible ways of setting up a frequency distribution for these data:

(1)	(2)	(3)
Classes	Classes	Classes
1–4	1–4	1–under 4
4–7	3–6	5–under 9
7–10	5–8	10–under 14
	7–10	

b. What do you think would be the best way to present these data?

2.15 The following table shows (a) percentage distribution of all companies classified in each enterprise industry category; (b) percentage distribution of sales/receipts, and (c) percentage distribution of employment of these companies in 1972.

	(a)	(b)	(c)
		Percentage distribution	
Category	Companies	Sales/receipts	Employment
Construction	17.8%	8.2%	9.1%
Minerals	0.4	1.2%	1.0
Selected service	33.3	5.7	11.4
Retail trade	36.7	22.9	25.2
Wholesale trade	6.5	21.1	7.2
Manufacturing	5.3	40.9	46.1

Source: U.S. Department of Commerce; Bureau of the Census, December 1979.

Construct a pie chart for each distribution.

2.16 One of the most frequently used statistics is *percentage of change*. We are often given two sets of data and asked to report the percentage of increase or decrease between the two. For example, in 1967 wholesale trade in the United States (in millions of dollars) equalled $459,467; in 1972, the comparable figure was $683,659. We calculate the *percentage change* by dividing the difference by the original figure and multiplying by 100. In this example:

$$\frac{683,659 - 459,467}{459,467} \times 100 = 0.49 \times 100 = 49 \text{ percent}$$

Calculate the percentage of change for each of the components of U.S. wholesale trade between 1967 and 1972.

Industry	Sales ($ millions)	
	1967	1972
Motor vehicles and automotive equipment	$46,122	$ 83,016
Lumber and construction materials	16,390	27,943
Metals and minerals (except petroleum)	33,704	43,488
Electrical goods	32,115	49,101
Machinery equipment and supplies	50,432	75,128
Drugs, chemicals, and allied products	27,795	39,288
Piece goods, notions, and apparel	21,280	27,933
Groceries and related products	74,458	109,781
Farm product- raw materials	38,148	52,401
Petroleum and petroleum products	33,373	46,284

Source: U.S. Department of Commerce; Bureau of the Census, December 1979.

2.17 Shown below is the deadweight tonnage (000) (carrying capacity of a ship in long tons [2,240 lbs]) of 39 oil tankers.

555	414	389
554	414	388
550	413	387
517	413	387
516	413	387
484	410	380
484	410	380
484	407	380
446	407	372
424	406	372

424	393	367
422	393	363
414	392	363

   a. Construct a frequency distribution, using approximately 7 classes.

   b. Construct a histogram, using relative frequency to represent the points on the vertical axis.

   c. Suppose you had 480,000 long tons of oil to ship. What are your chances of finding an oil tanker that can handle this load safely?

2.18 J.J. contends that driving on holiday week-ends is much more risky than on regular week-ends. To support his contention, he compares the number of highway fatalities over the most recent holiday week-end (e.g., Memorial Day) with the number occurring over the last (non-holiday) week-end. Finding almost twice as many fatalities, he uses this evidence to support his belief.

   Can you see any fallacies in his reasoning?

   Do you agree? Why, or why not?

2.19 The following table shows the number of accidental deaths and total miles traveled during 1978 by various modes of transportation.

Kind of transportation	Passenger miles (billions)	Passenger deaths
Passenger autos and taxis*	2,190.0	28,450
Buses.................................	78.1	135
Railroad passenger trains	10.2	13
Scheduled air transport (domestic)	188.0	13

* Drivers of passenger automobiles are considered passengers.

   a. Determine the death rate per billion passenger miles for each transportation mode.

   b. Construct a bar chart showing the comparison of death rates in various transportation modes.

2.20 The table below shows the total motor fuel consumption by state (in millions of gallons) in 1977 and 1978.

	1977	1978		1977	1978
Alabama	2,342	2,459	Montana	564	612
Alaska	265	272	Nebraska	1,063	1,082
Arizona	1,506	1,619	Nevada	511	553
Arkansas	1,462	1,520	New Hampshire ...	460	472
California	12,268	12,862	New Jersey	3,604	3,678
Colorado	1,592	1,675	New Mexico	900	961
Connecticut	1,501	1,523	New York	6,425	6,519
Delaware	334	327	No. Carolina	3,407	3,557
Florida	4,978	5,180	No. Dakota	503	517
Georgia	3,370	3,533	Ohio	6,022	6,126
Hawaii	334	350	Oklahoma	2,019	2,079
Idaho	590	608	Oregon	1,576	1,679
Illinois	5,869	6,026	Pennsylvania	5,798	5,968
Indiana	3,335	3,387	Rhode Island	416	415
Iowa	2,018	2,037	So. Carolina	1,799	1,897

	1977	1978		1977	1978
Kansas	1,570	1,625	So. Dakota	537	547
Kentucky	2,062	2,124	Tennessee	2,792	2,899
Louisiana	2,279	2,395	Texas	9,246	9,708
Maine	625	642	Utah	793	850
Maryland	2,123	2,208	Vermont	295	307
Massachusetts ...	3,560	2,624	Virginia	3,030	3,152
Michigan	5,224	5,444	Washington	2,121	2,222
Minnesota	2,384	2,479	W. Virginia	984	1,010
Mississippi	1,464	1,584	Wisconsin	2,596	2,698
Missouri	3,152	3,248	Wyoming	445	477

Source: Adapted from *The World Almanac and Book of Facts 1979,* p. 140; and *The World Almanac and Book of Facts 1980,* p. 125.

    a. Construct a frequency distribution for each year, using a width of *(a)* 1,000; *(b)* 1,500; *(c)* 2,000.

2.21 The table below shows the athletic shoe production (000 pairs) for each month for four consecutive years.

    a. Construct a relative frequency distribution for each year.

    b. Construct line charts corresponding to these relative frequency distributions and superimpose on the same chart.

	1973	1974	1975	1976
Jan	861	829	603	702
Feb	802	838	601	769
Mar	884	815	749	897
Apr	860	809	656	1,011
May	943	871	711	822
June	842	844	543	890
July	569	774	563	558
Aug	924	852	704	878
Sept	867	776	672	957
Oct	927	923	723	882
Nov	914	774	691	775
Dec	737	727	701	923
Total	10,130	9,832	7,917	10,064

Source: Adapted from *Business Statistics,* 1977, p. 141.

2.22 The following table summarizes the number of firms and employment for minority-owned businesses in the U.S. in 1972.

	Number of firms (000)	Employment (000)
Black	195	197
Spanish Origin	120	150
Other	67	109
Total	382	456

Source: Adapted from E. S. Buffa and Barbara A. Pletcher, *Understanding Business Today* (Homewood, Ill.: Richard D. Irwin, 1980), p. 105. © 1980 by Richard D. Irwin, Inc.

a. Construct relative frequency distributions.
b. Construct bar charts, using relative frequency on the vertical axis.

2.23 The following data present the number of gallons of water used per day for a 4-unit building in West Los Angeles.

875	1,052	1,146	723	822
797	1,122	1,052	650	645
853	1,334	956	711	608
907	1,140	1,006	722	650
985	1,018	1,217	806	724
1,177	916	1,067	1,085	799
1,128	877	842	935	650
903	993	981	920	941
				1,015

a. Using 100 as the width, set up a frequency distribution for these data.
b. Construct a histogram.
c. The Department of Water and Power has decided to impose certain penalties when usage exceeds 1,000 gallons per day. If this sample is representative of the true use of water in this building, what percentage of the time will this building be penalized?

2.24 Suppose you were provided with data that showed how various income level groups spend their money on such things as food, housing, transportation, recreation, etc.

a. Discuss ways in which this information can help you make business decisions.
b. What are some variables that might influence the distribution of spending patterns?

2.25 In 1977, home accidents resulted in approximately 24,000 deaths. The following shows the breakdown by category (rounded to nearest 100)

	f
Falls	7,500
Firearms	1,200
Fires, burns	5,200
Poison (solid, liquid)	3,000
Poison (gas)	1,000
Suffocation	2,700
Other	3,400
Total	24,000

Source: *The World Almanac and Book of Facts 1979*, p. 955.

a. Construct a relative frequency table for these data.
b. Construct a bar chart and a relative frequency bar chart.

Summarizing Statistical Data

Montana Rincon was deeply concerned. His audit of 1,000 accounts had uncovered an error rate of 6 percent. More importantly, he noticed that the typical error was extremely small and that the spread of scores was likewise small. He recalled that the Salami technique involved slicing barely noticeable and variable amounts off a tiny proportion of the total number of accounts. Such minute disparities are unlikely to be detected; if detected, unlikely to be reported; if reported, unlikely to spur an investigation.

The lack of extreme values continued to gnaw at him. His prior familiarity with audits had led him to expect at least an occasional extreme value—one so large that there would be a few embarrassed faces in the company. In later years, these errors become the topic of good-humored jokes, "Do you remember when. . . ?" and everyone would laugh. Not this time.

There was one other fact that disturbed him. All the errors were in one direction only—all favored the company. Not a single one favored the customer. Very strange.

Rincon's thinking had carried him to the core of the present chapter. To describe the typical error requires a measure of central tendency; similarly, to describe the spread or variability of these errors requires a measure of dispersion.

3.1 SUMMARY DESCRIPTIVE MEASURES

So far we have made some progress toward our goal of making sense out of masses of unorganized raw data. Through the use of such techniques as frequency distributions and graphic displays, we have been able to discern patterns in the data that were previously unrecognizable. However, these techniques provide only broad overviews of data. More often we are interested in pinpointing single descriptive measures that summarize some salient aspect of our data set. For example, imagine that you are a vice president in charge of manufacturing who is faced with the necessity of replacing obsolete equipment. You must choose between two competing models that are equally expensive to purchase and operate. In short, your choice boils down to the question of which model provides a greater product output. It would be difficult to compare and draw conclusions about their comparative outputs through the use of frequency distributions or charts. But if your plant manager informs you that a sample of Model A machines yielded an average daily output of 6,565 items and that Model B's yield was 5,105 items, you would have a single, immediately understandable basis for comparison.

As we indicated previously, our interest usually extends beyond the descriptive statistics themselves. We commonly wish to generalize from a sample statistic to a population parameter. For example, our decision to purchase Model A machines would be based on the assumption that the difference in the average output observed in the sample reflects differences that exist in the population. In other words, we would expect that, in the long run, the A models would provide an economic advantage over the B models.

We do not mean to imply that once you have calculated a sample statistic, you may then assume that you have also obtained the corresponding population parameter. Suffice it to say at this point, you have merely obtained an *estimate* of the parameter. In later chapters we examine procedures for determining the accuracy of this estimate.

In this chapter we look at ways of calculating numerical values that summarize and describe data sets. We shall be dealing primarily with two types of statistical measures—those that describe central tendency and those that describe the spread or variability of the data.

3.2 MEASURES OF CENTRAL TENDENCY

One of the most common descriptive measures is the arithmetic average or mean. Although it inevitably results in some loss of information, its appeal lies in the fact that it permits us to use a single number to summarize masses of data. In addition, means permit us to make direct comparisons of two or more sets of data even when these data sets are based on different numbers of observations.

For example, suppose you were told that the total personal income for February 1979 was $1,851.4 billion and for March 1979, $1,872.1 billion. You might be tempted to conclude that personal income improved from February to March. However, February's total is based on a 28-day month and March on 31 days. Dividing the income figures for each month by their respective numbers of days yields an average of $66.12 billion per day for February and $60.39 billion per day for March. Thus, the average daily income actually dropped in March.

Up to this point, we have been looking at one measure that is often used to describe the center of a data set—the mean. In actuality, there are several different ways of defining the central region of a distribution. Thus, there are several different measures of central location, each with a different definition. We refer to these indexes of central location collectively as *measures of central tendency*. In this text, we shall examine three of these measures: the mean, median, and mode.

measures of central tendency

Indexes that locate the center of a data set.

3.2.1 The arithmetic mean

The *arithmetic mean* is defined as the sum of the values of a variable divided by the number of observations. The definition is the same for both the sample and the population, although we use a different symbol to refer to each.

For the sample, the mean is defined symbolically as follows:

arithmetic mean

Sum of the values of a variable divided by the number of observations.

$$\overline{X} = \frac{\Sigma X}{n}$$ (3.1)[1]

---

[1] The formula for the mean lends itself to some interesting sidelights. Since $\overline{X} = \frac{\Sigma X}{n}$, it is also true that $\Sigma X = n\overline{X}$. There may be times when we know the value of the mean but wish to obtain the sum of the underlying measures. See the *Study Guide* for the rules of summation.

This tells us that the mean (X bar) is obtained by dividing the sum of the values of the variable X by the number of observations. Thus, if $X_1 = 8$, $X_2 = 14$, and $X_3 = 26$, $\overline{X} = \dfrac{8 + 14 + 26}{3} = 16$.

The Greek symbol μ (pronounced "mew") is used to represent a population mean. Thus,

$$\mu = \frac{\Sigma X}{N}$$

Note that we use N to represent the number in the population and n, the number in the sample.

Example A: Calculation of mean from ungrouped data

Between November 1948 and March 1975, there were six business recessions in the United States. The number of months that each recession lasted is shown in Table 3.1. As can be seen, the mean is 11. Thus, on the average, the six business recessions lasted for 11 months.

Table 3.1

Six economic slumps between the end of World War II and 1975

Recession	Number of months
Nov. 1948–Oct. 1949	11
July 1953–May 1954	10
Aug. 1957–April 1958	8
April 1960–Feb. 1961	10
Dec. 1969–Nov. 1970	11
Nov. 1973–Mar. 1975	16
Total	66

$$\Sigma X = 66; \ n = 6; \ \overline{X} = 11$$

We have seen that, for convenience, data sets are often cast in the form of a grouped frequency distribution. In calculating the mean, we should recall that grouping results in the loss of some information. Therefore, the group mean will often be somewhat different from a mean calculated from the individual measurements. The disparity will usually be quite small and must be regarded as the price we pay for the convenience of grouping.

Using the midpoint to represent each class, we find the mean by applying the following formula:

$$\bar{X} = \frac{\Sigma fX}{n},$$ (3.2)

where

X is the midpoint of each class, and
f is the corresponding frequency.

Note that multiplying the midpoint of each class by its corresponding frequency is the same as adding the midpoint f times. Thus, in Table 3.2, there are five cases in the first class. Rather than adding 12,500 five times, we merely multiply 12,500 by five to obtain 62,500.

Example B: Calculation of mean from grouped data

The classified ads in the *Los Angeles Times* for Sunday, August 3, 1980, under the section headed "Accounting," included opportunities for clerks, auditors, accountants, and controllers. The frequency distribution of starting salaries is shown in Table 3.2. Note that the ads are not representative of *all* opportunities available. Rather they constitute a sample of those for which starting salaries were listed.

Class limits	Midpoint (X)	f	fX
$10,000–under $15,000	12,500	5	62,500
15,000–under 20,000	17,500	16	280,000
20,000–under 25,000	22,500	13	292,500
25,000–under 30,000	27,500	9	247,500
30,000–under 35,000	32,500	7	227,500
35,000–under 40,000	37,500	0	0
40,000–under 45,000	42,500	1	42,500
Total		51	1,152,500

$$\bar{X} = \frac{\$1,152,500}{51} = \$22,598.04$$

Table 3.2
Frequency distribution of salaries offered in classified ads under the section headed "Accounting"

3.2.2 Properties of the mean

The mean is very much like the fulcrum of a teeter board. It is the point of balance in which the distances from the fulcrum times the weight on one side is equal to the distances times weight on the other. Thus, if a single measurement is two units away from the fulcrum on one side, and two measurements are one unit away from the other side of the fulcrum, the product of the weight and the distance is identical in both cases: $1 \times 2 = 2$ and $2 \times 1 = 2$ (see Fig. 3.1). For this reason, the mean is the point about which the sum of deviations is zero. To illustrate, the mean of 4, 12, 15, and 17

54

Figure 3.1

Like the fulcrum in a teeter board, the mean is the point of balance in a data set. Note that a single value two units away from the fulcrum can balance two values one unit away.

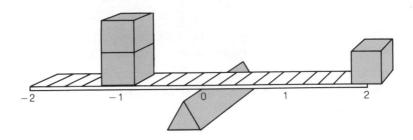

is 48/4 = 12. Subtracting the mean from each data point yields the deviations: −8, 0, +3, +5. Notice that the sum is equal to zero. Symbolically,

$$\sum_{i=1}^{n} (X_i - \overline{X}) = 0,$$

where the summation sign (Σ) directs us to sum the differences over all values of X from $i = 1$ through $i = n$.

Unlike some other measures of central tendency, every score contributes to the determination of the mean. This means that, at times, a single measurement can greatly influence the mean. Thus, as with the teeter board, an extreme value draws the fulcrum or mean toward it. For example, the mean of 2, 4, and 6 is 12/3 = 4, whereas the mean of 2, 4, and 27 is 33/3 = 11. At times, then, the mean does not adequately represent the central portion of a data set. Let's look at a real-life example of this property. Suppose you were interested in the mean number of shares traded on the New York Stock Exchange during the week of January 5, 1981. The volume for each day (in millions of shares) was:

Monday:	59
Tuesday:	67
Wednesday:	93
Thursday:	55
Friday:	50

The mean of these five values is 64.8. Note that there is one day on which the volume hit a record high. If we calculate the mean without this day's volume included, we obtain $\overline{X} = 57.75$ for the remaining four days.

At the same time that the mean is the point that makes the sum of the deviations from it equal to zero, it also makes the sum of the squares of the deviations from it minimal. Look at Table 3.3. Note that the squared deviations around the mean are less than the squared deviations around any other value, i.e., $\sum_{i=1}^{n} (X_i - \overline{X})^2$ = minimum. This fact takes on added significance as we progress through the course.

X	$(X-6)^2$	$(X-9)^2$	$(X-\bar{X})^2$	$(X-15)^2$
6 0		9	16	81
9 9		0	1	36
15 81		36	25	0
Total 30 90		45	42	117

Table 3.3
The squares and sums of the squares of deviations from the mean and from various scores in a distribution

$$\bar{X}=10$$

EXERCISES

NORMAL
DIST.?

3.1 For the following set of data: (3, 4, 5, 6, 7) demonstrate that
  a. The sum of deviations from the mean equals zero.
  b. The sum of the squared deviations from the mean is less than the sum of the squared deviations from any other data point.

3.2 Using the Montana Rincon data in Table 2.3, find the mean error of the 60 accounts.

3.3 Calculate the mean amount of error for the frequency distribution of Rincon's data shown in Table 2.5. Compare and comment on the values obtained in this and the preceding exercise.

3.2.3   The weighted mean

You are planning to bid on a construction job. You have five different classes of labor with the following hourly rates: $5.50, $6.25, $7.15, $8.30, and $12.90. You estimate that the job will require 1,250 hours of labor. You find the mean hourly rate by adding these five values together and dividing by five. The mean hourly rate would appear to be $8.02. You then multiply $8.02 by 1,250 hours to obtain a total estimated labor cost of $10,025. Are you satisfied that these procedures are correct?

As a matter of actual fact, they would be correct only if each category of labor contributed exactly the same number of hours. This is rarely the case. In construction, it is more likely that individuals with lower hourly wages contribute a greater number of hours than those at the upper end. The table below shows the numbers of hours worked at each hourly wage.

Hourly rate	Number of hours
$ 5.50	425
6.25	370
7.15	180
8.30	225
12.90	50
Total	1,250

To find the mean hourly rate for the job, we must multiply each hourly rate by its corresponding number of hours, sum the products, and divide by the total number of hours. Thus,

$$\bar{X}_w = \frac{(5.50)\,(425) + (6.25)\,(370) + (7.15)\,(180) + (8.30)\,(225) + (12.90)\,(50)}{1,250}$$

$$= \frac{8449.5}{1250} = 6.76.$$

weighted mean

Sum of the values of a variable multiplied by their respective weights and divided by the sum of the weights.

In the above example, the average shown is a *weighted mean*. We multiplied each rate by the number of hours worked (the weight), summed these products, and divided by the the sum of the weights.

In symbolic form,

$$\bar{X}_w = \frac{\Sigma wX}{\Sigma w} = \frac{8449.5}{1250} = 6.76. \tag{3.3}$$

Note how valuable the weighted mean can be. If you were told that the job was being expanded so that 3,150 hours of labor would be required, you could quickly multiply 3,150 by 6.76 to obtain a new estimate of $21,294 as the total labor cost. Note that, had we stayed with our original calculation of $8.02 per hour, we would have calculated the labor cost at $25,263—an overestimate of nearly $4,000.

Weighted averages are widely used in business and economics where various classes contribute differentially to totals. As we shall see, such is the case with index numbers and numerous seasonally adjusted economic indicators.

Box 3.1

STRAINING UNDER THE WEIGHTED MEAN

"I'll tell you what I'm gonna do," says the carpet salesman with the slicked-down hair and the Fu-Man-chu moustache. "I'm gonna make you an offer you can't refuse."

"Oh?"

"You want 3,000 square yards for that apartment house you're fixing up, right?"

"Right."

"You've selected five grades of carpeting—$9, $11, $12, $13, and $15. That's an average of $12 a square yard. Multiply that by 3,000 square yards and we get,

let's see . . . . $36,000. I'll give you the whole lot for $32,000. That's an 11 percent discount over our already low prices."

How does that sound to you? Is it really a good deal? It depends. Our slick friend has calculated a very deceptive mean. He just took the price tags and averaged them without regard to the quantities of each grade of carpeting you are buying. Here's the rub. The sum of the prices of these grades divided by five gives us the mean *only* when we buy equal quantities of each. But what if this isn't the case? Then we must calculate the weighted mean. Let's say you wanted to purchase the quantities of carpeting shown in the second column of the table:

Box 3.1 (*continued*)

X Price of carpet ($ per sq. yd.)	w Quantity (sq. yds.)	wX Total cost
$ 9	1,650	$14,850
11	500	5,500
12	350	4,200
13	250	3,250
15	250	3,750
Sums	3,000	$31,550

To find the mean here, you must multiply the price of each grade by the quantity ordered *(w)*, add them together, and divide by the sum of the weights. In other words,

$$\text{Weighted mean} = \frac{\text{Sum } wX}{\text{Sum } w}$$

$$= \frac{[(\$9)(1,650) + (\$11)(500) + (\$12)(350) + (\$13)(250) + (\$15)(250)]}{1,650 + 500 + 350 + 250 + 250}$$

$$= \$10.52.$$

Aha. The actual mean is $10.52, rather than the $12.00 represented by your carpet dealing carpet-bagger friend. Your initial cost estimate should have been $31,550. Four hundred and fifty dollars less than his magnanimous offer. Some deal! Chalk up the extra 450 bucks as the price of ignorance.

Weighted averages are used and abused much more than you might think in the shenanigans of daily living. If you bought 100 shares of stock at $5.00 a share, 50 at $6.00, and 10 at $7.00, your average price is not $6.00. It is

$$\frac{[(100)(5) + (50)(6) + (10)(7)]}{100 + 50 + 10} = \frac{870}{160} = \$5.44.$$

Similarly, if your dream machine guzzles an average of 20 MPG on a 300-mile trip, 23 MPG over 800 miles, and 25 MPG over 1,000 miles, your average number of miles per gallon is

$$\frac{(20)300 + 23(800) + 25(1,000)}{300 + 800 + 1,000} = \frac{49,400}{2,100} = 23.52$$

Source: Adapted from R. P. Runyon, *How Numbers Lie: A Consumer's Guide to the Fine Art of Numerical Deception.* (Lexington, Mass.: Lewis Publishing, 1981), pp. 106–8. Copyright © Lewis Publishing Co., Lexington, Mass.

EXERCISES

3.4 Investors use the concept of the weighted mean when they buy shares of a given stock at various prices and wish to determine the price at which they "average out." TVB Investment Associates bought 15,000 shares of a common stock at the following prices and amounts.

Price per share	Number of shares
$3.00	5,300
4.75	4,500
5.00	2,600
5.50	1,600
6.50	1,000
Total	15,000

a. Exclusive of commissions, what was the total price paid?

b. What was the average price per share (exclusive of commissions)?

3.5 Often an investor buys shares of stock at varying prices, only to find sudden shifts in the price. Referring to Exercise 3.4, let us imagine that the stock has dropped to $2.50 a share. TVB Investment Associates are confident that the stock will rebound. How many shares must it purchase to average out at $3.50 a share?

3.2.4 The median

As we have seen, the mean uses all of the data to arrive at a measure of central location. This is often advantageous since each and every value of the variable is included in the calculation. However, there are times when this is not to our advantage. For example, if there are extreme values in the distribution, the mean may actually lead to a misleading index of central tendency. The most common example of this is income. All we need are a few extremely high values and we find ourselves with a mean that is disproportionately high and does not at all reflect the underlying distribution.

median

The point in a distribution above and below which half the cases fall.

Another measure of central tendency—the *median*—by its very nature, circumvents this type of problem. The median in effect divides the distribution into two equal parts. Half of the data set falls below and half above. In a sense the median is blind to the specific numerical values on either side of it. All that is important is whether the values are higher or lower, not *how much* higher or lower.

When dealing with a relatively small data set, the median is easily obtained by arranging the numerical values in an array and pinpointing the middle value. If, for example, 2, 4, 8, 9, 10 represent our data set, the median value is 8 since two values fall above and two below. If the data set contains an even number of values, we estimate the median to be halfway between the two middle values. Thus, the median for the measurements 2, 4, 8, 9, 10, 11 would be halfway between 8 and 9 or 8.5.

It should be noted that, when there are indeterminate values, such as an open-ended class,[2] no mean can be calculated. Under these circumstances, the median is usually the statistic of choice.

Let us look at a few examples that may help provide an intuitive understanding of the median:

In 1977 the median rent of apartments in the United States was approximately $230. This means that, if the apartments available are representative of the nationwide stock of apartment units, and you picked an apartment at random, you had a 50-50 chance of renting an apartment for less than $230.

In 1980, the median age of the population was approximately 30. By the year 2000, projections indicate that the median age will approach 35. As the median age rises, the relative number of younger people declines.

In 1970, the median family income was $8,734. By the end of 1977, it had increased to $13,572. This means that a family at the median in 1970 would have required an increase of $4,838 just to maintain the same relative position.

Example C: Calculation of median from grouped data

Let's calculate the median error for the Montana Rincon data shown in Table 3.4.

Class limits	f	cf
1–under 5	3	3
5–under 9	9	12
9–under 13	9	21
13–under 17	10	31
17–under 21	14	45
21–under 25	5	50
25–under 29	3	53
29–under 33	4	57
33–under 37	3	60
	60	

Table 3.4
Grouped frequency and cumulative frequency distribution of errors in the Rincon data

By definition, the median is the middle observation (i.e., the one numbered $n/2$ in a set of n values of a variable). Since there are 60 observations of error in Rincon's sample, the median corresponds to the 30th observation (i.e., $60/2 = 30$). Now, looking at the cumulative frequency distribution, we see that the class 13 to under 17 includes the 30th observation. Thus, the median is at least as high as 13 and is located somewhere within the

[2] Precise values of a variable may be indeterminate for a number of reasons. For example, income figures often indicate an open-ended class at the upper end of the distribution (e.g. $50,000 per year and above). Under these circumstances, an annual income of $1 billion (an oil sheikh) carries no more weight than $75,000 earned by a junior executive in a multinational corporation.

class limits of 13 to under 17. To locate the median precisely within the class, we note that there are a total of 10 observations in that class. We also note that the 30th observation is the ninth observation in that class (i.e., 30 − 21). It is therefore, nine tenths of the way through a class that has a width of 4. Putting the information together, the median is 13 + (9/10) 4 = 16.6.

The following formula for the median helps to simplify the calculations.

$$Mdn = \frac{0.5n - F}{f} i + Lm, \tag{3.4}$$

where:

Mdn = Median
n = Number of observations in sample
F = Cumulative frequency of class immediately preceding the median class
f = Frequency in the median class
i = Width of the median class
Lm = Lower limit of the median class

In the preceding example, $n = 60$, $F = 21$, $f = 10$, $i = 4$, and $Lm = 13$. Substituting in Formula (3.4), we find:

$$Mdn = \frac{(0.5)\,(60) - 21}{10} (4) + 13$$

$$= \frac{9}{10} (4) + 13$$

$$= 16.6$$

3.2.5 The mode

In almost any distribution, there are certain values that occur more frequently than others. For example, more men wear size 10 shoes than any other size. In the Rincon data, more errors were in the class 17 to under 21 than in any other class. These examples illustrate a third measure of central tendency—the *mode*. The mode is the most frequently occurring value in the data set. We can usually obtain the mode by inspection: in a frequency distribution, the modal class is the one with the greatest associated frequency.

Although the mode has limited uses in statistical analyses, it is occasionally useful in making certain business decisions. For example, a shoe manufacturer has little interest in the mean or median since they may be values that do not exist in the real-world of shoe sizes. It is difficult to manufacture a size 9.38 even though this may be the average. In contrast, the mode does represent

mode

The value of the variable that occurs with the greatest frequency.

an actual shoe size and, in fact, may influence the manufacturer's production schedule.

3.2.6 Mean, median, and skewness

Charts of data sets may take on an unlimited number of different forms. Some may be symmetrical so that each side is the mirror image of the other. One of the most common of the symmetrical distributions is shown in Figure 3.2. We refer to this bell-shaped distribution as the *normal curve*. Not all symmetrical curves are bell-shaped. However, all symmetrical curves have an important characteristic in common—the values of the mean and median are identical.

In contrast, the measurements in many data sets tend to pile up at locations other than the center of the distribution. Such distributions are said to be *skewed*. As we have already seen, the mean is strongly influenced by extreme scores. In fact, it is drawn in the direction of the skew.

Figure 3.3 presents an example of a *positively skewed distribution*.

Broadly speaking, we should be alerted to the possibility that a data set will be positively skewed whenever there are limits at the lower end of the scale but no similar restrictions at the upper end. For example, life expectancies of everything from light bulbs to people might be expected to be positively skewed. Neither can survive less than zero time but, theoretically, there is no limit at the upper end. Light bulbs have been known to burn many years and an occasional human will survive the century mark. For similar reasons we might expect data sets of the following to be positively skewed: loan amounts, family income, savings deposits, family size, and time scores in athletic events.

Negatively skewed distributions are most likely to occur when the restrictions are on the upper end of the distribution. For example, if a bank places a limit of, say, $2,500 on personal loans, a large number of clients may be expected to apply for the limit. Relatively few would request loans in the low hundreds. Figure 3.4 illustrates a negatively skewed distribution.

One of the characteristics of skewed distributions is the relationship between the mean and the median. In a positively skewed distribution, the

normal curve

Frequency distribution with a characteristic bell-shaped form.

skewed distribution

Frequency distribution that departs from symmetry and tails off at one end.

positively skewed distribution

Frequency distribution with relatively fewer measurements occurring at the high end of the horizontal axis.

negatively skewed distribution

Frequency distribution with relatively fewer measurements occurring at the low end of the horizontal axis.

Figure 3.2

Symmetrical bell-shaped frequency curve. In this as well as other unimodal symmetrical distributions, the mean, median, and mode have identical values.

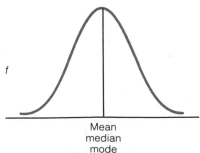

f

Mean
median
mode

Figure 3.3

Positively skewed distribution. A large number of establishments have few employees, whereas relatively few have large numbers of employees. Distributions of this sort are commonly used to present business and economic data. Note that the classes are *not* of equal width and that the last class is open-ended. Therefore, the horizontal axis should reflect these differences in class widths, and the midpoint of the last class cannot be precisely located.

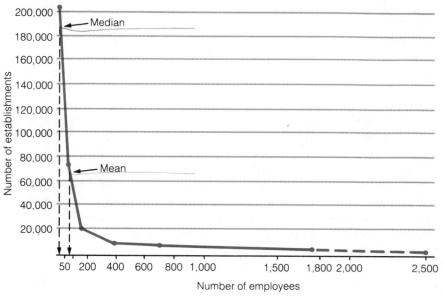

Source: Adapted from E. J. McCarthy, *Essentials of Marketing* (Homewood, Ill.: Richard D. Irwin, 1979), p. 133. © 1979 by Richard D. Irwin, Inc.

Figure 3.4

A negatively skewed distribution. The number of clients requesting various amounts of personal loans when a restriction of $2,500 is placed on the upper limit of the loan.

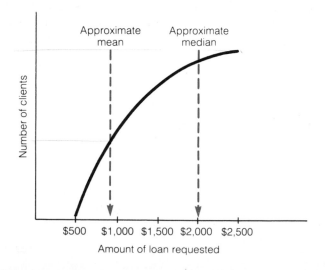

mean will always be higher than the median since it is being pulled toward the higher scores (see Fig. 3.3). In a negatively skewed distribution, the reverse is true (see Fig. 3.4). The mean will equal a value less than the value of the median. In fact, it is this relationship between the mean and the median that enables us to determine quickly whether a distribution is skewed and, if so, in which direction.

EXERCISES

3.6 In a survey of nine food markets, the price per head of lettuce was found to be: 26¢, 42¢, 21¢, 33¢, 54¢, 32¢, 29¢, 22¢ and 25¢.
- *a.* Calculate the mean.
- *b.* Calculate the median.
- *c.* Why is the mean a higher value than the median?

3.7 The owner of a carpet store is offered a "deal he can't refuse" by a sales representative. He may purchase 1,000 square yards of carpeting at a cost of $7,000. The usual cost of carpeting is $6, $7, $8, and $9 a square yard. Multiplying the mean ($7.50) by 1,000 square yards, the sales representative informs him that the usual cost is $7,500. The owner accepts the offer. Was it a wise business decision?

3.8 Referring to Exercise 3.7, the quantities of the various types of carpeting are shown below:

Cost per square yard	Square yards of each
$6	450
7	250
8	200
9	100

- *a.* How much would he have paid for the carpeting if he had taken the quantities of each grade into account?
- *b.* Find the weighted mean.

3.9 Suppose you work for a television research firm. You collect data on who watches what. One of the key variables that interest potential advertisers is the age of the viewer. What would you expect the following age distributions to look like?
- *a.* Viewers of Sesame Street.
- *b.* A Lawrence Welk special.
- *c.* A recent highly acclaimed film.

3.10 Given the following sales by size of retail establishments:

Sales	Number (000)
$ 0 to $ 9,999	217
$ 10,000 to $ 19,999	171
$ 20,000 to $ 29,999	127
$ 30,000 to $ 49,999	188

Sales	Number (000)
$ 50,000 to $ 99,999	304
$ 100,000 to $ 299,999	450
$ 300,000 to $ 499,999	106
$ 500,000 to $ 999,999	71
$1,000,000 to $1,999,999	38
$2,000,000 to $4,999,999	27
$5,000,000 or more	9

a. Graph this distribution and describe the form.
b. What would be the preferred measure of central tendency? Why?

3.11 Given the following measures of central tendency, indicate whether there is evidence of skew and, if so, in what direction?

a. $\overline{X} = 48$; median = 72.
b. $\overline{X} = 72$; median = 48.
c. $\overline{X} = 48$; median = 48.

3.3 MEASURES OF VARIABILITY

The volume of stock market transactions fluctuates from day to day; your moods shift like drifting sands; the price of precious metals swings like a pendulum in response to world crises; the number of cars on the road increases or decreases daily; the price of a head of lettuce may be 29 cents today and 34 cents tomorrow; the Dow Jones Industrial Average sometimes goes up and sometimes goes down.

If there is one thing we can count on, it is that variability is a fact of life. It may sometimes plague us, but at other times, it adds a touch of spice to daily life.

Suppose you are an employee of a manufacturing firm whose success is highly dependent on fulfilling customer orders in a timely fashion. Thus, you need to have production schedules that are consistent and predictable. You count on the expectation that a specific number of items will be produced each day and all your planning (shipping products, ordering supplies, etc.) is based on this figure. Then disaster strikes. There is dissatisfaction among your employees. You find that absenteeism suddenly has become highly unpredictable and variable, leading to tremendous variability in the number of items produced daily. Havoc!

Once you have variability, you lose the security that comes with knowing an event is consistent and, therefore, predictable. Variability introduces uncertainty—you can no longer rely on your carefully designed production schedule.

Variability also opens the doors to opportunity. For example, during periods of widely fluctuating changes in the market, an astute investor will take

Prime rate for 12 months of 1979 and 1980. Note the large variability in 1980 as contrasted   **Figure 3.5**
with 1979.

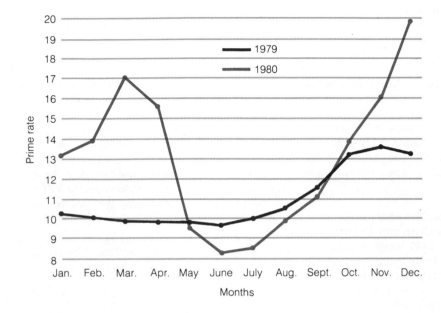

advantage of this variability. He or she will buy when prices are low and hope to cash in when prices are on the upswing.

To demonstrate the variability of business and economic data, Figure 3.5 displays two line charts—the average prime rate for each month of 1979 and 1980. Box 3.2 presents an amusing excerpt from *The Wall Street Journal* that comments on the tremendous variability in the prime rate in 1980.

3.3.1 Measures of range

The simplest measure of variability is the *range*. The range is merely the difference between the largest and the smallest measurements. Referring to Fig. 3.5, we see that in 1980 the range of prime rates was (19.91 − 8.31) = 11.60, whereas in 1979 the range was (13.53 − 9.79) = 3.74. Thus, the range confirms our observation of greater variability in 1980 than in 1979.

Stock market data, in general, lend themselves to the use of the range as a measure of variability. In addition to other information, individual stock prices are always reported in terms of "high" and "low" for the day and for the year. This range often plays a big role in investment strategies. Conservative investors often tend to focus their attention on stocks that show little fluctuation within a year or from year to year. Speculators are more likely to be attracted to stocks showing large fluctuations, particularly if they can buy when the price is at low ebb.

The obvious disadvantage of the range is that it is completely dependent

range

A measure of dispersion; the difference between the largest and the smallest score.

Box 3.2

*Wading Warily into
a Shallow Year*
_____

NEW YORK

Just about a year ago, President Carter's economists came out with their projection that 1980 would be a lackluster year, one in which real gross national product would average about 1% below 1979's physical output of goods and services. Indeed, the latest private estimates attest, real GNP last year did decline by around 1%. For 1981, the consensus favors a gain averaging about 1%.

Ho hum?

Not necessarily. Perhaps unavoidably in the aftermath of the lost election, the Carter crowd's commendably accurate forecast for 1980 brings to mind the tale of an economist who failed to return from a hiking vacation in unfamiliar territory. He had, it turns out,

been assured that the river he was about to wade across had an average depth of three feet. It did, but he was drowned anyhow.

A lot like that river, the 1980 economy was one in which seemingly placid averages masked some dangerous extremes. Interest rate gyrations have received the most attention, understandably so when the cost of business-sustaining credit reaches a record 20% annual rate, sinks by nearly half, and then doubles again to an even higher record level, all in one year.

Source: Reprinted by permission of *The Wall Street Journal,* January 19, 1981, © Dow Jones and Company, Inc., 1981.

---

fractile

A point below which some specific proportion or percentage fall.

quartile

Numbers that divide a distribution into four equal parts.

interquartile range

The score at the third quartile (Q_3) minus the score at the first quartile (Q_1); includes the middle 50 percent of observations.

on only two values in the distribution. For this reason, *fractiles* are often used to express variability in range.

A fractile is the point below which some specific proportion or percentage of measurements fall. We have already encountered the 50th fractile, which we identified as the median. Recall that 50 percent of the measurements fall below the median. Among the most common fractiles are the *quartiles,* which divide a distribution into four approximately equal numbers of observations. Twenty-five percent of the observations fall below the first quartile (Q_1), 50 percent below the second (Q_2), the median, and 75 percent below the third (Q_3). A range measure based on quartiles is the *interquartile range*. It represents the difference between the score at the third quartile and the first quartile, i.e., $Q_3 - Q_1$. Fifty percent of the cases fall within this range.

EXERCISES

3.12 Given the following data on net exports and imports for the four quarters of 1979, use the range to determine which shows the greater variability.

	$ billions			
	I	II	III	IV
Exports	238.5	243.7	267.3	280.0
Imports	234.4	251.9	269.5	287.7

Source: Survey of Current Business, January, 1980

3.13 The following table shows amounts spent in selected categories by family income level.

	Under $3,000	$3,000 to $3,999	$5,000 to $5,999	$8,000 to $9,999	$10,000 to $11,999	$12,000 to $14,999	$15,000 to $19,999	$20,000 to $24,999	$25,000 and over
Food	$ 739	$ 943	$1,126	$1,392	$1,563	$1,750	$2,010	$2,293	$2,651
Housing	1,310	1,539	1,709	2,128	2,342	2,591	3,027	3,495	4,682
Education	14	17	18	35	53	74	132	253	425
Alcohol	25	39	50	63	84	79	106	117	177

Find the range for each category.

3.3.2 Measures of deviation: Variance and standard deviation

Suppose you are a contractor involved in a highly competitive bidding situation for the construction of a major new shopping center. Your bid, of course, will be based on your estimate of costs—labor, materials, etc. Therefore, it is imperative that you be able to estimate these costs as accurately as possible. In addition, you need to know the variability to expect so that your bid can realistically reflect variations in your costs. If you do not take variability into account, you could, conceivably, lose money on what otherwise should be a lucrative deal. Moreover, in choosing suppliers, you will want to select those who have shown the most consistency or the least variability. Therefore, you need to have some way to measure variability.

The measures of variability based on range are easily calculated, readily understood, and useful as descriptive measures. However, when we are seeking precise quantitative measures that lend themselves to applications in inferential statistics, we must look elsewhere. We need not look far.

The variance and standard deviation, based on deviations from the mean, are important and widely used descriptive measures of dispersion. Moreover, their use is so pervasive in inferential statistics that they may well be considered its cornerstones.

The variance of a population is defined as the sum of the squared deviations of each score from the mean divided by the size of the population N. Symbolically,

$$\sigma^2 = \frac{\Sigma(X - \mu)^2}{N}$$

The standard deviation is simply the square root of the variance:

$$\sigma = \sqrt{\frac{\Sigma(X-\mu)^2}{N}}$$

Since we rarely have access to all the data in a population, most of the time we will be working with sample data. The formulas we use to define the sample variance and standard deviation are:

$$s^2 = \frac{\Sigma(X-\bar{X})^2}{n-1} \qquad (3.5)$$

$$s = \sqrt{\frac{\Sigma(X-\bar{X})^2}{n-1}} \qquad (3.6)$$

Note that we use italic letters to designate the three sample statistics:

$\bar{X}$ = mean of a sample
s^2 = variance of a sample
s = standard deviation of a sample

You may have noticed a difference in the denominator of the sample variance compared to the population variance. The use of $(n-1)$ in calculating the sample variance yields an unbiased estimate of the population variance. This means that, with repeated sampling from a population, the use of $(n-1)$ in the denominator will, on the average, provide a more accurate estimate of the population variance than using n in the denominator. Indeed, mathematicians have shown that the use of n in the denominator leads to consistent underestimates of the size of the population variance. It is, therefore, said to be biased. An unbiased estimate, on the other hand, yields the population variance as its average value.

Example D: Mean deviation method

Calculate the variance and standard deviation for the following sample data: 4, 5, 6, 7.

Solution

X	$X - \bar{X}$	$(X - \bar{X})^2$
4	−1.5	2.25
5	−0.5	0.25
6	+0.5	0.25
7	+1.5	2.25
Total 22	0	5.00

$$\bar{X} = 5.5$$

$$s^2 = \frac{\Sigma(X - \bar{X})^2}{n - 1} = \frac{5.00}{3} = 1.67$$

$$s = \sqrt{\frac{\Sigma(X - \bar{X})^2}{n - 1}} = 1.29$$

Raw score method. For most sample data, the mean has some fractional value. Consequently, subtracting the mean from each score and squaring can be both time consuming and fraught with the possibility of error. Fortunately, there is a way of obtaining $\Sigma(X - \bar{X})^2$ directly from the raw scores. It has been shown algebraically[3] that

$$\Sigma(X - \bar{X})^2 = \Sigma X^2 - \frac{(\Sigma X)^2}{n}. \tag{3.7}$$

Formula 3.7 directs us to add together the squares of each measurement (ΣX^2) and subtract the squared sum of the measurements $[(\Sigma X)^2]$ divided by n.

Once we have found $\Sigma X^2 - \dfrac{(\Sigma X)^2}{n}$ by use of the raw score method, we may substitute this numerical value for $\Sigma(X - \bar{X})^2$ in Formulas (3.5) and 3.6).

Example E: Raw score method

The table below, and continuing on page 70, illustrates the calculation of the mean, variance, and standard deviation from the data in Example D, using the raw score method.

X	X^2
4	16
5	25
6	36
7	49
Total 22	126

---

[3] $$\Sigma(X - \bar{X})^2 = \Sigma X^2 - 2\Sigma X \bar{X} + \Sigma \bar{X}^2$$

Since $\Sigma X = n\bar{X}$ and summing the mean square over all values of $\bar{X}$ is the same as multiplying by n,

$$\Sigma(X - \bar{X})^2 = \Sigma X^2 - 2n\bar{X}^2 + n\bar{X}^2$$
$$= \Sigma X^2 - n\bar{X}^2$$
$$= \Sigma X^2 - n\left(\frac{\Sigma X}{n}\right)^2$$
$$= \Sigma X^2 - \frac{(\Sigma X)^2}{n}.$$

$$\bar{X} = \frac{22}{4} = 5.5$$

$$\Sigma(X - \bar{X})^2 = 126 - \frac{(22)^2}{4}$$
$$= 126 - 121$$
$$= 5$$

$$s^2 = \frac{5}{3} = 1.67$$

$$s = \sqrt{\frac{5}{3}} = 1.29$$

Much of the data we encounter in business and economics is presented to us in the form of frequency distributions. As with the mean, grouping results in some loss of information and accuracy. Consequently, the variance and standard deviation calculated from a frequency distribution will differ somewhat from the same measures calculated directly from the raw scores.

The computational formulas for the variance, standard deviation, and $\Sigma f(X - \bar{X})^2$ for grouped data are:

$$s^2 = \frac{\Sigma f(X - \bar{X})^2}{n - 1} \qquad (3.8)$$

$$s = \sqrt{\frac{\Sigma f(X - \bar{X})^2}{n - 1}} \qquad (3.9)$$

$$\Sigma f(X - \bar{X})^2 = \Sigma f X^2 - \frac{(\Sigma f X)^2}{n} \qquad (3.10)$$

Example F: Grouped data

Many decisions we make in life are based on considerations of variability. Gas, electric, and water utilities, mass transit systems, telephone networks, bridges, to name but a few, are not designed for "average" use but must anticipate the extreme demands known as peak load. Otherwise the systems fail at crucial moments. Even in athletic competition, variability is a prime consideration. Imagine that a quarterback is faced with a third-down situation. He decides on a ground play, but which running back should he use? Both candidates have similar means, as can be seen in the table on page 71.

Net yardage	Runner A			Runner B		
	Midpoint	f	fX	Midpoint	f	fX
−4 under −2	−3	0	0	−3	5	−15
−2 under 0	−1	2	−2	−1	9	−9
0 under 2	1	12	12	1	10	10
2 under 4	3	30	90	3	4	12
4 under 6	5	4	20	5	13	65
6 under 8	7	2	14	7	11	77
Total		50	134		52	140

$$\bar{X}_A = \frac{134}{50} = 2.68 \qquad \bar{X}_B = \frac{140}{52} = 2.69$$

a. Calculate the variance and standard deviation for each player.
b. Which one should he call upon if two yards are needed for a first down?
c. Which one should he call upon if six yards are needed for a first down?

Solution

a.

Class limits	Runner A				Runner B			
	Midpoint	f	fX	fX²	Midpoint	f	fX	fX²
−4 under −2	−3	0	0	0	−3	5	−15	45
−2 under 0	−1	2	−2	2	−1	9	−9	9
0 under 2	1	12	12	12	1	10	10	10
2 under 4	3	30	90	270	3	4	12	36
4 under 6	5	4	20	100	5	13	65	325
6 under 8	7	2	14	98	7	11	77	439
Total		50	134	482		52	140	964

Runner A:

$$\Sigma f(X - \bar{X})^2 = 482 - \frac{(134)^2}{50}$$

$$= 122.88$$

$$s^2 = \frac{122.88}{49} = 2.51$$

$$s = \sqrt{\frac{122.8}{49}} = 1.58$$

Runner B:

$$\Sigma f(X - \bar{X})^2 = 964 - \frac{(140)^2}{52}$$

$$= 587.08$$

$$s^2 = \frac{587.08}{51} = 11.51$$

$$s = \sqrt{\frac{587.08}{51}} = 3.39$$

a. Although, on the average, both gain an amount sufficient to make a first down, A is more consistent. Most of his gains are for small amounts. B is far more variable, with some large gains as well as a number of losses. A should be chosen.

> *b.* Although A is consistent, he doesn't often gain as much as six yards (2/50 = 0.04). B, although variable, gains six or more yards with a greater relative frequency (11/52 = 0.21). B should be chosen.

A note of caution. A common error when using the raw score method to calculate s and s^2 is to confuse the similar appearing ΣX^2 (or $\Sigma f X^2$) and $(\Sigma X)^2$ [or $(\Sigma f X)^2$]. Keep in mind that the former is the sum of the squares of each individual score, whereas the latter represents the square of the sum of the scores. It is impossible to obtain a negative value for $\Sigma(X - \overline{X})^2$ since squaring eliminates all negative signs. In the event you obtain a negative value for $\Sigma(X - \overline{X})^2$, you have probably interchanged the two terms.

3.3.3   The interpretation of the standard deviation

The example involving the two running backs illustrated the fact that the standard deviation permits us to compare the variability of two or more distributions. The greater the standard deviation, the greater the variability.

The standard deviation takes on added significance when we wish to obtain the approximate proportion of measurements falling between various values of the variable. For example, suppose you have the responsibility for maintaining the weights of packages of soap. By law you are required to state the weight of the contents. However, there is bound to be variability among the different boxes (some will weigh more, some less). But as long as 75 percent of all weights fall between a specified range of values, you have satisfied the legal requirement. How would you go about specifying this range? Fortunately, a rather startling mathematical statement, known as *Chebyshev's theorem,* permits determinations of this sort.

The theorem states:

> No matter what the form of the distribution, the proportion of the measurements lying within k standard deviations of the mean is *at least* $1 - 1/k^2$, where k is any positive number larger than 1.

Thus, between the mean and ± 1.2 standard deviations, the proportion of area is at least

$$1 - \frac{1}{(1.2)^2} = 1 - 0.69 = 0.31.[4]$$

Similarly, between the mean and ± 2.0 standard deviations, the proportion of area is at least

$$1 - \frac{1}{(2.0)^2} = 1 - 0.25 = 0.75.$$

---

[4] Note that if we multiply the obtained proportion by 100, we obtain a percentage. Thus, 0.31×100 equals 31 percent

The proportion of area between the mean and ± 3.0 standard deviations is at least

$$1 - \frac{1}{9} = 1 - 0.11 = 0.89.$$

This is shown graphically in Figure 3.6.

Graphic illustration of Chebyshev's theorem. In any distribution *(a)* at least 31 percent of the area is between the mean and ± 1.2 standard deviations; *(b)* 75 percent is between the mean and ± 2.0 standard deviations; and *(c)* 89 percent is between the mean and ± 3.0 standard deviations.   **Figure 3.6**

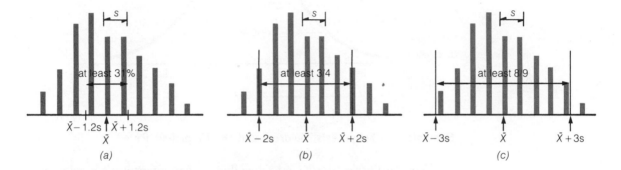

Let us return to the problem of the soap packages. Suppose your quality control department reports that the mean weight of all boxes is 425 grams with a standard deviation of 1.5. Using Chebyshev's theorem, you already know that 425 grams ± 2 standard deviations will include *at least* 75 percent of all the cases. Thus, the range of weights will be:

$$425 \pm 2(1.5)$$
$$= 425 \pm 3$$

or between 422 and 428 grams.

You may have noticed that we emphasized the words *at least* when stating Chebyshev's theorem. This is because the theorem is very conservative and, in many cases, the actual proportion will be greater than specified. Thus, this theorem applies to *any* distribution, regardless of its form.

It has often been observed that many distributions are bell-shaped and approach a form known as the normal curve. With this type of distribution, we may specify more precisely the proportion of area between the mean and various standard deviations above and below the mean. More specifically, when we know that our data come from approximately normally distributed populations, we are able to calculate intervals that are narrower than those calculated from Chebyshev's theorem:

74

> *Empirical Rule* (applies only to approximately normal bell-shaped distributions):
> The proportion of area between the mean and:
> ±1.0 standard deviation is approximately 0.68;
> ±2.0 standard deviations is approximately 0.95;
> ±3.0 standard deviations is approximately 0.997.

This is shown graphically in Fig. 3.7.

Figure 3.7
Graphic illustration of the Empirical Rule

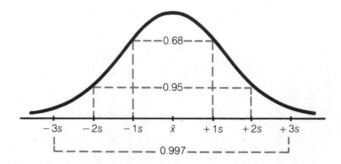

Example G: Chebyshev's Theorem and the Empirical Rule

> When airlines establish connecting flights, it is important that consideration be given to possible flight delays and transfer times at the terminal. A random sample of flight records reveals that the mean time lost through transfer times and flight delays is 16.5 minutes with a standard deviation of 4.5.
>
> a. Using Chebyshev's theorem, estimate the range of delays that will include 89 percent of all delays.
> b. Assume that the delays are normally distributed about the mean, estimate the range of delays that will include about 95 percent of all the delays. Through clerical error, you were given only seven minutes to make a connecting flight. Assuming that the connecting flight leaves on time, how confident are you that you will make a successful connection?

Solution

a. Using Chebyshev's theorem, at least 89 percent of all delays will be between the mean and ± 3 standard deviations. Thus, the minimum range of delays that will occur 89 percent of the time is 16.5 ± (3)(4.5), or between 3 and 30 minutes.
b. Using the Empirical Rule, 95 percent of the delays will be between the mean and ±2 standard deviations. On the assumption of a normal distribution, the range of delays that will occur 95 percent of the time is

16.5 ± (2)(4.5), or between 7.5 and 25.5 minutes. Since delays are rarely less than 7.5 minutes, your confidence in making the connection should be extremely low.

3.4 JUDGING NORMALITY OF A DISTRIBUTION

The use of the Empirical Rule requires that a data set approach the form of the normal curve. How do we know when this is, in fact, the case? To provide a precise answer to this question is beyond the scope of this book. Nevertheless, we may obtain a crude but useful judgment as to whether or not the data are approximately normal by use of the following two procedures: estimating the extent of skew; and finding the percentage of measurements falling between ±1, ±2, and ±3 standard deviations of the mean.

Judging skewness. In Section 3.2.6, we saw that the mean is to the right of the median when there is positive skew and to the left when there is negative skew. The absolute difference between the mean and median provides us with a numerical value representing the amount of skew but, in the absence of a standard for comparison, it is of little use. A useful quantitative estimate of skew is provided by Pearson's coefficient of skew *(sk)*. It expresses the deviation between the mean and median in terms of the standard deviation, i.e.,

$$sk = \frac{3(\overline{X} - \text{Median})}{s}.$$ (3.11)

Let us apply this formula to the array of measurements shown in Table 3.5.

1.7	3.0	3.4	3.6	4.0
2.2	3.0	3.4	3.6	4.0
2.2	3.1	3.4	3.6	4.0
2.3	3.1	3.4	3.6	4.3
2.5	3.1	3.4	3.7	4.5
2.7	3.1	3.4	3.8	5.1
2.8	3.1	3.5	3.9	5.1
2.9	3.1	3.5	3.9	5.4
2.9	3.3	3.5	3.9	5.6
2.9	3.3	3.6	4.0	5.8

$$\overline{X} = 3.52$$
$$\text{Median} = 3.40$$
$$s = 0.84$$

Table 3.5
Array of rates of motor vehicle traffic deaths per 100 million passenger miles in all 50 states during 1978

Using the mean, median, and standard deviation calculated from these data, we obtain the following index of skew:

$$sk = \frac{3(3.52 - 3.40)}{0.84} = 0.43$$

This indicates some positive skew. If a distribution is symmetrical, the mean and median are the same, so that sk = 0. However, it is generally accepted that data sets with indexes of skew ranging between plus and minus 0.50 may be considered symmetrical for all practical purposes.

Calculating proportions of area between ±1, ±2, and ±3 standard deviations. Recall that, if our distribution approaches a normal curve, the area included between $\overline{X}$ and $\pm 1s$ is 68 percent; between $\overline{X}$ and $\pm 2s$ is 95 percent; between $\overline{X}$ and $\pm 3s$ is 99.7 percent.

If the data shown in Table 3.5 are drawn from a population that approximates normality, we would expect to find the areas between the mean and ±1, ±2, or ±3 standard deviations to conform roughly to the Empirical Rule. Thus, approximately 68 percent (i.e., 34 measurements) should fall in the interval $\overline{X} \pm 1s$ or 3.52 ± 0.84, or between 2.7 and 4.4, rounded to the first decimal place. Counting the number of measurements that fall between these values, we find 39 cases or 78 percent. Similarly, 46 measurements (or 92 percent) fall between 1.8 and 5.2. Finally, we see that 100 percent of the measurements fall between ±3 standard deviations. Thus, the obtained values are crude approximations to expectations based on the Empirical Rule.

SUMMARY

In this chapter, we examined several statistics calculated from data in which single descriptive measures summarized some salient aspect of our data. Specifically, we presented and discussed several measures of central tendency and dispersion.

1. Under measures of central tendency, we looked at the mean, median and mode. Every measurement participates in the calculation of the mean.
2. The median is the point in the distribution that divides the distribution in half, so that 50 percent of the data set falls on either side of the median.
3. The mode is the measurement or the class with the highest associated frequency.
4. The mean is a valuable measure of central tendency, particularly when there is no pronounced skew. It cannot be calculated when there are indeterminate values. It is influenced by and drawn toward extreme measurements.
5. The median is often the measure of choice whenever data sets are skewed or when there are indeterminate values.
6. Variability is a fact of life. Things do not remain the same and fluctuations are the rule rather than the exception. The measures of dispersion examined in this chapter included range measures (the range and interquartile range) and deviation measures (the variance and standard deviation).
7. The variance and standard deviation provide precise quantitative measures of variability. Both are widely used in both descriptive and inferential statistics.
8. We examined two rules that permit us to make use of our knowledge of the mean and standard deviation to interpret data. Chebyshev's theorem

describes the minimum proportion of measurements between the mean and k standard deviations from the mean when k is greater than 1. It applies to any distribution, no matter what the form. The Empirical Rule, which applies to normally distributed variables, describes the proportions of area between the mean and ± 1, ± 2, and ± 3 standard deviations from the mean.

9. Finally, we examined two procedures for judging the normality of a data set. These involved calculating Pearson's index of skew and finding percentages of area between ± 1, ± 2, and ± 3 standard deviations from the mean.

TERMS TO REMEMBER

measures of central tendency (51)
arithmetic mean (51)
weighted mean (56)
median (58)
mode (60)
normal curve (61)
skewed distribution (61)

positively skewed distribution (61)
negatively skewed distribution (61)
range (65)
fractile (66)
quartile (66)
interquartile range (66)

EXERCISES

3.14 Two manufacturers state that the average life of their microwave ovens is six years. They base their claims on the following data that show the service life in years:

> I: 4,4,4,5,5,5,5,6,6,6,6,6,6,7,7,7,7,8,8,8
> II: 1,2,3,4,4,4,4,5,5,5,6,6,6,6,6,7,7,19,19,19

    a. What measure of central tendency was each manufacturer reporting?
    b. Which oven do you think would be a better buy in terms of longer average life?
    c. With which oven do you feel more confident in stating that the mean is a representative measure of central tendency?

3.15 The table below shows the percent changes in the Consumer Price Index (CPI) over a seven-year period for selected goods and services.

ungrouped

	1972	1973	1974	1975	1976	1977	1978
Food, drink	4.1	13.2	13.8	8.4	3.1	6.0	9.7
Housing	3.8	4.4	11.3	10.6	6.1	6.8	8.6
Apparel, upkeep	2.1	3.7	7.4	4.5	3.7	4.5	3.4
Transportation	1.1	3.3	11.2	9.4	9.9	7.1	4.9
Medical care	3.2	3.9	9.3	12.0	9.5	9.6	8.4
Entertainment	2.9	2.8	7.5	8.9	5.0	4.9	5.1
Other	4.2	3.9	7.2	8.4	5.7	5.8	6.4

Source: Adapted from *The World Almanac & Book of Facts 1980*, p. 41, copyright © Newspaper Enterprise Association, New York, N.Y. 10166.

Calculate the mean and standard deviation for each of the categories listed. Which shows the greatest variability?

3.16 There are 450 employees in the assembly line skilled in a particular aspect of tooling. The time required to complete this job is normally distributed with a mean of 68.5 and a standard deviation of 8.5.

A rush job comes in. The only way to meet the deadline is to allot 60 minutes for this particular procedure. What are the chances of success, if an employee is picked at random from this assembly line?

3.17 An investor finds that the mean price of a particular stock was $13.50 with a standard deviation of 0 during the last month. What was the price on the ninth day? On the sixteenth day?

3.18 Late in 1973 the national speed limit of 55 mph was enacted into law. An apparent side effect has been a drop in the number of highway fatalities. The accompanying table shows frequency distributions of fatalities per 100 million passenger miles during the years 1973 and 1978.

	f	
Classes	1973	1978
1.2–under 1.8	0	2
1.8–under 2.4	0	2
2.4–under 3.0	5	7
3.0–under 3.6	5	20
3.6–under 4.2	15	11
4.2–under 4.8	10	3
4.8–under 5.4	6	3
5.4–under 6.0	3	3
6.0–under 6.6	6	—
6.6–under 7.2	1	—
7.2–under 7.8	0	—

a. Calculate the mean rate of traffic fatalities for each year.
b. Calculate the standard deviation and variance for each year.
c. Calculate the median for each year.

3.19 Referring to Exercise 3.18:
 a. Use Chebyshev's theorem to estimate the values that include 75 percent of the area on both sides of the mean.
 b. Estimate the values including the middle 68 percent, using the Empirical Rule.

3.20 A testing laboratory subjected a random sample of 29 plastic toys to a stress test in which it recorded the number of pounds of pressure per square inch required to produce stress cracks. The results were as follows:

22.3	17.2	74.4	38.6	75.5	23.7	94.6	18.2	13.1	19.0
84.4	54.7	42.6	14.2	65.0	38.8	65.1	35.5	36.0	74.3
38.9	36.1	57.1	93.7	32.8	92.0	95.8	39.5	27.6	

 a. Find the mean, median, variance, and standard deviation.
 b. Using Chebyshev's theorem, find the values of the variable between which at least 75 percent of the measurements fall.
 c. Does the data set appear to approach the normal curve?

Probability: Introduction
to Basic Concepts

4.1 PROBABILITY IS PROBABLY EVERYWHERE

Uncertainty pervades all aspects of human endeavor. Probability is one of our most important conceptual tools because we use it to assess degrees of uncertainty and thereby to reduce risk.

Whether or not you have had formal instruction in this topic, you are already familiar with the concept of probability since it pervades almost all aspects of our lives. Without consciously realizing it, many of your decisions are based on probability. For example, when you study for an examination, you concentrate more on areas that you feel are likey to be covered on the test. You may cancel or postpone an outdoor activity if you believe the likelihood of rain is high. How often have you decided to try something because you felt there was a good chance that you would enjoy it?

In business, probability plays a key role in decision making. The owner of a retail shoe store, for example, orders heavily in those sizes that he or she believes likely to sell fast. The owner of a movie theatre schedules matinees only during holiday seasons because the chances of filling the theatre are greater at that time. Two competing companies decide to merge when they believe the probability of success is greater for the consolidated company than for either independently.

The following excerpts from newspaper articles further demonstrate the pervasive role of probability:

> Portrait of a murder: Odds are that it will take place in the second half of the year. The victim will be male and he and the killer will have known each other and and have been arguing before the slaying. There is a 20 percent chance that the killer and victim will be related.

> The odds of a major earthquake are high.

> Most printers don't expect a paper shortage, increased postal rates, or a recession to hurt this year's business much.

> Top executives are queasier about spending huge sums on products that history warns are likely to fail.

> According to a post-election poll sponsored by the New York Stock Exchange, "Fully 52 percent of all Americans say a depression in the next two or three years is either very likely or somewhat likely."

We usually express probability as a proportion, with values varying between 0 and 1. If it is impossible for something to occur, the probability is zero. On the other hand, if something is *certain* to occur, the probability is 1.00. Thus, for any given outcome, say O_i,

$$0 \leq p\ (O_i) \leq 1.00.$$

Probability may also be expressed as a percentage, or the number of chances in 100. Thus, if the probability of an event occurring is 0.80, we may expect this event to occur 80 percent of the time, or the chances that this event will occur are 80 in 100. When expressed as a percentage, probability values vary between 0 and 100 percent.

We may also express probability in terms of odds for or against (see Box 4.1).

Box 4.1

ODDS AND ENDS

Odds are the exact number we attach to a future possibility to state how likely it is to occur. If you shop every other day, the odds of your shopping on any given day are 50/50: in other words on a 50% chance the odds are even. If you shop only on every third day, we can find the odds of your shopping on any given day as the relationship between two days of non-shopping and one day of shopping. Thus, the odds of your going shopping on any given day are 2–1 against.

That's all the elements needed to compute the odds. First, you need to know the number of times an event actually occurs compared to the number of times it does not occur (in this case shopping vs. not-shopping), and you need a constant time frame for the two events (in this case three days).

The next principle of odds: For any given odd in favor of something happening there is a reverse odd in favor of it not happening. If we are discussing dogs and it is known that one out of 100 dogs bite, this can be discussed in one of two ways:

A. The odds that a dog won't bite, 99–1 for.
B. The odds that a dog bites, 99–1 against.

Here is how the odds break on your money and your job.

What are the odds on becoming a millionaire?

Thirty years ago there were just 13,000 millionaires in the United States; one for every 11,300 citizens. Today there are 520,000 people with a net worth of more than one million. The odds are based on 2.36 millionaires per thousand people.

The odds: 423.1 against.

What are the odds on hitting a jackpot?

Have you ever noticed how tantalizingly close you come to hitting the jackpot on a slot machine? They're all planned that way. There are 20 squares on each of the three wheels. Typically the middle one has nine bells out of the 20, meaning that the odds on your middle wheel turning up a bell are almost even. But the outer wheels have only one bell out of 20 alternatives. The calculation looks like this: $9/20 \times 1/20 \times 1/20$.

The odds: 889.1 against.

What are the odds on having to pay the IRS more money on your individual tax return?

Of the 87,386,000 individual returns filed for fiscal 1978, a total of 7,567,313 drew added assessments (amounting to $448 million dollars). This included 91,000 returns paid by bum checks.

The odds: against.

Total	10.5–1
Failure to pay	22.1–1
Tax due	29.1–1
Delinquency	122.0–1
Bad check	959.0–1
Negligence	1,323.0–1
Other (includes not reporting tips) . .	12,119,110–1
Fraud	13,041,168–1

What are the odds that your tax return will be examined?

The IRS examined 306,433 individual returns, 147,273 corporate returns and 27,579 partnership returns during fiscal 1978.

The odds:

Individual	284.0–1 against
Corporate	14.9–1 against
Partnership	42.7–1 against

What are the odds on a check bouncing?

According to the Federal Reserve Board, of the 32 billion checks written annually 0.67% bounce. It sounds like a small figure. It adds up to a large number of rubber checks, 214 million, almost one check for every man, woman and child in America.

Interestingly, Americans don't bounce checks where they most often (after banks) cash them: at the supermarket. The average supermarket cashes 1,707 checks each week yet just 1.5 bounce. What's more the average bounced check at $25.68 is well be-

Box 4.1 *(continued)*

	Women	Men
Union hiring hall	99.9–1	37.5–1
Civil service test	34.7–1	61.5–1

low the average of all checks cashed, $40.85. So if you're a supermarket operator the odds are 1,137–1 that any given check is OK.

What are the odds on men and women getting their jobs in various ways?

For both sexes about one-third of all workers get jobs they've applied for directly. Newspaper ads are next in importance for women, while men rely on friends to tell them about openings where the friends already work or elsewhere.

The odds: against.

	Women	Men
Applied directly to employer ...	1.9–1	1.8–1
Via friends	8.3–1	6.2–1
Via relatives	18.6–1	13.5–1
Answered newspaper ad	5.4–1	7.5–1
Employment agency (state or private)	6.6–1	10.4–1
School placement office	34.7–1	31.3–1

What are the odds of a business going bankrupt?

In 1978 a total of 6,619 businesses failed. The average debt when they went under was a hefty $355,946. The failure rate per 10,000 businesses has varied dramatically over the last five years as shown by the figures which follow.

The odds: against.

1978	416–1
1977	356–1
1976	285–1
1975	232–1
1974	262–1

Source: Reprinted by permission of G. P. Putnam's Sons from *The Odds on Virtually Everything* by the Editors of Heron House. Copyright © 1980 by Heron House, Inc.

4.2   THE BIRTH OF PROBABILITY THEORY

It was 17th-century France. An aristocrat by the name of Antoine Gombauld was in severe economic circumstances. The cost of maintaining his status as the playboy known as the Chevalier de Méré was digging deeply into the family coffers. Not equipped by education or inclination to seek gainful employment, he desperately sought a way out of his financial dilemma. Turning to the activity he knew best—gaming tables—he contrived a scheme that he was certain would reverse the financial drain. He reasoned as follows: "If I toss a single die once, the probability that it will turn up a six is one in six. If I toss it twice the chances of obtaining a six are two in six. A third toss will even the odds at three chances in six. A fourth toss, however, will turn the odds in my favor. The chances of a six are four in six. The odds are two to one in my favor. If I can get people to give me even odds that I will obtain at least one six in four tosses of a die, I will clean up."

And clean up he did. Oddly, though, he was right for the wrong reasons. If he had carried his reasoning two steps further, he would have surely detected a flaw. If he tossed five dice, would he then argue that the probability is five out of six in favor of obtaining a six? If six dice are tossed, are the chances six out of six of obtaining a six? That's certainty. Anyone familiar with dice will know that certainty is not one of their most dependable attributes. In truth, as we shall see, the probability of obtaining at least one six in six tosses of a die (or one toss of six dice) is $\frac{2}{3}$.

Nevertheless, the money continued to roll in. Fortunately for de Méré, the probability of obtaining a six in four toss is slightly better than 50–50. It was this slight advantage, repeated many times over in countless wagers, that maintained a positive cash flow.

Then de Méré either became greedy or he sought to add a new thrill to whet the appetites of his jaded clientele. Instead of a single die, he would toss a pair of dice. Pursuing the same reasoning as before, he concluded that the probability of getting a 12 (two 6s) is 1 in 36. On 24 tosses of a pair of dice, he reasoned that the chances of obtaining a 12 are 24 out of 36. "That's still two to one in my favor," he exulted. But suddenly he began to lose. Unbeknownst to him, the odds in the new game were ever so slightly in favor of the bettors. However, a small loss coupled with a rapid turnover can spell financial disaster, as de Méré soon discovered.

In desperation, he turned to the youthful mathematician, Blaise Pascal, for an answer to his quandary. Pascal, in turn, began a correspondence with one of the most eminent mathematicians of his day—Pierre de Fermat. Together they worked out the reasons for de Méré's failed venture. In the process, they gave birth to the theory of probability.

The Chevalier de Méré was not completely wrong. In fact, his overall conception was sound even if the details were faulty. He recognized that the behavior of a die is not chaotic and disorderly. Although the outcome of any particular toss may be unpredictable, the proportions in a large number of tosses are orderly and predictable. Probability theory is concerned with identifying and describing that order.

4.3 THREE APPROACHES TO PROBABILITY

4.31 Classical approach to probability

There are several different approaches to the development of probability theory. De Méré used what has come to be known as the *classical approach.* It starts out by posing a conceptual *experiment* in which the following types of questions are raised: What is the probability of obtaining a six on one toss of a theoretically perfect die? What is the likelihood of obtaining a head on a single trial (one toss) of a balanced coin? With the classical approach, no dice are actually thrown nor are coins tossed. The reason is that theoretically perfect dice and coins exist only in the mind and not in the real world. The structure of these hypothetically perfect populations is known. A coin has two sides, each of which is equally likely to appear on the face (it cannot get lost during the hypothetical toss nor land and remain on its edge). A perfect die has six sides, each also equally likely to appear. The *outcome* of any toss of a coin must be a head or a tail, and the outcome of throwing a die must be 1, 2, 3, 4, 5, or 6. The two outcomes with a coin and the six with a die include all possible outcomes. In other words, when we have enumerated all outcomes, we have exhausted all of the possibilities. We speak

classical approach to probability

The assignment of probability values on the basis of theoretical expectations in a hypothetical population.

experiment

The process by which we obtain measurements or observations of different outcomes.

outcome

A single possible result of an experiment; one that cannot be further subdivided.

84

exhaustive

Outcomes are exhaustive if the sum of their separate probabilities is equal to 1.00.

mutually exclusive

Two or more outcomes that cannot occur simultaneously.

of all possible outcomes as being *exhaustive* and the sum of their separate probabilities must equal 1.00.

Moreover, in this theoretical world of coin throws and die tosses, two different outcomes cannot be obtained at the same time on any given toss. Thus, we cannot obtain both a head and a tail on a single toss of a coin. In other words, each outcome is *mutually exclusive*. This characteristic is not true for all situations to which classical probability theory has been addressed.

Consider a deck of hypothetically perfect playing cards. The 52 cards include 13 different face values (from ace through king) and 4 different suits. On a single trial, it is possible to obtain, say, a king and a spade. In other words, the face value of the card and the suit are not mutually exclusive. However, if each card is considered a possible outcome, it is mutually exclusive of every other card. Thus, it is not possible to obtain both a king of spades and the king of diamonds on the same trial.

Having specified the complete structure of the hypothetical experiment, the definition of probability is quite straightforward. Since there are six mutually exclusive and equally likely outcomes of tossing a single die, the probability of obtaining one specific outcome (for example, a six), is defined by the following proportion:

$$p(6) = \frac{\text{Number of outcomes favoring a six}}{\text{Total number of possible outcomes}}$$
$$= \frac{1}{6}.$$

Similarly, the probability of obtaining a head *(H)* on a single toss of a perfectly balanced coin is $p(H) = \frac{1}{2}$.

4.3.2 Empirical or relative frequency approach to probability

The classical theory of probability is fine for dealing with games of chance (hypothetical ones, at that!). But how often do we know the complete structure of a population in the real world? How often are we able to specify all possible outcomes? Do we ever deal with theoretically perfect circumstances?

Consider the following real-world problems:

1. What is the probability of a defect in a silicon chip produced by a new assembly-line technique?
2. What is the probability that a student will be accepted for an internship in an accounting firm after failing his or her first course in accounting?
3. What is the probability that, during the open season for duck hunting, a whooping crane will be shot?
4. What is the probability that a contractor making the low bid on a building project will be awarded the contract?

In each of these cases, the structure of the population is unknown. We don't know nor can we expect to know all possible outcomes of silicon chip production, hiring practices of all accounting firms, the whooping crane toll of all hunters, and the bids of all contractors on all projects. But we can collect sample data from populations and determine the number of outcomes favoring a given result relative to the total number of outcomes observed. When we have done this, we have used the *empirical* or *relative frequency* approach to probability. Under this approach, the probability of A is ascertained by

empirical or relative frequency approach to probability

The assignment of probability values based on the observed relative frequency with which that outcome occurred in the past.

$$p(A) = \frac{\text{Number of observed outcomes favoring } A}{\text{Total number of observed outcomes}}. \qquad (4.1)^1$$

It is important to stress that the probability value obtained in this way (from observed or sample data) is an *estimated* probability, not a "true" population probability. Later we shall see how sample probabilities may be revised on the basis of new experimental evidence. In the meantime, using this approach, the obtained probability is our best estimate of the true probability in the population.

Thus, if 120,463 silicon chips are examined and 463 are found defective, we estimate the probability of a defective $[p(A)]$ as

$$p(A) = \frac{463}{120,463} = 0.0038$$

All other chips are assumed to be nondefective. Therefore, we have a two-category situation (defective versus nondefective). The second category is referred to as the *complement* of the first and is symbolized by $\overline{A}$, and called "not A." Since the two categories are mutually exclusive and exhaustive (meaning that one *or* the other must occur) and the p (certain event) = 1.00, it follows that

complement

The complement of an outcome is its opposite.

$$p(A) + p(\overline{A}) = 1.00$$

and

$$p(\overline{A}) = 1.00 - p(A)$$

---

[1] We may also define "empirical" probability as the limit of the relative frequencies:

$$p(A) = \lim_{n \to \infty} \left(\frac{x}{n}\right),$$

where

x = number of observed outcomes favoring A

n = total number of outcomes

Given the following values of x and n, we find $p(A)$:

x	n	p(A)
0	0	Undefined
0	> 0	0
$0 < x < n$	> 0	$0 < p(A) < 1.00$
$0 < x = n$	> 0	1.00

In the silicon chip example, the probability of obtaining a nondefective chip is, therefore,

$$p\,(\overline{A}) = 1.00 - 0.0038$$
$$= 0.9962$$

Note that the probabilities based on classical theory are as firm and solid as the Rock of Gibraltar (i.e., the theoretical Rock: the actual Rock could blow up any day). In contrast, empirical probabilities are often erected on shifting sands. For example, if the assembly line process developed a serious malfunction, the probability of obtaining a defective chip might increase dramatically. This is one of the reasons that quality control procedures are a continuing process. Can you see why? [Hint: Look back at Formula (4.1), which defines empirical probability.] Another example that illustrates how empirical probabilities can change dramatically is provided by the behavior of the prime rate. Prior to March 1980, the empirical probability of the prime rate reaching 20 percent was zero since it had never happened before. However, after that point (i.e., after the prime rate reached 20 percent), the empirical probability changed. There are some economists who, *prior* to March 1980, were predicting that the likelihood of the prime rate reaching 20 percent or higher was very high. How could they predict this when the empirical probability was zero? This leads us to the subjective approach to probability.

4.3.3 Subjective approach to probability

There are many occasions when we have little or no data on which to base our empirical probabilities. This is particularly the case when a new discovery, technology, or development bursts upon the scene. We simply have no records of past performance. A current example is nuclear technology. With the availability of fossil fuel sources of energy imperiled by the unstable world situation, some have hailed nuclear reactors as the energy source of the future. Others have responded with alarm at the prospects of sabotage, explosions, radioactive leaks, contamination of the environment by waste products, and the theft of fuel by radical groups for manufacturing atomic bombs. How do we assess the risks? It may not be possible to count the instances in which sabotage or explosions occurred; hence a relative frequency cannot be obtained. Therefore, we form *subjective probabilities*. Even though we are unable to confirm or defend their accuracy, it is often necessary to arrive at some estimate in order that decision-making processes may continue. An industrialist may learn that the latest economic indicators suggest a downturn in the economy but may go ahead with an expansion plan based on "feelings in the bones." We should not dismiss subjective probabilities out of hand. Of course, different people may arrive at different subjective probabilities. However, using all objective *and* subjective evidence available, we should understand that, at times, a given individual's belief or confidence (i.e., subjective probability) is the only reasonable way to assess a particular situation. There are numerous instances where subjective strategies could be, and are,

subjective approach to probability

The assignment of probability values based on the individual's belief or confidence that a particular outcome will occur.

used in determination of actions. For example, the director of personnel must assess the degree of success associated with various job application procedures; or the structure of a real estate deal may reflect anticipated mortgage rates one year from now; or the production supervisor of an automobile manufacturer may weigh the gas saving/riding comfort tradeoff in terms of the introduction potential for a new series of energy-efficient cars. Subjective probabilities play an important role in business and economic decisions and may indeed be based on objective evidence. We shall accept the concept of subjective probability and expect it to follow the same rules as probabilities obtained through the classical or empirical approach.

Example A: Putting it all together: Three approaches to probability

Let us use Montana Rincon's problem to see how the three approaches to probability may be integrated.

We shall define the population of interest as all accounts that contain errors. There are two possible outcomes for each error—either it favors the company or it favors the customer. Using the classical approach, we assume we are dealing with a hypothetical population in which chance and chance alone determines the structure of the population. Therefore, in this theoretical world in which only chance factors are operating, we expect that each outcome is equally likely to occur. In other words, the probability that an error, selected at random, will favor the company is ½. Likewise, the probability that a randomly selected error will favor the customer is ½.

On the other hand, if we take the empirical approach, we estimate the probability of a particular outcome from the observed relative frequency with which that outcome occurred in the sample data. Thus, since all the errors Rincon found favored the company, the empirical probability that an error will favor the company is 1.00. Similarly, since none of the observed errors favored the customer, the empirical probability of this outcome is 0.00.

Based on the sharp contrast between the theoretical probabilities and the empirical evidence, Rincon would strongly suspect that something was rotten in computerland. Coupled with the accounts of computer fraud he had recently read, he made a subjective assessment that the probability of fraud in this case was high.

EXERCISES

4.1 Identify each of the probability statements below as classical, empirical, or subjective.

    a. A pollster says that a candidate will probably collect more than 50 percent of the votes.

b. Mindy J., president of Consolidated Enterprises, says that the probability is high that the year will be profitable. She bases her statement on earnings through November.

c. Larry M. claims that there will be a world conflict within three years.

d. The probability of obtaining a royal flush, a straight flush, or five cards in the same suit is 0.002.

4.2 There are 2,598,960 different possible outcomes when dealt five cards in a poker game. Calculate the probability of obtaining the following hands given the number of outcomes favoring each hand. (For consistency, round to the seventh decimal place.)

Royal flush	4	Straight	10,200
Other straight flush	36	Three of a kind	54,912
Four of a kind	624	Two pairs	123,552
Full house	3,744	One pair	1,098,240
Flush	5,108	Nothing	1,302,540

Confirm that the probability of all the hands shown is exhaustive (allowing for rounding error).

4.3 In early 1980, gold reached, for the first time, a price of $875 an ounce. Prior to this, what was the empirical probability of gold at $875 an ounce?

4.4 The Interstate Commerce Commission (ICC) voted six to one to eliminate the per diem premium charged by railcars. These revenues must be spent on building, rebuilding, or leasing new cars.

Suppose you are interested in the record of one particular member of the ICC. What is the probability that his was the one dissident vote?

4.5 There are a total of 36 different outcomes when tossing one die twice (or two dice once). Given the frequency of occurrence of the following sums of these two dice, find the associated probabilities. (For consistency, round to four places.)

Sum	f
2	1
3	2
4	3
5	4
6	5
7	6
8	5
9	4
10	3
11	2
12	1

Confirm that the total of the probabilities of all outcomes shown is 1.

4.6 An article in the *Los Angeles Times* on January 28, 1981, stated:

In what Hollywood often considered a harbinger of the Academy Award nominations, the Directors Guild of America (DGA) announced Tuesday its own candidates for "outstanding directorial achievement in motion pictures." . . . In only two instances since the DGA awards began 33 years ago has there been a different Oscar winner.

a. What is the probability that the DGA winner will also win the Oscar? What approach to probability is the article based on?

b. On March 14, 1981, Robert Redford won the DGA award. On March 31, 1981, Redford won the Oscar. What is the probability that the 1982 DGA winner will also win the Oscar? Compare this probability with that obtained in *(a)*. What characteristic of empirical probability does this illustrate?

4.7 Find the probabilities of consecutive winning plays at the craps table given the following information. (For consistency, round to four places.)

Consecutive wins	Frequency
1	244 in 495
2	6 in 25
3	3 in 25
4	1 in 17
5	1 in 37
6	1 in 70
7	1 in 141
8	1 in 287
9	1 in 582

4.4 SAMPLE SPACE AND EVENT

Thus far we have looked at a simple experiment. We have enumerated all possible outcomes and raised questions concerning the probability of obtaining a specific outcome on any one trial. Thus, with the die experiment, we ascertained the probability of obtaining a six on one toss of a single die (the classical approach). Similarly, using the empirical approach, we raised questions concerning the probability of obtaining a defective chip when selecting a single chip from the daily output. In real life, the questions we raise are usually far more complex.

To deal with complex problems of probability it is helpful to introduce the concepts of *sample space* and *event*.

We have seen that an experiment yields various outcomes. We refer to all possible outcomes as the *sample space*. We may refer to each element in the sample space as a *sample point*. Thus, in tossing a single die, the sample space consists of the outcomes or sample points: 1, 2, 3, 4, 5, 6. Since each sample point refers to a specific result, we may define an *event* as a specific subset or collection of outcomes or sample points. For example, all odd numbers may constitute an event. Let us call this event *A*. The *complement* of event *A* is all outcomes in the sample space of positive whole numbers *not* included in *A*. Thus, the complement of event *A,* in this example, is all even numbers. Since *A* and *Ā exhaust* all outcomes in the sample space,

sample space

All possible outcomes of an experiment.

sample point

An element in the sample space.

event

A collection of specific sample points or outcomes.

$p(A) + p(\overline{A}) = 1.00$. In addition, these events are *mutually exclusive,* since an even number and an odd number cannot occur simultaneously on the same trial.

A convenient device for illustrating probability concepts is the *Venn diagram,* which graphically portrays each sample event as a sample point in a sample space (see Figure 4.1).

Venn diagram

A pictorial presentation of each sample event as a sample point in a sample space.

Figure 4.1

Venn diagrams illustrating the sample space for die tossing and various event relationships

(a) Sample space for die tossing

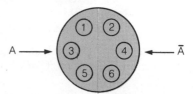

(b) Events that are mutually exclusive and exhaustive:
  Event A: All odd numbers
  Event $\overline{A}$: All even numbers

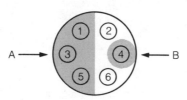

(c) Events that are mutually exclusive but not exhaustive:
  Event A: An odd number
  Event B: The number 4

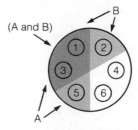

(d) Events that are not mutually exclusive or exhaustive:
  Event A: An odd number
  Event B: A number less than 4

tree diagram

A technique used to represent the sample space as branches or forks of a tree.

Another technique used to represent the sample space, particularly when there are various stages in the experiment, is the use of a *tree diagram.*

Each fork in a tree diagram shows the possible outcomes that may occur at a particular stage. For example, suppose you check the price of a stock two days in a row. On the first day, there are three possible outcomes: it goes up, it goes down, it does not change. These three outcomes are represented by the branches of the first fork in the tree diagram presented in Figure 4.2. On the second day there are the same three possible outcomes and these are shown as branches off the first three forks. Finally, we wind up with nine branches, each of which represents a possible outcome. These nine outcomes comprise the sample space: (up–up); (up–no change); (up–down); (no change–up); (no change–no change); (no change–down); (down–up); (down–no change); (down–down).

Later (in Section 4.7) we shall illustrate the use of tree diagrams in the solution of probability problems. In Chapter 9 we discuss the use of tree diagrams in decision-making situations.

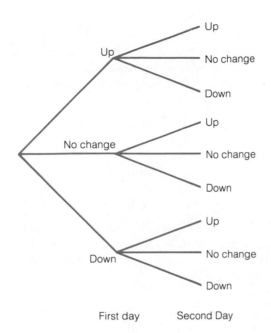

Figure 4.2
Tree diagram illustrating
the various outcomes
(sample space) when
observing the price of
a stock on two days

4.5 ADDITION RULE

4.5.1 General case (events that are not mutually exclusive)

Within the sample space for any experiment, we may define any number of events. These events may or may not be mutually exclusive. For example, in Figure 4.1 *(d)* we defined event *A* (an odd number) and event *B* (a number less than four). Suppose we were interested in the probability of *either A* occurring *or B* occurring. Note that the sample points comprising these two events overlap. In this instance, the sample points 1 and 3 overlap both events. Thus, to find the probability of *A or B,* we must be careful not to include the overlapping sample points twice. In other words, the number of sample points comprising *A,* an odd number, is 3 (1, 3, and 5). Therefore, $p(A) = \frac{3}{6} = \frac{1}{2}$. The number of sample points comprising *B,* a number less than four, is 3 (1, 2, and 3); $p(B) = \frac{3}{6} = \frac{1}{2}$. In order to determine the probability of obtaining either an odd number *or* a number less than four [$p(A \text{ or } B)$], we use Formula (4.2), which subtracts out the probability associated with the overlapping points. This formula represents the general addition rule for events that are *not* mutually exclusive.

$$P(A \text{ or } B) = p(A) + p(B) - p(A \text{ and } B) \qquad (4.2)$$

in which $p(A \text{ and } B)$ is the probability associated with the overlapping sample points.

In the present example, $p(A) = \frac{1}{2}$, $p(B) = \frac{1}{2}$, and $p(A \text{ and } B) = \frac{2}{6} = \frac{1}{3}$. Thus, $p(A \text{ or } B) = \frac{1}{2} + \frac{1}{2} - \frac{1}{3} = \frac{2}{3} = 0.67$.

Example B

There are six people qualified for selection to the position of deputy mayor. The office is to be filled by randomly selecting one person from the following list, which shows their sexes and numbers of years of experience in government:

Person 1 is a male with 8 years of experience.
Person 2 is a female with 7 years of experience.
Person 3 is a male with 12 years of experience.
Person 4 is a female with 4 years of experience.
Person 5 is a female with 8 years of experience.
Person 6 is a male with 4 years of experience.

a. What is the probability that the individual selected will be either female or a person with more than six years of experience?
b. What is the probability that the person selected will be either a male or a person with less than five years of experience?

Solution

a. We'll define A as a female. In this example, $p(A) = \frac{3}{6}$. Event B is a person with more than six years of experience. Since there are four such people, $p(B) = \frac{4}{6}$. However, there are two females with more than six years of experience. Therefore, $p(A \text{ and } B) = \frac{2}{6}$.

$$p(A \text{ or } B) = p(A) + p(B) - p(A \text{ and } B)$$
$$= \frac{3}{6} + \frac{4}{6} - \frac{2}{6} = \frac{5}{6}$$

Let us look at a picture of the sample space:

The circled points represent the event (A or B)

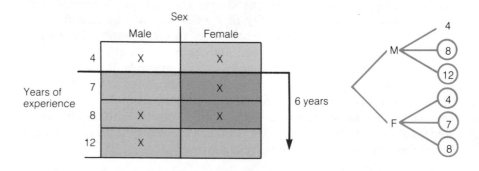

b. We'll define C as a male. In this example, $p(C) = \frac{3}{6}$. Event D is a person with less than five years of experience; since there are two such persons, $p(D) = \frac{2}{6}$. The number of males with less than five years of experience is 1; thus, $P(C \text{ and } D) = \frac{1}{6}$. Therefore, $p(C \text{ or } D) = \frac{3}{6} + \frac{2}{6} - \frac{1}{6} = \frac{4}{6}$ or 0.67.

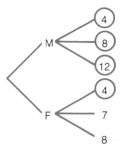

4.5.2   Special case (Events that are mutually exclusive)

A special case of the addition rule presented by Formula (4.2) occurs when we are dealing with *mutually exclusive* events. Since mutually exclusive events *cannot,* by definition, overlap, [i.e., $p(A$ and $B) = 0$] the probability of either event occurring is simply the sum of their separate probabilities. Or,

$$p(A \text{ or } B) = p(A) + p(B), \qquad (4.3)$$

where A and B are mutually exclusive.

Earlier we defined an event as a subset or collection of specific outcomes or sample points. Since the outcomes comprising any particular event are mutually exclusive, the probability of a given event is the sum of the probabilities of the sample points comprising that event.

For example, if event A is an odd number on a throw of a die, then event A occurs if either 1, 3 or 5 turns up on the die. That is (see Figure 4.1),

$$p(A) = p(1, 3, \text{ or } 5) = p(1) + p(3) + p(5)$$
$$= \tfrac{1}{6} + \tfrac{1}{6} + \tfrac{1}{6}$$
$$= \tfrac{3}{6} = 0.5.$$

Example C

An investor wants to pick two stocks from a total of four suggested by her broker. The four stocks have different yields; Y_1, Y_2, Y_3, Y_4. Suppose the names of these four stocks are written on poker chips. The chips are dropped in a hat and shaken up.

What is the probability of drawing (while blindfolded) the two chips that carry the name of the two stocks with the highest two yields or the two chips with the names of the two stocks with the lowest two yields?

Solution

Let 1, 2, 3, 4 represent the four stocks ranked from best (1) to poorest (4). Thus, the sample space consists of the following mutually exclusive sample points: (1,2); (1,3); (1,4); (2,3); (2,4) (3,4).

Since the two chips are selected at random, we assign each point a probability of $\frac{1}{6}$.

Let us define event A as the selection of the two most highly ranked stocks, and event B as the selection of the two stocks with the poorest yields. Thus, we are really asking: what is the $P(A$ or $B)$? Since these events are mutually exclusive, $p(A$ or $B) = p(A) + p(B)$.

The only sample point in event A is (1,2); therefore, $p(A) = \frac{1}{6}$. Similarly, the only sample point in event B is (3,4), and $p(B) = \frac{1}{6}$. Thus, $p(A$ or $B) = \frac{1}{6} + \frac{1}{6} = \frac{2}{6} = 0.33$.

EXERCISES

4.8 Nancy and Frank serve tables in a particular restaurant. Four people enter at the same time. If each server is randomly assigned two people, and we assume that each patron tips a different amount, what is the probability that one of them gets either the best tipper or the worst tipper? Use a Venn diagram to illustrate the sample space.

4.9 A study conducted by the National Society to Prevent Blindness reported that 35 percent of all eye injuries are caused by foreign bodies, 25 percent by open wounds, 25 percent by contusions, and 15 percent by burns.

Suppose a patient comes in with an eye injury, what is the probability that it was caused by either a foreign body or by an open wound?[2]

4.10 A committee consists of the following five members:

> Professor 1, a male
> Professor 2, a female
> Professor 3, a female
> Student 4, a male
> Student 5, a female

If one member is randomly selected to appear before the student council, what is the probability that:

    a. Either a female or a professor will be selected?
    b. Either a male or a student will be selected?

Use a tree diagram to represent the sample space. Circle the points corresponding to the events in *(a)* and *(b)* above.

4.11 A cereal manufacturer claims that the quart size package contains thirty-two ounces, plus or minus one-tenth of an ounce. An independent research firm randomly selects 1,000 packages from grocery shelves for testing. They find the following distribution of weights of the contents of the packages.

31.5–under 31.7	12
31.7–under 31.9	40
31.9–under 32.1	880
32.1–under 32.3	50
32.3–under 32.5	18
	1,000

[2] Source: *Los Angeles Times,* Aug. 29, 1980.

a. What is the probability that a single package, selected at random, will meet the manufacturer's specifications?
b. Contain less than specified?
c. Contain more than specified?
d. Be out of the specified limits?
e. Contain as much as or more than promised by the manufacturer?

4.12 Referring to Exercise 4.11, a consumer organization criticizes the cereal manufacturer for providing what it promised only 88 percent of the time. The cereal manufacturer replied, "To the contrary, 95 percent of the time, the purchaser gets an amount at least equal to what we promise." Comment.

4.6 JOINT, MARGINAL, AND CONDITIONAL PROBABILITIES

We are often interested in the probability that two or more events will *both* occur. For example, what is the likelihood of obtaining four heads in a row on four tosses of a coin? or what is the probability of obtaining three defective chips when randomly selecting from the daily output? or what is the probability that a given individual works at plant B *and* will be fired?

Questions such as these involve the *joint probability* of two or more events. In the case of two events, we are really asking: What is the probability that *both A* and *B* will occur? That is, what is $p(A \text{ and } B)$? We have already seen this expression when discussing the addition rule. We saw that, for mutually exclusive events, $p(A \text{ and } B) = 0$. That is, if two events are mutually exclusive, the probability is zero that *both* will occur. Thus, mutually exclusive events are *dependent* and not *independent;* independent events are not mutually exclusive. Let us look at an example to illustrate these relationships among events.

A bond broker has 40 different municipal issues available. Ten of these issues are from Harris County, Texas (even *A*). Of these ten, eight are general obligation bonds. A total of 16 general obligation bonds are available (event *B*). Let us summarize this information in a frequency table:

joint probability

The probability that two events will *both* occur.

dependence

Two events are dependent if the occurrence of one affects the probability of the occurrence of the other.

independence

Two events are independent if the occurrence of one does not affect the probability of the occurrence of the other.

	Harris County (A)	Not Harris County (Ā)	Total
General obligation (B)	8	8	16
Not general obligation (B̄)	2	22	24
Total	10	30	40

We may illustrate joint probabilities as follows:

If a bond is chosen at random from this group of 40, the joint probability that it will be from Harris County and a general obligation bond is

$$p(A \text{ and } B) = \frac{8}{40} = 0.20$$

Similarly, the probability that a bond, randomly selected, is from Harris County, but *not* a general obligation bond is

$$p(A \text{ and } \overline{B}) = \frac{2}{40} = 0.05$$

The probability that a randomly selected bond will *not* be from Harris County but will be a general obligation bond is

$$p(\overline{A} \text{ and } B) = \frac{8}{40} = 0.20$$

Finally, the probability that a randomly selected bond is *not* from Harris County and *not* a general obligation bond is

$$p(\overline{A} \text{ and } \overline{B}) = \frac{22}{40} = 0.55$$

We may summarize these joint probabilities in a joint probability table as shown below:

	Harris County (A)	Not Harris County ($\overline{A}$)	Marginal probabilities
General obligation (B)	0.20	0.20	0.40
Not general obligation ($\overline{B}$)	0.05	0.55	0.60
Marginal probabilities	0.25	0.75	1.00

You may recognize that a joint probability table is a relative frequency table (Chapter 2).

In addition to the joint occurrence of two events, we are often interested in how the probability of one event is influenced by whether or not another event occurs. First, let us look at the probability of an event that is *not* influenced by whether or not another event occurs. We refer to this as the *marginal probability* of an event. We use the totals in the margins to obtain these probabilities. For example, the probability that a randomly selected bond is from Harris County is $\frac{10}{40} = 0.25$. Or, $p(A) =$ marginal probability that A will occur. Similarly, the probability that a randomly selected bond is a general obligation bond is $\frac{16}{40} = \frac{2}{5}$, i.e., $p(B) =$ marginal probability that B will occur. We call these marginal probabilities since they are found in the margins of a joint probability table.

Another way to obtain these marginal probabilities is to recognize that

marginal probability

The probability that an event will occur regardless of other conditions that may prevail.

the cells in a joint probability table are mutually exclusive. We may then use the addition rule for mutually exclusive events to obtain the marginal probabilities.

For example, the marginal probability of being from Harris County has two mutually exclusive components: (1) the joint probability that a bond is from Harris County *and* is a general obligation bond plus (2) the joint probability that a bond is from Harris County *and* is *not* a general obligation bond. Stated symbolically,

$$p(A) = p(A \text{ and } B) + p(A \text{ and } \bar{B})$$
$$= 0.20 + 0.05 = 0.25$$

Marginal probabilities are sometimes called unconditional probabilities since they are not dependent or conditional on whether or not another event has occurred. However, we are often interested in the probability of one event, given that another event has occurred or is certain to occur. We call this *conditional probability* and symbolize it: $p(A|B)$ (which is read "the probability of event A given B"). The conditional probability of A given that B has occurred is found by dividing the joint probability of A and B by the marginal probability of B. Thus,

$$p(A|B) = \frac{p(A \text{ and } B)}{p(B)}.$$

(4.4)

conditional probability

The probability of an event given that another event has occurred. Symbolically, p(A|B) is the probability of A, given that B has occurred.

In our bond example, $p(A|B)$ refers to the question: What is the probability that a bond will be from Harris County, given that it is a general obligation bond? $p(A|B) = 0.20/0.40 = 0.50$.

In order to have a better understanding of Formula (4.4), let us obtain $p(A|B)$ directly from the frequency table. We see that there are 16 general obligation bonds; of these 16, 8 are from Harris County. Therefore, given that a bond is a general obligation bond, the probability that it will be from Harris County, i.e., $p(A|B) = \frac{8}{16} = 0.50$.

4.7 MULTIPLICATION RULE

Contrast the following two situations. You toss a theoretically perfect coin. What is the probability that you will obtain a head? It is 0.5. Now you toss it again. What is the probability that you will obtain a head? The probability remains the same: it is still 0.5. The reason that the probabilities remain the same is that outcomes on the two trials are said to be *independent*. One toss does not affect the other toss. Nor does the coin have a memory. It does not "try" to come up a tail the second time in order to conform to the so-called laws of chance. Stated another way, the probability of event A given that event B has occurred is still equal to the probability that A will occur, or $p(A|B) = p(A)$ for *independent* events. Likewise, $p(B|A) = p(B)$ when A and B are independent.

Now you select a card from a well-shuffled deck of cards. The probability is $\frac{4}{52}$ that you will select an ace. Imagine that fortune shines upon you and you obtain an ace. Without returning the ace to the deck, you now select a second card. What is the probability that this card will also be an ace? Is it still $\frac{4}{52}$? No. You have taken an ace so that there are only three remaining. In addition, there are only 51 cards left in the deck. The probability of selecting an ace on the second trial, *given that an ace has been selected on the first,* is now $\frac{3}{51}$. In other words, the probability on the second trial has been changed by what preceded it. The outcomes are no longer independent. When the occurrence of one outcome or event depends on the prior occurrence of another outcome or event, they are said to be *dependent.* Stated symbolically, $p(A|B) \neq p(A)$; or $p(B|A) \neq p(B)$ for *dependent* events.

When we are interested in the joint probability of two or more events, we must first determine whether or not the events are independent.

If we are dealing with dependent events, we may use the formula for conditional probability to determine joint probabilities:

Since

$$p(A|B) = \frac{P(A \text{ and } B)}{p(B)}$$

it follows that

$$p(A \text{ and } B) = p(B)p(A|B); \qquad (4.5)^3$$

This formula is called the multiplication rule for dependent events.

How do we determine the joint probability of independent events? Since $p(A|B) = p(A)$ for independent events, the formula for joint occurrence reduces to

$$p(A \text{ and } B) = p(A)p(B) \qquad (4.6)$$

A formula called the multiplication rule for independent events.

Example D.   Putting it all together: Dependent events

Let us imagine that Dynoair Industries has announced plans to lay off 600 employees. There are 2,500 employees in the total workforce. In the absence of any additional information, the chance that any one person will be laid off (Event A) is $600/2,500 = 0.2400$.

However, Dynoair has three plants with the following number of employees: Plant B: 400; Plant C: 800; Plant D: 1,200. Thus, the probability that an individual works at plant B is $500/2,500$, or $p(B) = 0.2000$. Likewise, $p(C) = 800/2,500 = 0.3200$, and $p(D) = 1,200/2,500 = 0.4800$.

---

[3] Formula (4.5) may also be written: $p(A \text{ and } B) = p(A)p(B|A)$.

Suppose the personnel director decides to lay off 200 employees at each location. Let us summarize this information in a frequency table:

	Laid off A	Not laid off $\bar{A}$	Marginal totals
Plant B	200	300	500
Plant C	200	600	800
Plant D	200	1,000	1,200
Marginal totals	600	1,900	2,500

There are six mutually exclusive and exhaustive sample points in our sample space. Figure 4.3 portrays this sample space in Venn and tree diagrams.

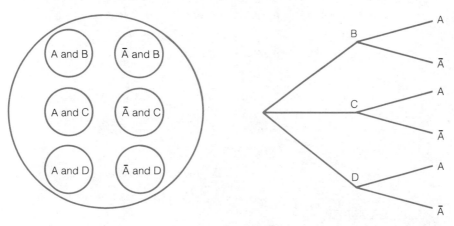

Figure 4.3
Venn and tree diagrams illustrating the sample space for the Dynoair example

We see that the probability of being fired is dependent upon where the individual works. These two variables are obviously related. The probability of being laid off [$p(A)$] is *not* the same as the probability of being fired given prior information as to where the individual works. The probability of being laid off, given that the person works at, say, Plant B, is a conditional probability; $p(A|B) = 200/500 = 0.4000$. Similarly, we may obtain the other conditional probabilities: $p(A|C) = 200/800 = 0.2500$; $p(A|D) = 200/1,200 = 0.1667$.

Figure 4.4 illustrates how a tree diagram may be used to obtain the probabilities associated with each sample point. Note that the first set of branches is composed of three mutually exclusive and exhaustive events: $p(B) + p(C) + p(D) = 1.00$.

The probabilities assigned to each segment of the second set of branches are conditional probabilities. For example, $p(A|B) = 0.4000$, $p(\bar{A}|B) = 0.6000$, $p(A|C) = 0.2500$, etc.

To obtain the probabilities associated with each sample point in our sample space, we multiply the probabilities along the branches leading to that sample point.

To illustrate,

$$p(A \text{ and } B) = p(B)p(A|B)$$
$$= (0.2000)\,(0.4000)$$
$$= 0.0800.$$

Finally, let us summarize these probabilities in a joint probability table:

	Laid off A	Not laid off $\overline{A}$	Marginal probabilities
Plant B	0.0800	0.1200	0.2000
Plant C	0.0800	0.2400	0.3200
Plant D	0.0800	0.4000	0.4800
Marginal probabilities	0.2400	0.7600	1.0000

Figure 4.4
Tree diagram illustrating the probabilities for all possible combinations of dependent events

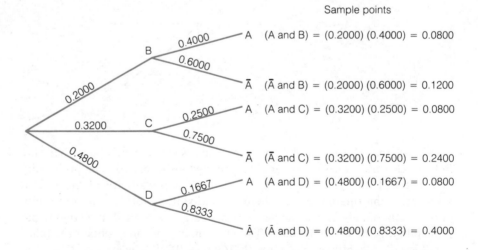

Sample points

A (A and B) = (0.2000) (0.4000) = 0.0800

$\overline{A}$ ($\overline{A}$ and B) = (0.2000) (0.6000) = 0.1200

A (A and C) = (0.3200) (0.2500) = 0.0800

$\overline{A}$ ($\overline{A}$ and C) = (0.3200) (0.7500) = 0.2400

A (A and D) = (0.4800) (0.1667) = 0.0800

$\overline{A}$ ($\overline{A}$ and D) = (0.4800) (0.8333) = 0.4000

The probabilities associated with the second set of branches are conditional probabilities. Since these events are related $[p(A|B) \neq p(A); p(\overline{A}|B) \neq p(\overline{A}),$ etc.], the probabilities differ for each set of branches. However, since each set of branches initiating from a single point represents mutually exclusive and exhaustive events, the total probability of each set is equal to 1.00.

Notice that the sum of the probabilities of all six sample points is also equal to 1.00 since these points are mutually exclusive and exhaustive.

Note also that the sum of the probabilities associated with the sample points comprising a particular event equal the probability of that event. For example, $p(A \text{ and } B) + p(\overline{A} \text{ and } B) = p(B)$, or $0.0800 + 0.1200 = 0.2000$.

Example E: Putting it all together: Independent events

Returning to the example of Dynoair Industries, a union committee points out the unfairness of dropping an identical number of individuals at each plant. Rather, they recommend that 24 percent be dropped from each plant. Recall that the number of employees at Plant B is 500, C is 800, and D is 1,200. Also, a total of 600 employees are to be laid off. Assuming that management accepts this recommendation:

a. Find the probability that a given individual will be dropped and works at Plant C. Are these events independent?
b. The formula for the probability of the joint occurrence of independent events may be easily extended to include any number of events. Thus, $p(A, B, C, \text{ and } D) = p(A)p(B)p(C)p(D)$. If two people are to be dropped on a given day, what is the probability that one will be from Plant B and the other from Plant C?
c. Construct a joint probability table showing the marginal probabilities as well.
d. Illustrate the use of a tree diagram in the solution of this probability problem.

Solution

Let us set up a frequency table to summarize the information:

	Laid off A	Not laid off $\overline{A}$	Marginal totals
Plant B	120	380	500
Plant C	192	608	800
Plant D	288	912	1,200
Marginal totals	600	1,900	2,500

a. Since 24 percent of all workers will be laid off regardless of where they work, i.e., the probability of being laid off, given that the individual works at Plant C, is the same as the marginal probability of being laid off. Or, $p(A|C) = p(A)$; therefore, the events are independent. Thus, $p(A \text{ and } C) = p(A)p(C) = (0.24)(800/2,500) = 0.0768$.

b. $p(A \text{ and } B) = p(A)p(B) = (0.24)(500/2{,}500) = 0.0480$.
   Thus, $[p(A \text{ and } B) \text{ and } p(A \text{ and } C)] = (0.0480)(0.0768) = 0.0037$.
c. Summarizing the probabilities we have calculated so far:

	Laid off A	Not laid off $\overline{A}$	Marginal probabilities
Plant B	0.0480		0.2000
Plant C	0.0768		0.3200
Plant D			
Marginal probabilities	0.2400	0.7600	1.0000

Although we could calculate the missing probabilities directly, we now have enough information to obtain the missing values through the use of the addition rule. For example, $p(B)$ has two mutually exclusive components: $(A \text{ and } B)$ and $(\overline{A} \text{ and } B)$. Thus,

$$p(B) = p(A \text{ and } B) + p(\overline{A} \text{ and } B)$$

Or,

$$\begin{aligned} p(\overline{A} \text{ and } B) &= p(B) - p(A \text{ and } B) \\ &= 0.200 - 0.0480 \\ &= 0.1520 \end{aligned}$$

Similarly,

$$p(A) = p(A \text{ and } B) + p(A \text{ and } C) + p(A \text{ and } D)$$

Thus,

$$\begin{aligned} p(A \text{ and } D) &= 0.2400 - 0.0480 - 0.0768 \\ &= 0.1152 \\ p(\overline{A} \text{ and } C) &= p(C) - p(A \text{ and } C) \\ &= 0.3200 - 0.0768 \\ &= 0.2432 \\ p(D) &= 1 - p(B) - p(C) \\ &= 1 - 0.2000 - 0.3200 \\ &= 0.4800 \\ p(\overline{A} \text{ and } D) &= p(D) - p(A \text{ and } D) \\ &= 0.4800 - 0.1152 \\ &= 0.3648 \end{aligned}$$

Using the multiplication rule for independent events, let us check our calculations:

$$p(\bar{A} \text{ and } D) = p(\bar{A})p(D)$$
$$= (0.76)\,(0.48)$$
$$= 0.3648;$$
$$p(\bar{A} \text{ and } C) = p(\bar{A})p(C)$$
$$= (0.76)\,(0.32)$$
$$= 0.2432;$$
$$p(\bar{A} \text{ and } B) = p(\bar{A})p(B)$$
$$= (0.76)\,(0.20)$$
$$= 0.1520$$

Finally, let us summarize our calculations in a joint probability table:

	Laid off A	Not laid off $\bar{A}$	Marginal probabilities
Plant B	0.0480	0.1520	0.2000
Plant C	0.0768	0.2432	0.3200
Plant D	0.1152	0.3648	0.4800
Marginal probabilities	0.2400	0.7600	1.0000

d.   Figure 4.5 summarizes the probabilities in a tree diagram.

Tree diagram illustrating the probabilities for all possible combinations of independent events. Since the events are independent, $p(A|B) = p(A)$, $p(A|C) = p(A)$, and $p(A|D) = p(A)$. Therefore, the probabilities associated with the second set of branches (conditional probabilities) are the same for each set.   **Figure 4.5**

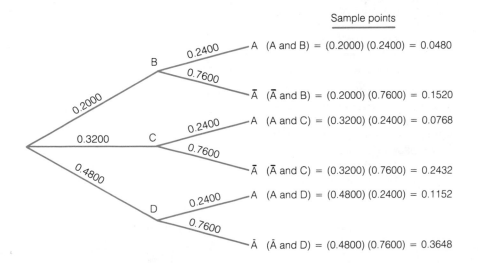

EXERCISES

4.13 Utility companies requesting rate hikes found that approximately 60 percent of the amounts requested were approved in 1979. In 1978, approximately 43 percent were approved. However, if an increase was approved in 1979, the chances are only 30 percent that another increase will be granted in 1980. If an increase was approved in 1978, the chances are 50 percent that another increase will be granted in 1980.

Which company has a better chance of getting an increase in 1980: one that received one in 1979 or one that received one in 1978?

4.14 In April and May of 1980, Pepsi-Cola had a 22.4 percent share of the market. If two people are selected at random, what is the probability both will choose Pepsi?

4.15 Many of the problems we face when dealing with non-independent events involve two variables, each of which has two independent and exhaustive categories. For example, Variable A: Good versus Bad; Variable B: Reject versus Not reject.

The following table gives the probabilities associated with the various possibilities:

	Not reject B_1	Reject B_2	Marginal probability
Good A_1	A_1 and B_1 0.92	A_1 and B_2 0.03	A_1 0.95
Bad A_2	A_2 and B_1 0.00	A_2 and B_2 0.05	A_2 0.05
Marginal probability	B_1 0.92	B_2 0.08	

Find the probability that an item is
  a. Good.
  b. Good and rejected.
  c. Rejected.
  d. Not rejected.
  e. Bad.
  f. Bad and rejected.
  g. Bad and not rejected.
  h. Good and not rejected.

4.16 Many samplings of a mass produced item has revealed that, on the average, 5 percent are defective. If an item is actually good, the probability that it will be erroneously rejected as defective is 0.032. If it is defective, the probability that it will be erroneously accepted is 0.00. Draw a decision tree to show the probabilities of the following events:
  a. A good item that is not rejected.
  b. A good item that is rejected.
  c. A bad item that is not rejected.
  d. A bad item that is rejected.

4.8   BAYES' THEOREM

The owner of a retail shoe store reads in his trade journal that a new style of shoe was a "hot" item in regions of the country where it had been introduced. He concludes that the probability is high that this item will do well in his store. Based on this belief, he orders heavily in that style. However, several months later he finds the shelves stocked with unsold quantities of this item. As a prudent businessman, he revises his assessment of the *prior probability* in the light of this new evidence. The new probability is referred to as *posterior* or *revised probability*.

As you might imagine, the availability of procedures that permit us to revise and alter prior probabilities on the basis of new experimental evidence puts a very powerful instrument in our hands.

This tool was provided by Thomas Bayes, an 18th-century British Presbyterian minister. He sought to prove mathematically the existence of God. His mental adventures carried him through the largely uncharted regions of probability theory. He sought a method that would permit him to reason backwards from effects to causes. In the course of so doing, he developed what has come to be known as Bayes' Theorem. This theorem is based on the concept of conditional probability and is the basic formula which permits us to determine the probability of one event (e.g. event A) based on information related to the probability of another event (e.g., event B). Recall that the basic formula for condition probability[4] [Formula (4.4)] is:

$$p(A|B) = \frac{p(A \text{ and } B)}{p(B)}$$

prior probability

The probability of an event before it is revised by additional information or evidence.

posterior (revised) probability

The prior probability that is modified in the light of new information. The revised probability is usually obtained by employing Bayes' theorem.

Stated verbally, the conditional (or revised) probability of A given B is the probability of the joint occurrence of A and B divided by the probability of B.

Let's illustrate the use of Bayes' Theorem. Imagine that you are a distributor who obtains one of your items from three different suppliers. Over the years, you have noted that all suppliers have occasionally shipped less than the number ordered but billed for the total order anyway. Their past records are shown in the table:

	Shipment without shortages B_1	Shipment with shortages B_2	Marginal frequency
Supplier A_1	486	18	504
Supplier A_2	576	72	648
Supplier A_3	612	36	648
Marginal frequency	1,674	126	1,800

---

[4] In Chapter 9 we develop a more comprehensive formula that is useful in decision theory.

You discover a shipment with severe shortages and are unable to identify the supplier since the return address label has been lost. Given this fact, what is the probability that A_1 was the supplier? A_2? A_3?

First, we convert the frequencies to probabilities. When this is done, the following joint probability table results:

	Shipment without shortages B_1	Shipment with shortages B_2	Marginal probability
Supplier A_1	0.27	0.01	0.28
Supplier A_2	0.32	0.04	0.36
Supplier A_3	0.34	0.02	0.36
Marginal probability	0.93	0.07	1.00

The marginal probabilities provide the prior probabilities of A_1, A_2, A_3. Thus, in the absence of any additional information, the probability that the shipment came from any one of the suppliers is:

$$p(A_1) = 0.28$$
$$p(A_2) = 0.36$$
$$p(A_3) = 0.36$$

However, given the information that there was a shortage, we may revise these probabilities in the light of this information. Thus, the revised probabilities are

$$p(A_1|B_2) = \frac{p(A_1 \text{ and } B_2)}{p(B_2)}$$

$$= \frac{0.01}{0.07} = 0.143$$

$$p(A_2|B_2) = \frac{p(A_2 \text{ and } B_2)}{p(B_2)}$$

$$= \frac{0.04}{0.07} = 0.571$$

$$p(A_3|B_2) = \frac{p(A_3 \text{ and } B_2)}{p(B_2)}$$

$$= \frac{0.02}{0.07} = 0.286$$

Note that the sum of these conditional probabilities (0.143 + 0.571 + 0.286) is equal to 1.00. This means that these three probabilities are exhaustive.

One of the suppliers is at fault. The one most likely guilty is A_2 with a probability of 0.571.

Example F

We have just inspected a shipment that we received several months ago. Because of too many defective items, the shipment is unacceptable. However, we have lost track of the supplier. From past records, we know that 10 percent of all shipments are unacceptable. Supplier ABC provides 30 percent of all shipments. However, past records indicate that 40 percent of the defective shipments were supplied by shippers other than ABC. What is the probability that the shipment in question was supplied by ABC? Identify the prior probability and the revised probability of this event.

Solution

Let us put the information we have in tabular form. Note that all of the information necessary to construct the joint probability table does not appear to be supplied. However, in a two-by-two table, all we need know is one marginal probability for each variable and one cell probability to fill in the remaining cell and marginal probabilities. To illustrate, we know that: the marginal probability of A_1 (not acceptable) is 0.10; the marginal or prior probability of B_1 (shipper ABC) is 0.30; and 40 percent of A_1 is in cell (A_1 and B_2), i.e., $(0.4)(0.10) = 0.04$. The known values are shown below:

	Supplier		(A) marginal
	ABC B_1	Other B_2	
Not acceptable A_1	A_1 and B_1	A_1 and B_2 0.04	A_1 0.10
Acceptable A_2	A_2 and B_1	A_2 and B_2	A_2
(B) marginal	B_1 0.30	B_2	

Since the A marginals are mutually exclusive and exhaustive, we may obtain $p(A_2)$ by subtraction. Thus, $p(A_2) = 1.00 - 0.10 = 0.90$. Similarly, $p(B_2) = 1.00 - 0.30 = 0.70$. Also, $p(A_1$ and $B_1) + p(A_1$ and $B_2)$ must add up to 0.10. Thus, $p(A_1$ and $B_1) = 0.10 - 0.04 = 0.06$. Similarly, $p(A_1$ and $B_1) + p(A_2$ and $B_1)$ must add up to 0.30. Thus, $p(A_2$ and $B_1) = 0.30 - 0.06 = 0.24$. Finally, $p(A_1$ and $B_2) + p(A_2$ and $B_2)$ must add up to 0.70.

Therefore, $p(A_2 \text{ and } B_2) = 0.70 - 0.04 = 0.66$. We may now construct our completed table of probability values.

	Supplier		(A) marginal
	ABC B_1	Other B_2	
Not acceptable A_1	A_1 and B_1 0.06	A_1 and B_2 0.04	0.10
Acceptable A_2	A_2 and B_1 0.24	A_2 and B_2 0.66	0.90
(B) marginal	0.30	0.70	1.00

It is now possible to find the conditional probability: i.e., given that the shipment is not acceptable (A_1), the probability that it was shipped by ABC (B_1) is

$$p(B_1|A_1) = \frac{p(A_1 \text{ and } B_1)}{p(A_1)}$$

$$= \frac{0.06}{0.10} = 0.60$$

Thus, the revised probability is 60 percent that the shipment came from supplier ABC as contrasted with the prior probability of 30 percent.

EXERCISES

4.17 Suppose that JAJ company has announced plans to hire 200 qualified applicants during the month of January. Because of pressure from a federal agency, management agrees to hire an equal number of men and women. Altogether 500 apply, of which only 150 are women. Nevertheless, management sticks to its promise.
   a. Given that you are a male, what is the probability that you will be hired?
   b. Given that you are a female, what is the probability of being hired?

4.18 In a given geographical region, IRS tax auditors found that 40 percent of all audited returns contained errors that benefitted the taxpayer. Two percent of the tax-payers have annual incomes of $100,000 or more. Among them, the error rate is 65 percent.
   Given that an auditor uncovers an error benefitting the taxpayer, what is the probability that it was committed by a taxpayer with
   a. An income of $100,000 or more;
   b. An income less than $100,000?

SUMMARY

In this chapter we looked at some of the rules and applications of probability theory.

1. We discussed three different approaches: classical, empirical, and subjective. Classical theory involves conceptual experiments in which the complete structure of the population is presumed to be known; empirical probabilities are obtained from actual experiments and define probability in terms of relative frequencies; subjective probabilities represent estimates based on each individual's degree of belief concerning the event.
2. We saw that the sample space (all possible outcomes of an experiment) may be represented by Venn and tree diagrams.
3. If we let A and B represent events and $p(A)$ and $p(B)$ the probability of their occurrence, then

    a. For events that are *not* mutually exclusive, the *general addition rule* is

$$p(A \text{ or } B) = p(A) + p(B) - p(A \text{ and } B)$$

    b. For *mutually exclusive* events, the *addition rule* is

$$p(A \text{ or } B) = p(A) + p(B)$$

    c. If A and B are *mutually exclusive*, the probability of their joint occurrence, $p(A \text{ and } B)$, is zero.
    d. If A and B are *exhaustive*, $p(A) + p(B) = 1.00$.
    e. If A and $\overline{A}$ are *complementary*, they are mutually exclusive [$p(A \text{ and } \overline{A}) = 0$] and exhaustive [$p(A) + p(\overline{A}) = 1.00$].
    f. If A and B are *independent*, $p(A|B) = p(A)$.
    g. If A and B are related or *dependent*, $p(A|B) \neq p(A)$.
    h. The multiplication rule for *dependent* events is

$$p(A \text{ and } B) = p(B)\ p(A|B)$$

    i. The multiplication rule for *independent* events is

$$p(A \text{ and } B) = p(A)\ p(B)$$

4. Bayes' theorem provides a basis for revising prior probabilities in the light of additional information or evidence by using the basic formula for conditional probability:

$$p(A|B) = \frac{p(A \text{ and } B)}{p(B)}$$

TERMS TO REMEMBER

classical approach to probability (83)
experiment (83)
outcome (83)
exhaustive (84)
mutually exclusive (84)
empirical or relative frequency approach
to probability (85)
complement (85)
subjective approach to probability (86)
sample space (89)
sample point (89)

event (89)
Venn diagram (90)
tree diagram (90)
joint probability (95)
dependence (95)
independence (95)
marginal probability (96)
conditional probability (97)
prior probability (105)
posterior (revised) probability (105)

EXERCISES

4.19 During the epic match between McEnroe and Borg at the 1980 U.S. open, McEnroe missed two consecutive cross-court forehands. The T.V. commentator, Tony Trabert, commented "If you give a guy ten shots like those two, he'll get at least eight." The color man, Pat Summerall replied, "He's got eight coming."

    a. What approach to probability theory was the commentator using?
    b. Comment on Pat Summerall's reply.

4.20 On a bill aimed at redistributing the $1.8 billion granted to persons with low incomes to help them pay for heating their homes the House of Representatives voted—215 yes, 199 no—to uphold a 1964 allocation that provides $150 million more for northern states.

    Suppose two representatives are picked at random, what is the probability (assuming that all votes are independent) that

    a. Both voted no?
    b. At least one voted yes?
    c. Could you improve your estimates of these probabilities if you had more information?

4.21 Suppose that the probability is 0.8 that a helicopter, exposed for two or more hours to a desert storm, is likely to remain operable. In a mission involving five helicopters and two hours exposure to a desert storm, what is the probability that all five helicopters will remain operable?

4.22 Walter Faintheart learns that the probability of boarding a plane that contains a bomb is 1 in a million. He reasons that the probability that two bombs are on board is $(1/1,000,000)(1/1,000,000)$, an incredibly small risk. Therefore, he carries a bomb on board on the grounds that "a second bomb on board is a virtual impossibility." Has he discovered a new formula for reducing aircraft risk?

4.23 If $p(A) = 0.7$, $p(B) = 0.4$, and $p(A \text{ and } B) = 0.15$,
    a. Are events A and B mutually exclusive? Explain.
    b. Are events A and B independent? Explain.

4.24 A total of 800 light bulbs were randomly selected from the inventory. Four hundred were burned continuously whereas the other 400 were used intermittently. However, both sets of bulbs were burned the same number of hours. The results were:

	Failed	Continued to function
Burned continuously	80	320
Burned intermittently	40	360

a. What is the probability that a given bulb failed?
b. What is the probability that a bulb burned continuously and failed?
c. What is the probability that a bulb burned intermittently and failed?
d. Given that a bulb failed, what is the probability that it burned intermittently?
e. Given that a bulb continued to function, what is the probability that it burned intermittently?
f. Summarize these probabilities in a tree diagram.

4.25 If $p(A) = 0.35$, $p(B) = 0.66$, and $p(A \text{ and } B) = 0.231$,
  a. Are events A and B mutually exclusive? Explain.
  b. Are events A and B independent? Explain.

4.26 If $p(A) = 0.80$, $p(B) = 0.40$, and $p(A \text{ and } B) = 0.00$,
  a. Are events A and B mutually exclusive? Explain.
  b. Are events A and B independent? Explain.

4.27 Lila, a real-estate broker, claims that she will sell a piece of property 80 percent of the time if she is given an exclusive listing for three months. She finds that 40 percent of her sales are consummated in the first month, i.e., she receives an offer and makes a sale. However, an offer is made in the first month only 50 percent of the time. What is the probability Lila will sell a piece of property, given that an offer is made during the first month? Is the sale of the property dependent on whether an offer is made the first month?

4.28 Steven S. sells jewelry door-to-door. Fifty percent of the time he makes a sale. He finds that 40 percent of the time he shows his merchandise and makes a sale. However, only 60% of the time does a potential customer give him the opportunity to show his merchandise. What is the probability that Steven will make a sale, given the opportunity to show his merchandise? Is a sale dependent on whether or not he gets this opportunity?

4.29 There are five workers in an assembly line ranked according to their productivity: 1, 2, 3, 4, 5 in which the rank of 1 is best.
  Suppose two workers are selected at random.
  a. List all the sample points in the sample space.
  b. List the sample points for the following events:
       $A = $ Best worker chosen.
       $B = $ Worst worker chosen.
  c. What is $p(A)$?
  d. What is $p(B)$?
  e. What is $p(A \text{ and } B)$?
  f. Are A and B independent?
  g. Are A and B mutually exclusive?
  h. What is $p(A \text{ or } B)$?

4.30 There are four contractors bidding on two jobs. Let 1, 2, 3, 4 represent the four contractors.

    a. Portray all sample points in the sample space in a Venn diagram.

    b. Let event A = one contractor gets both jobs. List the sample points and find $p(A)$.

    c. Let event B = two different contractors win the two bids. List the sample points and find $p(B)$.

    d. Are events A and B independent? Are they mutually exclusive?

    e. Illustrate events A and B in a Venn diagram.

4.31 Experience has shown that, in a particular area, 6 percent of the renters skip out on the last month's rent. That is, they don't pay the rent due on the first of the month but stay the full month anyway.

A new service for landlords guarantees that it is 80 percent accurate in correctly predicting good tenants on the basis of a credit score.

If 85 percent of all applicants for an apartment score favorably and are rented an apartment, find the probability that an applicant will *not* skip out on the rent if he or she is given the apartment based on the service's prediction.

4.32 A recent survey conducted by the National Association of Home Builders found that 30 percent of home-buyers were first-time buyers and of those, 68 percent had two incomes.

You have your house up for sale. Three different (and unrelated) couples make offers to buy.

    a. What is the probability that all three couples are first-time buyers and have two incomes?

    b. What is the probability that at least one of the couples is both a first-time buyer and has two incomes?

4.33 A recent Federal Home Loan Bank report shows that only 15 percent of potential home buyers can afford the monthly payments required for such a purchase. In a group of three unrelated potential buyers, what is the probability that

    a. All three cannot afford the monthly payments?

    b. At least two cannot afford the monthly payments?

    c. All three can afford the payments?

4.34 According to a survey by the University of California in Los Angeles, private preparatory and parochial schools supplied 31 percent of the 1979 freshman class at the nation's most selective colleges and universities.

UCLA defines "most selective" as private colleges and universities in which the students' combined math and verbal aptitude test scores average 1,175 or more.

By this definition, 98 colleges or universities out of the nearly 3,000 in the country qualify.

What is the probability that a student, selected at random, is from a private preparatory or parochial school and is a freshman at one of the nation's most selective colleges or universities?

4.35 In the 1980 competition for acceptance at one of the "most selective" universities (as defined in Exercise 4.34), there were 11,180 applicants. From these, 2,000 were accepted. One private boarding school in Massachusetts received 40 acceptances.

    a. What was the probability that a given applicant was accepted at this university?

    b.   Given that a student was accepted, what is the probability that he or she came from the private school in Massachusetts?

    c.   What is the probability that a student, selected at random, was accepted at this university and came from the private school in Massachusetts?

4.36   In a given geographical region with a population of approximately 4,444,000 people, 44,000 died of various causes during a year. One hundred thousand were involved in automobile accidents, of which four thousand resulted in fatalities.

    a.   What is the probability that a person died during that year?

    b.   Given that a person died, what is the probability that the death was due to an accident?

    c.   Given that a person was in an accident, what is the probability that he or she died?

4.37   Freeman Gosden, president of Direct Mail Marketing Club of Southern California, recently stated what he calls the "40/40/20" rule in direct marketing. He explained: "Forty percent of your success is who you are, what you're offering, and what the price and deal are *(A)*. The second 40 percent is hitting the right audience *(B)*. And the 20 percent is the creatives—the visuals and the copy *(C)*."[5]

    Are events *A, B,* and *C*

    a.   Mutually exclusive?

    b.   Exhaustive?

    c.   Complementary?

    d.   Independent?

What approach to probability theory is Gosden using?

4.38   One of the problems that small businesses face in the current economic climate is undercapitalization. Growth runs ahead of profits and so outstrips capital to finance continued growth.

    How to finance capital? A recent article in the *Los Angeles Business Journal* listed four options available to small businesses:

    A = A term loan from a bank.

    B = An SBA loan.

    C = A loan from a friend or relative.

    D = A second trust deed on a house.

    Suppose that the probabilities that a small business will choose a given option are: $p(A) = 0.25$; $p(B) = 0.20$; $p(C) = 0.30$; $p(D) = 0.25$.

    Two small businesses need to borrow money. Each selects one of the options available. Assume their selections are independent.

    Find the probability that

    a.   Both small businesses choose the same option.

    b.   Each selects a different option.

    c.   Each selects a different option, but neither chooses to borrow from a friend or relative.

4.39   Laurie J., president of Sunset Corp., is planning on marketing a new product. With no competition, she figures there is an 80 percent probability of success. With competition, she estimates the probability of success at 60 percent. She figures that there is a 50 percent chance that she will have competition. What is the probability that Laurie's product will be successful? Use a tree diagram to illustrate this problem.

---

[5] Source: *Los Angeles Business Journal,* September 1, 1980.

4.40 David and Scott both go into business right after graduation. The probability that David's business will show a profit the first year is 0.50. Since Scott's business is more competitive, the probability that he will be successful the first year is 0.30. Assume that the two business ventures are independent.

Find the probability that, during the first year,

    a. Both will be successful.

    b. Neither will be successful.

    c. At least one will be successful.

4.41 A case study of how even the most carefully laid plans can be spoiled by unexpected events:

Procter & Gamble spent 18 years and millions of dollars (including $10 million in free samples) to develop and promote its new Rely tampon.

That effort appears jeopardized by government allegations linking Rely with a mysterious, sometimes fatal, disease known as "toxic shock syndrome."

The Center for Disease Control linked tampon use with the syndrome. The agency said it found 70 percent of a sample of 50 women who contracted the disease in July and August had used Rely tampons.

The ailment is rare, affecting only about 3 in 100,000.

What is the probability that a woman, selected at random, used Rely and got the disease?

4.42 Refer to the data in Exercise 4.41.

Rely was introduced in 1975 but was not distributed nationally until early 1980. According to *The Wall Street Journal* (September 19, 1980), the product has done well and already has garnered at least 20 percent of the tampon market.

If this estimate of Rely's share of the market is accurate, what is the probability of getting toxic shock syndrome given that the individual uses Rely?

How would you evaluate a conclusion that stated, "the chances of getting this disease is 3.5 times as great for a Rely user"?

Random Variables and Probability

Distributions

5.1 RANDOM VARIABLES

It is an inescapable fact of life that decisions are always made under conditions of uncertainty. Should you buy a house today or wait until interest rates go down? Will they go down? When is the best time to borrow money? Will the prime rate go up or down tomorrow? Should you take the first job offered or wait for a better opportunity? Should you spend your advertising money on radio, TV, or newspaper ads? If you represent management, should you accept union demands or take a chance on losing production because of a strike?

In each of these cases, the measurement of interest may vary in a random manner. Thus, we refer to these measurements as *random variables*. A random variable may be defined as a quantitative variable that varies in a chance or random manner. However, suppose we could determine the probability of different values of a random variable. As we shall see in this chapter, we can derive *probability distributions* associated with a particular random variable; i.e., a theoretical distribution that represents the probability of each value of a random variable.

Recall our previous discussion of continuous and discrete variables (Section 1.4.1). A variable is said to be continuous within its range if, for any pair of real values with $p > 0$, there exist other values between these two for which $p > 0$. In other words, any nonzero interval within the range of the variable will have a probability greater than zero. To illustrate: Imagine it is possible to obtain an infinitely large group of people. Suppose further that we obtained the height of each on an incredibly accurate scale of measurement and expressed the value to any number of decimal places. Of course, there would be a limit to the range of values (no one would be as short as zero nor as tall as a mature elm tree). However, within the range of values that do occur, any given interval would have an associated probability greater than zero. In other words, at least *someone's* height would fall within any conceivable interval. Since the number of fractional values is unlimited, we cannot list all possible fractional values of a *continuous random variable* and calculate a probability value associated with each. Rather we discuss probabilities in terms of intervals of a continuous random variable. Thus, in this height example, we might describe the probability of obtaining a value between 6.157 and 6.158 feet rather than a specific value, such as 6.1575 feet.

Contrast this with a *discrete random variable*. Discrete variables are characterized by gaps in which no real values are found. Thus, there are either one or two or three or . . . . . children in a family. A random variable is said to be discrete if, for any pair of values for which there is a probability greater than zero, there exist other values between these two for which the probability is zero. Thus, if $p(a = 1) > 0$ (one child in a family) and $p(b = 2) > 0$ (two children in a family), there exist values between a and b for which $p = 0$ (e.g., 1.5 children, 1.7 children, etc.).

Therefore, a finite discrete random variable is one for which it is possible to list all values of the variable within its range and calculate the probabilities associated with each value. As we have seen, such is not the case with a continuous random variable. Figure 5.1 represents graphically the contrast between the distributions of discrete and continuous random variables.

Graphic representation of distinction between distribution of discrete *(a)* and continuous   **Figure 5.1**
(b) random variables. Note that the values of a finite discrete variable can be listed. However, values of a continuous random variable can be best represented by the probabilities associated with intervals of values of the random variable.

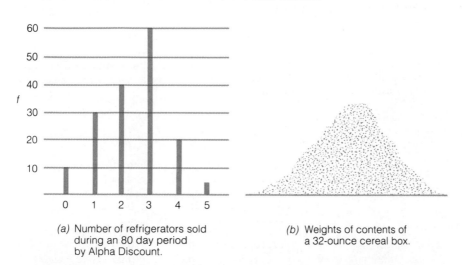

(a) Number of refrigerators sold during an 80 day period by Alpha Discount.

(b) Weights of contents of a 32-ounce cereal box.

In this chapter we shall examine several mathematical models that are widely used in statistical inference: the binomial, Poisson, and normal distributions. At the outset, it should be noted that the mathematical models are themselves purely theoretical. They are not facts of the real world that are either true or false. Rather they represent theories concerning possible relationships between probabilities and values of a variable. Thus, the binomial model is a theory concerned with describing probabilities when sampling from a discrete two-category variable. Similarly, the normal model represents a theory concerning possible relations between intervals of a continuous variable and probabilities of that variable. Normal distributions do not exist in the real world. However, we often find that real world distributions come close to fulfilling the requirements of a normal distribution. Consequently, practical applications of the theoretical model can be found. In this chapter, we examine several of these applications.

5.2 PROBABILITY DISTRIBUTIONS AND EXPECTED VALUE OF DISCRETE RANDOM VARIABLES

Let us imagine that you are the dispatcher of repair trucks for an automobile rental agency in a large metropolitan region. One of your problems is to anticipate the demand for repair services so that you have sufficient trucks and personnel available to meet the expected demand. You collect data on the number of requests per day over a 100-day period.

Table 5.1 shows this historical record. By dividing each frequency by *n* (the sample size), we arrive at a record of the probability of past demand on any given day. Note that this is an empirical distribution based on past records: the probabilities are therefore subject to change with changing times and conditions. However, if we may assume relatively stable circumstances in which present demand fairly accurately reflects the recent past, the probability distribution provides a better estimate of demand than sheer guesswork.

Table 5.1

Frequency and probability distributions of demand for repair services of an auto rental agency over a 100-day period. The probability of each *X* value, i.e., *p(X)* is obtained by dividing its associated frequency by *n*.

Demand *(X)*	f	p(X)
3	1	0.01
4	4	0.04
5	10	0.10
6	25	0.25
7	32	0.32
8	12	0.12
9	10	0.10
10	6	0.06
Total	100	1.00

Note that the demand for repair service has never been less than 3 nor greater than 10 on any given day. The events 3 through 10 are mutually exclusive and exhaust all outcomes during the period. Thus, the sum of the probabilities is equal to 1.00.

A concept of considerable importance in describing a probability distribution is that of expected value. Its origins are tied to games of chance when gamblers estimated what they might expect to win (or lose) in the long run by pursuing a given strategy or participating in a particular game. Contemporary statistics retains the spirit of this concept. The *expected value* of a random variable is the long-term average for that variable over many samplings. Symbolized by *E(X),* the expected value of a variable is the mean of the probability distribution. This is, in turn, the mean of the variable. Thus,

expected value

A form of weighted average in which the probabilities associated with each value of the random variable are used as weights; long-term average over many samplings.

$$E(X) = \Sigma Xp(X) = \text{mean of } X \text{ (i.e., } \mu) \qquad (5.1)$$

To obtain *E(X),* the expected value of *X,* we multiply each value of *X* by its associated probability and sum. In effect, we are weighting each value of X by its expected probability of occurrence. The column headed by $Xp(X)$ in Table 5.2 shows these weighted values. The sum of the column provides a sort of weighted average of the demand that you, as dispatcher, might expect in the future. It is the expected value of the random variable "demand for repair services." Therefore, on the average, you should expect approximately seven daily requests for repair.

Demand (X)	p(X)	Xp(X)
3	0.01	0.03
4	0.04	0.16
5	0.10	0.50
6	0.25	1.50
7	0.32	2.24
8	0.12	0.96
9	0.10	0.90
10	0.06	0.60
	1.00	E(X) = 6.89

Table 5.2
Calculation of expected demand for repair services

Example A

Maribeth, the owner of a retail store, is thinking of opening another branch. She plans to base her decision on her experiences with her first store. She has found that, in any given month, her income is equally likely to be $3,000, $4,000, $5,000, or $6,000. Regardless of her income, she also finds that her expense are equally likely to be $2,000, $3,000, $4,000, or $5,000. Determine the probability distribution of her net income (i.e., Income − Expenses).

Her cash flow situation is such that she requires an expected income of at least $500 a month to make this project feasible. Should she go ahead?

Solution

In this case, we are not given the distribution of the random variable of interest—net income. Therefore, in order to construct the required probability distribution, we must first list all the possible combinations of income and expenses.

There are four possible incomes and four possible expenses. Since each income can combine with each expense, 16 possible combinations can be generated, i.e., $3,000 income with $2,000, $3,000, $4,000, $5,000 expenses; $4,000 income with $2,000, $3,000, $4,000, $5,000; etc. The list of 16 possible combinations is shown in Table 5.3.

Table 5.3

Sixteen combinations of
four different incomes and
four different expenses

Income	Expenses	Net income (Income − Expenses)
$3,000	$2,000	$+1,000
3,000	3,000	0
3,000	4,000	−1,000
3,000	5,000	−2,000
4,000	2,000	+2,000
4,000	3,000	+1,000
4,000	4,000	0
4,000	5,000	−1,000
5,000	2,000	+3,000
5,000	3,000	+2,000
5,000	4,000	+1,000
5,000	5,000	0
6,000	2,000	+4,000
6,000	3,000	+3,000
6,000	4,000	+2,000
6,000	5,000	+1,000

Next we prepare a frequency distribution of the column labeled "net income." Table 5.4 shows the frequency distribution of net income, the probability distribution obtained by dividing each frequency by 16, and the product of income (X) times frequency.

Table 5.4

Frequency and probability
distributions of net income
and calculation of
expected value

Net income (X)	f	$p(X)$	$Xp(X)$
$-2,000	1	0.0625	−125
−1,000	2	0.1250	−125
0	3	0.1875	0
+1,000	4	0.2500	+250
+2,000	3	0.1875	+375
+3,000	2	0.1250	+375
+4,000	1	0.0625	+250
	16		$E(X) = 1,000$

Since her expected net income of $1,000 is more than the required $500, she should go ahead with her expansion plans.

EXERCISES

5.1 Based on the information in Example A (see p. 119), what is the probability that, in any given month,

   a. She will either make no money or will lose money?
   b. She will make at least $3,000?
   c. She will lose money?

Box 5.1

THE EXPECTATION OF THE VARIANCE AND THE STANDARD DEVIATION

An alternative procedure for obtaining the variance and standard deviation of a set of measurements is formulated in terms of expected values. Earlier in the chapter, we saw that the expected value of a probability distribution is the mean of that distribution. Thus, $E(X) = \Sigma Xp(X) = \bar{X}$. A comparable procedure exists with respect to the expected values of the variance and the standard deviation.

When working with sample data, the expectation formula for the unbiased estimate of the population variance is

$$s^2 = \{E(X^2) - [E(X)]^2\}\left[\frac{n}{n-1}\right]$$

The square root of this value yields the standard deviation. Thus,

$$s = \sqrt{\{E(X^2) - [E(X)]^2\}\left[\frac{n}{n-1}\right]}$$

The table below shows the calculation of s^2 and s for runner B in Example F, Chapter 3. As we shall see (Chapter 6), the expectation formula for the variance and standard deviation is of considerable importance when dealing with many theoretical distributions.

X	f	p(X)	X^2	$X^2p(X)$	$Xp(X)$
−3	5	0.096	9	0.864	−0.288
−1	9	0.173	1	0.173	−0.173
1	10	0.192	1	0.192	0.192
3	4	0.077	9	0.693	0.231
5	13	0.250	25	6.250	1.250
7	11	0.212	49	10.386	1.484
				$E(X^2) = 18.558$	$E(X) = 2.696$

$$s^2 = \{E(X^2) - [E(X)]^2\}\left[\frac{52}{51}\right]$$

$$= [18.558 - (2.696)^2]\left[\frac{52}{51}\right]$$

$$= [11.29]\left[\frac{52}{51}\right] = 11.51$$

$$s = \sqrt{11.51} = 3.39$$

5.2 Identify the following as either discrete or continuous random variables:
  a. Number of defectives in the daily output.
  b. Grade of ore expressed in ounces of gold per ton of ore.
  c. Number of kilowatts used daily by a four-unit apartment building.
  d. Number of work-days lost each month in the steel industry due to illness.
  e. Length of time to check out items at a counter in a supermarket.
  f. Number of children in a family.
  g. Number of television sets owned per family.
  h. Weight of trout caught by a fisherman on a given day.

5.3 Frances C., a buyer for a large retail outlet, is considering bidding on the entire stock of a bankrupt company. She feels that, if successful, her outlet may expect a profit of $125,000. From past experience with competing bidders, she estimates her chances of winning the contract at 0.30. However, the cost for preparing the bid is $15,000. Her company will be satisfied if the profit is $75,000. Should she bid on the stock of the bankrupt company?

5.4 Referring to Table 5.1, what is the probability that
  a. Demand, on any given day, will be 9 or more?
  b. Demand, on any given day, will be between 6 and 8?

    c. On two consecutive days, demand will be 8 or greater?

    d. On three consecutive days, demand will be 10?

5.5 An investment group is considering the purchase of a gas lease on a piece of land where seismic data suggest the possibility of natural gas. The cost for purchasing the option and drilling a well is $5,000,000. The expected return, if successful, is $200,000,000. The probability of success, based on experiences with similar sites in the region, is 0.15. If the expected return is positive, the investment group will further explore the possibilities. What will it do in this case?

5.3 DISCRETE VARIABLES: THE BINOMIAL

Bernoulli trials

Experiments that can result in only one of two mutually exclusive and exhaustive outcomes.

    The most elementary type of probability distribution is one in which there are two mutually exclusive and exhaustive classes of events—heads or tails, male or female, success or failure, no error or error, etc. Returning to Montana Rincon, if an account is selected at random for audit, one of two events must occur, namely, an error is found or no error is found. Each has some specific probability of occurring. Experiments that lead to one of two possible outcomes are referred to as *Bernoulli Trials*. The probability distribution of the two classes of events is called a Bernoulli distribution.

    It is traditional to call one of the two events a "success" and the other a "failure," with neither word connoting either desirable or undesirable results. We will use P to symbolize the probability of a success and $Q = 1 - P$ for the probability of a failure.

    Since events in Bernoulli trials are mutually exclusive and exhaustive,

$$P + Q = 1.00$$
$$P = 1.00 - Q$$
$$Q = 1.00 - P$$

    Let us assume that, in the population studied by Montana Rincon, the number of error-free accounts is 4.7 million out of 5 million. Thus, $P = 0.94$. Since Q equals $1 - P$, the probability of an error is 0.06.

    Now suppose we selected three independent observations from the population.[1] How many different sequences of Ps and Qs might we observe? The rule in the two-category case is 2^n. Thus, when $n = 3$, the total number of sequences is $2^3 = 8$. Shown below are all eight possible sequences as well as the probability of each.

---

[1] Actually, we would in all likelihood sample without replacement. Technically speaking, these observations would not be independent since the probabilities of an event change as each observation is withdrawn from the pool of possible observations. However, with a population N of 5 million and a sample n equal to 4, the difference between sampling with and without replacement is negligible.

All successes	$PPP = (0.94)(0.94)(0.94) = 0.830584$
Two successes and one failure	$PPQ = (0.94)(0.94)(0.06) = 0.053016$ $PQP = (0.94)(0.06)(0.94) = 0.053016$ $QPP = (0.06)(0.94)(0.94) = 0.053016$
One success and two failures	$PQQ = (0.94)(0.06)(0.06) = 0.003384$ $QPQ = (0.06)(0.94)(0.06) = 0.003384$ $QQP = (0.06)(0.06)(0.94) = 0.003384$
No success	$QQQ = (0.06)(0.06)(0.06) = \underline{0.000216}$ 1.000000

Note that the probability of any given number of successes in n Bernoulli trials depends only on P and the number of successes. In other words, the *sequence* of successes and failures has no effect on the probability that a given *number of successes* will occur.

In most real-world problems involving Bernoulli trials, we are not interested in the order or sequence of successes or failures. Rather, we are usually interested in the probability of a given number of successes in n trials. The number of successes is a discrete random variable. The distribution that yields the probability of a given *number of successes* in n Bernoulli trials is called the *binomial distribution*. The binomial distribution plays an important role in decision making situations in which there are only two alternative events—accept versus reject, elect versus fail to elect, compete versus not compete, innocent versus guilty.

Let's take a look at two different situations and see what they have in common.

> The leafcutter is an insect that attacks the leafy growth of cotton plants. Suppose that, without pesticide control, 10 percent of the plants are attacked. In one field, 20 plants are examined and 10 are found to be infested with the leafcutter. Is there reason to suspect that the infestation is higher than usual?

> It has been found that 60 percent of foreigners entering the country have not been administered polio vaccine. During the next 24 hours, four are expected to arrive in your city, where there has been a recent outbreak of polio. What is the probability that none has received the vaccine?

Each of these is an example of a situation to which the *binomial* applies. They share the following four characteristics:

☆ 1. The variable of interest consists of two mutually exclusive and exhaustive categories; e.g., infested versus noninfested; vaccinated versus not vaccinated.

☆ 2. Each trial is identical. For example, we do not use different criteria for judging infestation of the leafcutter for each observation (trial).

☆ 3. The probability of a success remains the same for each trial, i.e., each trial is independent of any other. For instance, the probability that any one individual has been vaccinated does not affect the probability that any other individual has been vaccinated.

binomial distribution

A theoretical distribution that yields the probability of a given number of successes in n Bernoulli trials.

binomial

A family of theoretical distributions which follow the same mathematical rule for relating probabilities to values of a discrete random variable. Each distribution differs in specific probabilities according to n and P.

4. While we assume that the chance of one success in one trial is known, our interest is directed toward ascertaining the chance of *x* successes in *n* trials or the number of times an observation falls in the *P* category. Similarly, *(n − x)* represents the number of times an observation falls in the *Q* category.

To illustrate, if the usual proportion of sales consummated by a particular salesperson is 0.33, what is the probability that he or she will make four sales out of the next six contacts? In this example, *P* = 0.33, *Q* = 0.67, *x* = 4, *(n − x)* = 2, and *n* = 6.

As we shall see, *p(x)* for a particular *P* is identical to *p(n − x)* for the corresponding *Q*. Thus, *p(x = 4)* when *P* = 0.33 is the same as the *p[(n − x) = 2]* for *Q* = 0.67.

Example B

An electronic surveillance device is constructed in which a given component appears in five separate locations. Under conditions of high humidity, the probability is 0.90 that a given component will function adequately. When in use during the monsoon season, what is the probability that all components will perform in a satisfactory manner? Does the binomial apply to this situation? If it does, specify the values of *P, Q, n, x,* and *(n − x)*.

Solution

The binomial applies if we may legitimately assume that the five components are independent, i.e., the failure of one component does not change the probability that other components will fail. The categories *function* and *fail* are mutually exclusive and exhaustive. The probability of success is the same for all five components. The criterion of success for each component is identical, namely, that it continues to function. *P* is the probability that a component will continue functioning; *Q* is the probability that a given component will fail, *n* is the number of components (5), and *x* = 5 since this is the event of interest. Also, *(n − x)* = 5 − 5 = 0.

Example C

Jerry J. is the owner of a mail-order company. From prior experience, he knows that the response rate to advertising material is usually 20 percent. He decides to try a new device—this time he includes a business reply card. He sends out 100 pieces and receives 30 replies. Does the binomial distribution apply to this situation? If so, specify *P, Q, n, x, (n − x)*.

Solution

Yes, this satisfies the requirements of a binomial distribution. The two mutually exclusive categories—"respond" and "not respond"—are exhaustive. Each trial is independent in that the response of one individual does not in any way affect the response of another. In this example, $P = 0.20$, $Q = 0.80$, $x = 30$, $(n - x) = 70$ and $n = 100$.

EXERCISES

5.6 You have a disagreement with a friend about a course of action to take. You suggest tossing a die. If it comes up even, you and your friend agree to pursue your suggested course. If it comes up odd, both will follow the friend's suggestion. Is this a situation to which the binomial distribution applies? If so, specify P, Q, n. x, $(n - x)$.

5.7 Past records show that the probability of a failure of a specific machine in the manufacture of carpeting is about 1 in 20 over the course of a year. There are 40 machines in service on a continuous basis. Is the binomial distribution appropriate for ascertaining the probabilities that five or fewer machines will fail? If so, specify P, Q, n, x, and $(n - x)$.

5.8 Suppose that, on an eight-hour flight, the probability of an engine failure on a four-engine aircraft is 1 in 10,000. What is the probability that two engines will fail? Is the binomial distribution appropriate for answering the question? If so, specify P, Q, n, x, and $(n - x)$.

5.3.1 The binomial expansion

Perhaps the simplest way to illustrate the probability distribution of a Bernoulli variable is to deal with a one-trial case, i.e., $n = 1$. In this case, we have only two possible outcomes, $x = 1$ or $x = 0$. That is, with one trial, the result is either a success or a failure. The probability distribution for a single trial is:

x	$p(x)$
1	P
0	Q

The probability distribution of a two-trial situation, $n = 2$, may be derived by applying the multiplication law for independent events:

$p(x = 2) = p(x = 1)p(x = 1) = P^2$ (probability of two successes out of two trials)

$p(x = 1) = p(x = 1)p[(n - x) = 1] = PQ$ (probability of a success followed by a failure)

$p(x = 1) = p[(n - x) = 1)]p(x = 1) = QP$ (probability of a failure followed by a success)

$$p(x=0) = p[(n-x)=1)]p[(n-x)=1] = Q^2 \quad \text{(probability of two failures out of two trials)}$$

The important point illustrated in this example is that the probability distribution of the binomial for any given n is obtained by expanding $(P + Q)^n$. Thus, when $n = 2$,

$$(P+Q)^2 = P^2 + PQ + QP + Q^2.$$

Note that we are not usually interested in distinguishing between the two events PQ and QP. Both indicate the probability of one success and one failure. Consequently, PQ and QP are treated as identical. Thus, we add them together to obtain

$$P^2 + 2PQ + Q^2.$$

Similarly, when $n = 3$,

$$
\begin{array}{r}
(P+Q)^3 = \quad P^2 + 2PQ + Q^2 \\
\times \quad P + Q \\
\hline
P^3 + 2P^2Q + PQ^2 \\
+ P^2Q + 2PQ^2 + Q^3 \\
\hline
1P^3 + 3P^2Q + 3PQ^2 + 1Q^3 \\
\uparrow \qquad \uparrow \qquad \uparrow \qquad \uparrow \\
\end{array}
$$

coefficients of the binomial

In this form of the binomial expansion, the exponents tell us the numbers of observations in both the P and the Q categories. Thus, P^3 involves three observations in the P category and none in the Q category, i.e., $x = 3$ and $(n - x) = 0$. Similarly, $3PQ^2$ involves $x = 1$ and $(n - x) = 2$.

Incidentally, the coefficients in the binomial expansion (1, 3, 3, 1) show the total number of different sequences of P and Q for a given value of x (see Box 5.2).

The probability of any given value of x may be obtained more conveniently by use of the following formula:

$$p(x) = \frac{n!}{x!(n-x)!} P^x Q^{n-x} \qquad (5.2)$$

where

$x =$ Number of observations in the P-category
$n - x =$ Number of observations in the Q-category
$n =$ Total number of trials or observations
$! =$ Factorial notation, which directs us to multiply the indicated number by all integers less than it, but greater than zero; e.g., $n! = (n)(n-1)(n-2)(n-3) \ldots (3)(2)(1)$

(By definition, $0! = 1$ and any value, other than zero, raised to the zero power $= 1$. For example, $(n - n)! = 0! = 1$, and $P^0 = 1$.)

Note that the first term to the right of the equality sign yields the coefficients of the binomial. The second term shows the value of P raised to the x power and Q raised to the $(n - x)$ power.

You may have noticed that, when each value of x is substituted in Formula

---

Box 5.2

PASCAL'S TRIANGLE

The French mathematician Blaise Pascal worked out a simple system for ascertaining the coefficients of the binomial. Known as Pascal's Triangle, it starts out with the coefficients for a two-category variable when $n = 1$. These are 1, 1 as in $1P^1$ and $1Q^1$. For $n = 2$, you place a 1 below and to the left of the coefficient above, another below and to the right of the coefficient below. The two coefficients are added together and placed below and midway between the two 1's. Thus, we obtain $1\,1\,2\,1\,1$. Continuing these procedures to any given n, we can obtain the coefficients of the binomial directly from the triangle. The figure

below shows Pascal's Triangle with coefficients through $n = 12$. The superscripts accompanying each coefficient show the corresponding exponent of P and Q respectively.

If we know n and wish to obtain the probability of a given x, we may go directly to Pascal's Triangle to obtain the appropriate term of the binomial. To illustrate, imagine that $n = 11$ and we wish to know $p(x = 9)$. We go to the row $n = 11$, and find the coefficient with exponents 9,2. We find the coefficient is 55. Thus $p(x = 9) = 55P^9Q^2$. All we need are the values of P and Q to calculate $p(x = 9)$.

Pascal's Triangle from $n = 1$ through $n = 12$. The boldface numbers show the coefficients of the binomial at each value of n. The superscripts show the exponents of P and Q, respectively, at varying values of n and x. To find the probability expression when $n = 10$, $x = 4$, and $(n - x) = 6$, we look across row 10 until we find the coefficient with 4,6 as superscripts. The coefficient is 210. Thus, $p(x = 4) = 210\ P^4Q^6$.

(Pascal's Triangle rows)	n	Number # of sequences 2^n
$1^{1,0}$ $1^{0,1}$	1	2
$1^{2,0}$ $2^{1,1}$ $1^{0,2}$	2	4
$1^{3,0}$ $3^{2,1}$ $3^{1,2}$ $1^{0,3}$	3	8
$1^{4,0}$ $4^{3,1}$ $6^{2,2}$ $4^{1,3}$ $1^{0,4}$ $(x, n-x)$	4	16
$1^{5,0}$ $5^{4,1}$ $10^{3,2}$ $10^{2,3}$ $5^{1,4}$ $1^{0,5}$	5	32
$1^{6,0}$ $6^{5,1}$ $15^{4,2}$ $20^{3,3}$ $15^{2,4}$ $6^{1,5}$ $1^{0,6}$	6	64
$1^{7,0}$ $7^{6,1}$ $21^{5,2}$ $35^{4,3}$ $35^{3,4}$ $21^{2,5}$ $7^{1,6}$ $1^{0,7}$	7	128
$1^{8,0}$ $8^{7,1}$ $28^{6,2}$ $56^{5,3}$ $70^{4,4}$ $56^{3,5}$ $28^{2,6}$ $8^{1,7}$ $1^{0,8}$	8	256
$1^{9,0}$ $9^{8,1}$ $36^{7,2}$ $84^{6,3}$ $126^{5,4}$ $126^{4,5}$ $84^{3,6}$ $36^{2,7}$ $9^{1,8}$ $1^{0,9}$	9	512
$1^{10,0}$ $10^{9,1}$ $45^{8,2}$ $120^{7,3}$ $210^{6,4}$ $252^{5,5}$ $210^{4,6}$ $120^{3,7}$ $45^{2,8}$ $10^{1,9}$ $1^{0,10}$	10	1,024
$1^{11,0}$ $11^{10,1}$ $55^{9,2}$ $165^{8,3}$ $330^{7,4}$ $462^{6,5}$ $462^{5,6}$ $330^{4,7}$ $165^{3,8}$ $55^{2,9}$ $11^{1,10}$ $1^{0,11}$	11	2,048
$1^{12,0}$ $12^{11,1}$ $66^{10,2}$ $220^{9,3}$ $495^{8,4}$ $792^{7,5}$ $924^{6,6}$ $792^{5,7}$ $495^{4,8}$ $220^{3,9}$ $66^{2,10}$ $12^{1,11}$ $1^{0,12}$	12	4,096

5.2, it yields results that are precisely the same as when we expanded the binomial. Thus, when $n = 3$, and $x = 3$, $p(x = 3) = \frac{3!}{3!0!} P^3 Q^0 = P^3$.

Similarly, when $x = 2$, $p(x = 2) = \frac{3!}{2!1!} P^2 Q^1 = 3P^2Q$; when $x = 1$, $p(x = 1) = \frac{3!}{1!2!} PQ^2 = 3PQ^2$ and when $x = 0$, $p(x = 0) = \frac{3!}{0!3!} P^0 Q^3 = Q^3$.

Example D

Before I go to pick up an arriving passenger at the airport, I call the airline for the ETA (estimated time of arrival). Surprisingly, 80 percent of the time, they give me misinformation—the plane comes in after their ETA. This week I have to pick up passengers arriving on three different flights.

What is the probability (according to information supplied by the airlines) that

a. All three will arrive on time?
b. None will arrive on time?
c. At least one will arrive on time?

Solution

$$P = \text{Arrive on time} = 0.20$$
$$Q = \text{Do not arrive on time} = 0.80$$
$$n = 3$$

a.
$$p(x = 3) = \frac{3!}{3!0!} (0.20)^3 (0.80)^0$$
$$= (0.20)^3 = 0.008$$

b.
$$p(x = 0) = \frac{3!}{0!3!} (0.20)^0 (0.80)^3$$
$$= (0.80)^3 = 0.512$$

c.
$$p(x = 1) = \frac{3!}{3!0!} (0.20)^1 (0.80)^2$$
$$= 3(0.20)(0.80)^2$$
$$= 0.384$$

$$p(x = 2) = \frac{3!}{2!1!} (0.20)^2 (0.80)^1$$
$$= 3(0.20)^2 (0.80)^1$$
$$= 0.096$$
$$p(x = 3) = 0.008$$

Therefore,

$$p(\text{at least one}) = p(x = 1) + p(x = 2) + p(x = 3)$$
$$= 0.384 + 0.096 + 0.008$$
$$= 0.488$$

Alternatively,

$$1 - p(x = 0) = 1.00 - 0.512 = 0.488$$

Example E

In a survey of 106,000 readers of *Cosmopolitan* magazine, 54 percent of the married women said they had had extramarital affairs. Assuming the sample of readers of *Cosmopolitan* is representative of the general population of married women, what is the probability that in a random sample of six women.

a. All have had extramarital affairs?
b. Five or fewer have had extramarital affairs?
c. One has *not* had an extramarital affair?

Solution

$$P = \text{yes extramarital} = 0.54$$
$$Q = \text{no extramarital} = 0.46$$
$$n = 6$$

a.
$$p(x = 6) = \frac{6!}{6!0!}(0.54)^6(0.46)^0$$
$$= 0.0248$$

b.
$$p(x \leqslant 5) = p(0) + p(1) + p(2) + p(3) + p(4) + p(5)$$
$$= 1 - p(6)$$
$$= 1 - 0.0248 = 0.9752$$

c. As stated previously, $p[(n - x) = 1]$ for $Q = 0.46$ is equivalent to $p(x = 5)$ for $P = 0.54$. Thus,

$$p[(n - x) = 1] = p(x = 5) = \frac{6!}{5!1!}(0.54)^5(0.46)^1$$
$$= 0.1267.$$

EXERCISES

5.9 A new type of checking account called NOW (for negotiable order of withdrawal) differs from traditional checking accounts mainly by paying interest on deposits.

The Federal Home Loan Banks (FHLB) were expressly authorized by Congress to provide this service to member savings and loan institutions that want it.

Eighty-three percent of California's regional FHLB are now offering NOW account servicing to their members.

If we randomly select four regional FHLB, find the probability that
- *a.* All offer NOW account servicing.
- *b.* At least two offer NOW.

5.10 A recent study of 1,500 randomly selected households in Tucson, Arizona,[2] reported that 69 percent of the households dine out monthly.

In a random sample of three households, what is the probability that
- *a.* At least two dine out monthly?
- *b.* All three dine out monthly?
- *c.* None dines out monthly?

5.3.2 The use of binomial probability tables

As you can see, calculating binomial probabilities can become quite laborious and subject to error as *n* increases. For this reason, Table I is included in Appendix C. This table shows the probabilities associated with various values of *x* for selected values of *P* and *n*. Let's look at an illustration.

Example E

Recent experiences with manufacturing a component used in a sophisticated solar tracking device has shown that approximately 25 percent are defective. A purchaser orders a lot of 30 and will be satisfied if no more than 9 are defective.

- *a.* What is the probability that exactly nine will be defective?
- *b.* What is the probability that nine or fewer will be defective?
- *c.* What is the probability of finding more than nine defectives?

Solution

a. Referring to Table I, we look under $n = 30$, column $P = 0.25$ and row $x = 9$: we find the probability of finding exactly nine defective components is 0.1298.

b. To find the probability that *x* is equal to or less than 9, we cumulate the probabilities from $x = 9$ through $x = 0$. Thus, $p(x \leq 9) = 0.1298 + 0.1593 + 0.1662 + 0.1445 + 0.1047 + 0.0604 + 0.0269 + 0.0086 + 0.0018 + 0.0002 = 0.8034$. Thus, the chances are about 8 in 10 that the shipment will contain nine or fewer defective components.

c. Here we take advantage of the fact that $p(x \leq 9)$ and $p(x > 9)$ are mutually exclusive and exhaustive. Thus, since $p(x \leq 9) + p(x > 9) = 1.00$, and $p(x \leq 9) = 0.8034$, it follows that $p(x > 9) = 1.00 -$

---

[2] Source: *Tucson Trends,* 1980.

0.8034 = 0.1966. The probability, then, is about 2 chances in 10 that a given shipment of 30 components will contain more than 9 defectives.

The probabilities shown in Table I are not approximate. They are exact, with accuracy to the third decimal place.

Example F

A co-op in the housewares-hardware industry is a group of independent retailers that form an association to take advantage of discounts available for quantity buying.

Suppose that 20 percent of the retailers in a particular area belong to a co-op and thus do their buying from one central source rather than from the local manufacturer's representative.

A manufacturer's representative randomly selects 12 retail accounts. What is the probability that

a. Five or fewer belong to the co-op?
b. At least seven do *not* belong to the co-op?

Solution

$$P = \text{belong to co-op} = 0.20$$
$$Q = \text{do not belong} \quad = 0.80$$
$$n = 12$$

a. The probability that x is less than or equal to some value (e.g., $x \leqslant 5$) may be obtained from Table I by cumulating the probabilities from $x = 5$ through $x = 0$. Thus, for $n = 12$, $P = 0.20$,

$$p(x \leqslant 5) = 0.0532 + 0.1329 + 0.2362 + 0.2835 + 0.2062 + 0.0687$$
$$= 0.9807.$$

b. As we indicated earlier, $p(x)$ for a particular P is identical to $p(n - x)$ for the corresponding Q. Thus, $p(n - x) \geqslant 7$ for $n = 12$ is equal to $p(x \leqslant 5)$.

In other words, the probability that seven or more do *not* belong is equal to the probability that five or fewer do belong. Thus, $p[(n - x) \geqslant 7] = p(x \leqslant 5) = 0.9807$.

Example G

According to an article in the *Los Angeles Times*, Aug. 29, 1980, approximately 5 percent of the U.S. population has some degree of difficulty seeing.

In a random sample of 25 people, what is the probability that at least 20 will have no difficulty seeing?

Solution

In order to use Table I, P must be ≤ 0.50. Thus, for convenience in the use of this table, we shall always let P correspond to the category in which the proportion is less than or equal to 0.50. Therefore, in some cases, P will be the category corresponding to "success" and, in other instances, P will be the category corresponding to "failure."

As long as we understand that $p(x)$ is related to the P-category, $p(n - x)$ is related to the Q-category, and define our categories appropriately, it makes no difference which is labeled P and which Q. In this example,

$$P = \text{difficulty seeing} = 0.05$$
$$Q = \text{no difficulty} \quad = 0.95$$
$$n = 25$$

Therefore,

$$p[(n - x) \geq 20] = p(x \leq 5) = 0.0060 + 0.0269 + 0.0930 + 0.2305$$
$$+ 0.3650 + 02774 = 0.9988.$$

EXERCISES

5.11 In 1980, Japan boasted a total of 968 citizens aged 100 or older. Of these, about 20 percent are male and 80 percent are female.

    a. If we were to select 10 at random, what is the probability that 5 or more would be males?

    b. What is the probability that exactly five would be male?

    c. What is the probability that none would be male?

5.12 Each week the *Los Angeles Times* lists the condominium developments or conversions that are opening that week in the southern California area.

Forty-five percent of the condominiums offered for sale beginning the week of Sept. 14, 1980, were priced at or above $150,000.

Suppose you randomly selected 15 condominiums that week; what is the probability that:

    a. All but two were priced at or above $150,000?

    b. At least 10 were under $150,000?

[**Hint:** $p(10$ or more$)$ in the Q-category $= p(5$ or less$)$ in the P-category.]

    c. Six or more were under $150,000?

5.13 Refer to Exercise 5.12. Eighty-five percent of the condominiums offered for sale that week had *less* than 2,000 square feet of living area.

Again, suppose you randomly selected 15 condominiums that week; what is the probability that

    a. At least 10 had less than 2,000 square feet?

    b. Six or more had less than 2,000 square feet?

5.14 Refer to the data in Exercises 5.12 and 5.13. Of those condominiums priced below $150,000, 16 percent had more than 2,000 square feet of living area.

    a. What is the probability of finding a condominium that was priced below $150,000 and that had more than 2,000 square feet?

b. ·Given that the condominium has more than 2,000 square feet, what is the probability it was priced below $150,000?

5.4   DISCRETE VARIABLES: THE POISSON DISTRIBUTION

Management is often faced with scheduling problems in which the variable of interest is discrete but is distributed over space or an interval of time. Let's look at a few illustrations.

> A furniture warehouse may have a total of four loading bays. The dispatcher may be concerned with estimating the probability that more than four trucks will arrive in any given hour.

> The committee preparing plans for a new obstetrics department in a hospital wishes to estimate the number of births so that adequate delivery room services will be provided.

> Based on experience, a manufacturer of video tape knows that, on the average, there are three flaws per 1,000 feet of tape. If a 400-foot length is selected at random, what is the probability that it will be without a flaw [i.e., $p(x = 0)$]?

5.4.1   The Poisson distribution

The *Poisson distribution* is ideally suited for dealing with problems of this sort. The requirements are: *(a)* A Poisson random variable must be discrete. However, unlike the binomial, which assumes a finite number of countable values, the Poisson assumes an infinite number of possible countable values. *(b)* The probability distribution is uniform.

The requirement of a uniform process carries with it several important implications. For example, *(a)* the probability of an event occurring during time period $2Y$ is twice the probability that it will occur during time period Y. Stated another way, the probability of an event is proportional to the length of time or amount of space during which a process is observed. If $p = 0.03$ that an event will occur in 5 minutes, the probability that it will occur in 10 minutes is 0.06. *(b)* The probability that a given number of events will occur within a given interval or amount of space must be independent of where the space or interval begins. For example, if we are looking at flaws in video tape, the probability that a given number of flaws will be observed over a given number of feet of tape must not depend upon where we begin looking at the tape. *(c)* The probability that a given number of events will occur in a given interval of space or time must not depend on the number of these events observed prior to the beginning of a specific interval. *(d)* The probability that two or more events will occur at precisely the same time or point in space is so low that it may be considered zero.

If we are examining variables such as demand for mass transit and demand for utility service (electricity, natural gas, telephone) we are not likely to be observing a process for which the Poisson is appropriate. This is due to

Poisson distribution

A theoretical distribution used to find the probability of the number of rare events observed in a given unit of time or space.

the fact that such demands are not usually uniform but rather, tend to bunch up during peak periods.

Let's illustrate the use of the Poisson distribution with the video tape problem.

The probability of an event x may be calculated from the formula

$$p(x) = \frac{(\mu^x)(e^{-\mu})}{x!} \qquad (5.3)$$

in which μ is the mean occurrence of an event per interval or the mean proportion of space occupied by an event, $e^{-\mu}$ is the base of the natural logarithm system raised to the negative mean power, and $x!$ is the event of interest multiplied by every integer less than x but greater than 0.[3] Fortunately, tabled values are available that allow us to obtain $e^{-\mu}$ for most values of μ.

The mean, μ, is in turn the product of the mean *rate* per unit times the number of units of space or the amount of time. To illustrate, imagine that the rate of defects per foot of tape, λ (the Greek letter lambda), is three per 1,000 feet, or 0.003. Assuming a uniform process, the expected mean in 400 feet is:

$$\mu = \lambda t \qquad (5.4)$$

in which

$\lambda =$ The mean *rate* of occurrence of an event
$t =$ The number of units of space or the amount of time

In 400 feet of tape, $t = 400$. Thus, $\mu = (0.003)(400) = 1.2$. Referring to Table II, we find that $e^{-\mu} = 0.301194$. Substituting this value in the formula, we obtain:

$$p(x = 0) = \frac{(\mu^0)(0.301194)}{0!}$$

Since μ^0 and $0!$ both equal 1, $p(x = 0)$ is $\frac{0.301194}{1} = 0.301194$. In other words, the probability is about 3 chances in 10 that there will be no flaws. This is another way of saying that the probability is 0.70 that at least one flaw will be found.

---

[3] Table III (Appendix C) gives the values of factorials of numbers 0 through 20.

Example H

A telephone order house receives, on the average, 24 orders per eight-hour shift. It has a sufficient number of lines and personnel to handle three orders per hour. What is the probability, during any given half-hour, that two or more potential customers will call?

Solution

This type of problem often arises when managers concern themselves with their firm's ability to handle different demands or loads. Since there are, on the average, 24 orders during an eight-hour shift, $\lambda = 24/8 = 3$. Thus, for a half-hour period (i.e., $t = 0.5$), $\mu = \lambda t = 3(0.5) = 1.5$. Looking in Table II under $\mu = 1.5$, we find $e^{-\mu} = 0.22313$. We may now calculate separately the probability that $x = 2$, $x = 3$, etc., and use the addition rule to find $p(x \geq 2)$. However, an alternative solution is available that makes use of the fact that the sum of all the probabilities is equal to 1.00. Since $p(x < 2)$ and $p(x \geq 2)$ are mutually exclusive and exhaustive, $p[(x < 2) + (x \geq 2)] = 1.00$. To find $p(x \geq 2)$ we need only calculate $p(x = 0)$ and $p(x = 1)$ and subtract from 1.00. Thus,

$$p(x=0) = \frac{(1.5)^0(0.22313)}{0!}$$
$$= 0.2231;$$
$$p(x=1) = \frac{(1.5)^1(0.22313)}{1!}$$
$$= 0.3347;$$
$$p(x \geq 2) = 1.00 - 0.2231 - 0.3347$$
$$= 0.4422.$$

Thus, the probability of receiving two or more potential orders within a 30-minute period is approximately 4 in 10.

5.4.2 Use of tables of Poisson distribution

Most problems involving the calculation of Poisson probabilities can be simplified by use of Table IV. This table shows $p(x)$ for various values of μ from 0.1 through 20. Let's illustrate its use on the preceding problem, in which $\mu = 1.5$ and we wish to find $p(x \geq 2)$. Under $p(x = 0)$ we find 0.2231, and under $p(x = 1)$ we find 0.3347. Therefore $p(x > 2) = 1 - 0.5578 = 0.4422$.

Note that care must be taken when applying the Poisson distribution to "switchboard" problems of this sort. Empirically, such distributions are not usually uniform. Rather, they tend to evidence peaks and valleys with demand alternating between highs and lows. In Chapter 11, we examine procedures

136

for ascertaining whether an empirical demand behaves according to a Poisson process.

5.15 On the average, 36 ships arrive at a given port during a 24-hour period. What is the probability that four or more will arrive during a given hour?

5.16 Which of the following *are not* likely to follow a uniform process?
   a. Passenger load on public transportation during a 24-hour period.
   b. Demand for electricity by industrial customers of an electric utility.
   c. Calls for police emergency services during a given night.
   d. Births in a hospital.
   e. Typing errors in a manuscript.

5.4.3   Poisson approximation to binomial

You may recall how difficult it is to calculate binomial probabilities as n increases. We saw that tables are useful, but even they are limited. Most reference tables show binomial probabilities for values of n up to 20 or 30. The 387-page manual issued by the National Bureau of Standards shows binomial probabilities only through values of $n = 49$. However, application of the binomial to business and economic problems often involve values of n in hundreds and even thousands.

Fortunately, the Poisson distribution may be used to approximate binomial values, thereby eliminating much of the chore of calculating binomial probabilities directly. Although no single guideline is generally accepted by statisticians, a helpful rule of thumb is that the Poisson probability distribution yields satisfactory approximations to binomial probabilities when n is large and nP is small. We recommend use of the *Poisson approximation* when either nP or $nQ \leqslant 7$. When nP or $nQ > 7$, another approximation procedure is available. This is discussed in Section 5.6.

Poisson approximation to binomial

A theoretical distribution used to obtain probabilities of events when n is large and P is small (nP or $nQ \leq 7$).

Example I

Phenyketonuria (PKU) is a rare genetic disease that affects the infant's ability to metabolize a class of amino acids. If the diet is not corrected early in life, the child suffers permanent retardation. Among Caucasians, its incidence is about 1 per 10,000 births. A medical insurance company is considering offering coverage to a group with a membership that includes 60,000 infants born over the past three months. What is the probability that five or more are afflicted with PKU?

Solution

Although this is a binomial problem, the sheer computational burden rules out the use of the binomial. Since $nP = (60,000)(0.0001) = 6$, and $n(1 - P) = 59,994$, we may use the Poisson approximation to binomial values.

First, we require μ. The mean of a binomial distribution (μ) is

$$\mu = nP \tag{5.5}$$

This is the value of μ we shall use when referring to the Poisson table (Table IV). Recall that the insurance company wishes to assess the risk that five or more children with PKU will be included in the population. Thus, we are interested in determining $p(x \geq 5)$ when $\mu = 6$. Since $p(x \geq 5) = 1 - p(x < 5)$, we may find the probabilities associated with values of x from $x = 0$ through $x = 4$ and subtract these from 1.00. Referring to the table, we find:

$$p(x = 0) = 0.0025$$
$$p(x = 1) = 0.0149$$
$$p(x = 2) = 0.0446$$
$$p(x = 3) = 0.0892$$
$$p(x = 4) = 0.1339$$

Therefore, $p(x \geq 5) = 1.00 - 0.2851 = 0.7149$. In other words, the chances are better than 70 percent that five or more children will be afflicted with phenyketonuria.

EXERCISES

5.17 A mail order company is testing a new, more expensive circular. Based on prior experience with other circulars, it expects a response rate of 3 percent. It sends 120 circulars to a sample randomly selected from its mailing lists. To break even, it estimates that it must receive six or more responses. Assuming the circular elicits the same response as previous ones, what is the probability that six or more responses will be obtained?

5.18 Rogers Tableware prides itself on the fact that 1 percent or less of its merchandise is defective. For example, in a shipment of 1,000 sets, there are usually 3 defectives (0.003). In fact, they are so confident of their record, they offer an extra discount if defectives ever exceed this amount.

Last week they sent out a shipment of 2,000 sets. The customer claimed that the shipment contained 15 defectives and demanded the discount. What is the probability that this really happened?

5.5   CONTINUOUS VARIABLES: THE NORMAL DISTRIBUTION

Consider the following:

A manufacturer of auto parts wants to decide whether to eliminate one step in an assembly-line production. A time-study analyst is called in and records the amounts of time spent to complete this particular step.

In a televised news conference on September 18, 1980, President Carter stated that the unemployment rate probably "won't vary much for the rest of the year." He also predicted that the inflation rate would remain below double-digit annual levels for the rest of the year.

Various metallic alloys are placed in a standardized situation in which the stress is gradually increased throughout the test period.

A critical factor in evaluating a gold find is the grade of the ore, that is, the number of ounces of gold per ton. A recent gold find near Sacramento is estimated to contain more than 6 million tons of ore, with a grade of about 0.17 ounce of gold per ton.[4]

All of these examples have one thing in common—the variables of interest are *continuous random variables.* Recall that, unlike discrete random variables, we cannot list all values of the variable. Hence, we cannot obtain probability values for any *specific* value of a continuous random variable. Rather, we describe the probability distribution in terms of probabilities associated with intervals of values of the random variable [see Fig. 5.1*(b)*]. Probability is therefore described in terms of the proportion of area included between any two points.

As indicated earlier in the chapter, the normal distribution is a theoretical mathematical model that has found widespread real-world applications because data produced by many random processes approximate the normal distribution. The approximation is never perfect.

For example, the normal curve approaches but never touches the *X*-axis at both ends. In other words, probabilities greater than zero range over *all* real numbers, both positive and negative. Empirical distributions, on the other hand, typically have values above and below which $p = 0$.

5.5.1   The standard normal distribution

The normal curve is a bell-shaped distribution in which the mean, median, and mode are identical. Recall that the median is a point in the distribution above and below which 50 percent of the cases fall. Since the median and mean are the same, it follows that half of the cases are above and half below the mean.

There is no single normal distribution. Rather, there is a family of normal

---

[4] *The Wall Street Journal,* August 28, 1980. © Dow Jones and Company, 1980.

There are an unlimited number of bell-shaped symmetrical distributions that may be described as normal. They differ from one another in terms of means (μ) and standard deviations (σ). **Figure 5.2**

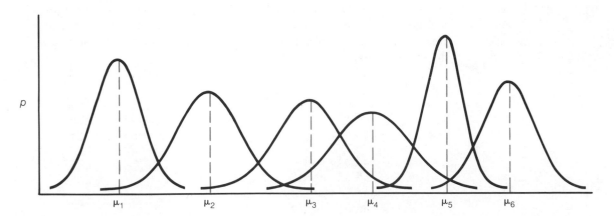

curves, all bell-shaped and symmetrical, that differ in terms of their respective means and standard deviations (see Fig. 5.2).

The importance of the normal curve in statistical analysis lies in the fact that there are *fixed* and *known* relationships between selected points along the *X*-axis and the proportions of area corresponding to these distances. We have already seen these relationships when we discussed the Empirical Rule (Section 3.3.3).

Recall that approximately 34 percent of the area is found between the mean and one standard deviation *above* the mean. Since the normal curve is symmetrical, about 34 percent of the area also falls between the mean and one standard deviation *below* the mean. These are useful benchmarks, but they don't help us to describe the area between the mean and a given score or between any two scores in the distribution. To accomplish these goals, we make use of the *z-score transformation*. The *z*-score represents the deviation of a score from the mean expressed in units of the standard deviation. Symbolically,

z-score transformation

The expression of any score in terms of standard deviation units.

$$z = \frac{X - \mu}{\sigma} \qquad (5.6)$$

As long as we know the mean and standard deviation, we may transform any distribution of scores into *z*-scores. However, only the *z*-scores of variables that are normally distributed will themselves be normally distributed. In other words, the transformation to *z*-scores does *not* in any way alter the original form of the distribution. If there is any reason to doubt the normality of a distribution, before applying this model, you might want to judge or check normality (see Section 3.4).

Why is the *z*-score transformation so important? Of all the possible normal distributions, the one we shall use is a *theoretical* continuous distribution called the *standard normal distribution*. This distribution has a mean (μ) of 0, a standard deviation (σ) of 1, and a total area equal to 1.00. Thus, when we transform any normally distributed variable into z- or *standard scores,* we can use one table (Table V), which provides the proportion of areas under the normal curve, regardless of the units of measurement in the original data.

Let's look at an illustration of the z-score transformation and the use of Table V.

The weights of packages of a brand of cereal are normally distributed with a mean of 32 ounces and a standard deviation of 1.3. What is the probability that a package, selected at random, will weigh between

a. 32 and 34 ounces?
b. 30 and 32 ounces?
c. 30 and 35 ounces?
d. 34 and 36 ounces?
e. 28 and 30 ounces?

In each case, it is helpful to make a rough sketch of the normal curve and shade in the desired area under the curve (see Figure 5.3).

Figure 5.3
Proportion of area between the mean and a score above the mean in a normal distribution with a mean of 32 and a standard deviation of 1.3

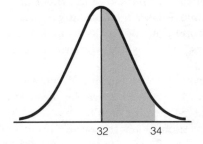

a. The z-score corresponding to $X = 34$ is

$$z = \frac{X - \mu}{\sigma} = \frac{34 - 32}{1.3} = 1.54$$

Referring to Table V, Column B, we see that the proportion of area between the mean and a z-score of 1.54 is 0.4382. Thus, the probability of randomly selecting a package weighing between 32 and 34 ounces is approximately 44 percent.

b. In Figure 5.4 the z-score corresponding to $X = 30$ is

$$z = \frac{30 - 32}{1.3} = -1.54$$

Since the normal curve is symmetrical about the mean, it is not necessary for Table V to show proportions of area corresponding to negative values of z. It is sufficient to indicate only the areas corresponding to positive z-

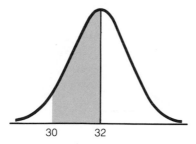

Figure 5.4
Proportion of area
between the mean and a
score below the mean

scores. Negative z-scores will have precisely the same proportions of area as their positive counterparts.

Thus, the proportion of area between the mean and $z = -1.54$ is the same as we previously obtained for $z = +1.54$. It is 0.4382.

c.   Table V (Column B) presents proportions of area under the normal curve with the mean as the point of reference. To find the area between two measurements when they are on different sides of the mean (see Figure 5.5), we must first obtain each z-score separately. We then find the corresponding proportions of area.

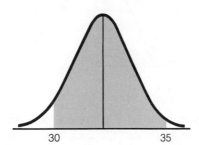

Figure 5.5
Proportion of area
between two scores, one
above and one below the
mean

Thus, z for $X = 35$ is $\dfrac{35 - 32}{1.3} = 2.31$. The proportion of area (Column B) is 0.4896.

The z for $X = 30$ is -1.54. The proportion of area is 0.4382.

Since we are dealing with two *mutually exclusive* events (the same package cannot weigh both more than and less than the mean *at the same time*), we simply use the addition rule for mutually exclusive events.

Thus, the probability of selecting a package that weighs between 30 and 35 ounces is $0.4896 + 0.4382 = 0.9278$, or approximately 93 percent.

d.   First, we find the z-scores and the corresponding proportions of area (Column B).

$$X = 36, \ z = 3.08; \ \text{proportion of area} = 0.4990;$$
$$X = 34, \ z = 1.54, \ \text{proportion of area} = 0.4382.$$

We can see from Figure 5.6 that the proportion of area between these two scores can be obtained by subtraction:

Figure 5.6
Proportion of area
between two scores,
both above the mean

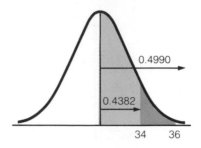

$$0.4990 - 0.4382 = 0.0608.$$

Thus, the probability of randomly selecting a package that weighs between 34 and 36 ounces is approximately 6 percent.

e. As you might expect, the procedures used in determining the proportion of area between two scores below the mean (Figure 5.7) are essentially the

Figure 5.7
Proportion of area
between two scores,
both below the mean

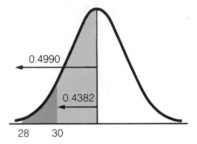

same as in the preceding example. You find the proportion of area corresponding to each *z* and subtract to find the area between the two scores. Thus,

$$X = 28, \ z = -3.08;$$
$$X = 30, \ z = -1.54.$$

Therefore, the probability of selecting a package weighing between 28 and 30 ounces is $0.4990 - 0.4382 = 0.0608$.

Thus far, all of our examples have utilized Column B of Table V in which the mean is used as the point of reference. Column C of Table V provides the proportion of area beyond a given value of *z*.

Let's look at an example.

Example J

Landlords have been concerned about accepting tenants who own waterbeds, even though the tenant may meet all usual standards of acceptability.

Thomas A. is the owner of an older apartment house that fails to meet many present-day building codes. A prospective tenant satisfies all of the usual standards of acceptability except that he owns a waterbed. The landlord fears that the weight of a waterbed will cause the floor to collapse. The possibility of spending endless days in court defending himself against lawsuits is not an appealing prospect. However, before making the decision, Thomas seeks out data.

On researching the question, he learns that a king-size waterbed without pedestals, when filled with 235 gallons of water, exerts a pressure of 48.29 pounds per square foot. However, the amount of water used to fill the mattress is quite variable. Consequently, the pressure is normally distributed about a mean of 48.29 with a standard deviation of 4.21. Structural engineers have told Thomas that the floors of his apartments can safely withstand a pressure of 59 pounds per square foot. What is the probability that a given king-size waterbed, when filled, will equal or exceed this limit?

Solution

$$z = \frac{59 - 48.29}{4.21} = \frac{10.71}{4.21} = 2.54$$

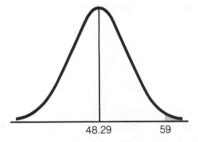

Looking in Column C, Table V, we find the proportion of area beyond a z of 2.54 is 0.0055. Thus, the risk does not appear to be too great.

Example K

There are many occasions when statistical analysis is used as the basis for establishing business, financial, or production goals. For example, automobile manufacturers are presently engaged in a competitive race to produce cars with high mileage ratings. The city miles-per-gallon ratings of compact cars are normally distributed with a mean of 25.49 and a standard deviation of 2.36.

An automobile manufacturer wants to design a car so that its mileage ratings are better than 99 percent of compact cars presently on the road. What mileage rating should the manufacturer establish as a production goal?

Solution

We must first find the z below which 99 percent of the area falls. Since 50 percent of the area is below the mean, we want to ascertain the value of z that includes 49 percent of the area between it and the mean. Looking up 0.4900 in Column B of Table V, we find that a z of 2.33 includes 0.4901 of the area between it and the mean. This is sufficiently close to 0.4900.

Since X (the desired mileage rating) is the unknown, we solve Formula 5.6 for X.

Thus, since

$$z = \frac{X - \mu}{\sigma}$$
$$X = \mu + z\sigma$$

(5.7)

Substituting known values of μ, z, and σ in this formula, we obtain:

$$X = 25.49 + (2.33)(2.36)$$
$$= 30.99$$

This is the mileage rating that the automobile manufacturer should set as a goal if he wants to beat 99 percent of the compact cars.

EXERCISES

5.19   The diameter of ¾-inch bolts produced by a foundry is normally distributed with a mean of 0.75 inch and a standard deviation of 0.02. If the diameter of a bolt is as small as 0.72 inch or as large as 0.78 inch, it will not perform properly. Out of a daily production in the hundreds of thousands,

- *a.*   What is the probability that a single bolt, selected at random, will be too wide?
- *b.*   What is the probability that it will be too narrow?
- *c.*   What is the probability that it will be either too wide or too narrow?
- *d.*   What is the probability that two out of two defective bolts are selected?
- *e.*   What is the probability that a single bolt, selected at random, will be within specifications?
- *f.*   What is the probability that two out of two bolts will be within specifications?

5.20   Past experience with a particular model of refrigerator has shown that the life of the compressor is approximately normally distributed with a mean of 8.45 years and a standard deviation of 2.51.

- *a..*   If the manufacturer is considering a two-year unconditional warranty on the compressor, what proportion of failures should it anticipate to be under warranty?
- *b.*   If 86,000 refrigerators are delivered to a specific service area, approximately how many compressors will require warranty service?

c. If the manufacturer wishes to reduce the risk to 0.0020, how long should the warranty be?

d. If the manufacturer is willing to increase the risk to 0.0250, what would be the length of the warranty period?

5.6 NORMAL APPROXIMATION TO BINOMIAL DISTRIBUTION

We have seen that the binomial distribution is the model that best describes the probability distributions of two-category variables which are mutually exclusive and exhaustive. However, in many real-life applications, the value of n is simply too large to use the binomial tables to determine the desired probability values.

In Section 5.4.3 we demonstrated the Poisson approximation to binomial values when n is large and nP or nQ is equal to or less than 7. Fortunately, binomial probability distributions begin to approach the form of the normal curve as n becomes larger. For any given n, the approximation improves as both P and Q approach 0.50 (See Figure 5.8). A widely accepted rule of thumb is that, when nP or nQ exceeds 5, the normal approximation may be used.

normal approximation to binomial

A theoretical distribution used to approximate binomial probabilities when n is large and P approaches 0.5. When nP and nQ exceed 5, the normal distribution provides a reasonable approximation to binomial values.

$$\mu = nP$$

Example L

According to a recent study by an Energy Department advisory group, many low-income households will spend 30 percent or more of their income on home energy costs.

This is leading to problems for the utility companies as well as the consumers.

Peoples Energy Corp. says that out of 900,000 natural-gas customers in the Chicago area, the number of "seriously delinquent" accounts this year is approximately 26,000.[5]

Suppose we randomly draw a sample of 175 natural-gas customers from a district in the downtown area and find 12 "seriously delinquent" accounts. Is there reason to believe that delinquency in this area is more severe than expected?

Solution

In this problem, $P = 0.03$, $Q = 0.97$, and $n = 175$. In order to use the normal approximation to the binomial, we must first find the mean and standard deviation.

The mean (μ) of a binomial distribution is [Formula (5.5)] $\mu = nP$. In this example, $\mu = (175)(0.03) = 5.25$.

---

[5] Source: *The Wall Street Journal*, 18 September 1980. © Dow Jones and Co., 1980.

Figure 5.8

Probability distributions of the binomial for selected values of *n* and *P*. Note that, when *n* is small, the form of the binomial distribution bears little resemblance to the normal curve, particularly when *P* is low (see *a*). However, as *n* increases, the binomial distribution more closely approaches the form of the normal distribution. This is so even when *P* has a low value (see *c*). Note that, even as *P* begins to approach 0.5, the approximation to the normal curve is better at a given n (compare *a* with *d*, *b* with *e*, and *c* with *f*). When n = 49 and *P* = 0.25, the approximation is quite close. As a rule of thumb, whenever *nP* and *nQ* equal or exceed 5, the normal distribution provides a reasonable approximation to the binomial probability distribution.

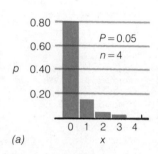

(a)

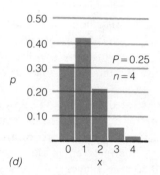

(d)

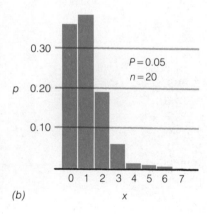

(b)

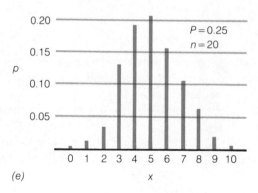

(e)

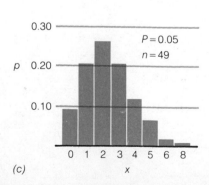

(c)

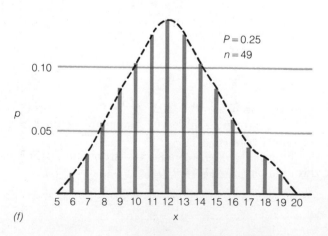

(f)

The standard deviation (σ) of a binomial distribution is

$$\sigma = \sqrt{nPQ} \qquad (5.8)$$

Therefore,

$$\sigma = \sqrt{(175)\ (0.03)\ (0.97)}$$
$$= 2.26$$

Now, we find the z-score corresponding to $x = 12$:

$$z = \frac{12 - 5.25}{2.26} = 2.99$$

Looking in Column C of Table V, we find that the probability is 0.0014. Thus, finding this many delinquent accounts is an extremely unlikely occurrence.

Example M

Recall the saga of Montana Rincon. Up to this point, he has a strong suspicion that all is not well in computerland. Six percent of the scores he audited contained errors, all of the errors favored the company, and they all involved small amounts that would ordinarily go unnoticed and, if noticed, would go unchallenged. Backtracking through the records, he noticed that there were 10 different keypunch operators, each responsible for 500,000 accounts. The following table gives *n*, the number of observations on each operator; *x*, the number of correct accounts; and *(n − x)*, the number containing errors.

Operator	n	x	n − x
A	105	105	0
B	93	73	20
C	110	110	0
D	100	100	0
E	107	83	24
F	99	99	0
G	95	95	0
H	100	100	0
I	90	74	16
J	101	101	0
	1,000	940	

Using the normal curve approximation to binomial values, calculate the probabilities associated with *x* of each operator.

Are any of the results sufficiently unusual to arouse the suspicions of Montana?

Solution

First, for each keypunch operator, we find the mean and the standard deviation of the binomial distribution. We obtain a z corresponding to each of the various values of x. We then find the probability of an event as unusual or more unusual than the observed x.

In all cases where $x = n$, $z = 0$. Thus, the results with seven of the operators were well in accord with our expectations of well-trained and careful operators. They made no errors. Such is not the case with operators B, E, and I. When we calculate z and probabilities associated with their numbers of successes (x), we obtain:

$$\text{Operator B: } \mu = 87.42, \ \sigma = 2.29$$

$$z = \frac{73 - 87.42}{2.29} = -6.30$$

$$\text{Operator E: } \mu = 100.58, \ \sigma = 2.4566$$

$$z = \frac{83 - 100.58}{2.4566} = -7.16$$

$$\text{Operator I: } \mu = 84.6, \ \sigma = 2.2530$$

$$z = \frac{74 - 84.6}{2.2530} = -4.70$$

The probabilities associated with zs for these three operators are so low that they do not appear on the table of areas under the normal curve. The probability that these errors occurred by chance is, therefore, extremely remote. Rincon concluded that either they are careless operators or they are in a conspiracy to commit fraud. The fact that previous analyses had shown all the errors to be in one direction raised serious doubts that the errors were random. Indeed, the evidence was quite consistent with a conspiracy hypothesis.

EXERCISES

5.21 "An alarming 4 out of 10 Americans are highly fearful they will become victims of violent crimes—murder, rape, robbery, or assault," according to a study called "Figgie Report on Fear of Crime: America Afraid."[6]

In a community of 200 people, what is the probability that fewer than 70 are fearful of violent crimes?

5.22 Refer to Exercise 5.21. The same study reported that, "to protect their homes, the sweeping majority (nearly 9 out of 10) always lock their doors and always identify people before letting them in."

In a community of 200 people, find the probability that a burglar will find at least 10 homes unlocked.

---

[6] Study made by Research and Forecasts, Inc. for the N.Y. Public Relations firm of Ruder and Finn.

5.23 Suppose an intensive anti-crime campaign is waged by law enforcement officials. Much publicity is directed to the danger of leaving doors unlocked. As a result of this campaign, 95 percent of the people keep their doors locked.

In a community of 200 people, find the probability that a burglar will find at least 10 homes unlocked. Compare your answer with the one obtained in Exercise 5.22.

SUMMARY

In this chapter, we looked at random variables and discrete and continuous probability distributions.

1. For a continuous random variable, any nonzero interval within its range will have an associated probability greater than zero. It is not possible to list all values of the variable.
2. For a finite discrete random variable there are gaps in which no real values are found. It is possible to list all values within its range and calculate probability values.
3. The Binomial and Poisson distributions are probability models used with discrete random variables. The Bernoulli is a finite, and the Poisson infinite, discrete random variable.
4. The normal probability distribution is most commonly used as the model for describing the probability distribution of continuous variables.
5. We examined the use of the expected value of a discrete random variable as the basis for decision making in business and economics.
6. A Bernoulli variable is a discrete finite variable in which there are only two possible values of the variable. When sampling from a Bernoulli variable, the binomial expresses the probability of obtaining various numbers of successes.
7. We examined the criteria for deciding when we have a binomial situation.
8. The probability distribution of the binomial is obtained by expanding $(P + Q)^n$.
9. As n increases, the calculation of binomial probabilities becomes increasingly laborious. The use of binomial tables simplifies the entire process.
10. The Poisson distributions involve infinite discrete random variables. It is particularly useful when the variable of interest is distributed over space and time. The Poisson assumes a uniform distribution of events over space and/or time.
11. As with the binomial, tables of Poisson probabilities considerably reduce the tedium involved in calculating probability values of events that interest us.
12. The Poisson distribution may also be used to calculate binomial probabilities when n is large and/or P is small.
13. The normal distribution is commonly used with continuously distributed variables. It is a bell-shaped distribution in which the mean, median, and mode are identical.

14. The value of the normal curve is that there are known and fixed relationships between selected points along the X-axis and proportions or area corresponding to these values.
15. The standard normal distribution has a mean of 0 and a standard deviation of 1. Converting a normally distributed variable to z-scores consists of expressing these scores in terms of the standard normal distribution.
16. We examined several illustrations of the use of the z-score transformations in business and economics.

TERMS TO REMEMBER

random variable (116)

probability distribution (116)

continuous random variable (116)

discrete random variable (116)

expected value (118)

Bernoulli trials (122)

binomial distribution (123)

binomial (123)

Poisson distribution (133)

Poisson approximation to binomial (136)

z-score transformation (139)

standard normal distribution (140)

standard scores (140)

normal approximation to binomial (145)

EXERCISES

5.24 If a tenant does not pay the rent, the landlord may go through the procedure of an "unlawful detainer." The probability, if one goes through this procedure, of getting a tenant out in approximately 22 days is 95 percent. A firm charges $75 for this service. However, there are ways that the tenant may delay these proceedings and stay as long as 60 days. Only 5 percent of tenants usually do this. The charge for a contested action such as this is $200. How much security deposit should a landlord collect on a $300 per month apartment to protect himself or herself from a loss?

5.25 Refer to Exercise 5.24. It is September, 1980. The marshal's office is experiencing an unusual number of unlawful detainers. Since action is now being delayed because of inability or lack of personnel to serve the tenants, the probability of getting a tenant out in 22 days has been reduced to 5 percent. There is a 65 percent chance that the action will go smoothly and the tenant will be evicted in 30 days. More tenants are fighting this action. Thus, the probability is now 30 percent that it will take 60 days to remove a tenant. The service charge for an uncontested service is still $75, and $200 for a contested action.

Now, how much security deposit should the landlord collect on a $300-a-month apartment to protect himself or herself from any losses?

5.26 Response time is a vital consideration among agencies required to provide emergency services. On September 25, 1980, a businessman was informed that the silent burglar alarm in his home had been tripped, probably by an intruder. The police were simultaneously informed. The businessman raced home and arrived before the police. When the police arrived, approximately 40 minutes after being told of the alarm, they found that the businessman had been murdered by the burglar (L.A. Times, 9/26/80).

Suppose that, in a large urban area, an average of 230 crimes are committed in the seven-hour period between 10 P.M. and 5 A.M. Assuming that the distribution of crimes is uniform:

    a. If the police are able to handle eight calls within a 20-minute period, what is the probability that they will be overloaded during any given period of 20 minutes?

    b. If the police are able to handle eight calls within a 10-minute period, what is the probability that they will be overloaded during any given period of 10 minutes?

    c. If the police are able to handle eight calls within a 5-minute period, what is the probability that they will be overloaded during any given period of 5 minutes?

5.27 The shelf life of a brand of alkaline battery is normally distributed with a mean equal to 583 days and a standard deviation equal to 50.

    a. If a battery has been on the shelf for 630 days, what is the probability that it will not function?

    b. If a battery has been on the shelf for 430 days, what is the probability that it will not function?

    c. If you want to be 90 percent sure that a battery will operate, how old a battery should you accept?

5.28 Burt Bach, noted attorney and author of the column "Bach's Score" in the *Los Angeles Daily Journal,* commented, "Ninety-five percent of civil actions for money damages are resolved before going to trial. Moreover, about 94½ percent of the actions result in a money settlement to the plaintiff. When they are resolved before trial, 99 percent of the cases result in a money settlement to the plaintiff."

    a. What is the probability that a given case, selected at random, will be resolved before going to trial and result in money damages to plaintiff?

    b. Given that the plaintiff wins a money settlement, what is the probability that it was resolved before going to trial?

5.29 Refer to Exercise 5.28.

Suppose an attorney is currently representing 10 different plaintiffs in actions of this sort. What sorts of questions can we answer? If we can treat the variables involved as two mutually exclusive and exhaustive categories with independent trials, etc., we can use the binomial model.

    a. Let the two categories be resolved versus not resolved before trial.

        i. What is the probability that all 10 cases will the resolved before trial?

        ii. What is the probability that at least six will be?

    b. Let's say that the attorney has taken these 10 cases on a contingency basis and gets paid only if the plaintiff wins a money settlement. Let us also say that this will make the expenditure of time and effort worthwhile only if the attorney wins money damages for at least four of the clients. What is the probability that it has been worthwhile to accept these cases on a contingency basis? (Let the two categories be: plaintiff receives a money settlement or plaintiff does not.)

    c. The attorney's costs are much greater if the case goes to trial. If the cases go to trial, at least five of the clients must win money judgments to make it worthwhile to have accepted the cases on a contingency basis. Given that the cases went to trial, what is the probability that at least five of the plaintiffs won money judgments?

5.30 Acceptance sampling is a quality control procedure in which the binomial figures prominently. A sample of a fixed size is selected from a large inventory. Imagine, for example, that a large shipment of small appliances is received by a warehouse. Sixty are tested. If more than two are found defective, the entire shipment is rejected. By variations of this technique, a purchaser can cut the risk to some predetermined probability level. To illustrate, what is the probability of accepting a shipment if the true proportion of defectives is:

    a. 0.20?
    b. 0.10?
    c. 0.04?
    d. 0.01?

5.31 Zeta Cable sells and installs cable TV. Service calls are free to its customers. Records show that the time between a customer's complaint and service is normally distributed with a mean of 4 days and a standard deviation of 1.

Suppose you are a Zeta Cable subscriber and you complain of a malfunction. There is a program on in two days that you are extremely anxious to see. What is the probability that you will be serviced in time to see this program?

5.32 Refer to Exercise 5.31. Zeta Cable would like to advertise that 95 percent of the time service is performed within a certain period of time. What should they tell their customers?

5.33 FBI statistics[7] show that the incidence of rape of women in the population is about 0.06% (or 0.0006).

In a random sample of 1,000 women, what is the probability that
    a. None will be raped?
    b. At least four will be raped?

5.34 During the two-hour period between 7 and 9 A.M., it is anticipated that 600 automobiles will pass through a newly constructed toll booth. If the station can handle 10 cars a minute, what is the probability that it will be overloaded during a given one-minute period? (Assume a uniform distribution of traffic).

5.35 Refer to Exercise 5.34. When the facility was opened, the traffic load was three times as great as anticipated. What is the revised probability that it will be overloaded during a given one-minute period?

5.36 A concept utilized by many businesses is the price-earnings ratio, or P/E ratio, as it is sometimes called. This refers to the multiplier that is applied to earnings per share in order to determine current value. The higher the P/E, the more positive the future looks.

Suppose a large conglomerate has been following the earnings and sales growth of one particularly successful company and finds that its P/E ratio is normally distributed with a mean of 26 and a standard deviation of 1.8. The president of the conglomerate decides that he is interested in buying out the stock of this company. Since the quality of management is one of the factors that influences the P/E ratio, he decides that if the P/E ratio reaches 28 he will retain present management. If, however, it falls below 25, he will replace management with his own team.

    a. What is the probability present management will be retained?
    b. What is the probability that new management will be brought in?

---

[7] Source: *Los Angeles Times*, September 19, 1980.

5.37 Teri F. has $10,000 to invest. She has a real estate broker working on a real estate deal. As soon as the negotiations are completed, she will have to use her $10,000 as a down payment. She estimates the probability that this will occur in one, two, or three months is about 5 percent for each of these three possibilities. She figures there is a 15 percent chance she will need her money in four months, 20 percent chance for five months, and a 50 percent chance that the deal will be completed in six months.

While she is waiting, she wants to make money on her $10,000. She narrows it down to two choices:

1. She can place her money in a regular passbook account that pays 5.5 percent simple interest.
2. She can buy a six-month money market certificate currently paying 10.5 percent simple interest. However, federal regulations that went into effect June 2, 1980, place a penalty on early withdrawal—a loss of 90 day's interest.
   Which alternative should she choose?

5.38 David F. estimates that he needs $25,000 a year to afford his current style of living. He holds a job as a supervisor in a large company with several different levels of management, and his salary is $18,000 a year. He learns that wages for supervisory employees, who report to middle management, are normally distributed with a mean of $22,000 and a standard deviation of $1,800.

    a. What is the proportion of supervisors who earn more than he does?

    b. Suppose he looks for a similar job with another company. What is the probability that he will earn a salary equal to his needs?

5.39 During a five-year period, a major corporation has had to replace 40 key personnel. What is the probability that it will have to fill a) 0; b) 1; c) 2; d) 3; e) 4 or more positions during the next six months? (Assume a uniform distribution.)

5.40 It is September 1980. The prime rate is now 12.5 percent. Stuart I. has a decision to make. Should he borrow money now or wait and take a chance that the interest rate will go down? After consulting with various economic experts, he comes up with the following probability distribution:

Prime rate (X)	p(X)
11.0%	0.05
11.5	0.05
12.0	0.15
12.5	0.25
13.0	0.25
13.5	0.20
14.0	0.05

What does he conclude?

5.41 Certain supermarkets have stopped marking prices on individual items in a test the industry hopes will show that item-pricing is unnecessary in stores with automated checkout equipment.

A crucial aspect of this test is customer acceptance. If 80 percent of the customers accept the new system, industry executives feel they stand a good chance to eliminate individual item-pricing. Consumer advocates contend that item-pricing is a useful shopping tool and estimate that only 50 percent of the customers will accept the new system.

If 500 customers are surveyed and the consumer advocates' estimate is correct, what is the probability that 80 percent or more of the sample will express acceptance?

5.42 Luis M. has to decide between two job offers. Job 1 pays a commission of $600 for each sale. No leads are provided and Luis figures he can make 10 customer calls a week. The probability of making all 10 sales is 0.30. Job 2 pays a commission of $400 for each sale. Five qualified leads are provided each week. The probability of making all five sales is 0.50. Luis calculates that, regardless of which job he takes, it will cost approximately $100 for each sales call.

Which job should he take?

5.43 "Something is wrong with this shipment of cutlery," says Mimi. "We've been in business for 10 years and have always had approximately 1 percent defective or returned merchandise. Suddenly today, out of 10 sold, two were returned as defective."

Is Mimi right? Is something wrong, or could this be due to chance?

5.44 A survey shows that the time spent on home chores weekly by most married women working for big concerns is normally distributed with a mean of 10 hours and a standard deviation of 1.6. What is the probability that a woman, selected at random, working for a large company, spends at least 10 hours a week on household chores?

5.45 Recall that Montana Rincon found 60 errors when he examined 1,000 randomly selected credit card accounts. When he examined the 60 errors, he discovered that all favored the company and none favored the customer. On the assumption that the errors represent a random process in which the probability was equal that an error favored the company ($P = \frac{1}{2}$) or the customer ($Q = \frac{1}{2}$), find the probability that all favored the company.

5.46 A recent marketing survey in the Los Angeles area revealed that the majority of the family's grocery shopping is done by the female household head in 71.6 percent of all households, whereas the male head does most of the shopping in only 15.5 percent of the households.[8]

In a random sample of 100 households, what is the probability that

    a. At least 60 female household heads do the shopping?

    b. Between 70 and 80 females do the shopping?

    c. Between 20 and 25 men do the shopping?

5.47 Charles S. went into a business without researching the market. He assumed he could fulfill orders for ladders at $2.50 below the market. He figured this item to be a "loss leader." His costs far exceeded his expectation, and he lost $1.50 on each sale. However, he figured he could make it up on another item—laundry carts on which his profit was $3.50 for each one sold. He went into business on the assumption that he would sell four laundry carts for each ladder sold.

However, he priced himself so much below the market that the loss leader proved so "successful" his orders ran four ladders for every laundry cart.

    a. If his original projections were correct, what would his expected profit have been?

    b. Given what actually happened, can you see why Charles S. went bankrupt?

---

[8] *Los Angeles Times,* September 21, 1981.

5.48   Approximately 35 percent of the nation's largest companies now have art collections.[9]

In a random sample of 20 of the nation's largest companies, what is the probability that

a. None has an art collection?
b. Four or less have art collections?
c. Ten or more do *not* have art collections?

5.49   If the average household expenditure for groceries during a typical week is normally distributed with a mean of $44.41 and a standard deviation of $2.50, what is the probability that a household, selected at random,

a. Will spend less than $40?
b. Will spend between $40 and $50?
c. Will spend between $35 and $40?
d. Will spend more than $40?

5.50   Many companies have now instituted a system whereby they pay workers for good ideas. A survey of 219 business firms reveals that total awards paid in 1979 were normally distributed with a mean of $228,310 and a standard deviation of $30,000.

What is the probability that a company, selected at random, will pay out more than $200,000 in incentive awards this year?

---

[9] Source: *The Wall Street Journal,* 18 September 1980. © Dow Jones and Co., 1980.

Sampling and Sampling Distributions

Joan B., owner of Waterbeds Galore, felt a tingle of excitement, as she always did when she sorted through the daily mail. Who knows, maybe this is the day she receives notice that she has won an all-expense trip to Hawaii, or maybe a certified check from a long-overdue account. She quickly sorted the mail into three piles—first-class mail, bills, and junk mail. The last of these she hastily dispatched to the circular file.

One of her salespeople looked at her quizzically and then asked, "Why did you throw these away without even opening them?"

"Junk mail."

"Mind if I look at the one from the trade association?"

"Which one?"

"The Waterbed Manufacturers Association."

"Help yourself, Fran."

"This is very interesting, Joan. It is the results of a survey of the waterbed industry for 1979."

"Survey, shmurvey, they never tell you anything."

"Well, I think this is interesting. Do you know that the industry grossed an estimated $671 million last year?[1] A flash in the pan, huh?"

"That much? I'm impressed."

"Would you mind telling me how much we grossed last year?"

"No secret. About $268,000. Why do you ask?"

Fran took out her pocket calculator, quickly pushed a few buttons, and looked up, "Do you realize that we account for about four hundredths of one percent of the gross sales of the entire industry?"

"Really. Wonder what it would take to get us up to a bigger percentage share of the industry?"

"Listen to this. In our region, there has been about a 29 percent increase in the number of retailers displaying hybrids. Beds, that is, not plants."

"Now that's really telling me something. I thought they would never fly. That's why we don't have any on display. Perhaps we should rethink our position."

"Also, five percent of the gross income derives from the sale of conventional furniture. Things to go along with the waterbeds."

"Say. Don't throw that survey away. After you finish with it, I'm going to take it home and read it this evening. Certainly beats TV fare."

6.1 WHY SAMPLE?

In this brief scenario, we are introduced to one of the most important tools in the decision making process of the business community—the survey. Although often treated with disdain, it can be highly informative and of enormous assistance to everyone from the multinational executive officer to the small retailer in Oshkosh, Wisconsin. Perhaps, the survey represents the outstanding example of using sampling techniques to make inferences about populations.

---

[1] "The 1980 Industry Survey Results," *Industry Magazine*, p. 10.

But why sample? Why not target a particular population in which we are interested and then conduct a census of this population? Certainly it would remove the uncertainty that inevitably accompanies drawing inferences from samples.

Example A

Statistics on the U.S. population have been collected and published decennially (every 10 years) ever since 1790, generally with increasing amounts of detail. The decennial census attempts to count *every* person, household, and so forth, in the population. It is designed to be the most complete census of the entire population. Perhaps because it is the biggest, it best illustrates the problems with census taking and why sampling is probably, even here, the best way to collect data.

Let's look at the 1980 census and some of the problems associated with an attempt to measure *every* member of a population and, thus, obtain the actual parameters, not just estimates from samples.

Because the population is so very large, and, in some cases, scattered, the cost of the 1980 census will well exceed $1 billion! The massive undertaking officially got under way in the spring. On March 28, 1980, census forms were sent through the mail. Let's look at some of the problems that arose from the system used to collect the data. First, many citizens reported that they never received the form in the March mailing. The U.S. Postal Service said that the mailing lists were filled with errors. For example, 13 percent of the census forms in Manhattan were undeliverable because of bureau errors. Approximately 85 percent of the mailed forms were filled out and returned. Since the census attempts a complete count of all citizens, over 300,000 workers were recruited and paid to follow up mailed forms and to contact residents of out-of-the-way places. Think of the monetary expenses involved here!

Timeliness, as we know, is a key factor. Unfortunately, the 1980 census found itself embroiled in legal difficulties. Thus, not only was there a significant time delay, but one can well imagine the impact these litigations will have on the final costs.

Then, of course, census officials encountered the usual types of problems associated with any census. Many members of the population were inaccessible, unavailable, or simply unresponsive. Although the law provides penalties for not giving the bureau required information, one census bureau supervisor reported, "It's very difficult to count people who don't want to be counted." He indicated that such people include illegal aliens, welfare cheats, tax evaders, and others who feel the census invades their privacy. This particular type of problem—dealing with the part of the population that is *not included*—has already led to various legal skirmishes. For example, a federal judge upheld Detroit's contention that its residents had been undercounted. New York city officials claim that the 1980 census missed at least 800,000 of its citizens. In recent years, big cities have asserted that the bureau has missed huge numbers of blacks and Hispanics because its approach to counting citizens is antiquated. As of September 1980, the bureau faced lawsuits filed by cities (other than Detroit) such as New York, Philadelphia, Chicago, and Newark. According to *The Wall Street Journal,* all the suits are aimed at a readjustment of the 1980 figures.

Thus, we see, even the biggest census of them all has its problems. Later, we shall examine some of the errors associated with these government censuses. The director of the National Opinion Research Center at the University of Chicago said, "Maybe [the Census Bureau] can count noses, but the more people they have to count, the less chance there is that they'll do it right."

Let's look at a few additional examples that illustrate why we sample.

The cost per minute of television advertising is determined by such factors as the time of the day, the day of the week, and the size of the estimated audience viewing a particular program. With approximately 98 percent of all U.S. homes owning at least one TV set, it is unthinkable that the TV viewing practices of the residents of all 75,000,000 homes could be obtained. The population is simply too large. Even if it were not, the cost would be prohibitive and the time required to collect and analyze the data might make the results as timely as last week's newspapers.

In order to assess the risk of serious injury or death to drivers and passengers in a given model compact car, the car is driven into a concrete wall at 35 miles per hour. A dummy, wired with sensitive electronic equipment, provides data on the possible consequences of the impact. Since the test involves the destruction of the vehicle, few cars would be available at the dealership if the entire population were tested in this way.

A government agency authorizes a study to determine the principal causes of mine disasters. The total population is not available since the sites of some of the disasters may be buried under tons of dirt and water. Or, a college is interested in the income and occupations of its alumni. For a variety of reasons, it may be impossible to query some members of this population. They may be dead, institutionalized, or inaccessible because they hold high political offices or for various other reasons. Hence, a sample must suffice to represent the population.

We have attempted to illustrate that there are a number of factors that make it necessary for us to sample rather than to attempt to measure the entire population.

In brief, the reasons for sampling are:

1. *Economy.* The larger and less accessible the population of interest, the greater the cost of obtaining a census vis à vis collecting a sample. There are times, however, where the population is sufficiently small and accessible that cost differences are negligible.

2. *Timeliness.* Generally, it requires less time to collect, analyze, and report the data from a sample than to perform similar operations with a census. If the survey is required in order to make a timely decision, the sample

survey has a distinct advantage. Again, if the population is sufficiently small and accessible, this advantage will be diminished. It should be emphasized that inferential statistical procedures are not required following the collection of census data. A mean or proportion *is* a parameter and is *not* an estimate.

3. *Destruction of item.* When testing destroys or impairs the operation of an item, there is no real choice but to sample.

4. *Infinite population.* When the population is unlimited, such as that resulting from an ongoing production process, sampling must be undertaken since a census is impossible.

5. *Accuracy.* Paradoxically, a sample may be more accurate than a census. This is especially the case in surveys that require much information from the respondents. Because of cost and time factors, a sample may permit detailed probing whereas a census frequently involves yes-no types of responses.

6.2 SOURCES OF ERROR

In order to use samples to make inferences about populations, we must first clearly define the target population. This population may vary according to our objectives. For example, J&B Scotch announced a plan to spend $12 million on advertising over a 12-month period. The "typical" scotch drinker is their target population. According to *The Wall Street Journal,* this population consists of "college-educated people, most often male, aged 25 to 44, who earn at least $22,000 yearly." Based on survey results that show southerners are losing their taste for bourbon, J&B has added southerners as a special target population.

Sometimes the population of interest can be small and narrowly defined; at other times, it can be large and broadly defined. Moreover, the population does not necessarily consist of people: it may be measurements, items in inventory, interest rates, prices of stocks, commodities, etc.

Once we have clearly defined the population of interest, the single most important consideration is to select a sample in such a way that the statistics resulting from repeated sampling, were it undertaken, would center on the parameter of interest. If the sample is collected in a careless and haphazard fashion, the resulting statistics are subject to sources of error that cannot be estimated. Increasing the sample size will not balance out these errors. Errors that consistently lead to statistics that center on any value other than the parameter of interest are known as *bias.*

bias

A systematic type of error that does not balance out with continued sampling.

For example, when testing for defects in a large rail shipment of items, we would not select the entire sample from a single location in the boxcar. The interior of a boxcar is subject to all manner of bumps and stresses during transit. Therefore, the proportion of defects in any single location is unlikely to center on the proportion of defects in the entire interior of the boxcar.

Similarly, we would not attempt to obtain national TV ratings of a network program by sampling from a single region of the country, nor would we sample only Caucasians if we wished to predict the results of a national election.

Bias is not the only source of error. There will always be a certain amount of *sampling error* due to the presence of a large number of uncontrolled variables, which we usually refer to as chance factors. For example, any given sample may include a disproportionate number of very high or very low values. This is the sort of error that causes sample statistics to differ from one another, even when selected from the same population. It is also a source of error whose probability *can* be assessed. Unlike errors due to bias, increasing the size of the sample is our best insurance against sampling error. The effects of uncontrolled factors tend to balance out in the long run.

sampling error

A random type of error that tends to balance out with continued sampling.

Example B

Seasonally adjusting data is a technique used by the government to minimize the effect of sampling error due to haphazard or random changes. In fact, many of these seemingly haphazard or random changes tend to show a consistency over long periods of time. Some of these consistencies seem to coincide with certain "seasons" or even months (e.g., winter season or the month of December) and thus, the government has devised a system whereby some statistics are adjusted for expected seasonal variations.

The unemployment rate is a good example of data that is seasonally adjusted. Data gathered over long periods of time have shown that the labor force tends to be higher at certain times of the year (e.g., summer months—many students enter (and thus swell) the labor force during these months). If the number of jobs available is not adequate to meet this increased need, the unemployment rate will show an increase. However, this particular example reflects a seasonal change that is predictable and expected. Thus, economists adjust the unemployment rate to reflect seasonal changes that could distort economic trends.

One other kind of error that can lead to nonrepresentative sampling may be caused by the manner in which the data are collected or the way in which the observations are made. This is sometimes called *nonsampling error*. If we are measuring the diameter of screws, for example, the fault may lie with the inaccuracy of the measuring instrument. In many jurisdictions, jury lists are compiled from voter registration records. Thus, unregistered citizens are systematically excluded from exercising this responsibility of citizenship. Questionnaire data may contain leading questions (e.g., "Why do you favor Brand A over Brand B?") or may restrict the choice of alternative responses. To illustrate, in 1980 certain Los Angeles supermarkets were involved in a test to measure consumer response to the elimination of individual item-pricing in stores with automated checkout equipment. Councilwoman Peggy Stevenson criticized the industry-designed test because it limited customer response to just three problem areas: the form asks whether the price charged

nonsampling error

An error that is caused by the way in which the observations are made.

was different from the shelf price, whether the shelf price was missing on an unmarked item, and whether the receipt tape was difficult to read.[2]

Not only do errors plague the samples we obtain, but they also crop up when we attempt to conduct a census. Let's return for another look at the major government decennial census. One estimate places a 2.5 percent error (undercounting) nationwide even in the 1970 census *(The Wall Street Journal)*.

Many of the errors in the 1980 census were attributed to the way in which the data were collected. Because of a variety of different reasons, many of the results were contaminated by the insidious effect of bias. For example, New York City claimed that drug addicts on methadone mainte- nance were hired to work in certain ghetto areas. Many felt that these employ- ees may have, perhaps unwittingly, or purposely, distorted or miscounted people in certain categories. Affidavits filed in New York City charge that census workers deliberately changed answers on census forms to expedite their work. For example, a census worker was paid $2.50 an apartment for basic housing information that could be obtained from the landlord in seconds. Contrast this with the $4.00 a unit they were paid for actually conducting interviews that were extremely time-consuming.

In some areas, census workers faced actual risks, while being paid $4.25 an hour. For example, one enumerator was told at the entrance to a building that if he tried to count the residents, "They'll cut your face." So, he classified this building as vacant.

Because minority groups tend to cluster in urban areas and are commonly undercounted, many cities hope the Census Bureau will alter the way counts are made, since many federal grants are tied to population figures. They hope to see head counts replaced by random sampling of the population.

Random sampling has already been incorporated as a key aspect of the census, at least for certain kinds of data. Four out of five people received a short form, which contained 19 questions. The rest received a long form, which asked an additional 46 questions. It will be interesting to see how the Census Bureau resolves all of the difficulties inherent in massive data collections of this sort. We look forward to seeing how much of a swing the bureau takes toward sampling as a means of obtaining information in response to some questions. The Constitution requires the taking of a census every 10 years; but it does not require that the same degree of detail be obtained from every respondent.

EXERCISES

6.1 The Public Service Commission of a state with nuclear electric power has received heated requests that it place a moratorium on planning and constructing nuclear power plants. The chairperson of the commission requests that the Concerned Citizens against Nuclear Power undertake a questionnaire to ascertain voter reactions to this proposal. What are some of the risks in this suggestion?

---

[2] *Los Angeles Times,* September 18, 1980.

6.2 Big Jim's Gym in Glendale, California, has experienced a drop in memberships from 500 to about 300 in six months, with profits dropping proportionately.[3]

Big Jim wanted to determine the reasons for this membership drop and decided to use a questionnaire. What sorts of questions should he include? Define the population of interest. Who should be asked to fill out this questionnaire? Is a census possible or would it be best to select a sample? Can you think of any pitfalls that should be avoided?

6.3 Comment: An economist is studying the personal income of Americans over a 12-month period. He finds an unexpected rise in the month of July. Upon investigating, he learns that this increase was the result of a cost-of-living increase given that month by the government to recipients of social security, veterans' benefits, etc.

6.4 For the following example, specify the population of interest.

A college recently sent out a questionnaire to its alumni requesting information on such things as current occupation and income. On the basis of the questionnaires that were filled out and returned, the college published statistics on mean income, types of positions, etc. in their alumni newsletter.

On the basis of the data collected, would you redefine the population? Comment on some of the problems inherent in this kind of data collection.

6.5 Suppose that you are interested in studying stock prices on the New York Stock Exchange (NYSE) over a three-year period. Specify the population of interest. Would it be more appropriate to conduct a census or to select a sample? What are some of the difficulties you might encounter?

6.3 TYPES OF SAMPLES

We have seen why it is necessary to sample. We have looked at the kinds of error that can affect sampling and, in turn, the inferences that we make about the population. By exercising care in the way in which we select our sample, we can reduce the effects of bias and sampling error. There are many considerations that will determine the way in which we choose our sample. There are also many different ways to select samples. In the following sections, we shall look at several types of nonrandom and random samples.

6.3.1 Nonrandom Sampling

Let's look at a familiar example of nonrandom sampling. A congressman has been accused of misappropriating funds for his own personal use. At a press conference, he announces, "I am completely innocent. Eighty-five percent of my mail supports my innocence. At least my constituency knows me for what I am—their honest and hardworking representative in Congress."

Can we conclude, with the congressman, that 85 percent of his constituency really believe in his innocence? Voluntary responses (such as letters, telegrams,

---

[3] R. B. Chase and N. J. Aquilano, *Production and Operations Management,* 2d ed.(Homewood, Ill.: Richard D. Irwin, 1977), p. 640. © 1977 by Richard D. Irwin, Inc.

and phone calls) are referred to as *convenience sampling*. They are rarely, if ever, representative of the general population. For example, the congressman is likely to receive a disproportionate number of letters from personal friends, financial backers, and members of his own party. Other legislators may be overwhelmed with an outpouring of letters from small, determined, but highly organized pressure groups. Some such groups have been known to write hundreds of letters in the hope of influencing legislation.

Convenience sampling need not be voluntary. Occasionally, in order to obtain a quick answer to a question, a survey will be conducted by telephone. Such a survey is unlikely to be representative of the population of interest since it leaves out people without telephones, those with unlisted numbers, and people who are not at home when the call is made. Those who are at home may be different in many ways from those who are not. For example, if calls are made during the day, we may obtain a disproportionately high number of unemployed individuals. If made during the evening hours, we will miss those who spend their evenings away from home—at restaurants, theaters, and meetings of civic and community groups. However, convenience sampling is inexpensive. With a severely limited budget, it may be the only way we can hope to obtain information bearing on the question of interest. But we should be acutely aware of its limitations so that we do not claim more than is justified.

Another type of nonrandom sample consists of *judgmental sampling*. To illustrate, imagine that the head of a marketing department is asked to conduct a survey of some of the leading manufacturers in the semiconductor industry. From past experience, she knows that some corporations do not respond to surveys. She also knows that others are extremely cooperative and usually informative. With this wealth of prior experience in mind, she looks over a list of semiconductor manufacturers and checks off those that should be included in the sample. Depending on the astuteness of the person making the judgments, the results of such a survey may range from useless to extremely beneficial.

The Consumer Price Index (CPI) is an example of an important economic indicator that, to some extent, uses judgmental sampling. It measures the average change in prices over time in a fixed market basket of goods and services. Thus, it is based on prices of food, clothing, shelter, fuels, transportation fares, charges for medical and dental services, and other goods and services we require on a day-to-day basis. The decision as to what items to include (e.g., what specific foods and what beverages) is largely a matter of judgment. However, the CPI is probably one of the most widely used economic indicator; consequently, much care and expertise go into the calculation of this index.

A note of caution is in order regarding all nonrandom sampling techniques. Since there is no random model against which to interpret the results of the survey, there is no way to estimate sampling error. As we shall see, our ability to estimate sampling error is one of the major advantages of random sampling procedures.

convenience sampling

Selecting a sample where expediency is the primary consideration.

judgmental sampling

Selecting a sample based on prior experiences with the population of interest.

6.3.2 Random sampling

On September 14, 1940, prior to the United States' entry into World War II, the first peacetime draft was approved. A huge cylinder, containing the names of those eligible for the draft, was put on public display. The cylinder was rotated a number of times so as to thoroughly mix the names. Then, one by one, they were selected from the cylinder. Assuming that the mixing was thorough, a random sample was obtained by selecting the names in such a way that each member of the population (each name in the cylinder) was equally likely to be selected. This is referred to as a *simple random sample*. There were few complaints about the procedures; most people accepted them as fair and equitable. Contrast this with a similar exercise in 1970. Then, capsules containing all birthdates were placed in a hopper. The order in which the birthday was chosen established the order of priority in drafting the young men. Those whose birthdays were drawn first received the highest eligibility ratings. But then—sudden bedlam!

Statisticians all over the country, amateur and otherwise, noted that December dates received unusually low numbers, meaning they were to be among the first selected. January dates were at the other end of the spectrum. What happened? Apparently, when the months were placed into the hopper, December was inserted last. Later investigations revealed that the capsules were not mixed together. Thus, there was a built-in bias in favor of selecting December dates and against selecting January dates, the first placed in the hopper.

Simple random sampling. Perhaps this lottery method was used because it is more immediately understood by the general population. A more accurate and statistically acceptable method was available—use of a table of random digits. Such a table is generated in such a way that each digit in the table is equally likely to precede and follow any other digit. Table VI (Appendix C) is a table of random digits generated by a random process at the RAND Corporation. Its use in selecting samples might best be shown by use of an illustrative example.

Imagine that you are in charge of quality control in an industrial plant that manufactures steel shafts. After production, the shafts move along a continuous belt, and they eventually arrive at a packaging facility. During an eight-hour shift, 5,760 of these shafts are produced. You wish to select 25 for testing. Conceptually, you assign the number one to the first shaft, two to the second, and so forth right up through the 5,760th shaft. You next refer to Table VI. Both the page you select and the column on that page should be entered in a random fashion. To illustrate, close your eyes and place your finger on the page. Wherever it lands is your starting point. Since the table is random both right or left, up or down, you should decide beforehand the direction in which you will read.

Using these procedures, we started on row 20 of the fourth column of digits. The first four digits are 4,616. Thus, the 4,616th shaft is included in our sample. Reading down, the next four digits are 7029. Since there is no shaft corresponding to this number, we ignore it. The next shafts are those

---

simple random sample

A sample, selected in such a way, that each member of the population has an equal chance of being chosen.

numbered 3,297, 1,286, 4,021, 5,192, etc. We stop when we have collected a sample of 25.

Simple random sampling constitutes one of the most accurate methods of selecting samples from populations. However, it is often inconvenient, time consuming, and expensive. For example, imagine you randomly selected a sample of individuals from a large metropolitan area who are to be interviewed in their homes. If a person is not at home, random sampling would require that you return to the residence until the person is interviewed. It is not difficult to imagine that these procedures could seriously delay the collection of data and rapidly exhaust the budget allocated for the project. For these reasons, other random sampling techniques are often used.

The systematic random sample.   It is not always convenient to choose a sample through the use of a table of random digits. It is a fairly long and complex procedure that may not be completely understood by the individual selecting the sample.

A simpler way of selecting a random sample is to choose the elements according to some predetermined system. We might select every 20th or 50th unit, for example, *after* the first unit is selected randomly. This method is known as *systematic random sampling*. To illustrate, imagine that you are the Personnel director of a firm with 10,000 employees. You wish to study employee attitudes about a new incentive system. You decide to select a random sample of 500. Suppose all the employees are listed on the computer in alphabetical order. First, you randomly select the first member (perhaps through the use of a table of random digits) and then select every 20th employee until you reach your desired sample of 500. In this method, only the first member (the starting point) is selected randomly. After that, every 20th person is included in the sample.

> **systematic random sample**
>
> A sample, selected by choosing elements at evenly spaced intervals, after the first element is selected randomly.

One caution should be mentioned with respect to this kind of sampling. It might just happen that there is a bias that can occur with this method. For example, suppose there were a large number of people from a particular minority group whose names all happen to start with one particular letter. Choosing the sample in this way might tend to include a disproportionate number of people from that group. This kind of bias is usually very subtle and often not picked up by the data collector.

Stratified sampling.   Sometimes we are dealing with a population that has various subgroups. If we have reason to believe that data collected will differ among these various groups, then we cannot afford to ignore their existence. We refer to the different subgroups as *strata*. Sampling from these strata is known as *stratified sampling*. Income, education, gender, ethnic group are but a few of the variables that might distinguish the different subgroups we wish to represent.

> **stratified sample**
>
> A sample, selected by randomly selecting from separate subgroups (or strata) within a population.

The procedure we use for stratified sampling is to divide the population into a number of different strata and then randomly select samples from each stratum.

How do we determine how many elements to select from each stratum? One method is simply to select the same proportion as exists in the population. This is known as proportional allocation. Let's look at an example.

Suppose we wanted to get some idea of the profits generated by the eleven and a half million proprietorship enterprises operating in the United States in 1977. Since we know there will be differences among the different kinds of industries, we divide our population into 10 different kinds of industries. Each of these industries contributes a different proportion to the total (e.g., service industries account for approximately 28 percent of the total). Let's say we wanted our total sample to number approximately 15,000 firms. The sample should be apportioned to the industries in the proportion that each industry bears to the total. There are (0.28) (11,500,000) = 3,200,000 service

Figure 6.1

A road map of Tucson, Arizona. Simple random sampling would involve identifying every household in the city and selecting a random sample. If each household were identified and numbered, a systematic procedure could be used. This might involve a listing of households, selecting the first one at random, and then selecting the remainder according to a systematic plan (such as every 25th household on the list). The large numbers in the figure identify different economic strata in Tucson. Stratified sampling would involve selecting random samples from within each stratum. Finally, each city block could be identified. Blocks could then be selected by a random procedure. Each household in the selected blocks would be interviewed. These procedures would constitute cluster sampling.

firms. Thus, there should be 3,200,000/11,500,000 $\times$ 15,000 = 4,200 service firms in the sample. Therefore, we would randomly select approximately 4,200 service industries. After randomly selecting from each of 10 categories, we would then compile our results separately for each industry and present the results accordingly.

Cluster sampling. Suppose you receive a large shipment of flatware that is case-packed, 24 sets per case. You wish to engage in acceptance sampling but balk at the prospects of opening every case, identifying each set, and then selecting a random sample from the entire collection of flatware sets. Clearly the cost, time requirements, and inconvenience would be enormous. Instead, you regard each case as an element in the population and you randomly select several cases from the total shipment. Then you inspect every flatware set in each case for flaws and record the number that are rejected. This technique is known as *cluster sampling*. In this example, each case would constitute a cluster.

Cluster sampling is commonly employed in surveys involving geographical divisions. For example, a city may be subdivided into blocks. The blocks may be listed and a random procedure used to select a sample of blocks. Then a person in every residence on the block is interviewed.

Our assumption in cluster sampling is that each cluster is representative of the population as a whole. Thus, each case is representative of the entire shipment of flatware and the block is representative of the households in the city. This may not always be precisely correct so that considerable thought should precede the use of cluster sampling. When appropriate, cluster sampling usually provides a distinctive advantage over simple random sampling in convenience, time, effort, and cost.

Figure 6.1 summarizes simple random sampling, systematic, stratified, and cluster sampling.

cluster sample

A sample in which the population is divided into naturally occurring groupings or clusters.

Example C

A manufacturer's representative organization represents a company that manufactures jewelry. So far sales have mainly been confined to a small group of specialty merchandisers. However, the company decides it wants to expand its market to include retail outlets such as department stores, boutiques, drugstores, and ladies' dress shops. It develops a new line especially designed for this expanded market.

It asks its representatives to sample these markets in order to determine whether to go into full-scale production on the new line.

Let's look at some of the ways the manufacturer's representatives sample this new population in order to comply with the company's request. In each case, identify the target population and the type of sampling technique used.

a. One of the representatives, Walter, familiar with department stores and the way they buy, chooses several that he knows from prior experience, are the "trendsetters" in the retail field. Based on their reactions, he feels he can estimate how successful the new line will be in most department stores.

b. Another representative, Cyndi, in charge of drugstores, wrestles with the problem that her accounts are scattered over a fairly broad geographical area. She decides that she will drive over to one particular area that happens to contain a few drugstores that are not too far from each other. She will sample these accounts and then make a few selected phone-calls to a number of other drugstores. She hopes that this sample will enable her to estimate accurately how the new line will "go" in drugstores.

c. Perry, a relatively new addition to the organization, devises a way to obtain sample reaction from boutiques. He divides his territory into shopping areas, and then randomly samples a few boutiques in each shopping center.

d. Aki, another new addition to the organization, decides to compile a list of all the ladies' dress shops in his territory. Then, after randomly selecting the first, he checks off every 10th one and schedules visits to each of these shops.

e. Norma, in charge of the organization, decides to test sample across-the-board. She divides the total territory into a number of different areas. She then randomly selects a few of these areas and visits every department store, boutique, drugstore, and ladies' dress shop in that area.

What kind of sampling technique has been left out? Can you see any reasons why this technique was not used?

Solution

a. Walter's target population consists of all department stores in his territory. He is using judgmental sampling. He has drawn upon his prior experience to select those accounts that he feels best represent his population.

b. Cyndi's target population: all drugstores in her territory. Although her technique is principally convenience sampling, she has probably used judgmental sampling as well in her choice of "selected" phone calls.

c. Perry's population: all boutiques in his territory. Stratified sampling is the procedure he chose.

d. Aki's population: all ladies' dress shops in his territory. He has used systematic random sampling as his method.

e. Norma's population: all retail outlets in her territory. (All of the others were subpopulations of Norma's). She has chosen cluster sampling to obtain her estimates.

Simple random sampling is the method that has been excluded. The representatives have decided that this technique would require too much time and effort. It probably would result in higher costs as well since the chances are the sample would be scattered over a broad geographical area. In this example, simple random sampling would certainly not be the most desirable method.

EXERCISES

6.6 Which of the following use judgmental, convenience, or random sampling? If random sampling, identify the type.

    a. The 30 stocks constituting the Dow Jones Industrials were selected by experts as representative of all industrial stocks.

    b. In order to check conformity to affirmative action, personnel files were subdivided into ethnic groups and random samples were selected from each group.

    c. To obtain overnight ratings of a highly advertised television special, each of a number of interviewers was instructed to make 25 phone calls from numbers they found in a phone book.

    d. To check on the quality of paper clips, 20 boxes were randomly selected from a shipment and all clips in each box were inspected.

    e. During a 32-year period of inflationary pressures, the penny gumdrop has resisted all upward price pressures. However, synthetic gums have replaced natural gums and solid cores have given way to hollow centers. Reasoning that the price stability is illusory, penny gumdrops are not included as a food item in the CPI.

    f. To assess employee reaction to an anticipated change in car-pooling incentives, the computer randomly selects 50 names from the 6,500 that are in the personnel files.

6.7 Comment: Each day, the news department of a local radio station conducts a person-in-the-street interview of five typical citizens. The interviews are conducted at noon at the shopping mall.

6.8 Comment: To assess the present status of the candidates in a national election, a public opinion analyst stands on a street corner in the financial district and interviews everyone who will talk to him. When possible sources of bias are pointed out, he replies, "I know, but I balance it out in sheer numbers."

6.4 THE CONCEPT OF SAMPLING DISTRIBUTIONS

Consider the following problem:

A proprietor of a firm that manufactures plastics has a factory with many machines. Each of these machines has a number of different parts that are known to break down at various times. He wants to decide whether to wait until the parts break down before replacing or whether to institute a program of routine preventive maintenance. He reasons that if he can anticipate when a machine will break down, routine preventive maintenance will, in the long run, save him time and money.

He takes a sample of 30 key parts and subjects them to various stress tests. He obtains a variety of descriptive measures relating to the length of life of these parts—mean, standard deviation, and the type of distribution. Can he make a decision based on the results obtained with this sample?

The answer to this question is "not really." The results he obtained were measures of the sample. But it is not the sample that he is interested in. His primary concern is in the population—*all* of the key parts. However, he cannot test the total population without destroying all of his machines. Therefore, he must somehow use the data obtained from the sample to make inferences about the population.

Suppose he took another sample of 30 parts. Would he obtain *exactly* the same mean and standard deviation? Probably not.

If he continued to take samples of the same size and calculate means and standard deviations for each sample, he would certainly find that these statistics differ from sample to sample. Assuming that he didn't go out of business first, he could obtain a distribution of all the sample means he collected. The resulting distribution would approach the *sampling distribution of the mean*. It would be a theoretical probability distribution—the probability associated with specific values of the sample mean.

Since it obviously would not be practical to obtain this sampling distribution empirically, we may conceive of a theoretical experiment in which we draw all possible samples of a fixed size and obtain a *sampling distribution*. This is really the heart of statistical inference—the sampling distributions, which are nothing more than theoretical probability distributions of the relevant statistics. There is a sampling distribution for every statistic—mean, standard deviation, proportion, etc. In most real-life situations, we obtain only one sample and make inferences based on these sampling distributions.

Let's talk a little more about results obtained from samples. Do you think that the mean $(\overline{X})$ of a given sample is *exactly* equal to the mean (μ) of the population from which it was drawn? How about the mean of a second sample? The answer is "probably not," but we are usually satisfied if we have reason to believe our sample mean is pretty close to the true (or population) mean. As a matter of fact, if we were to continue taking samples and then calculated the mean of *all* the sample means we obtained, we might be pleasantly surprised to find that *this* value is, in fact, pretty close to the actual population mean.

sampling distribution of the mean

A *theoretical* probability distribution of sample means that would result from drawing *all* possible samples of a given size from some population.

sampling distribution

A *theoretical* probability distribution of any sample statistic that would result from drawing *all* possible samples of a fixed size from some population.

6.5 THE SAMPLING DISTRIBUTION OF THE SAMPLE MEANS: NORMAL POPULATION

Imagine it is possible to select all samples of a fixed sample size from an infinite, normally distributed population. If we were to construct a probability distribution of these sample means, what would we discover?[4]

First, we would see that the sampling distribution of sample means, like

---

[4] Recall that, when the population is infinite or very large, sampling with or without replacement yields essentially identical results. Therefore, we don't distinguish here between sampling with and without replacement.

the population from which they were drawn, is normally distributed. Moreover, we would find that the mean of the probability distribution $(\mu_{\bar{x}})$ is equal to the mean of the population. We would also discover that this distribution is more compact and less dispersed than the distribution of scores in the original population. This is another way of saying that the standard deviation of the sample means $(\sigma_{\bar{x}})$ would be less than the standard deviation of the population.

The relationship between the standard deviation of a sampling distribution of means and the standard deviation of the population is quite precise. It is

$$\sigma_{\bar{x}} = \frac{\sigma}{\sqrt{n}},$$ (6.1)

in which $\sigma_{\bar{x}}$ is the standard deviation of the sampling distribution of means and is referred to as the *standard error of the mean.* As you can see, both the sample size *(n)* and the value of the population standard deviation affect the size of the standard error of the mean. Figure 6.2 shows how the dispersion decreases with increasing sample size. Figure 6.3 shows the sampling distributions of the mean for a fixed sample size *(n)* when the population standard deviations differ.

standard error of the mean

Standard deviation of the sampling distribution of sample means.

Effect of increased sample size on the sampling distribution of sample means drawn from a normally distributed population in which $\mu = 4.00$ and $\sigma = 1.00$. Note how the variability decreases as the sample size *(n)* increases.

Figure 6.2

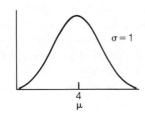

Population distribution

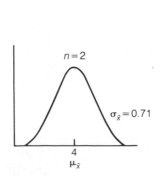

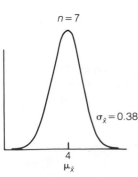

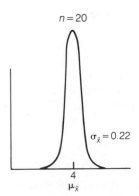

Figure 6.3

Sampling distributions of sample means drawn from normally distributed populations with different standard deviations. In both cases, the sample size is the same ($n = 49$). Note that when the variability in the population is high, so also is the variability of the sample mean from that population.

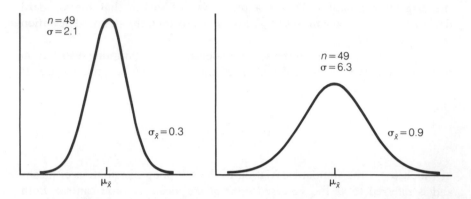

The standard error of the mean ($\sigma_{\bar{X}}$) has all the properties of any standard deviation of normally distributed variables. We can convert any sample mean into a z-score (Section 5.5) and, by reference to areas of the standard normal curve, ascertain (a) the proportion of area between any sample mean and $\mu_{\bar{X}}$; (b) the proportion of area between two sample means; and (c) the proportion of area beyond any sample mean. These proportions also represent probabilities.

Before we look at an example, we want to be sure you understand the difference between the standard deviation of the population (σ) and the standard error of the mean ($\sigma_{\bar{X}}$). Although they are related, they are really quite different. Refer to Figure 6.2. The top distribution is the actual distribution of scores in the population. The standard deviation of this distribution is σ. The three figures below represent probability distributions of sample means drawn from this population. The standard deviation of each of these distributions is $\sigma_{\bar{X}}$. Although both σ and $\sigma_{\bar{X}}$ are standard deviations, it is probably best to always refer to $\sigma_{\bar{X}}$ as the *standard error of the mean* to help avoid any possible confusion.

Example D

A milling operation produces boards in which the widths are normally distributed with $\mu = 0.625$ and $\sigma = 0.012$. Using acceptance sampling, an entire lot is rejected if the mean width of the sample is either less than 0.610 or greater than 0.640. What is the probability of rejecting a lot if the sample size is (a) 1? (b) 5? (c) 25?

Solution

a. This involves a familiar problem—ascertaining the probability that a single board, selected at random, will fail to meet acceptance standards.

For this solution, we use the population standard deviation. Thus,

$$z = \frac{0.640 - 0.625}{0.012} = 1.25$$

and

$$z = \frac{0.610 - 0.625}{0.012} = -1.25$$

In Column C, Table V, we find the area beyond a z-score of 1.25 is 0.1056. Therefore, $p(X > 0.640 \text{ or } X < 0.610) = 0.1056 + 0.1056 = 0.2112$. There are about 21 chances in 100 that a single board will be rejected. Figure 6.4 portrays this relationship graphically.

Relationship between z and area. Since $z = \dfrac{X - \mu}{\sigma}$, it follows that $z\sigma = X - \mu$ (difference   **Figure 6.4**

between the value of a random variable and the mean).

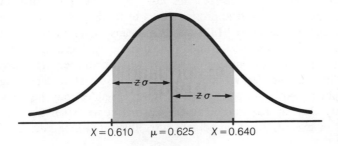

b. $\sigma_{\bar{X}} = 0.012/\sqrt{5} = 0.0054$. Therefore, $z\,(\bar{X} > 0.640) = 0.015/0.054 = 2.78$ and $z\,(\bar{X} < 0.610) = -2.78$. We find the area beyond a z-score of 2.78 equals 0.0027. Therefore, $p(\bar{X} > 0.640 \text{ or } \bar{X} < 0.610) = 0.0027 + 0.0027 = 0.0054$. The chances of rejecting the lot are now less than 1 in 100.

c. $\sigma_{\bar{X}} = 0.012/\sqrt{25} = 0.0024$. Thus, $z\,(\bar{X} > 0.640) = 0.015/0.0024 = 6.25$. This value of z is so large that there is no tabulated value. The probability of rejecting the lot, based on an $n = 25$, is extremely small.

6.6 THE SAMPLING DISTRIBUTION OF THE SAMPLE MEANS: POPULATION NOT NORMAL

We have noted that, when we draw samples of a fixed size from a normally distributed population, the distribution of sample means is normal with $\mu_{\bar{X}} = \mu$ and the standard error of the mean, $\sigma_{\bar{X}}$, equal to $\sigma/\sqrt{n}$. This feature of the sampling distribution of means permits us to use the standard normal curve as a model against which to assess the results of sampling experiments.

Unfortunately, many variables used in business and economics do not approximate a normal distribution. Such variables as wages, expected life of equipment, and time required to achieve compliance with a governmental regulation are often skewed and occasionally distributed in an assortment of different shapes including U, uniform, and J-shaped distributions. Consequently, it would appear that the normal curve could not possibly serve as a model against which to evaluate the results of sampling experiments when such nonnormally distributed variables are used. This would indeed be the case were it not for a rather remarkable mathematical postulate known as the *central limit theorem*. This theorem specifies a relationship of surpassing importance in statistical inference:

central limit theorem

If random samples of a fixed *n* are drawn from any population (regardless of the form of the population distribution), as *n* becomes larger, the sampling distribution of means approaches normality.

No matter what the form of the population distribution, the sampling distribution of sample means approaches normality as the sample size increases.

This characteristic of sampling distributions is illustrated in Figure 6.5. Note that, as the sample size reaches $n = 30$, the sampling distributions of means are extremely good approximations to the bell-shaped normal curve. What does all of this mean? In a word, it means that, as long as the sample size is sufficiently large, the normal curve may be used as a model to describe sampling distributions of means no matter what form the parent distribution takes. What sample size is sufficiently large? Most statisticians agree that the sampling distribution of means of most variables may be considered a sufficiently close approximation to the normal curve when the sample size, *n*, equals or exceeds 30. This point is also illustrated in Figure 6.5.

Example E

The mean tread life of a premium-brand tire is known to be 22,560 miles with a standard deviation of 2,153. However, the distribution is positively skewed so that it cannot be considered normal. The owner of a small taxi fleet wishes to purchase 36 tires but he estimates that they must average more than 21,500 miles tread wear to justify the cost.

a. What is the probability that a randomly selected sample of 36 tires will achieve a mean tread wear less than 21,500 miles?
b. What is the probability that the mean tread wear will be as high as 23,500 miles?

Figure 6.6 presents each of these questions in graphical form. Note that, even though the population of tread wears is not normally distributed, the sampling distribution of sample means, when $n = 36$, approaches normality.

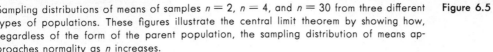

Sampling distributions of means of samples $n = 2$, $n = 4$, and $n = 30$ from three different types of populations. These figures illustrate the central limit theorem by showing how, regardless of the form of the parent population, the sampling distribution of means approaches normality as n increases.

Figure 6.5

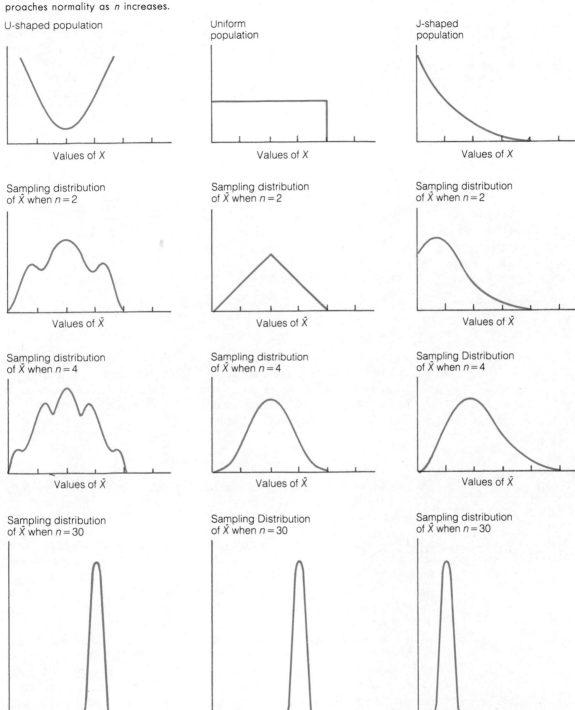

U-shaped population

Values of X

Uniform population

Values of X

J-shaped population

Values of X

Sampling distribution of $\bar{X}$ when $n = 2$

Values of $\bar{X}$

Sampling distribution of $\bar{X}$ when $n = 2$

Values of $\bar{X}$

Sampling distribution of $\bar{X}$ when $n = 2$

Values of $\bar{X}$

Sampling distribution of $\bar{X}$ when $n = 4$

Values of $\bar{X}$

Sampling distribution of $\bar{X}$ when $n = 4$

Values of $\bar{X}$

Sampling Distribution of $\bar{X}$ when $n = 4$

Values of $\bar{X}$

Sampling distribution of $\bar{X}$ when $n = 30$

Values of $\bar{X}$

Sampling Distribution of $\bar{X}$ when $n = 30$

Values of $\bar{X}$

Sampling distribution of $\bar{X}$ when $n = 30$

Values of $\bar{X}$

Figure 6.6
Finding the probability
under the sampling
distribution of means
corresponding to sample
means with tread wears
beyond selected values

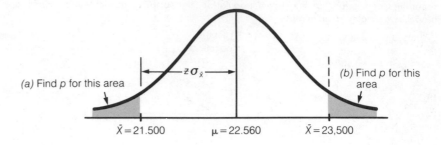

(a) Find p for this area $-z\sigma_{\bar{x}}-$ *(b) Find p for this area*

$\bar{X} = 21.500$ $\mu = 22.560$ $\bar{X} = 23.500$

Solution

a.   Since $\sigma = 2{,}153$ and $n = 36$,

$$\sigma_{\bar{X}} = \frac{2153}{\sqrt{36}} = 358.83$$

The z corresponding to 21,500 miles is

$$z(\bar{X} < 21{,}500) = \frac{21{,}500 - 22{,}560}{358.83} = -2.95$$

Referring to Column C in Table V, we find that p equals 0.0016. Thus,
he may feel quite secure that he will achieve the minimum tread wear
required to justify the expense. The probability is less than one in a
hundred that he will obtain a mean lower than 21,500.

b.   The z corresponding to 23,500 miles is

$$z(\bar{X} > 23{,}500) = \frac{23{,}500 - 22{,}560}{358.83} = 2.62$$

In Column C of Table V we find that p equals 0.0044. It is unlikely
that the fleet owner will obtain a mean tread wear as high as 23,500
miles.

The solution to Example E presumes that we know the value of the population standard deviation. In the real world, it is rare that we know parameters:
for infinite populations, the parameters can never be known. For all practical
purposes, the standard deviations of very large populations may also be regarded as unknowable. How, then, do we obtain the standard error of the
mean when the population standard deviation is unknown?

In a word, we estimate it from the sample standard deviation. Just as
the mean of a sample provides a point estimate of the population mean, so
also does the sample standard deviation provide a point estimate of its corresponding parameter, the population standard deviation. Moreover, the larger
the sample size, the more likely that a given sample standard deviation will
approach the population standard deviation. To estimate the standard error
of the mean from a sample statistic, we use the formula:

$$\text{Est. } \sigma_{\bar{X}} = s_{\bar{X}} = \frac{s}{\sqrt{n}} \qquad (6.2)$$

Note that $s_{\bar{X}}$ estimates $\sigma_{\bar{X}}$ and s estimates σ in this formula. It is important to remember that the italic letters tell us that we are using sample statistics to estimate parameters.

Let's now look at an example of the use of z when the standard error of the mean is based on the use of a sample standard deviation. For illustrative purposes, we shall use a large sample since we know that the sampling distribution of the mean approaches the normal curve when n is large. In Chapter 7, we shall look at another sampling distribution that is used when n is small (i.e., $n < 30$).

Example F

Forty new automobile batteries are randomly selected from an inventory numbering in the thousands. They are run through a series of tests simulating normal conditions of use. They are found to have a mean life of 1,277.5 days with a standard deviation of 85.4.

You know that one of your most important customers demands that the mean life of battery lots that he purchases exceed 1,250 days. What is the probability of failure if the customer purchases lots of *(a)* 40 batteries? *(b)* 60 batteries? *(c)* 120 batteries?

Solution

a. First we find the estimated standard error of the mean:

$$s_{\bar{X}} = \frac{85.4}{\sqrt{40}} = 13.50.$$

Therefore,

$$z(\bar{X} < 1250) = \frac{1250 - 1277.5}{13.5}$$

$$= -2.04$$

In Column C of Table V, we find the area beyond z is 0.0207. The estimated risk would appear rather low, about 2 chances in 100.

b. When $n = 60$,

$$s_{\bar{X}} = \frac{85.4}{\sqrt{60}} = 11.02.$$

Thus,

$$z(\bar{X} < 1250) = \frac{1250 - 1277.5}{11.02}$$

$$= -2.50$$

We find the area beyond z is 0.0062. The estimated risk is now less than 1 in 100.

c. When $n = 120$,

$$s_{\bar{X}} = \frac{85.4}{\sqrt{120}} = 7.80$$

Therefore,

$$z(\bar{X} < 1250) = \frac{1250 - 1277.5}{7.80}$$

$$= -3.53$$

This value does not appear in Table V. However, by inspection of the table, it is clear that $p < 0.0002$. The risk is now about 2 in 10,000. Moral of the story? There is more than one reason you should convince your customer to buy in as large lots as possible.

6.7 THE SAMPLING DISTRIBUTION OF THE SAMPLE MEANS: POPULATION SMALL

As we have just noted, many populations with which we deal in business and economics are either infinite or so large that, for all practical purposes, they may be treated as infinite. Under these circumstances, the difference between sampling with and without replacement is negligible. However, when the population of interest is finite and relatively small, there is a distinct difference between these two sampling procedures, particularly as the sample size approaches the size of the population. More specifically, when sampling without replacement, the formula for calculating the standard error of the mean overestimates the true standard error. This is not the case when sampling with replacement. However, as a matter of actual practice, most sampling in business and economics uses sampling without replacement (why test the same item twice?). Consequently a correction should be applied whenever relatively large samples are drawn without replacement from a small population. This correction consists of multiplying the standard error of the mean by the *finite correction factor (fcf)*,

finite correction factor (fcf)

A term used to reduce the size of the uncorrected standard error of the mean, when the ratio of sample size to population size (n/N) is greater than 0.05.

$$\text{fcf} = \sqrt{\frac{N-n}{N-1}} \tag{6.3}$$

Thus, when σ is known and the population small, the standard error of the mean multiplied by the correction factor is

$$\sigma_{\bar{X}} = \left(\frac{\sigma}{\sqrt{n}}\right)\left(\sqrt{\frac{N-n}{N-1}}\right) \tag{6.4}$$

When the standard error of the mean is estimated from a sample standard deviation, the standard error of the mean becomes

$$s_{\bar{X}} = \left(\frac{s}{\sqrt{n}}\right)\left(\sqrt{\frac{N-n}{N-1}}\right) \tag{6.5}$$

The effect of the finite correction factor is to reduce the size of the standard error of the mean. To illustrate, if $s_{\bar{X}} = 15$, $n = 20$, and $N = 100$, the corrected standard error of the mean is

$$s_{\bar{X}} = 15\sqrt{\frac{100-20}{99}} = (15)\,(0.8989) = 13.48$$

The finite correction factor is particularly important when the sample size is a relatively large proportion of the population. For example, if $n = 5$ and $N = 100$, the correction factor is:

$$\sqrt{\frac{N-n}{N-1}} = \sqrt{\frac{95}{99}} = 0.9796$$

However, if $n = 40$ and $N = 100$, the correction factor becomes:

$$\sqrt{\frac{N-n}{N-1}} = \frac{60}{99} = 0.7785$$

In the second instance, the finite correction factor has a considerable impact on the size of the standard error of the mean. A widely accepted rule of thumb is that the finite correction factor should be used whenever the sample size is more than 5 percent of the population from which it was drawn. Figure 6.7 illustrates why this is so. Note that, when n is small relative to N, the correction is negligible (i.e., the fcf is close to 1.00). However, as the ratio of n to N increases, the correction factor takes on increased significance in the calculation of the standard error of the mean. Note also that this statement holds true for relatively large as well as relatively small population sizes.

Example G

Monitoring equipment revealed that, during a two-hour period in the automated production of submarine electrical circuits, there had been wide fluctuations in the voltage supplied to the production facility. It was feared that this interruption in the power supply may have adversely affected the operating characteristics of the circuits.

Suppose 40 were selected for intensive testing and the mean output was found to be 600 milliamperes with a standard deviation of 27.

Find the estimated standard error of the mean, if, in the two-hour period, the number of circuits produced had been (a) 500? (b) 200? (c) 100?

Figure 6.7

An illustration of the effect of the ratio n/N on the size of the finite correction factor. As the ratio increases (i.e., sample size becomes a larger proportion of population size), the fcf decreases.

The lower the fcf, the greater the impact on the size of the standard error of the mean. Note that the size of the population (e.g., $N = 100,000$ versus $N = 100$) plays only a minor role. It is the size of the ratio that is the critical factor.

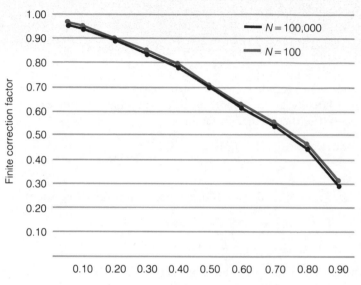

Solution

The uncorrected standard error of the mean is $27/\sqrt{40} = 4.27$.

a. The finite correction factor is

$$\sqrt{\frac{500 - 40}{499}} = 0.9602$$

Thus, the corrected standard error of the mean is $(4.27)(0.9602) = 4.10$.

b. The finite correction factor is

$$\sqrt{\frac{200 - 40}{199}} = 0.8967$$

The corrected standard error of the mean is $(4.27)(0.8967) = 3.83$.

c. The finite correction factor is

$$\sqrt{\frac{100 - 40}{99}} = 0.7785$$

giving us a corrected standard error of the mean of (4.27) (0.7785) = 3.32.

This example illustrates that, as the proportion of the sample size *(n)* to the size of the population *(N)* [i.e., n/N] increases, the correction factor has a greater impact on the size of the standard error of the mean. Thus, in *(a)* the ratio $n/N = 0.08$, and the corrected standard error of the mean changes little; in *(b)* the ratio $n/N = 0.20$, and the effect of the correction factor starts to take on added significance; in *(c)* the ratio $n/N = 0.40$, and the effect on the standard error of the mean is now quite considerable.

EXERCISES

6.9 Cinemascore™, a syndicated newspaper and a national radio and TV feature, measures audience reactions to new movies. In 1979, they obtained a sample of 112 movies and found a mean running time of 107.2 minutes (or approximately 1 hour 47 minutes) and a standard deviation of 7.15 minutes. (Running time is measured from the first frame to the end of the story, including opening logos and titles, but not including credits after the story has ended.)[5]

Suppose we wish to use these sample results to estimate the true mean running time (μ) of all movies, whether showing in 1979 or not. What is the probability that the sample mean $(\overline{X})$ will be

    a. Within two minutes of μ?
    b. Within one and one-half minute of μ?
    c. Within one minute of μ?
    d. Within one-half minute of μ?

6.10 A builder is considering a new condominium project. However, a candid discussion with the loan officer at a savings and loan reveals that interest rates to his buyers will be anywhere from 12 to 14 percent.

He has tentatively priced his condominiums at $85,000 each. With a 20 percent down payment, an annual income of $33,600 would qualify the buyers for a 12 percent loan. But, at 14 percent, the qualifying income is $38,688.

He decides to conduct a survey to estimate the mean annual income of his potential buyers. Suppose he samples 100 potential buyers and finds $(\overline{X}) = \$35,000$, $s = \$4,000$.

What is the probability that these buyers can afford the condominiums if

    a. Interest is 12 percent?
    b. Interest is 14 percent?

6.11 A trade journal for morticians reports that the present cost of an adult, full-service funeral is normally distributed with a mean, μ, of $1,700, and a standard deviation, σ, of $500.

A random sample of $n = 25$ funerals is chosen.

    a. What is the probability that the obtained $\overline{X}$ will differ from μ by more than $100 in either direction?
    b. What is the probability that $\overline{X}$ will be greater than $2,000?
    c. This population mean figure has been vigorously disputed. There are some

---

[5] Source: *Cinemascore,* 1980.

who place the average cost for a full-service funeral at \$2,500. Assuming the same population standard deviation, what is the probability $\overline{X}$ will be greater than \$2,000 if $\mu = \$2,500$?

6.12 Refer to Exercise 6.11. Suppose that an independent consumer group does not believe the figures reported by the trade journal. It wishes to obtain an independent estimate of the true "cost of dying" (μ). They will use the mean of a sample of $n = 25$ funerals as an estimate. Assuming that the population standard deviation is \$500,

    a. Find the probability that the estimate will lie within \$100 of the actual population mean.

    b. Suppose the sample size was increased to $n = 100$. Now find the probability that the estimate lies within \$100 of the true mean.

Compare your answer with *(a)* and comment on the effect of increased sample size upon the reliability of $\overline{X}$ as an estimate of μ.

6.13 The Department of Health Services has conducted studies measuring the level of carbon tetrachloride, a suspected carcinogen, in a number of different wells.

    a. A random sample of 36 wells is taken and the mean level of carbon tetrachloride found to be 4.5 parts per billion (ppb). Previous studies showed $\sigma = 1.5$. If the health department uses this sample mean to estimate the population mean level, what is the probability that the estimate will be within ± 0.75 ppb of the true mean?

    b. Suppose that the department selects another sample of 36 wells, this time concentrating on wells that previously had problems with TCE, another suspected carcinogen. This sample yields the same sample mean, $\overline{X} = 4.5$. However, previous studies on wells with a similar history showed increased variability, $\sigma = 2.9$. Find the probability that this estimate will lie within ± 0.75 ppb of the true mean.

Compare your answer with *(a)* and comment on the effect of increased variability in the population upon the reliability of $\overline{X}$ as an estimate of μ.

6.14 Refer to Exercise 6.13. Although no federal standards have yet been set, health officials have adopted 5 ppb as the maximum acceptable level of carbon tetrachloride. In other words, the health department wants to close water wells down if they have levels greater than 5 ppb. What is the probability that the wells studied in Exercise 6.13 *(a)* and *(b)* will exceed the accepted level?

6.8 PUTTING IT ALL TOGETHER

Let's look at an hypothetical example of a small population and construct a sampling distribution of means for various sample sizes. Note that the example is used only to clarify the key concepts relating to sampling distributions. In actual practice, sampling would not be used with such a small population.

Let us imagine that the plant manager of a carpet mill is conducting an inventory of various pieces of equipment and their ages. He wishes to calculate their mean ages. There are six carpet looms with the ages shown in Table 6.1.

Loom	Age (X)	(X−μ)	(X−μ)²
A	3	−1	1
B	6	2	4
C	4	0	0
D	4	0	0
E	5	1	1
F	2	−2	4
Sums	24	0	10

The mean population age is $\mu = \dfrac{\Sigma X}{N} = 4.0$.

The variance of the population is $\sigma^2 = \dfrac{\Sigma(X-\mu)^2}{N} = \dfrac{10}{6} = 1.667$.

The standard deviation is $\sigma = \sqrt{1.667} = 1.2910$.

Now imagine that the plant manager wished to estimate the parameters, μ and $\sigma_{\overline{X}}$, selecting a single sample of $n = 2$. Sampling without replacement, there are 15 different samples of $n = 2$ that could be drawn. These are shown in Table 6.2.

Sample no.	Looms	Ages	$\overline{X}$
1	A,B	3,6	4.5
2	A,C	3,4	3.5
3	A,D	3,4	3.5
4	A,E	3,5	4.0
5	A,F	3,2	2.5
6	B,C	6,4	5.0
7	B,D	6,4	5.0
8	B,E	6,5	5.5
9	B,F	6,2	4.0
10	C,D	4,4	4.0
11	C,E	4,5	4.5
12	C,F	4,2	3.0
13	D,E	4,5	4.5
14	D,F	4,2	3.0
15	E,F	5,2	3.5

Now we may construct a frequency and probability distribution of the 15 sample means. These are shown in Table 6.3.

Here we see that the expected value of the sample means is μ (i.e., $E(\overline{X}) = \mu$), the mean of the population from which the samples were drawn. It is the long-term average of the sample means over many samplings. Note that the mean of the sampling distribution (i.e., $60/15 = 4.00$) equals the population mean and the expected value of the sample means.

Finally, Table 6.4 shows the calculation of the variance and the standard error of the mean of the sampling distribution. Note that the standard error of the mean, calculated directly from the sampling distribution, yields a value identical to $\sigma_{\bar{X}} = \frac{\sigma}{\sqrt{n}} \sqrt{\frac{N-n}{N-1}}$.

Table 6.3
Frequency distribution of means and calculation of the expected value of a sample mean

$\bar{X}$	f	$p(\bar{X})$	$\bar{X}p(\bar{X})$	$f(\bar{X})$
2.5	1	0.0667	0.1668	2.5
3.0	2	0.1333	0.3999	6.0
3.5	3	0.2000	0.7000	10.5
4.0	3	0.2000	0.8000	12.0
4.5	3	0.2000	0.9000	13.5
5.0	2	0.1333	0.6665	10.0
5.5	1	0.0667	0.3667	0.5
	15	1.0000	$E(\bar{X}) = 3.9999*$	$\Sigma f(\bar{X}) = 60.0$

* Disparity of 0.0001 due to rounding error.

Table 6.4
The calculation of the variance and standard error of the mean directly from the sampling distribution of means in which the sample size is 2

$\bar{X}$	f	p	$\bar{X}^2$	$\bar{X}^2 p(\bar{X})$
2.5	1	0.0667	6.25	0.4169
3.0	2	0.1333	9.00	1.1007
3.5	3	0.2000	12.25	2.4500
4.0	3	0.2000	16.00	3.2000
4.5	3	0.2000	20.25	4.0500
5.0	2	0.1333	25.00	3.3325
5.5	1	0.0667	30.25	2.0176
				$E(\bar{X}^2) = 16.6667$

We have previously seen that $E(\bar{X})$ is $\mu = 4.0$. Thus,

$$E[(\bar{X})]^2 = 16$$

Hence,

$$\sigma_{\bar{X}}^2 = E(\bar{X}^2) - [E(\bar{X})]^2$$
$$= 16.6667 - 16.000$$
$$= 0.6667$$

Therefore, the standard error of the mean is

$$\sigma_{\bar{X}} = \sqrt{0.6667} = 0.8165$$

Using the formula for the standard error of the mean that includes the finite population correction factor, we obtain

$$\sigma_{\bar{X}} = \frac{\sigma}{\sqrt{n}} \cdot \sqrt{\frac{N-n}{N-1}}$$

$$= \frac{1.2910}{\sqrt{2}} \cdot \sqrt{\frac{6-2}{6-1}}$$

$$= (0.9129)(0.8944)$$

$$= 0.8165$$

Figure 6.8 ties the preceding discussion together. It shows a plot of the population values, the probability of each, and the standard deviation of the population. It also shows the plot of the sample means when $n = 2$, the probability of randomly selecting each mean (without replacement), and the standard error of the mean.

Plot of the population values μ and σ, and their associated probabilities (above horizontal line). Plot of sample means $(n = 2)$, $\mu_{\bar{X}}$, and their associated probabilities (below horizontal line). **Figure 6.8**

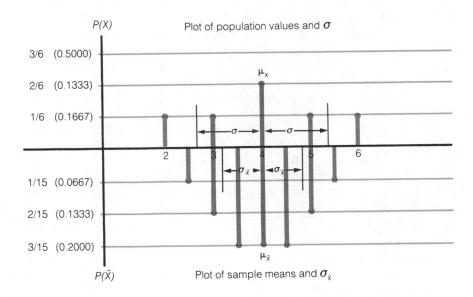

SUMMARY

In this chapter we looked at some of the reasons that we sample, the kinds of errors we can make, the types of samples, and the concept of sampling distributions.

1. We saw that there are several reasons that we sample rather than make a census of the whole population:

   a. The population is simply *too large* for every member to be examined.

b. It is much *less expensive* to measure a sample than to measure the whole population.

c. It *takes less time* to sample than to measure the whole population.

d. In some cases, *testing destroys the sample;* by testing a small sample, the losses are kept to a reasonable level.

e. In some cases, sampling is *more accurate* than trying to conduct a census.

2. There are different types of error that can occur as we measure our sample or population:

a. Those due to *bias* in the selection of the sample, making it nonrepresentative.

b. *Sampling errors* due to uncontrolled or chance conditions.

c. *Nonsampling errors* resulting from the way in which the data are collected or the observations are made.

3. Several different types of sampling techniques were discussed:

a. *Nonrandom sampling,* which includes convenience and judgmental sampling.

b. *Random sampling,* in which every element of the population has an equal chance of being selected. There are various ways to achieve random sampling:

i. *Simple random sampling,* which is perhaps the most thorough, since the selection of any one element is purely a chance process; usually, tables of random digits are used in the selection process.

ii. *Systematic random sampling,* in which the first element is randomly selected and the remainder are chosen by picking at regular intervals (such as every 10th or 50th candidate).

iii. *Stratified sampling,* in which the population is divided into certain meaningful subgroups (the strata) and then candidates are randomly selected from each of the strata.

iv. *Cluster sampling,* in which the population is divided into areas or groups, then certain areas are randomly selected and a census is made on all the elements in those areas.

4. A sampling distribution is a *theoretical* probability distribution of a statistic obtained by drawing *all* possible samples of a given size.

5. Mainly, we concentrated on the sampling distribution of the mean. We saw that the mean of this sampling distribution ($\mu_{\bar{x}}$) is equal to the mean of the population from which the samples were drawn. The standard deviation of this distribution is known as the *standard error of the mean* ($\sigma_{\bar{x}}$).

6. The *central limit theorem* tells us that, regardless of the form of the population distribution, the sampling distribution of the mean approaches normality with increasing sample size.

7. Finally, we saw that when the population is small, but particularly when the ratio of sample size to population size *(n/N)* is large, we must use the *finite correction factor* to correct the size of the standard error of the mean.

TERMS TO REMEMBER

bias (161)

sampling error (162)

nonsampling error (162)

convenience sampling (165)

judgmental sampling (165)

simple random sample (166)

systematic random sample (167)

stratified sample (167)

cluster sample (169)

sampling distribution of the mean (172)

sampling distribution (172)

standard error of the mean (173)

central limit theorem (176)

finite correction factor (fcf) (180)

EXERCISES

6.15 For each of the following examples, define the population of interest and specify whether a census or a sample would be more appropriate. Explain your answers.

a. In order to estimate the durability of a new line of luggage, the company conducts a series of tests, including dropping the luggage from varying heights onto a concrete pad and hitting it with various size sledge hammers on the top and bottom surfaces and on each of the four sides.

b. A medical researcher wishes to investigate the relationship between sexual relations during the last month of pregnancy and the incidence of illness and infection in the newborn infant.

c. In two months, voters will be asked to decide on a referendum that specifically sets up separate areas for smoking and nonsmoking in all public places. Suppose you want to predict the outcome of this vote.

d. The automobile industry wishes to conduct a consumer survey of all registered automobile owners in order to determine when they intend to purchase a new automobile.

6.16 For each of the following, specify the population of interest and comment on any possible sources of error in the selection of the sample or collection of the data.

a. A bank is interested in determining the average monthly checking account balance of its customers. For convenience, it selects its sample from among those accounts that show little activity over the course of the month.

b. A committee consisting of three people, all members of a minority group, conduct a survey of the effects of a new affirmative action program within a specific industry.

c. A union is involved with direct negotiations with management. Management presents its offer to union officials. In order to assess union members' reaction to the proposed contract, members are asked, "Will you vote to accept the wage increases management is offering, even though they are less than those we proposed?"

d. A questionnaire, prepared to study attitudes on safety measures in a particular firm, is mailed to all employees of that firm. The questionnaire is received the day after a major accident at one of the plants.

6.17 Jerry J. is a manufacturer's representative who primarily calls on general merchandise suppliers to direct salesmen. This is a fairly specialized (albeit high-volume) market and there are only about 50 accounts in this category. Whenever he receives a new item, he must decide whether or not to include it in his current line of merchandise. He decides to accept an item only if he can project a minimum volume (μ) of $50,000. He figures that his 10 percent commission rate for this minimal volume will just about cover his costs. The usual standard deviation (σ) associated with the total volume is $1,000.

Jerry decides to sample some of his accounts in order to estimate the mean total volume on a new item. He wants his estimate to fall within ±$500 of μ.

What is the probability that the estimate will meet the desired accuracy if, he samples:

    a. Five accounts?
    b. Ten accounts?
    c. Twenty-five accounts?

6.18 Refer to Exercise 6.17. Based on sampling only five accounts, Jerry finds the probability that he can estimate his total volume to within ±$500 is almost 80 percent. He is satisfied since sampling only five accounts is relatively easy for him.

Find the probability that his estimate will meet the desired accuracy if his population of accounts increased from 50 to

    a. 100.
    b. 500.

(Assume the same $\sigma = \$1,000$).

6.19 Repeat Exercise 6.18, this time using $n = 10$ and

    a. $N = 100$.
    b. $N = 500$.

6.20 What kind of error can be minimized by increasing the sample size? What kinds are unaffected by increased sample size?

6.21 The operations officer at one of the Bank of America branches finds that traffic is lightest during the morning hours from 10 to 11:30 A.M. and heaviest during lunch-time from 12 to 1:30 P.M. She finds that waiting time is normally distributed with $\mu = 3.5$ minutes and $\sigma = 0.75$ for the morning hours, and $\mu = 7.5$, $\sigma = 1.5$ for lunchtime hours.

She now wants to see whether day of the week is a factor. Random samples of $n = 20$ are taken for the two time periods for different days of the week.

If we assume that the above results are the true population parameters, what is the probability that:

    a. The morning sample will differ by more than a half-minute from the original mean?
    b. The lunch-time sample will differ by more than a half-minute from the original mean?

Explain why the probabilities differ in *(a)* and *(b)*.

6.22 Give examples of situations in which a sample would be preferable to a census because of the following reasons:

    a. Population too large.
    b. Costs too large.
    c. Timeliness.
    d. Process involves destruction.
    e. Inaccessibility of members of the population.

6.23 Give examples of situations in which the following types of error could occur in the selection of the sample or the collection of the data:

    a. Bias leading to the nonrepresentative sample.

b. Sampling error due to uncontrolled factors.

c. Nonsampling error due to the way the observations are made.

6.24 Comment: The Labor Department reported a drop in wholesale prices for September 1980. This was the first time in more than four years that this index has dropped. However, Labor Department officials revealed that the decline resulted from a change in the way the index was computed. For the first time, discounts given on cars and trucks at the end of a model year were included. Without this change, there would have been a 0.4 percent increase in wholesale prices instead of the reported 0.2 percent decrease.

The Commissioner of labor statistics said that the change stemmed from a decision "to use the actual prices paid to producers for their finished goods."[6]

6.25 Comment: In September of 1980, the jobless rate in Los Angeles County "surged" to 7.5 percent. A spokesman for the California Employment Development Department attributed the rise in the county rate to "strikes in the motion picture and construction industries and to lagging manufacturing production."[7] The county rate, unlike the national and state figures (to which it is normally compared) is *not* adjusted to reflect seasonal changes.

6.26 In the following situations, indicate the kind of random sampling procedures that should be used. Comment on your choice.

a. A credit card company wishes to estimate the average number of times their customers use the card over the course of a month.

b. The Air Quality Management District wants to compare the ozone concentrations in the air during the summer and winter months. They are interested in sampling from a variety of areas and at different times of day.

c. A supermarket is interested in assessing consumer reaction to a new pricing system. Because of a limited budget, the sample will be obtained from interviewing customers at the store.

6.27 Unpublished authors often glue pages of their manuscript together to learn if the entire manuscript has been read. One frustrated author accosted the editor who had just rejected his manuscript. "I glued pages 8, 9, and 10 together. They came back still glued together. Now I know how carefully you read manuscripts from unknown authors."

The editor replied, "I don't have to smell a whole limburger cheese to know it stinks."

Is this a fair comment?

6.28 Comment: The vice president of a gasoline additive company states: "Letters from our satisfied customers prove that you can expect to get up to 30 percent better mileage by using our additive with carbuferate."

6.29 Comment: A school board wishes to obtain voter sentiment on plans to construct a new elementary school in the district. Each child is given a questionnaire to take home to his or her parents.

6.30 Comment: The state agricultural department wishes to estimate the cotton yield on farms stretched along the Colorado River on the western border of Arizona.

---

[6] Source: *Los Angeles Times*, October 4, 1980.

[7] Ibid.

Since members of the Colorado River Indian tribes own most of the land and, in turn, lease it to private bidders, the tribal council is asked to conduct the survey.

6.31 A direct-selling organization finds itself experiencing the usual problems associated with this type of business—the extremely high turnover rate. The mean time employees spend on the job is 82 days ($\mu = 82$) with a standard deviation of $\sigma = 15$.

In order for the organization to realize a profit, employees must stay at least 75 days. If they stay less than 60 days, the company loses money. New employees are hired. What are the probabilities that the company will make money or lose money on this group if

    a. $n = 10$?

    b. $n = 25$?

6.32 Another direct-selling organization, run in a slightly different way, (Exercise 6.31), finds that employees, before they move on, spend a mean time of 75 days ($\mu = 75$), with the same standard deviation, $\sigma = 15$.

The same time standards apply for profit and loss in this organization.

What are the probabilities that this company will make or lose money on a new group of employees if:

    a. $n = 10$?

    b. $n = 25$?

6.33 Suppose you were interested in comparing the wages of government employees in a certain classification (e.g., data processing analysts) with their counterparts in private industry. Assume all the data from these populations are available to you.

Suggest how you might go about obtaining the following kinds of samples:

    a. Simple random.

    b. Systematic random.

    c. Stratified.

    d. Cluster.

6.34 A construction firm wishes to sample the durability of a new, fairly expensive line of tools, to be sure these tools can stand up to all kinds of stresses and still come out "passable" in three months. The employees currently pay for their own tools through salary deductions. The firm wishes to make sure the tools are all paid for before new ones are needed. They have determined that a pay-out period of 12 weeks is hardly noticed by the employees. However, if the tools need to be replaced before the old ones were paid, this could create problems.

    a. They sample 36 of the tools and find the mean time under the "passable" criterion $(\overline{X})$ with $s = 10$.

        1. Suppose they want to be 95 percent sure that the tools will last at least 90 days. What value of $\overline{X}$ would satisfy this condition?

        2. Suppose they want to be 99 percent sure; what value of $\overline{X}$ would satisfy this condition?

    b. Suppose they increase their sample to $n = 100$. Answer (1) and (2) above.

One of the ways to increase the precision of the obtained results would be to increase the sample size. Can you suggest any reasons why the firm might be reluctant to do this?

6.35 Beverly, the owner of a restaurant, reviews her financial statement and decides she is not satisfied with the profits she is making. She knows she can raise her prices, but would rather avoid going this route.

Suppose she calls an independent group in to study the problem. They identify one area of major concern—her turnover rate at the tables. They know that restaurants are geared for different turnover rates—from the extreme of one seating per table an evening, to the fast-food places.

After consulting various studies on similar situations, they find that similar restaurants have achieved a minimal turnover rate of three seatings every two hours. The profitable restaurants in this category show a normal distribution of table-time with $\mu = 35$ minutes and $\sigma = 9$ minutes.

They decide to conduct a sampling study in order to estimate the mean time customers spend at the tables at Beverly's restaurant.

    a.   Suggest some sampling techniques they might employ. Discuss your choices.

    b.   Suppose they sample 100 customers and the mean table-time $\overline{X}$ is found. What is the probability that $\overline{X}$ will

        1.   Fall at or below 34 minutes?

        2.   Lie between 30 and 40 minutes?

        3.   Lie between 36 and 37 minutes?

        4.   Fall above 40 minutes?

    c.   By decreasing the clean-up time between customers, the optimal goal of three seatings every two hours can be achieved if $\mu = 38.5$ and $\sigma = 9$. Recalculate the four probabilities in *(b)*.

Estimation

7.1   POINT AND INTERVAL ESTIMATION

Montana Rincon's investigations had carried him to the point where he was convinced he could build a good prima facie case for collusion to commit fraud. Now he wished to estimate the extent of the alleged crime. He began to ask himself some far-reaching questions—questions that required answers that went beyond the sample statistics themselves. "What is my best estimate of the true proportion of falsified accounts in the population of 5 million accounts? What are the largest and the smallest proportions that would seem reasonable in the light of the observed sample proportion? Among the accounts that contain errors, what is my best estimate of the mean error of the population? Within what interval is it reasonable to estimate that the population mean falls?"

These questions carried Montana to the core of statistical estimation. Recall that he had originally selected, at random, 1,000 accounts from a population of 5 million. He calculated the proportion in the sample that contained errors. He found this to be 60/1,000 or 0.06. He also calculated the mean error in the 60 accounts containing errors. Now he wanted to obtain *estimates* of parameters, using the sample statistics as *estimators*.

In this he was not unlike any intelligent planner and decision-maker. Consider the following examples:

> The convention manager of a large premium show needs to estimate the average area required by each exhibitor.
>
> The executive compensation director of an aerospace firm wishes to investigate the possibility of using travel as an incentive to increase productivity. She wants to estimate the mean cost per trip (for two people).
>
> A survey by American Management Associations reported that traveling sales reps are costing their companies 12 percent more this year. The sales manager of a firm with numerous traveling reps must estimate this year's travel budget in order to determine whether cut-backs are required.
>
> A flour mill receives many orders for large sacks of flour from its customers. To avoid the cost of individually weighing each sack, the manager decides to bill on the basis of an estimate of the mean weight per sack.

In each of these examples, decision makers are interested in estimating a particular parameter and will take certain actions based on the value of that parameter. When we wish to obtain an estimate of a parameter, we must rely on statistics calculated from samples. For example, to estimate the mean weight of the flour sacks we could randomly select a sample from the warehouse, and use the sample mean as our estimator.

There are two kinds of estimates—*point estimates* and *interval estimates*. A point estimate is a single number that is calculated on the basis of sample data. For example, suppose we obtained $\overline{X} = 250.3$ for the mean weight of the flour sacks. With point estimates such as this, we have no idea how closely our estimate approximates the true mean. That is, it is impossible to tell whether the true mean is very close to 250.3 or far away from this estimate.

estimate

The numerical value of the statistic used to estimate a parameter.

estimator

A statistic used to estimate or make inferences about a parameter. Estimators include sample means, variances, standard deviations, and proportions.

point estimate

A single numerical value of a statistic used to estimate a parameter.

interval estimate

A range of values within which the parameter presumably lies.

On the other hand, interval estimates do provide such information. We can construct an interval estimate so that we have a specified amount of confidence that this interval will contain the desired parameter. Moreover, we can attach a specific probability value to this confidence. For example, we could construct an interval estimate of the mean weight per flour sack and establish a probability level of, say, 95 percent that this interval will include the population mean weight.

This procedure of estimating intervals is quite familiar to all of us. If you are asked to estimate a person's weight, what do you say? Usually, you provide a range of values. You might say, "Around 200 pounds, give or take 10." The 200 pounds is your point estimate and the "give or take 10" is your interval estimate. Your interval estimate says, in effect, that you would not be at all surprised if the person's weight were really as low as 190 or as high as 210. This is really little different from the statement by the chairman of the board, "We expect profits to be up between 5 and 8 percent over the last quarter." Or the economic advisor to the president who announces, "By the end of the year, we expect the rate of inflation to be somewhere between 7 and 8 percent." The boundary values to these interval estimates are saying two things: we feel confident that the true value is within these limits and we doubt that the parameter is less than the lower limit or higher than the upper limit.

A word of caution is in order. The population parameter is *not* a random variable—it is a constant. Therefore, it makes no sense to say that the probability is 95 percent that the population parameter has a value between the upper and lower limits of the interval. Either it is in an interval or it is not. If repeated samples of a given size are drawn from a population, 95 percent of the interval estimates will include the population mean.

7.1.1   Criteria in choosing an estimator

When we randomly select a sample from a given population, there are a variety of statistics that may be calculated. How do we decide which of these statistics to use as an estimator of the parameter of interest? For example, if we want an estimate of the mean weight of the flour sacks, we could consider any measure of central tendency, such as the mean, median, or mode. Our choice will depend on which estimate is closest to the population parameter. However, since we don't know the value of the parameter, how do we decide which estimator, on the average, will be closest? We find the answer to this question by examining the sampling distribution of each of these statistics. In short, we seek answers to three questions: Is it an unbiased estimator? Is it efficient? Is it consistent?

Lack of bias.   In Chapter 6 we discussed a type of bias that results from selecting a nonrepresentative sample from the population of interest. When we talk about an *unbiased estimator,* however, we are alluding to a somewhat different type of bias. Even if the sample is randomly selected and is representative of the population, some estimators do not yield estimates that are, on

unbiased estimator

An estimator which equals, on the average, the value of the parameter.

the average, equal to the population parameter. For example, we previously defined, in Formula (3.4), the variance of a sample as

$$s^2 = \frac{\Sigma(X - \bar{X})^2}{n-1}$$

It has been mathematically proven that this provides an unbiased estimate of the population variance since, on the average, the mean of the sampling distribution of s^2 will equal σ^2. However, had we used n in the denominator instead of $n - 1$, we would have a biased estimator. On the average, $\Sigma(X - \bar{X})^2/n$ would underestimate the population variance.

The sample mean is an unbiased estimator since, as we saw in the previous chapter, the expected value of the mean equals the mean of the sampling distribution, which, in turn, equals the population mean (i.e., $E(\bar{X}) = \mu_{\bar{X}} = \mu$).

efficient estimator

One estimator is said to be more efficient than another when the variability of its sampling distribution is less.

Efficiency. In addition to being unbiased, an estimator must also be *efficient*. This means that it has a smaller variance and standard deviation than other estimators. It follows, then, that a good estimator should have a higher probability of being close to the population value as the sample size increases. Compare the sampling distributions in Figure 7.1. If we were to select a single sample mean from each sampling distribution, which one would probably be closer to the population mean? Since B shows less variability, we would expect sample means drawn from this distribution to be closer to the population mean, on the average, than those drawn from distribution A. Thus, sample means drawn from sampling distribution B would be classified as "more efficient" than those drawn from A.

Figure 7.1

Two sampling distributions of means with identical means but differing in variability. The standard error of the mean in B is less than in A. Therefore, sample means from B provide more efficient estimates of the population mean.

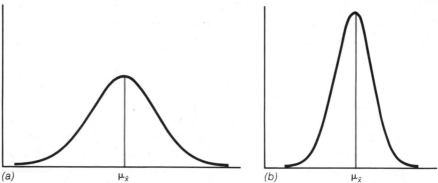

(a) $\mu_{\bar{x}}$ (b) $\mu_{\bar{x}}$

Consistency. The final criterion of a good estimator is whether or not it is *consistent*. As we just saw, the lower the variability in the sampling distribution, the more likely is the estimate to approach the parameter. We have previously seen that the variability of the sampling distribution of the mean, $\sigma_{\bar{x}}$, is dependent upon the variability in the population and the size of the sample, i.e., $\sigma_{\bar{x}} = \sigma/\sqrt{n}$. Thus, as we increase n, we decrease the value of the standard error of the mean. In other words, by increasing n, we also increase the consistency of the estimate. However, as Figure 7.2 shows, increasing n does not affect consistency in a linear fashion. In fact, beyond a certain point, large increases in n are necessary in order to bring about relatively minor decreases in the standard error of the mean. Since increasing sample size often has economic implications, we are usually faced with striking a balance between increased consistency versus increased costs.

consistent estimator

If an estimator is consistent then as you increase n, you increase the probability that the estimate approaches the value of the parameter.

The relationship between increasing sample size and decreasing standard error of the mean in a population where $\sigma = 10$. Note that this is a curve of diminishing returns. Beyond a certain point, increasing n has relatively little effect on the standard error of the mean.

Figure 7.2

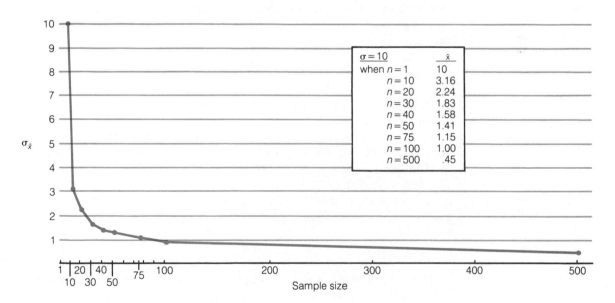

$\sigma = 10$	$\bar{x}$
when $n = 1$	10
$n = 10$	3.16
$n = 20$	2.24
$n = 30$	1.83
$n = 40$	1.58
$n = 50$	1.41
$n = 75$	1.15
$n = 100$	1.00
$n = 500$	.45

7.2 INTERVAL ESTIMATION OF THE MEAN

Both point and interval estimates, used in concert, play important roles in the decision-making process. For example, when trying to establish production schedules for the manufacture of new cars, Chrysler Corporation must obtain an estimate of the most likely dollar volume that will be ordered within a given period. Beyond that, however, it is imperative that the range of possible error also be estimated. Thus, the estimate might read as follows: The mean dollar volume per month for cars ordered over the coming quarter is $72 million, give or take $3 million. In other words, the true dollar volume per month is expected to fall between $69 million and $75 million. If the parameter (the actual dollar volume) falls short of this interval, the financial consequences may be quite serious. If it falls within the interval, the company is well prepared to handle the many factors involved in the timely production and delivery of automobiles ordered. Planning proceeds on the basis of the assumption that the parameter (the actual dollar volume) will be in that interval. Good management practices will assure a favorable financial position as long as the parameter is not, in reality, below the lower bound of the interval. Clearly, if the estimate turns out to be incorrect and the dollar volume is actually below the lower bound, the financial consequences could be considerable, depending on the extent of the error.

There are two main objectives when establishing an interval estimate: *(a)* to make it as small or as precise as possible while retaining a high degree of confidence that the interval actually includes the desired parameter, and *(b)* to provide a probability value associated with our confidence that the interval encloses the parameter.

The interval estimate is called the *confidence interval*. The upper and lower boundaries constitute the *confidence limits* and the probability value associated with this interval is our *confidence coefficient*.

The general formula for a confidence interval when σ is known and the sample is large is

$$\overline{X} \pm z\sigma_{\overline{X}} \tag{7.1}$$

in which z is the value that includes the desired proportion of area surrounding the mean.

To illustrate, since 95 percent of the area under a normal curve falls between $\pm 1.96\ \sigma_{\overline{X}}$, the 95 percent confidence interval is $\overline{X} \pm 1.96\ \sigma_{\overline{X}}$.

If we wish to obtain the 99 percent confidence limits, we substitute $z = \pm 2.58$: $\overline{X} \pm 2.58\ \sigma_{\overline{X}}$.

If σ is not known but the sample is large, we define the confidence interval as

$$\overline{X} \pm z\, s_{\overline{X}} \tag{7.2}$$

Thus, the 95 percent and 99 percent confidence intervals when σ is unknown and n is large becomes

$$95 \text{ percent interval} = \overline{X} \pm 1.96\ s_{\overline{X}}$$
$$99 \text{ percent interval} = \overline{X} \pm 2.58\ s_{\overline{X}}$$

When we define the confidence limits, we are saying that 95 (or 99) percent of the time, μ will lie within the obtained interval. We are *not* saying that the probability is 95 (or 99) percent that μ is equal to a value between the upper and lower limits. Remember that μ is a constant. It is either in the specified interval or it is not. Figure 7.3 illustrates this point.

7.2.1 Interval estimate of the mean—Large samples

A flour mill receives numerous orders for large sacks of bleached flour from its customers. Because of the high costs of labor each sack is not individu-

confidence interval

Interval which has a certain probability of containing the population parameter.

confidence limits

The upper and lower boundaries of the confidence interval.

confidence coefficient

The probability that the confidence interval will contain the population parameter. The most commonly used confidence coefficients are 0.95 and 0.99.

Figure 7.3

Estimating the confidence interval of the mean from a number of samples of a fixed size drawn from the population. Note that the sample means are variable as are the confidence intervals. In fact, each sample yields a different confidence interval. The population mean, μ, is fixed. In this illustration, all except one of the confidence intervals include the population mean within their limits. In actual practice, when the 95 percent confidence interval is used, 95 percent of the confidence intervals will include the population mean. Similarly, 99 percent of the intervals will enclose the population mean when the 0.99 confidence coefficient is used.

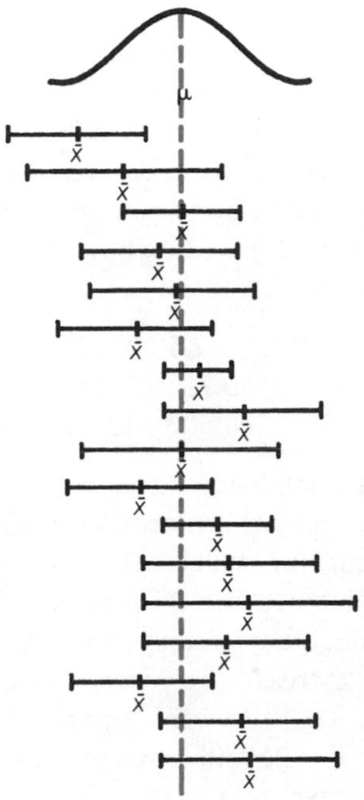

ally weighed. Rather they are filled by automated equipment set to deliver approximately 250 pounds. For billing purposes, it is decided to randomly select 50 sacks from the warehouse, weigh them, and then use the mean as the best estimate of the average weight per sack. The results are as follows:

253	231	246	247	250
247	254	251	253	253
250	249	254	252	250
249	250	249	248	251
251	246	248	251	247
250	250	253	251	253
252	251	250	247	251
248	253	251	250	249
254	249	247	253	252
253	248	250	249	251

$$\bar{X} = 250.3; \quad s^2 = 4.8265; \quad s = 2.1969.$$

Suppose we wish to estimate the 95 percent confidence interval. Estimated 95 percent confidence interval is:

$$\bar{X} \pm 1.96 \, s_{\bar{X}}$$
$$= 250.30 \pm (1.96)\left(\frac{2.20}{\sqrt{50}}\right)$$
$$= 250.30 \pm (1.96)(0.31)$$
$$\text{or } 249.69 \text{ to } 250.91$$

Thus, the probability is high that the interval 249.69 to 250.91 contains the true mean weight of the plant's output.

If we wanted an even higher degree of confidence that we have successfully bracketed the population mean, we could use a more demanding confidence coefficient. For example, if we wished to establish the 99 percent confidence interval, we substitute $z = \pm 2.58$ in the above formula. Thus, the estimated 99 percent confidence interval is

$$\bar{X} \pm 2.58 \, (0.31)$$
$$= 250.30 \pm 0.80$$
$$= 249.50 \text{ to } 251.10$$

Note that the 99 percent confidence interval is wider than the 95 percent confidence interval. In general, the more demanding we are in our choice of the confidence coefficient, the wider will be the resulting interval.

Imagine that, having obtained the 99 percent confidence interval, we are not satisfied that it is sufficiently precise, i.e., the width of the interval is too large. How might we increase the precision of the estimate? In a word, increase the sample size. To illustrate, suppose that we had used 100 as the sample size, obtaining $\bar{X} = 250.30$ and $s = 2.20$. The estimated 99 percent confidence interval would then be

$$\bar{X} \pm (2.58)\left(\frac{2.20}{\sqrt{100}}\right)$$
$$250.30 \pm (2.58)(0.22)$$
$$249.76 \text{ to } 250.87.$$

Thus, by doubling the sample size, the width of the confidence interval has been narrowed from 1.60 to 1.11.

Example A: Point and interval estimate of total quantity

Often our interest is directed toward finding both a point estimate and an interval estimate of the total quantity of the variable of interest. For example, the manager of the flour mill may wish to estimate both the total number of pounds of flour processed within a month and the confidence interval that includes the true total.

Suppose that, during a period of one month, the mill processes a total of 6,000 sacks of flour.

> *a.* What is the estimated total weight that will be processed within a month?
>
> *b.* What is the 99 percent confidence interval for the total weight?

Solution

a. The point estimate of weight per sack was 250.3. To estimate the total weight processed during a given period of time, multiply the point estimate by the total number of sacks filled during that time. In the example, estimated total output is $(250.30)(6,000) = 1,501,800$ lbs. or, dividing by 2,000, it is 750.9 short tons.

b. We have previously obtained the 99 percent confidence interval of 249.50 to 251.10. Multiply each limit by 6,000 to obtain the 99 percent confidence interval of the total output. Thus, 99 percent confidence interval for total weight $= (6,000)(249.50)$ to $(6,000)(251.10) = 1,497,000$ to $1,506,600$ pounds. Expressed in short tons, the interval is 748.5 to 753.3.

7.2.2 Interval estimation of the mean: *t*-distributions

We have seen that when σ is known, we may use the standard normal curve as the basis for establishing interval estimates of the mean. This is also true for large sample sizes because of the central limit theorem. Thus, we are able to use the sample standard deviation in calculating our estimated standard error of the mean. What do we do when σ^2 is estimated on the basis of sample data and the sample size is small, i.e., $n < 30$? At the turn of the century, William Gosset, who published under the name Student, observed that the approximation of s to σ is poor for small samples. In fact, with small samples, s will tend to underestimate σ more than half the time. Student described a family of symmetrical distributions, called *t-distributions,* that permit interval estimates when σ is unknown and the sample size is small. Although generated by mathematical rules that are different from those governing the normal curve, the *t*-distributions assume that the population distribution is normal. Like the normal curve, they are unimodal and bell-shaped. However, they are more spread out in the tails and slightly flatter in the middle. Each of these distributions varies as a function of n or, more specifically, *degrees of freedom (df).* Let's examine the concept of degrees of freedom in more detail.

The term *degrees of freedom* refers to the values of the variable that are free to vary once we have placed certain restrictions on our data. To illustrate, let us suppose that we have four numbers upon which we have placed the following restriction: their sum must equal 20. Note that we could select the first three numbers out of a hat, if we wanted, but the last number would not be free to vary. It would be fixed by the three numbers previously drawn and by the restriction that the sum must equal 20. Thus, if the three numbers are 7, 5, and 6, the last number *must* be 2. Thus, the restriction has limited the number of values that are free to vary. In fact, we may

t-distributions

Theoretical sampling distributions with a mean of zero and a standard deviation that varies as a function of sample size.

degrees of freedom (df)

The number of values that are free to vary after certain restrictions have been placed on the data.

generalize: For a sample on which we have placed a single restriction, the number of degrees of freedom (df) is equal to $n - 1$. Hence, if our sample size is 16, df $= 15$. Figure 7.4 shows how the t-distributions vary as a function of degrees of freedom.

Figure 7.4

Proportion of area beyond $z = \pm 1.96$ in various t-distributions and the standard normal curve (i.e., df $= \infty$). Note that as df increases, the proportion of area approaches that of the standard normal curve, i.e., 0.05. In fact, when df equals infinity, there is no difference between the two.

In the t-distributions for df $= 3$ and df $= 12$, note the t-value required to mark off a total of 0.05 of the area.

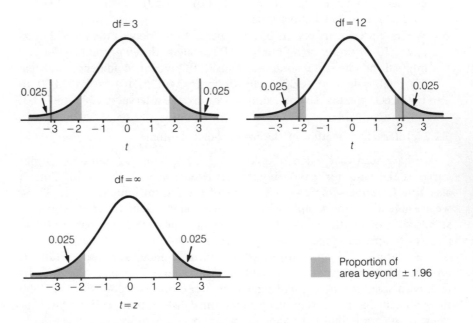

The table of values of the t-distributions (Table VII) is quite different from Table V, which shows areas of the standard normal curve. Ideally, there should be a separate table for each number of df, showing the areas between the mean and various values of t. However, to do so would produce an excessively lengthy set of tables. Instead, Table VII shows the values of t with related proportions of area beyond that value.

The row at the bottom of Table VII (Appendix C) shows selected confidence coefficients. To find the t corresponding to a given coefficient, we move up the column until we locate the tabulated value corresponding to the appropriate degrees of freedom. For example, if df $= 15$, the t-value corresponding to the 0.95 confidence coefficient is 2.131.

The formula for a confidence interval when σ is unknown and the sample is small is

$$\bar{X} \pm t_{df}s_{\bar{X}} \qquad (7.3)$$

where t_{df} is the value of t for a given number of degrees of freedom.

Example B: Interval estimation of the mean, small samples

In a time and motion study of a single stage in an assembly line operation, a random sample of 10 time scores yielded the following values, in seconds.

128	150
135	125
105	132
111	117
142	124

$$\bar{X} = 126.90$$
$$s = 13.73, \text{df} = 9$$

Calculate the point estimate of the population mean and the 95 percent confidence interval.

Solution:

$$\Sigma X = 1269;$$

$$\bar{X} = \frac{1269}{10} = 126.9$$

$$s_{\bar{X}} = \frac{13.73}{\sqrt{10}} = 4.34$$

Thus the 95 percent confidence interval is

$$\bar{X} \pm (t_{df})(4.34)$$
$$= 126.90 \pm (2.262)(4.34)$$
$$= 117.08 \text{ to } 136.72$$

The probability is high that the confidence limits bracket the population mean number of seconds.

Example C: Point and interval estimation of the total, small samples

Referring to Example B, during the course of a week, 1,095 items are produced.

a. What is the total estimated time that this stage will require each week?

> *b.* What is the 95 percent confidence interval for the total time required for this stage?

Solution

a. We multiply the mean estimated time by the number of items produced. Thus, our best estimate is that the stage will consume $(1,095)(126.9) = 138,955.5$ seconds or 38.60 hours.

b. We multiply each confidence limit by 1,095 to obtain the 95 percent confidence interval for the total time required by this stage. Thus, 95 percent confidence interval for total $= (117.08)(1,095)$ to $(136.72)(1,095)$ seconds $= 35.61$ to 41.59 hours.

Therefore, we estimate that the average time required for the process, μ, falls in the interval 35.61 to 41.59. The 95 percent confidence coefficient implies that, with repeated sampling from the population under similar conditions, 95 percent of the confidence intervals would bracket the population mean.

7.2.3 Interval estimation of the mean—Small populations

The procedures are the same as finding point and interval estimates for large or infinite populations. However, if the sample size is 5 percent or more of the population size, the finite population correction factor should be used when finding the estimated standard error of the mean $(s_{\bar{x}})$. Let's look at an example.

Example D

A total of 250 cattle were moved by train from the ranch to feeding lots. To obtain an estimate of the weight of the shipment, 29 head were randomly selected and weighed. The results were as follows:

1350	1463	1644
1808	1705	1404
1475	1395	1752
1735	1501	1492
1526	1480	1540
1242	1490	1510
1495	1585	1475
1520	1523	1429
1409	1475	1525
1610	1504	

Find:
a. The point estimate for the population mean.
b. The 95 percent confidence interval.

c. The point estimate for the total weight.
d. The 95 percent confidence interval for the total weight.

Solution

a. The point estimate for the population mean is

$$\bar{X} = \frac{44,062}{29} = 1,519.39$$

b. The standard deviation is 121.77. The uncorrected standard error of the mean is $121.77/\sqrt{29} = 22.61$. When multiplied by the finite correction factor:

$$s_{\bar{X}} = 22.61 \sqrt{\frac{221}{249}} = 21.30$$

The 95 percent confidence interval of the mean is

$$\bar{X} \pm (t_{df})(21.30) = 1519.39 \pm (2.048)(21.30)$$
$$= 1475.77 \text{ to } 1563.01$$

c. The point estimate of the total weight is $(250)(1,519.39) = 379,847.5$ pounds, or 189.92 short tons.
d. The 95 percent confidence interval of the total weight is $(250)(1,475.87)$ to $(250)(1,562.91) = 368,967.5$ to $390,727.5$. Expressed in short tons, the interval estimate is 184.48 to 195.36.

EXERCISES

7.1 Which of the following is an estimate and which an estimator?
    a. A sample mean.
    b. A sample standard deviation equals 4.63.
    c. The sample median is 47.62.
    d. A sample median.
    e. The sample standard error of the mean is 1.75.
    f. The standard error of the mean.

7.2 Is an estimate or an estimator a random variable?

7.3 Do any of the following illustrate lack of bias, efficiency, or consistency? If so, indicate which.
    a. The mean of a sampling distribution of standard deviations equals the population standard deviation.
    b. The sampling distribution of the median has a standard error of 6.15, compared to the sampling distribution of the mean with a standard error of 4.95.

    c. A sample mean of $n = 10$ deviated from the population mean by 6.23. Another mean, based on $n = 50$, deviated from the population mean by 3.10.

7.4 Explain circumstances under which point and interval estimates of the total quantity may be more useful than point and interval estimates of the mean.

7.5 The convention manager of a large premium show needs to estimate the average area required by each exhibitor. He takes a random sample of 100 exhibitors at similar shows and finds $\bar{X} = 213.75$ square feet with $s = 37.5$.
    a. Find the 95 percent confidence limits for μ.
    b. Find the 99 percent confidence limits for μ.
    c. What is the effect of the increase in the level of confidence?

7.6 Refer to Exercise 7.5. Suppose the convention manager decides to sample 400 exhibitors. Assume he obtains the same $\bar{X}$ and s. Find the
    a. 95 percent confidence limits for μ.
    b. 99 percent confidence limits for μ.
    c. Compare your results with those obtained in Exercise 7.5 for $n = 100$. What is the effect of increasing sample size?

7.7 Swap meets have become extremely lucrative for many merchants who view them as a steady business, supplying an integral part of their income. One enterprising young merchant decides to see for himself just how profitable this business can be. He randomly selects steady merchants and finds that their mean daily income, $\bar{X} = \$450$, $s = \$110$. Find the 95 percent confidence limits for μ if
    a. $n = 16$.
    b. $n = 25$.
Comment on the effects of increased sample size.

7.8 Refer to Exercise 7.7 Assume the same $\bar{X} = \$450$, $s = \$110$. Find the 99 percent limits for
    a. $n = 16$.
    b. $n = 25$.
Compare your results with those obtained in Exercise 7.7. What is the effect of the increase in the level of confidence?

7.9 Transit worker absenteeism is becoming a problem in many cities. In a city with 3,000 transit workers, a municipal employee decided to study the number of days bus drivers stayed home over the past few years in an attempt to predict absenteeism for the coming year. Taking a sample of 400, she found that transit workers' absences increased yearly 17.5 percent with $s = 3.0\%$. If the number of days absent per worker this year is 18, find the 95 percent confidence limits for
    a. The number of days absent per worker next year.
    b. The total number of days of absence next year.

7.3 INTERVAL ESTIMATION OF THE DIFFERENCE BETWEEN INDEPENDENT MEANS

There are numerous occasions when we wish to evaluate the effectiveness of some change of procedures on an outcome of interest. For example, we may wish to learn the answers to such questions as: Does a new additive affect the miles-per-gallon performance of automobiles? Will a new filament design extend the life of incandescent bulbs? Does a new tread design affect the wear characteristics of tires? Does a change in personnel practices lead to improvement in productivity?

One way to answer questions of this sort is to set up two experimental conditions, one reflecting the changed procedure and the other maintaining the status quo. We wish to obtain a point estimate of the difference between population means ($\mu_1 - \mu_2$) and find the confidence limits at a given level of confidence.

Recall that, when finding a point estimate and the confidence interval of a mean, we conceptualize the sampling distribution of the mean as the probability distribution of all sample means, of a fixed size, drawn from a given population. Similarly, when dealing with the sampling distribution of the difference between means, we conceptualize the selection of *pairs* of samples of a fixed size. We then obtain the difference between sample means. The sampling distribution would consist of a probability distribution of all *differences* between means. The mean of this sampling distribution would equal the difference between the population means from which the samples were drawn, i.e., $\mu_1 - \mu_2$. If the samples were drawn from populations with the same mean (or the same population), the mean of the sampling distribution is zero. The *standard error of the difference between means* (the standard deviation of the sampling distribution) is estimated by pooling the two sample variances. Thus,

standard error of the difference between means

The standard deviation of the sampling distribution of differences between sample means.

$$s_{\bar{X}_1 - \bar{X}_2} = \sqrt{\frac{n_1 s_1^2 + n_2 s_2^2}{n_1 n_2}} \qquad (7.4)[1]$$

in which

$s_{\bar{X}_1 - \bar{X}_2}$ is the estimated standard error of the difference between means.
s_1^2 is the sample variance of one sample.
s_2^2 is the sample variance of the other sample.

When n_1 and n_2 are equal, the formula simplifies to

$$s_{\bar{X}_1 - \bar{X}_2} = \sqrt{\frac{s_1^2}{n_1} + \frac{s_2^2}{n_2}} \qquad (7.5)$$

Because of the central limit theorem, we may use the standard normal curve as long as the n in each sample equals or exceeds 30. However, the use of the pooled estimate requires that the sample variances not be greatly different from one another.

---

[1] Formula 7.4 uses the unbiased estimator of the sample variance $\left(\text{i.e., } s^2 = \dfrac{\Sigma (X - \bar{X})^2}{n-1}\right)$.

7.3.1 Interval estimation of the difference between means—Independent large samples

The research department of United Electric claims that it has developed a new filament that increases the life of an incandescent light bulb. Forty old-style and 40 new-style bulbs were randomly selected from inventory. Since the selection of items for one sample was not dependent on the selection of items for the second sample, the two samples are said to be independent. The bulbs were burned continuously under controlled laboratory conditions. The length of time each bulb burned was recorded. The results are summarized below.

	Sample 1 (new filament)	Sample 2 (old filament)
Mean	742	667
Standard deviation	125	138
n	40	38

The point estimate for the difference between population means ($\mu_1 - \mu_2$) is $\overline{X}_1 - \overline{X}_2 = 742 - 667 = 75$. Since the sample sizes are equal, we may use Formula 7.5. Thus, the estimated standard error of the difference between means is

$$s_{\overline{X}_1 - \overline{X}_2} = \sqrt{\frac{(125)^2}{40} + \frac{(138)^2}{40}}$$

$$= 29.44$$

The 95 percent confidence interval becomes

$$(\overline{X}_1 - \overline{X}_2) \pm (1.96)(29.44) = 75 \pm 57.70$$
$$= 17.30 \text{ to } 132.70$$

The chances are high that these limits encompass the true population difference between means. Note that we may also draw another inference: It appears that the bulbs with the new filament enjoy a longer mean life since a zero difference between means is not included within the 95 percent boundaries. Chapter 8 takes up this aspect of statistical inference in greater detail.

7.3.2 Interval estimation of the difference between means—Independent small samples

When degrees of freedom are small (less than 30), the Student t-distribution should be used. In two group experiments, the number of degrees of freedom is $n_1 + n_2 - 2$. Let's look at an example.

The U.S. Department of Agriculture wished to evaluate the effectiveness of a wide-spectrum pesticide on the production, per acre, of cotton. Thirty plots, of two acres each, were randomly selected, and 15 of them were randomly assigned for treatment by the new pesticide. The remaining 15 plots received standard pesticide control procedures. The results were as follows:

	Cotton production, pounds per acre	
	Sample 1 (new pesticide)	Sample 2 (standard pest control)
$\bar{X}$...........	1,425	1,395
s	200	195
n	15	15

The point estimate for the difference between population means is $\bar{X}_1 - \bar{X}_2 = 1,425 - 1,395 = 30$ lbs. The estimated standard error of the difference between means is

$$s_{\bar{X}_1 - \bar{X}_2} = \sqrt{\frac{(200)^2}{15} + \frac{(195)^2}{15}}$$
$$= 72.12$$

The 99 percent confidence interval, when df $= 28$, is

$$\bar{X}_1 - \bar{X}_2 \pm (2.763)(72.12) = 30 \pm 199.27$$
$$= (-169.27) \text{ to } (+229.27)$$

We estimate that the true population difference lies between -169.27 and 229.27. The new pesticide does not appear to have produced a substantial advantage, if any, since a true difference of zero is included within the bounds of the confidence interval.

EXERCISES

7.10 A commercial real estate specialist decides to investigate the mean difference in office space rents between city and suburban office buildings. Since this mean difference is critical to certain plans, he decides he wants to find the 99 percent confidence limits.

He randomly samples 100 city and 100 suburban locations and obtains the following results, in dollars per square foot:

	City	Suburban
$\bar{X}$	$23.50	$16.00
s	$ 4.00	$ 3.00

Estimate $(\mu_1 - \mu_2)$, the difference in mean rental rates, and find the 99 percent confidence limits.

7.11 In order to achieve compliance with federal highway speed requirements, many states are experimenting with a variety of different techniques, e.g., setting up a greater number of speed traps, or ticketing a greater number of cars for minor speed-limit infractions. In several jurisdictions, the shells of cars are painted to resemble highway patrol cars and are deployed along the roadway to simulate a speed trap. The hope is to deceive motorists into compliance.

In one study, the speeds of randomly selected automobiles were checked by radar under two different conditions: the dummy patrol cars present and dummy patrol cars absent. The results were as follows:

Dummy patrol cars	
Absent (X_1)	Present (X_2)
66	52
60	48
58	58
62	59
69	55
70	53
80	57
65	51
71	61
67	70
70	60
63	50
69	52
61	56

Find
  a. The point estimate of the difference between means.
  b. The interval estimate of the difference between means, using the 0.95 confidence coefficient.

7.4 POINT AND INTERVAL ESTIMATION OF THE POPULATION PROPORTION

One of the most common types of decision requires that we make a choice or distinction between two alternatives—candidate A versus candidate B in an election; defective versus non-defective item; loosening the money supply versus tightening money controls; favoring a contract negotiated by a labor union versus not favoring the contract. You will recognize these as involving two-category variables. Although it is theoretically possible to establish confidence limits of a two-category variable by expanding the binomial, the procedures are complex and time consuming. Fortunately, as we have seen, as *n* increases the binomial increasingly takes on the form of a normal distribution. When *n* is sufficiently large, we may use the normal curve for interval estimation. Most statisticians agree that the normal approximation of the binomial is adequate when $n > 30$ and both nP and nQ equal at least 5.

When we do not know the population proportion, we must estimate it on the basis of sample results. The estimated proportion is $\bar{p} = x/n$, where x is the number of successes or the number of events in the P category. Thus, if we sample 80 union members and find that 46 favor a strike vote, $\bar{p} = 46/80 = 0.5750$.

The standard error of the proportion is

$$s_{\bar{p}} = \sqrt{\frac{\bar{p}\,\bar{q}}{n}} \qquad (7.6)[2]$$

in which

$\bar{p} =$ the proportion in the P category

$\bar{q} = 1 - \bar{p}$ or $\dfrac{n - x}{n}$

The formula for a confidence interval for a proportion is

$$\bar{p} \pm zs_{\bar{p}} \qquad (7.7)$$

Let's look at an example.

Example E

Amalgamated Workers Local No. 8 has a total membership of 6,000. The labor contract expires in 60 days. To date, management has not acceded to union demands. The leaders in the local wish to take the pulse of the membership in the event that no progress is made in negotiation with management. A total of 105 randomly selected voting members are polled by secret ballot. Fifty-seven oppose a strike and 48 favor.

a. Find the point estimate of the population proportion favoring a strike vote.

b. Find the 95 percent confidence interval of the proportion.

c. Assuming the entire membership votes, estimate the 95 percent confidence interval for the strike vote of the total membership.

---

[2] To be completely accurate, $(n - 1)$ should be used in the denominator since $s_p^2 = \bar{p}\,\bar{q}/n$ tends to underestimate the population variances. However, in most business applications, this slight correction for bias is of little consequence. Hence, most business statistics books omit the correction.

Solution

a.

$$\bar{p} = \frac{48}{105} = 0.4571$$

$$\bar{q} = \frac{105 - 48}{105} = \frac{57}{105} = 0.5429$$

Our best estimate is that approximately 46 percent of the membership presently favor a strike vote.

b. The standard error of the proportion is

$$s_{\bar{p}} = \sqrt{\frac{(0.4571)(0.5429)}{105}} = 0.0486$$

Therefore, the 95 percent confidence interval is

$$\bar{p} \pm (1.96)(0.0486) = 0.4571 \pm 0.0952 = 0.3618 \text{ to } 0.5524$$

c. 95 percent confidence interval for the total $= (0.4571)(6{,}000)$ to
$$(0.5524)(6{,}000)$$
$$= 2171 \text{ to } 3314.$$

EXERCISES

7.12 Nine cases of leukemia were found among 3,200 soldiers exposed to low levels of radiation during a 1975 Nevada nuclear bomb test.

The total population of soldiers who participated in the test was approximately 250,000.

Because the doses of radiation were apparently quite low (the equivalent of an X-ray breast examination or gastrointestinal study), there are implications for the American population as a whole.

    a. Find the 95 percent limits for the proportion.
    b. Find the 95 percent confidence limits for the number of exposed soldiers who have leukemia.
    c. Find the 99 percent confidence limits for the proportion.
    d. Find the 99 percent confidence limits for the number of exposed soldiers who have leukemia.

7.13 In a sample of 1,000 accounts, Montana Rincon found 60 errors.
    a. Find the 90 percent confidence interval of the proportion.
    b. Find the 90 percent confidence interval for the total number of errors.

7.5 ESTIMATING SAMPLE SIZE

One of the most frequent questions directed toward statisticians is, "How large should my sample be?" The answer is that it depends on several considerations. How precise do you want the estimate to be? Do you know the population standard deviation? If not, do you have any sample data that provide an estimate of the population standard deviation? Then there are the inevitable economic questions. How much does each observation cost? What is your budget?

Let us put economic questions aside, and direct our attention to obtaining the degree of precision we want, regardless of cost. If, following this estimate, cost is an overriding concern, then we may adjust our goals with respect to precision in order to bring costs more in line with the realities of available funds.

7.5.1   Interval estimation of the mean

Let's look at the rationale behind estimating the sample size required to achieve a specified degree of precision. Imagine that Bulkwheat Inc. is packaging a cereal product that requires a label indicating the average net weight of the contents of its packages. The population standard deviation is known to be 1.2. The firm wishes to keep the sample within 0.25 ounce of the population mean with a probability of 0.95. Recall that, with sufficiently large n, the sample means are normally distributed with a ~~mean~~ equal to $\sigma/\sqrt{n}$. Thus, we want the allowable error, e, to equal: *standard error of the mean*

$$e = z\, \sigma_{\overline{X}}$$

$$= z\frac{\sigma}{\sqrt{n}}$$

in which

$e =$ Allowable error $(\overline{X} - \mu)$
$z =$ The value of a standard normal variable that includes the desired proportion of area surrounding the mean.

Solving in terms of n as the unknown, we obtain

$$\sqrt{n} = \frac{z\, \sigma\, \overline{x}}{e}$$

Thus,

$$n = \left(\frac{z\sigma}{e}\right)^2 = \frac{(z\sigma)^2}{e^2} \tag{7.8}$$

Let's apply Formula 7.8 to the Bulkwheat packaging problem. Since we are estimating the average weight per package and we wish our allowable error to be no more than 0.25 ounce, we would expect that, with repeated sampling, 95 percent of the sample means will be within 1.96 $\sigma_{\overline{X}}$ of the population mean. Thus,

$$0.25 = 1.96 \, \sigma \, \bar{x}$$

$$= 1.96 \frac{\sigma}{\sqrt{n}}$$

Therefore,

$$n = \frac{(1.96 \, \sigma)^2}{(0.25)^2}$$

$$= \frac{[(1.96)(1.2)]^2}{0.0625}$$

$$= \frac{5.53}{0.0625}$$

$$= 37.63 \quad 88.51$$

Thus, we would require an n equal to about 38 in order to achieve a 95 percent probability that our confidence interval will include the population mean.

In the above example, we knew the population standard deviation. In most real-life situations, σ is unknown. Therefore, we must estimate it. If we have prior sample data, we may use the standard deviation of the sample as our best available estimate of the population standard deviation. Lacking this, but having some estimate of the range of the values of the variable, we may obtain a crude estimate of the standard deviation by dividing the range by 4. Thus, if our range was found to be 6.2, our approximation of the standard deviation would be 1.55. Thus,

$$n = \frac{[(1.96)(1.55)]^2}{0.0625}$$

$$= 48.61$$

This sample size of 49 would not be precise since the estimate of the standard deviation is so crude. For this reason, the estimate should be regarded as a minimum sample size. If time and funds permit modest increases in n, they should be made. However, once the first testing is done, you will have a sample standard deviation that will provide a better basis for estimating the sample size in all future experiments.

Lacking any prior information, we should consider taking a pilot sample, i.e., a small, randomly selected sample the purpose of which is to obtain some estimate of the population standard deviation. We may use the pilot standard deviation as the basis for obtaining an estimate of the sample size required to obtain the desired degree of precision.

7.5.2 Interval estimation of a proportion

The general manager of an automobile agency wants to estimate the proportion of the hourly labor rate for repair service that goes for labor. The remainder is for overhead and profit. A survey by the Associated Press shows that approximately 33 percent or 0.33 of the customer's hourly charge goes for labor. The general manager wants the probability to be 0.95 that the sample proportion differs from the population proportion by no more than 0.05. How large should the sample be? Suppose he wants to be 99 percent sure, what size sample should he take?

Since we wish our error to be no more than 0.05, we would expect that with repeated samples, 95 percent of the sample proportions will be within 1.96 $s_{\bar{p}}$. In other words, we wish 0.05 to equal 1.96 $s_{\bar{p}}$. Thus:

$$0.05 = 1.96 \sqrt{\frac{\bar{p}\bar{q}}{n}}$$

$$(0.05)^2 = (1.96)^2 \frac{(\bar{p})(\bar{q})}{n}$$

$$n = \frac{(1.96)^2}{(0.05)^2} \bar{p}\,\bar{q}$$

$$= (1,536.64)(0.33)(0.67)$$

$$= 339.75$$

Thus, he needs a sample size of approximately 340 to be 95 percent sure that his interval estimate of the population proportion will be off by no more than 0.05.

To be 99 percent sure:

$$n = \left(\frac{2.58}{0.05}\right)^2 (0.33)(0.67)$$

$$= 588.69 \text{ or approximately } 589$$

When we have no prior knowledge of P, the safest approach is to take the worst possible case, i.e., assume the population proportion, P, is actually 0.5 and $1 - P = 0.5$. This procedure would guarantee a sufficiently large n no matter what the actual values of P and $1 - P$. Thus, at the 99 percent interval,

$$n = \left(\frac{2.58}{0.05}\right)^2 (0.50)(0.50)$$

$$= 665.64 \text{ or approximately } 666$$

EXERCISES

7.14 A survey of 282 companies that used travel as an incentive to their sales force programs reported that the mean cost per trip (for two people) was $1,235 with $s = \$500$.

    a. Find the 95 percent confidence limits for μ.
    b. Find the 99 percent confidence limits for μ.

c. Suppose you wanted to be 95 percent confident that your estimate was off by no more than $50. What size sample should you use?

7.15 In a heated congressional race between two candidates, a pollster wishes to estimate the 95 percent confidence interval for the proportion of voters favoring Candidate A with an error not to exceed 2 percent. A previous poll showed 47 percent of the registered voters favoring her candidacy.

a. What sample size should be used in the survey?

b. If the pollster is willing to tolerate an error as high as 5 percent, what *n* should be used?

7.16 Why is the *n* estimated in the preceding exercise only an approximate value? Explain.

SUMMARY

Estimation in an essential ingredient of both planning and decision making. This chapter covered point and interval estimates of a mean, differences between means, and proportions.

1. Point estimates are single values that are intended to estimate the value of the corresponding parameter. Statistics such as the mean, median, standard deviation, and proportion are common estimators of corresponding population parameters.

2. Interval estimates involve the specification of an interval in which we have a given degree of confidence that the parameter of interest is included.

3. The three criteria in selecting an estimator are that: the mean of its sampling distribution should equal the parameter that is being estimated (lack of bias); the estimator must be efficient, i.e., the lower the variability in the sampling distribution of the estimator, the more efficient the estimate; finally, as we increase the sample size, the sample statistic should provide increasingly close approximations to the parameter of interest.

4. Interval estimation involves *(a)* specifying the degree of confidence we wish an estimate to achieve; *(b)* constructing a confidence interval; and determining the confidence limits. The probability associated with this interval is the confidence coefficient.

5. Interval estimates of the mean were shown when $n > 30$, $n < 30$, and when the population is small. In the world of business and management, point and interval estimates of the total quantity are frequently of critical importance. Procedures for making such estimates were demonstrated.

6. Interval estimates of the difference between means were demonstrated with small and large sample sizes.

7. Decisions commonly involve making a choice between two alternatives. Procedures for obtaining both point and interval estimates of a proportion were demonstrated.

8. Estimating the sample size is both a statistical and economic decision. The statistician is typically concerned with degree of precision of an estimate. The text demonstrated procedures for estimating the sample size necessary to approximate a desired degree of precision.

TERMS TO REMEMBER

estimate (196)

estimator (196)

point estimate (196)

interval estimate (196)

unbiased estimator (197)

efficient estimator (198)

consistent estimator (199)

confidence interval (200)

confidence limits (200)

confidence coefficient (200)

t-distributions (203)

degrees of freedom (df) (203)

standard error of the difference between means (209)

EXERCISES

7.17 A real estate firm decides it wants to set up a "hot line" to answer real estate questions. In order to make time, equipment, and personnel plans, it decides to model itself after an existing service in Chicago. Lois, the director of the project, decides she needs to know the average volume of calls to expect per day. She randomly selects 50 days from the Chicago service's records and finds a mean, $\overline{X} = 200$ calls daily with $s = 30$.

What are the 95 percent confidence limits for the mean volume of calls?

7.18 Refer to Exercise 7.17. In looking over Chicago's records, Lois finds a great deal of variability in the length of the call. She randomly selects 50 calls and finds a mean time, $\overline{X} = 11.5$ minutes with $s = 3$.

What are the 95 percent confidence limits for the mean length of call? Would it be correct to say that, of all incoming calls, 95 percent will be between these time limits?

7.19 Refer to Exercises 7.17 and 7.18. It is now a year later. The "hot line" appears to be functioning smoothly. However, Lois suspects that the mean time per call is higher for her service than for the one in Chicago. She randomly samples 50 calls from each service and finds:

	Chicago	The "hot-line"
$\overline{X}$	11.5	12.8
s	3.0	2.5

Estimate the difference $(\mu_1 - \mu_2)$ in mean time and place bounds on the error of estimation. We wish to be 95 percent confident.

7.20 According to a survey by the Associated Press, business is booming at most auto repair service departments. In fact, most dealers have a healthy backlog of service appointments.

Marcy G. wants to estimate the mean time between a call for service and the designated appointment. She randomly samples nine customers at a particular dealership and finds that the mean "waiting" time, $\overline{X}$, is 12.5 days and $s = 2.5$.

a. Find the 95 percent confidence limits for μ.

b. Find the 99 percent confidence limits for μ.

Comment on the effect of increasing the level of confidence.

7.21   Refer to Exercise 7.20. Suppose Marcy finds another dealer nearby who can service her car. She decides to estimate the mean difference in waiting time between the two dealerships. She randomly samples nine customers from the second dealer and summarizes her results:

	Dealer 1	Dealer 2
$\bar{X}$	12.5	8.5
s	2.5	3.5
n	9	9

Estimate the difference $(\mu_1 - \mu_2)$ in mean waiting time and find the 95 percent confidence limits.

7.22   A study of 100 companies revealed that their travel budgets increased 8.2 percent with $s = 1.9\%$. Find the 99 percent confidence interval for μ.

7.23   Refer to Exercise 7.22. What size sample is needed to estimate the true mean percentage increase for the travel budgets to within 0.25 percent at the 95 percent level of confidence?

7.24   Find the n required to estimate the proportion of defectives to within 4 percentage points with a confidence coefficient of 0.95 when:
   a. $\bar{p} = 0.10$.
   b. $\bar{p} = 0.25$.
   c. $\bar{p} = 0.40$.
   d. $\bar{p} = 0.50$.

Generalize: What is the effect of $\bar{p}$ on the sample size required to maintain a given degree of precision?

7.25   An analysis of the arrival times of 20 city transit buses revealed that they were, on the average, 15.72 minutes late with $s = 5.02$. Determine the confidence interval of the mean using a confidence coefficient of
   a.   0.95.
   b.   0.99.

7.26   One of the eligibility requirements for federal highway funds is that the drivers in a state be in compliance with the 55 miles-per-hour speed limit. As of October, 1980, several states are already threatened with a cut-off of funds based on sample studies that indicated failure to comply. In one study, the speed of 22 automobiles was randomly determined by a radar device. The speeds recorded were as follows:

58	50
52	57
73	61
56	70
59	63
65	55
56	51
68	60
58	59
55	68
62	60

Find
 a. The point estimate of mean speed per vehicle.
 b. The interval estimate at the 95 percent confidence level.
 c. The 99 percent confidence limits.

7.27 Many cars manufactured in the United States carry new car warranties of one year. The manufacturer is required to repair or replace warranteed parts if they develop defects or wear out. In a sample of 5,000 cars by one manufacturer, it was found that 112 developed fluid leaks in the automatic transaxles on the car.
 a. Find the 95 percent confidence interval for the proportion of cars developing defects.
 b. If 750,000 cars are produced within a year, estimate the interval that will contain the total number of transaxle defects during the year, using a confidence coefficient of 0.95.

7.28 The manager of a firm was concerned that production appeared to be slowing down. He noticed that more people were taking 70-minute lunch-hours and quitting a few minutes earlier to make a dental appointment or catch a child's Little League game. He called in an independent research group to estimate the number of minutes people actually worked. The group observed a sample of 25 employees and found the mean number of minutes worked, $\bar{X} = 435$, $s = 10$. Suppose you headed the research group, estimate the mean number of minutes worked so that you are 95 percent confident of your estimate.

7.29 Refer to Exercise 7.28. You now learn that the total number of employees in this firm is 200. Based on the results obtained for the 25 employees, find the 95 percent confidence limits for the number of minutes worked. Explain the difference between these results and those of Exercise 7.28.

7.30 A pharmaceutical house has developed a new nonprescription sleeping tablet. The marketing manager wishes to estimate the 99 percent confidence interval of the market potential with an error of no more than 5 percent. The market manager estimates that the product should achieve 20 percent market penetration. Approximately what sample size should be used?

7.31 Referring to Exercise 7.30, in the actual field tests, a sample of 450 users of sleep-aids were asked to try the new compound. Sixty-eight said they preferred the new compound to the one they were currently using.
 a. Estimate the 90 percent confidence interval of the proportion.
 b. If there are 30 million potential customers in the country, find the point and interval estimate of the total sales potential, using the 0.90 confidence coefficient.

7.32 Refer to Exercises 7.30 and 7.31. Recall the manager's estimate of 20 percent market penetration. Does his estimate seem to be supported?

7.33 In any given field there are a large number of textbooks available for professors. Their choice depends upon a variety of factors. A new textbook, written by well-known authors in the field, receives extremely good reviews. In a sample of 70 professors teaching this particular course, 19 select this new text. Suppose the total population of professors teaching this course is 250. Find the
 a. 95 percent confidence limits for the proportion.
 b. 95 percent confidence limits for the total number of professors adopting this book.

7.34 The marketing manager of a supermarket chain approaches the owner with a great advertising gimmick: "Come to our store—buy any 25 items. Go to any competitor—buy the same 25 items. Compare the tapes. If the competitor's is lower, we will refund double the difference."

"It sounds like a great idea," says the owner, "but how much could it cost us? I want to be very sure we don't lose more than we will gain in added business. I want you to estimate the mean difference and be 95 percent confident of your estimate."

The marketing manager sends out a comparison shopper and instructs him to select 25 items from our store and the same 25 items from competing stores. He repeats this process 10 times, each time selecting a different competitor. The following results are obtained:

	Our store	Competitors
$\bar{X}$	$50.75	$45.65
s	$ 4.50	$ 6.00
n	10	10

Based on the results, the owner is wary. "We may lose money." The marketing manager assures him that very few people actually go through the trouble of doing this. At worst, previous markets that have used this gimmick have found that 100 customers actually made the comparison. In the meantime, he estimates increased business from the advertising will total at least $2,200.

    a.   Find the 95 percent confidence limits of the difference in the mean.

    b.   Find the 95 percent confidence limits of the total refunds if 100 customers actually make this comparison. What do you conclude?

7.35 According to a survey by Cinemascore™, 81 percent of the 2,345 people polled gave *Star Trek—The Motion Picture* an A or B rating.[3]

    a.   Find the 95 percent confidence limits for the proportion.

    b.   Find the 99 percent confidence limits for the proportion.

7.36 *Rocky II* was a 1979 sequel to the extremely successful movie *Rocky*. According to a survey by Cinemascore™, 98 percent of the 502 people polled gave *Rocky II* an A or a B rating.[4]

    a.   Find the 95 percent confidence limits for the proportion.

    b.   Find the 99 percent confidence limits for the proportion.

---

[3] Source: *Cinemascore,* 1980.
[4] Ibid.

Hypothesis Testing

8.1   BASIC CONCEPTS

We have seen that, in making decisions, we always have to deal with uncertainty. We can never be *absolutely* sure that any decision we make is correct (except perhaps after the fact—when it may be too late). Although intuitive feelings and hunches often play a role in managerial decisions, we need to have a more objective manner to reduce the uncertainty. Unfortunately, as we shall see, we can never completely eliminate uncertainty. We can, however, remove some of the guesswork and make our decisions in a more objective manner.

In this chapter we shall examine the role of hypothesis testing in decision making. We shall develop rules that will help us control and minimize the probability of error in choosing among alternatives. To illustrate some of the basic concepts in hypothesis testing, let us look at an example.

Ever since the Arab embargo on oil following the Yom Kippur War in 1973, numerous products have entered the market claiming to substantially increase the miles-per-gallon ratings of the family "dream-turning-nightmare" machine. Some involve additives that are mixed with the gasoline of the car, whereas others are devices that must be installed along the fuel line or attached directly to the carburetor. All claim significant improvements in gasoline mileage. You have undoubtedly seen the advertisements, "Enjoy 10 percent to 40 percent improvement in your car's mileage ratings." On what basis are these claims made? Unfortunately, they often represent nothing more than testimonials by "satisfied" users, usually in the form of letters written to the manufacturer. We know that convenience sampling of this sort is rarely representative of the population of interest (i.e., all purchasers of the additive or device). Moreover, few consumers are capable of conducting or motivated to conduct the objective and systematic research necessary to test the claims. Additionally, there is a strong psychological tendency for consumers to look for and find improvements in order to justify their expenditure of funds.

Fortunately, there are a number of federal and private agencies that are set up specifically to evaluate claims of this sort. Let's follow through both the logic and the procedures for testing the mileage claims. To begin with, the claim is either true or it is false: an additive either increases mileage or it does not. It is our role as researchers to establish procedures for deciding between these two alternatives. Stated in formal terms, these alternative possibilities are:

1.   The additive has no effect on gasoline mileage.
2.   The additive has an effect on gasoline mileage.

Since the first is often an hypothesis of "no effect," it is referred to as the *null hypothesis (H$_0$)*. As we shall see, the null hypothesis may also specify a particular value of the parameter of interest. But characteristically it denies the effect we are seeking to evaluate. The second hypothesis, called the *alterna-*

null hypothesis (H$_0$)

A statement that specifies hypothesized values for one or more of the population parameters. In a statistical test, the null hypothesis states the hypothesis to be tested. It often, but not necessarily, involves the hypothesis of "no effect."

tive hypothesis (H_1), asserts that the null hypothesis is false. Note that these two hypotheses are *mutually exclusive* and *exhaustive*. They cannot both be true or false at the same time. One *must* be true and one *must* be false. It is the role of our data collection and statistical procedures to decide between these two alternatives. But, if you think for a moment, you'll realize how difficult it is to conceive of proving the null hypothesis. Except for taking a complete census, which is not feasible considering the millions of cars on the road, we must rely on drawing samples from a population. We have already seen that these sample statistics distribute themselves about a central value, such as the population mean or the population proportion. The sample means virtually never equal the population mean, and the sample proportions virtually never equal the population proportion. In addition, they almost always differ from one another and by variable amounts. How then can sample statistics be used to prove no difference? In a word, they cannot.

However, since H_0 and H_1 are mutually exclusive and exhaustive, rejecting H_0 allows us to assert H_1. Note that the proof is indirect. We assert H_1 by rejecting H_0. Also, statistical proof is always one-way. We may reject H_0 and assert H_1, but we cannot reject H_1 and, thereby, assert H_0. In brief, just as we cannot prove H_1 directly neither can we disprove it. For example, if two sample means happen to be identical, this does not prove no difference in the population parameters. In other words, we have not proved that they come from the same population, or from two different populations in which the population means are the same. We know that sample statistics, even if drawn from different populations, can occasionally be the same. It should be noted that the null and alternative hypotheses are always stated in terms of population parameters rather than in terms of sample statistics, although we use sample statistics to test the hypotheses.

Let's briefly summarize the logic of hypothesis testing up to this point. First, we set up two mutually exclusive and exhaustive hypotheses.

alternative hypothesis (H_1)

A statement that specifies that the population parameter is a value other than that specified in the null hypothesis.

1. The null hypothesis (H_0):
    a. Cannot be proved. We cannot prove the mileage additive or device does not work.
    b. Can be rejected. If rejected, we assert H_1. We can reject the hypothesis that the additive or device does not work and thereby assert that it does work.
2. The alternative hypothesis (H_1):
    a. Cannot be proved directly. We cannot directly prove that the mileage additive or device works.
    b. Cannot be rejected directly. We cannot directly reject the possibility that the additive or device works.

Only by rejecting H_0 *(b)* can we assert our "indirect proof" of the alternative hypothesis (see Box 8.1).

Box 8.1

POSSIBILITY AND PROBABILITY

We have stated that the logic of statistical inference asserts that the null hypothesis can never be proved. In other words, we can never make an absolute definitive statement that we have *proved* the null hypothesis. Let us examine the logic behind this assertion.

In order to prove something is true, we must be able to state with *absolute certainty* that it will occur (or has occurred). Stated in terms of probability, we are saying the probability is *exactly* 1.00 that the statement is true. However, if the *possibility* exists that the statement may *not* be true (even if that is highly unlikely), then we no longer have *absolute proof.* The best we can say is that the probability is very high that the statement is true.

Looked at another way, if it is *impossible* for something to occur, the probability of its occurrence is *exactly* zero. But if even the slightest chance exists that it may occur, we cannot say with *absolute certainty* that it will not occur. Thus, the occurrence of an event may be possible but not probable.

It is this distinction between *possibility* and *probability* that underlies the logic of statistical inference.

Let's look at an example to illustrate the relationship between possibility and probability. Suppose we set up two mutually exclusive hypotheses:

H_0: The price of gold will be $1.00 an ounce.
H_1: The price of gold will not be $1.00 an ounce.

The probability that H_0 is true is extremely low; for all intents and purposes, the probability is almost zero. *However*—is it possible? Herein lies the crux of *absolute* proof. Although we would all agree that it may be possible, it is highly improbable. But as long as some possibility exists that it may occur, we cannot say we have *disproved* the null hypothesis. If the occurrence of an event is unlikely (but *not* impossible), the associated probability of its occurrence is low. Therefore, in this instance, since the probability that H_0 is true is extremely low, we would choose to conclude that it is false. That is, we would reject H_0 and assert H_1.

On the other hand, let us examine two other mutually exclusive and exhaustive hypotheses:

H_0: The former president Lyndon B. Johnson will be the next president.
H_1: The former president Lyndon B. Johnson will not be the next president.

Here we are dealing with a situation in which H_0 is not only improbable, it is impossible (at least until someone figures out a way to elect a dead man president). Thus, since H_0 is impossible, we may state with absolute certainty that H_0 is false. Since H_0 deals with an *impossible* occurrence, we may finally say we have *proved* that H_0 is false and H_1 is true.

Thus, absolute proof is possible only when *no* possibility exists that H_0 may be true. As long as even the smallest possibility exists, we must make our conclusions in terms of probabilistic rather than absolute statements.

But how do we go about rejecting H_0? This is where probability theory enters the arena. First, it should be emphasized that in hypothesis testing we assume that the null hypothesis is the *true* distribution. If, in the sampling distribution of a statistic under the null hypothesis, a particular result would have a low probability of occurrence, then one of two possible conditions has occurred. Namely, either the null hypothesis is true, and we obtained a result which had a very small probability of occurring, or the null hypothesis is not true. We *choose* to accept the latter condition, i.e., we reject H_0 and assert H_1.

How do we define low probability? The definition is arbitrary but not capricious. Some researchers are willing to reject H_0 when the obtained statistic would have occurred 5 percent of the time or less in the appropriate sampling distribution. This criterion for rejecting H_0 is variously referred to as the 5 percent *significance level,* the 0.05 *level of significance,* or simply the 5 percent or 0.05 level. The criterion of rejection is typically represented by the Greek letter α *(alpha).* Thus, when we use the 0.05 significance level, $\alpha = 0.05$.

Other researchers set up a more stringent criterion for rejecting H_0 and asserting H_1, namely the 0.01 or 1 percent significance level (i.e., $\alpha = 0.01$). Only when the obtained statistic would have occurred 1 percent or fewer times in the sampling distribution of interest would the researcher be willing to reject H_0 and assert indirect proof of H_1. These two levels of significance are commonly used, although we will occasionally encounter other values, such as 0.10 or 0.001.

The level of significance that is set is not merely a matter of preference among different researchers or different statisticians. The choice has to do with the consequences of making one of two types of error—mistakenly rejecting a true H_0 or failing to reject a false H_0. The *same* researcher or statistician may use different levels of significance in different experiments.

Note that *statistical proof is not absolute proof in any sense of the word.* If our tests allow us to reject the null hypothesis that the additive or device does not work, we have not demonstrated beyond all reasonable doubt that it does work. As we have noted repeatedly, statistical analysis and probability theory help to reduce uncertainty. They do not eliminate it altogether. Indeed, analysis of the logic of statistical inference reveals that there are two types of errors we may commit:

1. We may reject the null hypothesis when it is true. Thus, we may falsely reject H_0, that the additive or device does not improve mileage, and assert H_1, that it does improve mileage. Such an error of falsely rejecting H_0 is known as a *Type I* or *Type α error.* The probability of this type of error is given by α. Thus, if $\alpha = 0.05$, we will mistakenly reject a true null hypothesis approximately five percent of the time. Hence, our claim to have demonstrated an effect of changed conditions (such as better mileage when mixing an additive with the gasoline) will be wrong about 5 times in 100. Some might consider this risk of error too high. That is why some investigators use $\alpha = 0.01$ or less. They are more conservative about their willingness to claim an effect where there may not be one.

2. In the second type of error, we fail to reject H_0 when it is actually false. This class of error is known as *Type II* or *Type β error.* If the device or additive actually improved mileage and we did not reject H_0, we would have failed to claim an effect when there was one. Note that we do not claim to have proved H_0 but merely that we failed to reject it. We should not minimize the importance of a Type II error. Many promising lines of research have undoubtedly been abandoned prematurely because the results of preliminary investigations were not encouraging.

An ideal situation is one that results in a balance between the two types of errors. In this ideal situation, we should be able to state in advance the probability of making both a Type I and a Type II error.

As we have seen, we may state the probability of a Type I error in terms of the value of α we employ. The probability of a Type II error is represented

---

significance level

A probability value that is considered so rare or unusual in the appropriate sampling distribution that one is willing to reject H_0 and assert that nonchance factors are operating. The most commonly used significance levels are 0.05 and 0.01.

alpha (α)

The probability that nonchance factors are operating.

Type I (Type α) error

The rejection of H_0 when it is actually true. The probability of a Type I error is set by the α-level.

Type II (Type β) error

The acceptance of H_0 when it is actually false. The probability of a Type II error is β.

by β. How can we evaluate this probability? We can determine β only when H_0 is false and we know the true value of the parameter under H_1. Since this is rarely the case, it is difficult to evaluate this probability. However, procedures are available that permit us to estimate β even when the parameter is not known. These procedures are beyond the scope of this text.

There are certain strategies, however, that may be employed to reduce the probability of a Type II error. For example, the lower we set our α, the greater the likelihood of a Type II error. Thus, in general, if $\alpha = 0.05$, β will be less than if $\alpha = 0.01$. Another way to reduce β is to increase the sample size. The larger the sample, the smaller the probability of a Type II error.

Table 8.1 summarizes the type of error as a function of the true status of H_0 and the decision we have made. Note that a Type I error can be made only when H_0 is true, and a Type II error only when H_0 is false. We see that $(1 - \alpha)$ is the probability of accepting H_0 when it is true, and α is the probability of rejecting a true H_0. Thus, if $\alpha = 0.05$ *and* H_0 is true, the probability of accepting H_0 equals $1 - 0.05 = 0.95$. Or, stated another way, if the null hypothesis is true, there is a 95 percent chance that it will be accepted.

Table 8.1

The type of error made as a function of the true status of H_0 and the decision we have made

Decision	True status of H_0	
	H_0 true	H_0 false
Accept H_0	Correct $(1 - \alpha)$	Type II error (β)
Reject H_0	Type I error (α)	Correct $(1 - \beta)$

How can we tell whether we are making a Type I or Type II error? Following the logic of statistical inference, we can't. Since we rarely know the value of the population parameters, we really never know whether H_0 is true or false. Unless we know (for sure) the true status of H_0, we can never be *absolutely* sure that our decision is correct. Once again we must deal with uncertainty and the fact that our knowledge is probabilistic and not absolute. However, since it is difficult to evaluate the probability of a Type II error, an acceptance of the null hypothesis is not as strong a conclusion as the rejection.

One final word on the logic of statistical inference. Both the null hypothesis and the alternative hypothesis may be *directional* or *nondirectional*. An alternative hypothesis is directional if it asserts the *direction* of the change or difference expected to result from our research procedures. To illustrate, we may be interested only in examining the claim that a device or additive *increases* gasoline mileage. Stated symbolically, $\mu > \mu_0$, in which μ_0 is the value of the population mean under the null hypothesis. Thus $H_1: \mu > \mu_0$ states that the sample using the additive comes from a population in which

directional hypothesis

An alternative hypothesis that states the *direction* that the parameter differs from that hypothesized under H_0.

nondirectional hypothesis

An alternative hypothesis that states merely that the parameter is different from the one stated under H_0.

the mean mileage rate is *greater* than that achieved without the additive. This H_1 is called a directional hypothesis since it asserts an expected difference in the *direction* of the effect, i.e., μ is *greater than* μ_0. If H_1 is directional, so also is the null hypothesis. Thus, our null hypothesis in this example would read: H_0: $\mu \leq \mu_0$. In order to reject H_0, the sample statistic would have to be significantly *greater* than expected under H_0 (not just different).

At other times, we may have little or no basis for predicting the direction of the effects of changed conditions. In this case, our alternative hypothesis would read: $\mu \neq \mu_0$. The null hypothesis would be: $\mu = \mu_0$. Thus, if our sample statistic is either significantly greater than or significantly less than the value hypothesized under H_0, we would reject the null hypothesis. Table 8.2 illustrates several directional and nondirectional null and alternative hypotheses that might be used in evaluating the results of changed conditions.

Examples of changed conditions of employment and directional and nondirectional hypotheses used in the evaluation of the effects of the changes. Note that H_1 states that we expect our changed conditions to have an effect. In the case of a directional H_1, we even predict the direction of the expected change. The null hypothesis is a purely statistical hypothesis, a straw man, set up in order to ascertain whether or not we can attribute effects to changed conditions.

Table 8.2

Changed conditions	Null and alternative hypotheses		Direction of hypotheses
Greater involvement of labor in management decisions	H_0:	Productivity unchanged or reduced; $\mu \leq \mu_0$	Directional
	H_1:	Productivity increased; $\mu > \mu_0$	
Elimination of grievance procedures	H_0:	Morale unchanged or improved; $\mu \geq \mu_0$	Directional
	H_1:	Morale decreased: $\mu < \mu_0$	
A mixed bag of changes in labor-management relationships	H_0:	Productivity unchanged; $\mu = \mu_0$	Nondirectional
	H_1:	Productivity changed; $\mu \neq \mu_0$	

two-tailed probability value

Probability value associated with nondirectional hypotheses. The proportion of area in both tails of the distribution is taken into account.

one-tailed probability value

Probability value associated with directional hypotheses. The proportion of area in only one tail of the distribution is taken into account.

The reason for distinguishing between directional and nondirectional hypotheses is that the nature of our alternative hypothesis determines which tail of the sampling distribution to refer to when evaluating the probability of an obtained result. The rule is simple and straightforward. If H_1 is nondirectional, we look at *both tails* of the sampling distribution to obtain the probability value associated with the result. If directional, we look at only *one tail* of the sampling distribution.

Figure 8.1 uses the normal probability model to define *critical regions* for rejecting directional and nondirectional null hypotheses at the $\alpha = 0.05$ and $\alpha = 0.01$ levels. If a statistic falls anywhere in the critical region, we reject H_0.

Figure 8.1

Critical regions for rejecting directional and nondirectional null hypotheses at $\alpha = 0.05$ and $\alpha = 0.01$ under the normal probability distribution. Note that a nondirectional test uses both tails of the distribution to assess probability values. Thus, a two-tailed test at $\alpha = 0.05$ uses the area beyond both -1.96 (0.025) and $+1.96$ (0.025). Together, the two areas add up to 0.05. However, for a directional test at $\alpha = 0.05$, we examine only one tail. We find a value of z beyond 1.645 includes 5 percent of the total area, as does a value of z beyond -1.645.

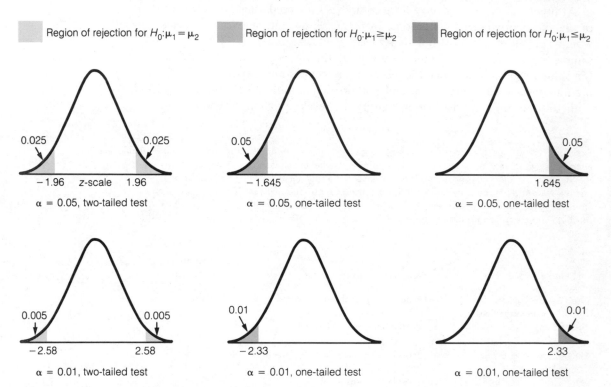

Region of rejection for $H_0 : \mu_1 = \mu_2$ Region of rejection for $H_0 : \mu_1 \geq \mu_2$ Region of rejection for $H_0 : \mu_1 \leq \mu_2$

0.025   0.025
-1.96 z-scale   1.96
$\alpha = 0.05$, two-tailed test

0.05
-1.645
$\alpha = 0.05$, one-tailed test

0.05
1.645
$\alpha = 0.05$, one-tailed test

0.005   0.005
-2.58 2.58
$\alpha = 0.01$, two-tailed test

0.01
-2.33
$\alpha = 0.01$, one-tailed test

0.01
2.33
$\alpha = 0.01$, one-tailed test

critical region

That portion of the area under a curve which includes the values that lead to rejection of the null hypothesis.

critical value

In a statistical test, the value that marks off the region of rejection at a specific α-level.

Note that the regions shown for rejecting null hypotheses illustrated in Figure 8.1 also show *critical values* that delineate a region. Thus, in a two-tailed test of significance, a value of z equal to or greater than $+1.96$ or equal to or less than -1.96 is a critical value for rejecting a nondirectional hypothesis at $\alpha = 0.05$. The normal probability table (Table V—Appendix C) provides exact probability values for varying values of z. Many tables that we use do not provide exact probability values. Rather, they provide *critical values* that define the region of rejection at various levels of α. The table of the t-distributions is a case in point (Table VII—Appendix C). The values shown in the body of the table are critical values required to reject H_0 at various α-levels, for one- and two-tailed tests at varying numbers of degrees of freedom. A portion of Table VII is excerpted in Table 8.3. Note that the critical value for rejecting a directional H_0 at $\alpha = 0.01$ and df = 6 is 3.143. To reject H_0 at $\alpha = 0.01$, two-tailed test and df = 6 would require a t equal to or greater than 3.707. We shall return to the t-distributions when illustrating tests of significance of means and difference between means when *n* is small.

For any given df, the table shows the values of *t* corresponding to various levels of probability. Obtained *t* is significant at a given level if it is equal to or *greater than* the value shown in the table.

Table 8.3

Critical values of *t*

df	Level of significance for one-tailed test					
	0.10	0.05	0.025	0.01	0.005	0.0005
	Level of significance for two-tailed test					
	0.20	0.10	0.05	0.02	0.01	0.001
1	3.078	6.314	12.706	31.821	63.657	636.619
2	1.886	2.920	4.303	6.965	9.925	31.598
3	1.638	2.353	3.182	4.541	5.841	12.941
4	1.533	2.132	2.776	3.747	4.604	8.610
5	1.476	2.015	2.571	3.365	4.032	6.859
6	1.440	1.943	2.447	3.143	3.707	5.959
7	1.415	1.895	2.365	2.998	3.499	5.405
8	1.397	1.860	2.306	2.896	3.355	5.041
9	1.383	1.833	2.262	2.821	3.250	4.781
10	1.372	1.812	2.228	2.764	3.169	4.587

Source: Excerpted from Table VII, Appendix C.

EXERCISES

8.1 In each of the following indicate the type of error, if any, that has been made. For pedagogical purposes, we shall suppose that we know the true status of H_0. (In the following exercises p = probability.)

a. A head of personnel reports that, following computer matching of a person to the job, the turnover rate of personnel has declined significantly. Using $\alpha = 0.05$, one-tailed test, he found $p \leq 0.02$, one-tailed value. True effect of computer matching: decrease personnel turnover.

b. After introducing a new procedure on the assembly line, the plant manager reported a significant decline in the time required to complete an assembly. Using $\alpha = 0.05$, one-tailed test, he found $p \leq 0.01$, one-tailed value. True effect of new procedures: no effect.

c. Using trained pigeons to detect flaws in printed circuits, a researcher finds no significant difference from humans doing the same job. He had hypothesized that pigeons, because of better visual acuity and greater concentration on repetitive tasks, would perform better than humans. Using $\alpha = 0.01$, one-tailed test, he obtained $p \leq 0.03$, with the pigeons obtaining better scores.

True difference between pigeons and humans on this task: pigeons superior.

  d. After adding trace amounts of molybdenum to a metallic alloy, a materials engineer found no statistically significant difference in wear characteristics. Using $\alpha = 0.05$, two-tailed test, he found $p = 0.40$. True effect of molybdenum: no effect.

8.2 Classify the following as directional or nondirectional alternative hypotheses.
  a. Improved working conditions in the plant will lead to lower absentee rates.
  b. Tightening the money supply will decrease the rate of inflation.
  c. Altered working conditions in the plant will lead to changed absentee rates.
  d. Decontrolling an industry will lead to changed prices for its products.
  e. An increased threat of an IRS audit will lead to fewer errors on tax returns.

8.3 In evaluating the hypotheses shown on the alternatives in Exercise 8.2, indicate whether a one- or a two-tailed test of significance would be used.

8.4 Comment: When testing the null hypothesis, Michael B. obtained $p = 0.08$, one-tailed value, when using $\alpha = 0.05$, one-tailed test. The results were in the direction specified under H_1. Michael concluded, "I have proved that the changed conditions produce no effect."

8.5 Comment: An investigator conducts a study to test his hypothesis that a concerted program of safety education will reduce time losses due to accidents on the job. Citing a *p*-value of 0.006 in favor of his hypothesis, he asserts, "I have proved beyond a shadow of a doubt that safety education works."

8.2 HYPOTHESIS TESTING INVOLVING A SINGLE MEAN

Let's return to the problem of evaluating the claim that a gasoline additive improves the miles-per-gallon performance of a car. Suppose an automobile manufacturer wishes to test this claim in order to make recommendations to purchasers of its cars. It is a common practice to randomly select automobiles as they come off the assembly line and to run them through a series of performance tests. Included in these tests are mileage ratings. Within a few months of introducing new model cars, considerable information has already been collected concerning the mileage ratings of these cars. For example, it might be found that the mean mileage rating is 26.54. The testing department could now select a random sample of cars off the production line, place the additive in the gasoline, and run the mileage test on this single sample of cars. The mileage performance of these cars could then be statistically compared with the standard established by prior testing. A statistical test of this sort is referred to as the one-sample case since a single sample statistic is used in the statistical test.

In a single-sample case, the standard of comparison is provided by the results of prior testing or it is some hypothetical value selected for theoretical or practical reasons. For example, suppose the opposition party wishes to show that the economic advisors to the president are misinformed. They take the position that *whatever* the advisors say is wrong. Thus, if we expressed this position in terms of null and alternative hypotheses, the null hypothesis would be whatever value the economic advisors are predicting, e.g., the inflation rate will be 10 percent. The alternative hypothesis would simply be that the true value is different from the one hypothesized. Symbolically,

$$H_0: \mu = \mu_0 = 10 \text{ percent}$$
$$H_1: \mu \neq \mu_0 \neq 10 \text{ percent}$$

8.2.1 Hypothesis testing involving a single mean: Large samples

The testing department of the automobile manufacturer randomly selected 60 cars off the production line for testing. The additive was placed in the gasoline mixture and the automobiles were run in the standard mileage test. The results were as follows: $\overline{X} = 27.05$, $s = 5.04$, $n = 60$.

Since n is large, the central limit theorem permits us to use z as the *test statistic* in deciding whether the additive significantly increased gasoline mileage over the previously established standard of miles per gallon. The test statistic, z, is

test statistic

A statistic used to test the null hypothesis.

$$z = \frac{\overline{X} - \mu_0}{s_{\overline{X}}} \tag{8.1}$$

in which

$\overline{X} =$ The sample mean of the cars with additive and is a point estimate of μ, the population mean of cars with additive

$\mu_0 =$ The population miles-per-gallon under the null hypothesis, i.e., $\mu_0 = 26.54$

$s_{\overline{X}} =$ The estimated standard error of the mean, $s\sqrt{n}$

Let us set up this problem in formal statistical terms.

1. *Null hypothesis.* H_0: The population mean after the additive is equal to or less than the population mean prior to mixing the additive, i.e., $\mu \leq \mu_0$.

2. *Alternative hypothesis.* H_1: The miles-per-gallon ratings on cars with the additive in the gasoline mixture are greater than the previous mileage standard established by the car, i.e., $\mu > \mu_0$.

3. *Statistical test.* The z-statistic is used since the test involves means and $n > 30$.

4. *Significance level.* $\alpha = 0.01$. A rigorous standard for rejecting H_0 was adopted by the automobile manufacturer to decrease the risk of a Type I error. It does not wish to take the chance of recommending an additive that really does not work.

5. *Sampling distribution.* The normal probability distribution is chosen because, with $n > 30$, the sampling distribution of means may be considered as approaching the form of the normal curve.

6. *Critical region.* Since H_1 is directional, the critical region consists of all values of $z \geq 2.33$. One percent of the area falls beyond $z = 2.33$ in one tail of the normal curve.

In the present example, a z corresponding to $\overline{X} = 27.05$ is

$$z = \frac{27.05 - 26.54}{\dfrac{5.04}{\sqrt{60}}} = \frac{0.51}{0.65} = 0.78$$

Since the obtained z is less than the critical value required for significance at $\alpha = 0.01$, one-tailed test, we fail to reject H_0. In fact, referring to Table V, we find that $z \geq 0.78$ will occur by chance approximately 22 percent of the time. The automobile manufacturer would be wise to withhold endorsement of the additive (see Figure 8.2).

It is important to note that decisions concerning the null hypothesis, alternative hypothesis, statistical test, significance level, sampling distribution, and critical region are all made during the planning stage of an investigation. We do *not* collect data first, and then, after the fact, decide the rules of the game.

8.2.2   Hypothesis testing involving a single mean: Small samples

The head of the shipping department in a New England company receives complaints from customers in the west that the orders are being received approximately two weeks after they are placed. He decides to check out another shipper. He randomly selects 16 orders from the next batch going to these customers and places them with a different carrier. He records the shipping time and finds $\overline{X} = 11$ days, $s = 3$. Is the new shipper's delivery time different from the old?

Let us set this problem up in formal statistical terms.

1. *Null hypothesis.* H_0: The mean of the population from which the sample was drawn equals 14, that is, $\mu = \mu_0 = 14$.

2. *Alternative hypothesis.* H_1: The mean of the population from which the sample was drawn does not equal 14, that is $\mu \neq \mu_0$. Note that H_1 is nondirectional since we are only asking whether the two shippers differ, not *how* they differ. Consequently, a two-tailed test of significance will be employed.

3. *Statistical test.*   Assuming that the population measurements are normally distributed, the Student t-ratio is chosen because we are dealing with a small sample. The test statistic,

$$t = \frac{\overline{X} - \mu_0}{s_{\overline{X}}} \tag{8.2}$$

In the sampling distribution of sample means ($n = 60$), the obtained $\bar{X} = 27.05$ ($z = $ **Figure 8.2** 0.78) does not lie in the critical region for $\alpha = 0.01$, one-tailed test. Therefore, we fail to reject H_0 and conclude that this sample mean could have been obtained by chance from the distribution specified under H_0.

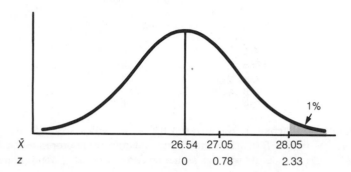

$\bar{X}$	26.54	27.05	28.05
z		0 0.78	2.33

4. *Significance level.* $\alpha = 0.05$. If the difference between the sample mean and the hypothesized population mean is so extreme that its associated probability of occurrence under H_0 is equal to or less than 0.05, we will reject H_0.

5. *Sampling distribution.* The sampling distribution is the Student t-distribution with df $= 15$.

6. *Critical region.* $t \geq |2.131|$. Since H_1 is nondirectional, the critical region consists of all values of $t \geq 2.131$ and $t \leq -2.131$. In the present example, $\bar{X} = 11$, $s_{\bar{x}} = 3/\sqrt{16} = 0.75$. Thus, $t = (11 - 14)/0.75 = -4.00$.

Decision. Since the obtained $t = -4.00$ falls within the critical region (i.e., $-4.00 < -2.131$), we may reject H_0 at the 0.05 level of significance. Thus, we may conclude that the sample was drawn from a population in which the mean shipping time is different from 14 days.

EXERCISES

8.6 Robbin has just purchased a 50-unit apartment building. In reviewing the records, she finds that many of the tenants do not pay their rent when due (on the first of the month). The rents tend to come in rather sporadically. Over the last two years, she finds rent payments are normally distributed with $\mu = 6.0$ days late and $\sigma = 3.0$ days late. She decides to institute a new system whereby a tenant is penalized when the rent is more than five days late. Her justification for this action is that the bank charges her a late fee if she is tardy with her mortgage payment.

The first month that the late penalty goes into effect, she finds the mean time rent is received, $\bar{X} = 4.0$ days. Has she succeeded in reducing her waiting time?

Set this problem up in formal statistical terms and draw the appropriate conclusions. Employ $\alpha = 0.01$ and decide whether a one- or two-tailed test is appropriate. (Although apartment buildings normally experience some vacancy rate, for purposes of this example, we shall use $n = 50$.)

8.7 In a five-year period from 1975 to 1979, the ratio of swimming pool sales to spa/hot tub sales nationally was $\mu = 11.20$. In the same five-year period, the ratio for sales in California was:

	Swimming pools
	Spa/hot tub
1975	7.58
1976	4.84
1977	1.54
1978	0.69
1979	0.05

Was California different from the rest of the nation with respect to this ratio?

Set this problem up in formal statistical terms and draw the appropriate conclusions. Employ $\alpha = 0.05$ and decide whether a one- or two-tailed test is appropriate.

8.3 HYPOTHESIS TESTING INVOLVING DIFFERENCES BETWEEN INDEPENDENT MEANS

Which of two different marketing strategies produces greater gains in sales?

Does an alloy containing a minute amount of a rare metal differ in hardness from the alloy without the rare metal?

Illumina Incorporated has two factories that produce fluorescent bulbs. It compares the life expectancies of two randomly selected samples of bulbs, one drawn from each factory.

Each of these examples has a point in common. Two independent samples are drawn from populations representing different conditions. We are interested in learning if these differences make a difference. Stated in another way, we wish to ascertain whether there is a statistically valid basis for concluding that the samples were drawn from different populations. Then, if we reject H_0, we may assert that the different conditions were responsible for the differences in sample means.

We shall examine the tests of significance for differences between means when the n in each group equals or exceeds 30 (large samples) and when the sample sizes are less than 30.

8.3.1 Hypothesis testing involving differences between independent means: Large samples

Because the sampling distribution of differences between means is approximately normal when each sample size is 30 or more, even if the underlying distribution is not normal, we may use z as the test statistic and the normal probability distribution for evaluating the significance of the difference. Thus, the test statistic is

$$z = \frac{\bar{X}_1 - \bar{X}_2}{s_{\bar{X}_1 - \bar{X}_2}} \qquad (8.3)$$

in which

$$s_{\bar{X}_1 - \bar{X}_2} = \sqrt{\frac{n_1 s_1^2 + n_2 s_2^2}{n_1 n_2}}$$

or if $n_1 = n_2$,

$$s_{\bar{X}_1 - \bar{X}_2} = \sqrt{\frac{s_1^2}{n_1} + \frac{s_2^2}{n_2}}$$

Example A

In an attempt to establish uniform product standards, Illumina Inc. wished to learn if there are differences in the life expectancies of fluorescent bulbs produced at its west- and east-coast factories. Two samples of $n = 60$ were randomly selected from the warehouse of each factory and were subjected to the same standard tests of life expectancy. The results were as follows:

$$\bar{X}_1 = 5{,}365 \qquad \bar{X}_2 = 5{,}128$$
$$s_1 = 850 \qquad s_2 = 925$$
$$n_1 = 60 \qquad n_2 = 60$$

a. Should we use a directional or a nondirectional alternative hypothesis?
b. Set the problem up in formal statistical terms, and test H_0 using $\alpha = 0.05$. Draw the conclusion appropriate to the findings.

Solution

a. Since there is no prior basis for expecting the life expectancies of bulbs to be greater at one factory than another, a two-tailed test of significance is appropriate.
b. 1. *Null hypothesis (H_0):* There is no difference in the population mean life expectancies of bulbs at the two factories, i.e., $\mu_1 = \mu_2$ or $\mu_1 - \mu_2 = 0$.
   2. *Alternative hypothesis (H_1):* There is a difference in the population mean life expectancies of bulbs at the two factories, i.e., $\mu_1 \neq \mu_2$ or $\mu_1 - \mu_2 \neq 0$.
   3. *Statistical test:* Since we are comparing two large samples randomly selected from the output of two different factories, the z-statistic for two independent samples is appropriate.
   4. *Significance level:* $\alpha = 0.05$.
   5. *Sampling distribution:* The standard normal probability distribution.

6. *Critical region:* Since H_1 is nondirectional, the critical region consists of all values of $z \geq 1.96$ or $z \leq -1.96$.
In the present example,

$$s_{\bar{x}_1 - \bar{x}_2} = \sqrt{\frac{(850)^2}{60} + \frac{(925)^2}{60}}$$
$$= 162.18$$

Therefore,

$$z = \frac{5365 - 5128}{162.18}$$
$$= 1.46$$

Decision: Since the value of the test statistic is not in the critical region for rejecting H_0, the null hypothesis cannot be rejected. Thus, there is no statistically valid basis for concluding that there was a difference in the mean life expectancies of bulbs produced at the two factories.

8.3.2 Hypothesis testing involving differences between independent means: Small samples

In October of 1980, economists were still trying to evaluate economic indicators to predict whether the recession had ended or even slowed down. The outlook for retailers at this point still appeared gloomy. Mass merchandisers reported mediocre gains (and some losses) from the previous year. In contrast, department-store companies had an improved (although still not spectacular) profit picture.

The following presents the percent change in sales from the previous year for six mass merchandisers and five department-store companies.

Mass merchandisers	Percent change	Department stores	Percent change
Montgomery Ward	+4.4	Federated Dept. stores	+8.5
Sears Roebuck	−3.9	May Dept. Store Co.	+8.1
J. C. Penney	+2.3	Carter Hawley Hale	+7.1
F. W. Woolworth	+4.6	Dayton-Hudson	+5.9
K mart	+4.0	Mercantile Stores	−2.0
Zayre	+11.8		

Is there a difference? Let us set this problem up in formal statistical terms.
1. *Null hypothesis (H_0):* There is no difference in the population means of the mass merchandisers and the department-store companies with respect to percent change in sales, i.e., $\mu_1 = \mu_2$, or $\mu_1 - \mu_2 = 0$.

2. *Alternative hypothesis (H_1):* There is a difference in the population means of the two groups with respect to percent change in sales, i.e., $\mu_1 \neq \mu_2$. Since we have no basis for predicting the direction of the difference, our alternative hypothesis is nondirectional. Therefore, we shall employ a two-tailed test of significance.

3. *Statistical test:* We use the t-ratio because we are dealing with small samples. The test statistic,

$$t = \frac{\overline{X}_1 - \overline{X}_2}{s_{\overline{X}_1 - \overline{X}_2}} \qquad (8.4)$$

4. *Significance level:* $\alpha = 0.05$.

5. *Sampling distribution:* The sampling distribution is the t-distribution with df $= n_1 + n_2 - 2$, or $6 + 5 - 2 = 9$.

6. *Critical region:* Since H_1 is nondirectional, the critical region consists of all values of $t \geq |2.262|$ or $t \geq 2.262$ or $t \leq -2.262$.

In this example, $\overline{X}_1 = 3.87$, $s_1 = 5.03$; $\overline{X}_2 = 5.52$, $s_2 = 4.32$;

$$s_{\overline{X}_1 - \overline{X}_2} = \sqrt{\frac{6(5.03)^2 + 5(4.32)^2}{30}} = 2.86$$

Therefore,

$$t = \frac{3.87 - 5.51}{2.86} = -0.58$$

Decision. Since our obtained t does not fall in the critical region, we fail to reject H_0. There is no statistically valid basis for asserting a difference between mass merchandisers and department-store companies with respect to percent change in sales.

EXERCISES

8.8 The creative department of a marketing firm has produced two different package designs for a product soon to be released for over-the-counter sales in drug departments. It wishes to ascertain which package design leads to greater sales. The country is divided into various test marketing regions and 300 stores are randomly selected for the test. One hundred fifty stores display one package design and the other 150 stores display the other design. The number sold within a specified time period is recorded. The results were:

$$\begin{array}{ll}
\overline{X}_1 = 27.45 & \overline{X}_2 = 37.14 \\
s_1 = 8.30 & s_2 = 9.44 \\
n_1 = 150 & n_2 = 150
\end{array}$$

a. Should a directional or a nondirectional alternative hypothesis be used?

b. Set the problem up in formal statistical terms and test H_0, using $\alpha = 0.01$. Draw the appropriate conclusion.

8.9 In 1976 Congress struck down a strong federal regulation requiring motorcyclists to wear helmets. Subsequently, states began repealing or weakening laws requiring the use of helmets. Motorcycle fatalities nationwide are up 47 percent since 1976.

A federal motorcycle-safety specialist claimed, "The single most significant factor in the increase in fatalities is the repeal of helmet laws."

In an attempt to test this assertion, the authors randomly selected four states that repealed their stringent helmet regulations and four states that continue to require helmets. We calculated the average annual increase (or decrease) in motorcycle fatalities during this period and obtained the following results:

No helmet required	Helmet required
$\bar{X}_1 = 0.2221$	$\bar{X}_2 = 0.1036$
$s_1 = 0.0211$	$s_2 = 0.0346$
$n_1 = 4$	$n_2 = 4$

On the basis of these results, can we conclude that there is a significant difference in annual rate of increase in motorcycle fatalities between states that do or do not require helmets? Set this problem up in formal statistical terms. Employ $\alpha = 0.05$ and decide whether a one- or two-tailed test is appropriate.

8.4 HYPOTHESIS TESTING INVOLVING DIFFERENCES BETWEEN MEANS: DEPENDENT OR PAIRED MEASURES

As we have seen, when we conduct a comparison involving independent means, we randomly select our sample from the population and then randomly assign members of the sample to each of the comparison groups. However, there are numerous occasions when the members are not randomly assigned to the comparison groups:

Before-after comparisons. Measures are taken on each member of the sample both before and after the introduction of changed conditions. Several examples are: Mortgage funds available from a savings and loan institution prior to and after changes in federal money policies; miles-per-gallon ratings of cars before and after mounting steel-belted radial tires; or employee morale before and after changes in personnel policies.

Matched measures. Before-after measures often introduce problems in interpreting research results. For example, if attitudes of employees are measured a second time, any changes may reflect their reactions to the second administration of the questionnaire. Or they may remember their responses to the first administration and, in order to be consistent, answer in the same way to the second administration. Another possibility is that some extraneous variable, occurring between first and second administrations, may have affected response measures. For example, a strike may override any reactions to improved personnel practices.

One way to overcome the problem of before-after measures on the same sample is to use two samples in which each member in one group is matched or paired with a comparable measure of the second group. For example, if you wanted to evaluate two different tread designs on motorcycle tires, you might put a tire of one design on the front wheel and one of the other design on the rear wheel. These tires are matched in the sense that they are exposed to the same driving conditions. Of course, it would be necessary to carefully balance the assignment of tires to the front and rear wheel since position on the motorcyle might also affect tread wear.

Before-after and matched-measures designs are called dependent since the assignment of a member to one condition is not independent of the assignment of a member to another condition. Thus, if one member of a pair is randomly assigned to condition A, the second member must be assigned to condition B.

When $n < 30$, we use the t-ratio for dependent measures. In a dependent-measures design,

$$t = \frac{\bar{d}}{s_{\bar{d}}} \tag{8.5}$$

in which
$\bar{d} =$ The mean difference, $(\Sigma d)/n$
$s_{\bar{d}} =$ The standard error of the difference

The standard error of the difference is, in turn, defined as follows:

$$s_{\bar{d}} = \frac{s_d}{\sqrt{n}} \tag{8.6}$$

in which
$s_d =$ The standard deviation of the difference and
$n =$ Number of pairs

Finally, the standard deviation of the difference, s_d, is defined as follows:

$$s_d = \sqrt{\frac{\Sigma d^2 - \dfrac{(\Sigma d)^2}{n}}{n-1}} \cdot \tag{8.7}$$

By combining Formulas 8.6 and 8.7, we may arrive at a single computational formula for the standard error of the difference:

$$s_{\bar{d}} = \sqrt{\frac{\Sigma d^2 - \dfrac{(\Sigma d)^2}{n}}{n(n-1)}} \cdot \tag{8.8}$$

Let's look at an example of a dependent-samples design.

Example B

Beverly M., president of a savings and loan association, claims that too much emphasis is placed on the amount of money each client has on deposit. Equally important is the length of time the money remains on deposit. She designed an index that multiplies the amount of money on deposit by the time on deposit, with adjustments made for additional deposits or withdrawals. She then introduced a premium incentive plan designed to discourage withdrawals. She randomly selected 15 accounts and compared the index scores before and after introducing the plan. The results were as follows:

Account	After	Before	Difference (d)
1	10,540	10,020	520
2	780	720	60
3	9,453	9,105	348
4	1,573	1,062	511
5	3,962	3,905	57
6	4,673	4,401	272
7	8,205	8,100	105
8	12,458	12,011	447
9	959	847	112
10	7,444	6,853	591
11	4,982	4,602	380
12	8,831	8,452	379
13	648	182	466
14	6,969	6,740	229
15	2,403	2,378	30

a. Should a directional or nondirectional hypothesis be used in assessing her claim?
b. What is the appropriate statistical test?
c. Using $\alpha = 0.01$, evaluate her claim.

Solution

a. Since we are interested only in evaluating her claim that premium incentives increase the deposit $\times$ duration measure, we use a directional test.
b. Since $n < 30$ and we are using a before-after design, we use the t-ratio for dependent measures.
c. Setting up the problem in formal statistical terms, we obtain:
   1. *Null hypothesis (H_0):* The population mean difference is equal to or less than zero, i.e., $\mu_d \leq 0$.
   2. *Alternative hypothesis (H_1):* The population mean difference is greater than zero, i.e., $\mu_d > 0$.
   3. *Statistical test:* Since $n < 30$, we use the t-test as the test statistic. In the case of dependent measures, n refers to the number of pairs.
   4. *Significance level:* $\alpha = 0.01$, one-tailed test.
   5. *Sampling distribution:* The t-distribution with df $= n - 1 = 14$ (the degrees of freedom is the number of pairs minus one).

6. *Critical region:* The critical value for df = 14, one-tailed test: $t \geq 2.624$.

When calculating the various measures, we find:

$$\bar{d} = \frac{\Sigma d}{n} = \frac{4,507}{15} = 300.47$$

$$s_d = \sqrt{\frac{\Sigma d^2 - \dfrac{(\Sigma d)^2}{n}}{n-1}} = \sqrt{\frac{1,864,655 - \dfrac{(4,507)^2}{15}}{14}} = 190.95$$

$$s_{\bar{d}} = \frac{s_d}{\sqrt{n}} = \frac{190.95}{3.87} = 49.34.$$

Therefore,

$$t = \frac{\bar{d}}{s_{\bar{d}}} = \frac{300.47}{49.34} = 6.090, \text{df} = 14$$

The critical value at $\alpha = 0.01$, one-tailed test when df = 14, is 2.624. Since $t = 6.090$ exceeds the critical value, we may reject H_0. Apparently, Beverly M.'s premium incentive plan does increase the measure based on deposits and duration of deposits.

EXERCISES

8.10 Swap meets (or "flea markets") generally take place on Saturdays and Sundays. One merchant decides to see whether the *day* makes a difference in the mean net income (revenue − cost of goods sold − expenses such as booth rental). He randomly selects 16 "steady" merchants and finds their daily net income to be (rounded to the nearest dollar):

Merchant	Saturday	Sunday
1	540	600
2	282	364
3	510	600
4	460	420
5	180	265
6	350	410
7	500	540
8	460	500
9	700	680
10	400	375
11	310	380
12	540	500
13	400	480
14	125	160
15	325	465
16	720	900

Does day of the week make a difference? Use $\alpha = 0.05$ and decide whether a one- or two-tailed test is appropriate. Set this problem up in formal statistical terms and make the appropriate conclusions.

8.5 HYPOTHESIS TESTING INVOLVING A SINGLE PROPORTION

The one-sample significance test of a proportion is conceptually the same as a one-sample test of a mean. A sample proportion is obtained, the standard error of the proportion is calculated from the sample data, and the statistical significance of the sample proportion is tested against a prior known or hypothesized proportion. When the sample size is equal to or less than 30, the binomial table provides direct probability values for a given value of x or $(n - x)$ (See Section 5.3.2). When $n > 30$, x is divided by n to yield the sample proportion. Since the sampling distribution of a proportion approaches normality with increases in n, z may be used as the test statistic. Thus,

$$z = \frac{\bar{p} - P_0}{s_p} \tag{8.9}$$

in which

$$s_{\bar{p}} = \sqrt{\frac{\bar{p}\,\bar{q}}{n}} \; ;$$

$P_0 = $ The proportion expected under H_0
$\bar{p} = $ The obtained proportion
$\bar{q} = (1 - \bar{p})$

The significance test of a proportion, one-sample case, is commonly used to evaluate the results of a survey in which the respondents have only two choices—candidate A versus B, strike versus no strike, like versus dislike. In many cases, the proportion hypothesized under H_0 is $P_0 = 0.50$. This is due to the fact that the decision is based on the rejection of the hypothesis of an equal split. Thus, a pollster would say that a candidate is leading if the proportion favoring the candidate is significantly greater than 0.50, or that a strike will ensue if more than 50 percent of the membership votes for a strike.

Example C

In order to encourage carpooling, a large supplier of software for the computer industry has instituted a policy of providing privileged parking spaces for employees who carpool. Because of a mixed bag of vocal opposition as well as support for the policy, a random sample of 300 employees was polled in order to take the pulse of the 8,000 employees. The results were as follows:

> 180 employees favored the policy
> 110 employees disliked the policy
> 10 were undecided

Thus, the total sample of those responding was 300 ("undecided" is considered a response). The proportion in the sample who favored the policy is $\bar{p} = 180/300 = 0.60$, and the proportion in the sample who were either undecided or disliked the policy is $\bar{q} = 0.40$.

a. What is the appropriate test of significance?
b. Is the alternative hypothesis directional or nondirectional?
c. Conduct the appropriate analysis, using $\alpha = 0.01$, and draw the conclusion warranted by the data.

Solution

a. Since we are dealing with a one-sample case involving the comparison of an obtained proportion with a hypothesized value ($P_0 = 0.50$), the one-sample test of a proportion is appropriate.
b. The test of significance is nondirectional. Management is attempting to assess the impact of the new policy because it has no clear idea of employee reactions.
c. 1. *Null hypothesis (H_0):* The sample was drawn from a population in which the proportion favoring the policy is 0.50, i.e., $P = P_0 = 0.50$.
   2. *Alternative hypothesis (H_1):* The sample was drawn from a population in which the proportion favoring the policy is not 0.50, i.e., $P \neq P_0$.
   3. *Statistical test:* Since the binomial distribution approaches the normal probability distribution with a large n, z is the appropriate test statistic.
   4. *Significance level:* $\alpha = 0.01$.
   5. *Sampling distribution:* The normal probability distribution.
   6. *Critical region:* $z \geq 2.58$ or $z \leq -2.58$.

The standard error of the proportion, $s_{\bar{p}}$, is

$$s_{\bar{p}} = \sqrt{\frac{(0.60)(0.40)}{300}}$$
$$= 0.0283.$$

Therefore,

$$z = \frac{0.10}{0.0283} = 3.53$$

Since z is in the upper critical region, it appears that the employees are not equally divided on the issue. Rather, the majority seems to favor the new policy.

8.6 HYPOTHESIS TESTING INVOLVING DIFFERENCES BETWEEN PROPORTIONS

A plant manager must replace an obsolete machine that produces ball bearings for washing machine motors. The choice has been narrowed to two competing machines. In fact, the decision will be based on estimates of the proportion of defective ball bearings produced by each machine.

The executive committee of a union is preparing its platform for negotiating a new contract. It wishes to learn if the west coast membership differs from the east coast membership in the proportion who would prefer direct salary increases rather than fringe benefits.

Two companies can both supply a certain type of flashbulb for photographic supply firms. Since the prices are the same, the choice will be determined by randomly selecting samples of flashbulbs from each company and subjecting these flashbulbs to various tests. Of 150 flashbulbs tested from Company 1, 7 were defective, whereas 20 of 300 were defective for Company 2. Can we conclude that Company 1 supplies a better flashbulb?

In each of these cases, we collect two separate samples and we wish to know whether the sample proportions are representative of the same population. The null hypotheses would be, respectively: there is no difference in the proportions of defective ball bearings produced by the two machines; there is no difference in the proportions of the membership at two different locations who prefer direct salary increases; and there is no difference in the proportions of defective flashbulbs supplied by the two companies. In this two-sample case, when H_0 is $P_1 = P_2$, a pooled estimate of the standard error of the difference between proportions should be obtained. This is a two-stage process: the first involves the calculation of a combined proportion using each sample proportion to estimate the overall proportion. Thus,

$$p = \frac{n_1 \bar{p}_1 + n_2 \bar{p}_2}{n_1 + n_2} \tag{8.10}$$

in which

$p =$ The pooled estimate of the population proportion

$\bar{p}_1 =$ The proportion of interest in sample 1

$\bar{p}_2 =$ The proportion of interest in sample 2

$n_1 \bar{p}_1 =$ The number of items of one type in sample 1

$n_2 \bar{p}_2 =$ The number of items of another type in sample 2

The formula produces a pooled estimate of the proportion of items of one type in the two populations, on the assumption that the proportion in the first population equals the proportion in the second population.

The second step involves the calculation of the pooled standard error of the difference in proportions. This is

$$s_{\bar{p}_1-\bar{p}_2} = \sqrt{pq\left(\frac{1}{n_1}+\frac{1}{n_2}\right)} \qquad (8.11)$$

in which

$$q = 1 - p$$

Therefore, the test statistic

$$z = \frac{\bar{p}_1 - \bar{p}_2}{s_{\bar{p}_1-\bar{p}_2}} \qquad (8.12)$$

Let's look at an example.

Example D

In any recession, companies may cut or omit dividends, but usually as a last resort. In fact, some companies may maintain dividends as an outward sign of good health. A survey by Standard & Poor compared the percentage of companies that cut or omitted dividends during the first nine months of 1980 to the percentage for a similar period in 1975. Based on 1,537 companies in 1975, 24.14 percent decreased or omitted dividends; whereas in 1980, of 2,026 companies, 9.62 percent did likewise. Is there a significant difference between 1975 and 1980 with respect to slashing of dividends?

These findings are summarized below:

$$\begin{array}{cc} 1975 & 1980 \\ \bar{p}_1 = 0.2414 & \bar{p}_2 = 0.0962 \\ n_1 = 1537 & n_2 = 2026 \end{array}$$

Our pooled estimate of the population proportion is

$$p = \frac{(1537)\,(0.2414) + (2026)\,(0.0962)}{1537 + 2026} = 0.1588$$

Therefore,

$$\begin{aligned} s_{\bar{p}_2-\bar{p}_2} &= \sqrt{(0.1588)\,(0.8412)\left(\frac{1}{1537}+\frac{1}{2026}\right)} \\ &= \sqrt{(0.1352)(0.0011)} \\ &= \sqrt{0.0001487} \\ &= 0.0122. \end{aligned}$$

Solution

In formal statistical terms we have:

1. *Null hypothesis (H_0):* Both sample proportions were selected from the same population, i.e., $P_1 = P_2$ or $P_1 - P_2 = 0$.
2. *Alternative hypothesis (H_1):* The two sample proportions were selected from different populations, i.e., $P_1 \neq P_2$. The alternative hypothesis is nondirectional since there is no clear-cut basis on which to predict the *direction* of the difference.
3. *Statistical test:* Since both n_1 and n_2 exceed 30, the z-statistic may be used.
4. *Significance level:* $\alpha = 0.01$.
5. *Sampling distribution:* The standard normal probability distribution.
6. *Critical region:* Since H_1 is nondirectional, any value of $z \geq 2.58$ or $z \leq -2.58$ will lead to rejection of H_0.

In this example, the test statistic z is

$$z = \frac{0.2414 - 0.0962}{0.0122}$$
$$= 11.90.$$

Since the test statistic is in the critical region, we reject H_0. There was significantly less slashing of dividends in 1980 than during a comparable period in 1975.

EXERCISES

8.11 A survey of doctorate recipients conducted by the National Research Council reported that, of 64,800 humanists (e.g., history, English) who earned their Ph.D.s in the period 1936–78, 25 percent were female. Yet of the 1,423 persons from this group who were unemployed and looking for work, 741 were female. Are there significantly more unemployed females than expected?

Set this problem up in formal statistical terms and state the appropriate conclusions. Employ $\alpha = 0.01$ and indicate whether a directional or nondirectional H_1 should be used.

8.12 The same survey (Exercise 8.11) reported that, of 308,800 Ph.D.s in science and engineering, 11 percent were female. Of 2,852 unemployed and seeking employment, 32 percent were female.

  a. Does this represent a significant departure from the proportion in the population?
  b. Is there a difference in the proportion of unemployed females with science and engineering Ph.D.s as compared to females with Ph.D.s in the humanities?

Set these problems up in formal statistical terms and state the conclusions warranted by the data. Employ $\alpha = 0.01$ and decide whether one- or two-tailed tests should be used.

SUMMARY

In this chapter we looked at hypothesis testing involving means, differences between means, proportions, and differences between proportions.

1. A statistical test of sigificance involves two mutually exclusive and exhaustive hypotheses, the null hypothesis (H_0) and the alternative hypothesis (H_1).
2. The null hypothesis specifies an expected value for the population. It usually takes the form of "no effect" or "no difference."
3. The alternative hypothesis denies the null hypothesis.
4. Statistical proof is both indirect and probabilistic. By rejecting the null hypothesis, we assert the alternative hypothesis.
5. Rejection of the null hypothesis involves a judgment based on probability. If the obtained result would have rarely occurred in the sampling distribution of the statistic, we reject H_0 and assert H_1.
6. "Rarely" is, in turn, defined by probability. The 5 percent significance level means that the result would be obtained by chance 5 percent of the time or less. Similarly, using $\alpha = 0.01$, H_0 is rejected when the result would be obtained by chance 1 percent of the time or less.
7. Statistical proof is not absolute. Two types of error that may be made are Type I or Type α error and Type II or Type β error.
8. A Type I error consists of falsely rejecting H_0; i.e., rejecting H_0 when it is true. The probability of this type of error is α.
9. A Type II error occurs when we fail to reject a false H_0.
10. Both the null and alternative hypotheses may either be nondirectional or directional. When nondirectional, the critical region is found in both tails of the sampling distribution. When directional, the critical region is only one-tailed.
11. Hypotheses involving a single mean are usually evaluated against a known prior mean or a hypothetical mean. When $n \geq 30$, z may be used as the test statistic. When $n < 30$, the Student t-ratio is used.
12. Tests involving two sample means usually involve two different conditions; i.e., there are *two samples*, which are drawn from *two populations*. The usual H_0 is that the mean of the first population equals the mean of the second population. Rejection of H_0 permits us to infer that the conditions produced different results.
13. Dependent tests of significance are used when the measurements are paired in some way. This may be accomplished by using before-after measures on the same persons or objects or by matching them on some known basis.
14. Hypothesis testing involving a single proportion is used to evaluate a sample proportion against a known or hypothetical proportion. It is used in binomial situations when $n > 30$.

15. Hypothesis testing involving two sample proportions is conceptually similar to the test of significance of the difference between means. H_0 is typically that the proportion of the first population equals the proportion of the second population. Rejection of H_0 permits us to infer differences in the populations of the variable of interest.

TERMS TO REMEMBER

null hypothesis (H_0) (224)
alternative hypothesis (H_1) (225)
significance level (227)
alpha (α) (227)
Type I (Type α) error (227)
Type II (Type β) error (227)
directional hypothesis (228)

nondirectional hypothesis (228)
two-tailed probability value (229)
one-tailed probability value (229)
critical region (230)
critical value (230)
test statistic (233)

EXERCISES

8.13 It is almost always difficult to get directly through to an airline's information line. If the line isn't busy, then you are usually answered by a recording, which tells you to hold on until the first available agent will service your call. When one or more airlines are on strike, the phone traffic increases to the airlines still operating.

How long will people wait on "hold"?

Harriet, in charge of public relations for an airline, decides to try two different systems and see whether she can increase the time people will wait until serviced. She finds that, normally, people will wait for an airline agent a mean time of six minutes. Once a person hangs up, there is no way to know whether they will call back or try a different airline.

    a. Harriet installs a taped information system so that the waiting customer hears information about special rates, offers, discounts, etc. during the waiting period. A random sample of 100 of these calls finds customers wait $\bar{X} = 7.2$ minutes, $s = 2.1$. Has she succeeded in increasing waiting time?

    b. Harriet instructs some of the operators handling these calls to make contact with the party every minute or so, saying "An agent will be with you shortly. Thanks for waiting." She randomly samples 100 of these calls and finds $\bar{X} = 8.1$ minutes, $s = 2.2$. Has this system succeeded in increasing waiting time?

    c. Is there a difference in mean waiting time between her two new systems?

Set these problems up in formal statistical terms and state the appropriate conclusions. Employ $\alpha = 0.05$ and indicate whether directional hypotheses are justified.

8.14 Recall that Montana Rincon found 60 errors when he randomly selected and audited 1,000 accounts from the records of a credit card company (Section 2.1). When researching through the literature, he finds three different estimates of error, namely,

    a. 0.03.
    b. 0.04.

c. 0.05.

Using each in turn as the hypothesized population proportion, test H_0: $P = P_0$. Use $\alpha = 0.01$.

8.15 Rincon also found all 60 errors were in one direction only—favoring the credit card company. He reasoned that a random variable should produce roughly an even split between positive and negative errors and, in the long run, a mean error of zero. He obtained a mean error of 16.23 cents with $s^2 = 69.20$.

    a.   What test of significance should he use?

    b.   Should the test be directional or nondirectional?

    c.   Set up in formal statistical terms, calculate the appropriate test statistic, and draw the conclusions warranted by the data.

    d.   Has Rincon proved computer fraud?

8.16 A survey of 1,500 randomly selected households in Tucson revealed that 77 percent of the homes were heated by gas. Does this represent a decrease from the 1977 figure of 86 percent? Use $\alpha = 0.05$. Set this problem up in formal statistical terms. Conduct the appropriate statistical test and state your conclusions.

8.17 The same survey (Exercise 8.16) of 1,500 randomly selected Tucson households found 35 percent of all female heads working in 1979 as compared with 31 percent in 1977. Does the 1979 figure represent a significant departure from 1977? Use $\alpha = 0.01$, two-tailed test. Set this up in formal statistical terms, calculate the appropriate test statistic and state your conclusions.

8.18 Refer to Exercise 8.17. Suppose we were only interested in whether or not there had been a significant *increase* in working women over the two-year period. Discuss the differences in the statistical test.

8.19 In the summer of 1980, much of the midbelt of the country, from Kansas to Texas, was hit by a devastating heat wave and drought. In order to ascertain if there were likely to be significant losses in the Texas cotton crop, the weight of the first pickings of 25 one-acre plots was obtained. The results were: $\overline{X} = 720$ lbs; $s = 140$; $n = 25$.

Based on prior experience, the same acreage would be expected to produce 825 pounds per acre.

    a.   What test of significance should be used?

    b.   Should the test be one- or two-tailed?

    c.   Conduct the appropriate statistical test, using $\alpha = 0.01$, and draw the conclusion warranted by the data.

8.20 A plant manager selected random samples of the output of each of two machines that produce ball bearings and tested them for defects. Out of 220 produced by machine 1, 15 were defective. A total of 18 out of 340 produced by machine 2 was found defective.

    a.   What is the appropriate test of significance?

    b.   Is H_1 directional or nondirectional?

    c.   Set up in formal statistical terms, test H_0, and draw the conclusion warranted by the data. Use $\alpha = 0.05$.

8.21 Companies wishing to explore for and mine uranium are complaining about the resistance to ore drilling in many northeastern states in the United States. Claiming eastern ore is much higher grade than western ore, these companies argue that only by mining in the northeastern states can we hope to stem the declining output of uranium oxide. In one study comparing the ore drillings of 23 western sites with those of 20 eastern sites, the following sample means and standard deviations were obtained. Assume the population values are normally distributed.

Eastern sites	Western sites
$\bar{X}_1 = 0.60\%$	$\bar{X}_2 = 0.12\%$
$s_1 = 0.15\%$	$s_2 = 0.05\%$
$n_1 = 20$	$n_2 = 23$

    a. What test of significance should be used?
    b. Should the test be directional or nondirectional?
    c. Conduct the appropriate statistical test, using $\alpha = 0.01$, and draw the conclusion warranted by the data.

8.22 According to the United States League of Savings Associations, mortgage loan deliquency was 0.80 percent (or 0.0080) in July 1980, based on a survey of 1,092 savings and loans associations. During the 1974–75 recession, delinquencies peaked at 1.02 percent (or 0.0102).

    Is the July 1980 figure significantly less than the level recorded during the last recession? Set this problem up in formal statistical terms. Employ $\alpha = 0.05$ and state the conclusion warranted by the data. Is the one- or two-tailed test appropriate?

8.23 In October of 1980, President Carter signed a bill that substantially deregulated the railroad industry. As a result, railroads were given considerable leeway to raise freight rates free of Interstate Commerce Commission control. Although deregulation in other transportation industries has usually led to slashing of some air fares and truck rates, rail deregulation has led to increased rail rates and profits.

    Some commodities have been harder hit than others insofar as rate increases. One furniture manufacturer in Atlanta complained that he has suffered greater increases than shippers of other commodities. He now pays $598 for shipping 12,000 pounds of furniture by rail to Charlotte, N.C. instead of $437. This represents an increase of approximately 37 percent. To support his claim, he randomly samples 40 rail increases on commodities such as grain, rock, logs, and cotton products. He finds that the average increase is 32 percent with $s = 11$ percent. Is his claim valid? Is a one-tailed test appropriate? Explain. Use $\alpha = 0.05$.

8.24 The personnel director of a large corporation has been charged with the responsibility of designing a program to reduce absenteeism. After designing an incentive program, she wishes to evaluate its effectiveness prior to implementation. She randomly selects the names of 30 employees from the personnel files, obtains the attendance records of each, and then forms pairs on the basis of the similarity of these attendance records. Thus, the two with the poorest records would constitute one pair, the next two would make up the second pair, etc. One member of each pair is randomly assigned to the incentive condition (experimental group) and the other to the control condition. Following the administration of the incentive to the experimental group, the following results were obtained during the ensuing six-month period.

	Days absent	
Pair	Control	Experimental
1	20	14
2	17	12
3	14	11
4	16	13
5	12	11
6	8	11
7	11	9
8	9	10
9	12	7
10	7	5
11	6	8
12	5	3
13	8	6
14	3	0
15	4	1

a. What is the appropriate test of significance?

b. Should the test be one- or two-tailed?

c. Set the problem up in formal statistical terms, calculate the appropriate test statistic and, using $\alpha = 0.05$, draw the conclusion warranted by the data.

8.25 The state agriculture department has begun a program to upgrade its system for detecting pesticide residues, starting with purchases of new equipment. An article in the *Los Angeles Times* reported that most laboratories are incapable of detecting more than 30 percent of the pesticides.

A spokesman for pesticide regulation in California said that even with the new equipment, the state laboratories will still not be able to detect more than 90 of the 279 pesticidal chemicals.

In a test of the new versus the old equipment, the following means and variances of percentage of detection were obtained:

New equipment	Old equipment
$\bar{X}_1 = 33.05$	$\bar{X}_2 = 30.14$
$s_1^2 = 16.43$	$s_2^2 = 14.99$
$n_1 = 35$	$n_2 = 31$

a. What is the appropriate test of significance?

b. Should a directional or nondirectional H_1 be used?

c. Conduct the appropriate statistical test, using $\alpha = 0.01$, and draw the conclusions warranted by the data.

8.26 In an effort to gain greater compliance with federal highway speed limits, an experiment was conducted in which the speed of randomly selected automobiles were checked by radar under two different conditions: simulated patrol cars visible along the highway versus patrol cars absent. A total of 14 observations were made under both conditions. The results were as follows:

Patrol cars absent	Simulated patrol cars
$\bar{X}_1 = 66.50$	$\bar{X}_2 = 55.86$
$s_1^2 = 32.27$	$s_2^2 = 32.13$
$n_1 = 14$	$n_2 = 14$

    a. What is the appropriate test of significance?

    b. Should a directional or nondirectional H_1 be used?

    c. Conduct the appropriate statistical test, using $\alpha = 0.01$, and draw the conclusion warranted by the data.

8.27 A cigarette manufacturer is considering the introduction of a new ultra-low-tar cigarette. It has narrowed its choice to two alternatives: a cigarette with 0.1 milligram of tar versus a cigarette with about 0.15 milligram of tar. Its marketing department decides to set up a taste comparison test. A panel of volunteer subjects was divided randomly into two groups: the first group smoked the 0.1 mg. cigarette and the second tried the 0.15 mg. cigarette. The subjects were asked, "Would you buy this low-tar cigarette in preference to your present brand?" A total of 40 percent of group 1 responded positively ($n_1 = 140$). Fifty-five percent of the subjects in group 2 indicated a willingness to switch ($n_2 = 180$).

    a. Is the preference of group 1 significantly different from 50 percent? Is H_1 directional or nondirectional? Use $\alpha = 0.01$.

    b. Is the preference of group 2 significantly different from 50 percent? Is H_1 directional or nondirectional? Use $\alpha = 0.01$.

    c. Do the preferences of the two groups differ significantly from one another? Use $\alpha = 0.01$.

8.28 Gina sells advertising space for a local newspaper chain. She decides to try to sell swap-meet merchants on the idea of advertising in her local daily. The largest part of Gina's audience is located an average of 25 miles from the swap meets. Gina claims that customers will travel 25 miles or more to come and shop. To justify her claim, she takes a random sample of 100 customers and records the distance traveled. She obtains $\bar{X} = 27$ miles and $s = 9$. What can we say about Gina's claim? Use $\alpha = 0.05$. Is a directional hypothesis warranted?

8.29 Union leadership conducted a poll of randomly selected members on the east and west coasts concerning their preference for direct wage increases as opposed to "hidden" increases in the form of fringe benefits. A total of 485 of 900 east coast union members favored direct increases; 510 of 880 west coast members favored direct increases.

    a. Test the hypothesis that the east coast members are evenly split on this issue. Use $\alpha = 0.05$. Is the hypothesis directional or nondirectional?

    b. Test the hypothesis that the west coast members are evently split. Use $\alpha = 0.05$. Is the hypothesis directional or nondirectional?

    c. Determine whether or not the east coast and west coast members differ on the issue. Is the hypothesis directional or nondirectional? Use $\alpha = 0.05$.

8.30 A comparison shopper randomly selects 21 items from a list of available grocery supplies. She visits two supermarkets and obtains the prices of identical items at each market. She finds:

Item	Market 1	Market 2
1	6.75	6.50
2	2.39	2.30
3	0.48	0.52
4	7.04	6.54
5	1.05	0.95
6	0.98	0.89
7	0.72	0.79
8	1.45	1.32
9	3.00	3.40
10	4.90	4.45
11	1.50	1.20
12	0.25	0.24
13	0.58	0.63
14	1.33	1.19
15	2.98	2.80
16	1.35	1.30
17	0.98	1.05
18	4.44	4.14
19	5.49	5.24
20	3.97	3.84
21	0.88	0.96

a. What is the appropriate test of significance?

b. Should H_1 be directional or nondirectional?

c. Formulate the problem in formal statistical terms, conduct the appropriate test of significance, $\alpha = 0.05$, and draw the conclusion warranted by the data.

8.31 *Recession-proof spender* is a term used to describe the individual who spends all of his or her income and seems virtually unaffected by any downturn in the economy.

Marketing analysts claim that 40 percent of the country's population seem to fall in this category. They come from the ranks of the wealthy, the single, childless two-income households, and those whose incomes are adjusted with inflation.

One economist from California, after looking at the demographics and income of 1,500 randomly selected people in California, concluded that 59 percent of the population in California are recession-proof spenders.

Do these results provide sufficient evidence to conclude that California residents are significantly different from the rest of the country? Use $\alpha = 0.05$. Is a directional hypothesis justified? Set this problem up in formal statistical terms, conduct the appropriate statistical test and state the conclusion warranted by the data.

Decision Theory

9.1 BASIC CONCEPTS

Imagine yourself in each of the following situations.

Your are faced with the choice of two jobs. How do you decide which one to take?

You cannot decide whether to rent an apartment or buy your own home.

You are at the crossroads of your educational career. Should you study for real estate? The stock market? Data processing?

You have a few dollars. What is the best way to invest it?

There is a bulletin board in our office that announces: "When confronted by two good choices, choose both." Unfortunately, life is never quite so simple. Some decisions are easier to make than others. This is particularly true when nothing is left to chance—when the outcome of each decision is known with complete *certainty*. Under these conditions, we only have to sift through all the alternatives in order to come up with the best decision.

However, most of the time we must make a decision even though we are *uncertain* about the outcome. We can choose only *one* job; we can either rent an apartment *or* buy a house, but not both (at this time); we can major in only *one* field; we have only a limited choice of investment opportunities.

You may have to make decisions regarding the kind and amount of inventory to carry; you may have to decide how many new cars, machines, etc., you can or should produce in a given period of time; you may be involved in deciding the size and capacity of a new plant; you may have to decide whether to install alternative forms of energy; you may have to decide whether to lay off a certain number of people within a specified period of time.

All of these examples have certain things in common—they all involve *decisions* (alternatives to choose from; actions). The decisions themselves are usually under our control. In contrast, various other factors may be operating over which we exercise little or no control. If we are fortunate, we can elucidate these factors and perhaps make our decision in the light of "What would happen if . . . ?"

For example, in choosing between two jobs, there may be differences in the way the money is earned (e.g., commission versus salary); there may be differences in the kind of future they offer (one may promise more rapid advancement than the other); and there may be geographical differences. In some cases, we can express these differences in monetary terms. However, some differences may not be strictly comparable on the basis of purely monetary considerations. To illustrate, the geographical implications may be more than just financial (e.g., they might involve health, comfort, or leaving family, friends, etc.). We shall refer to this "measure of value" as *"utility."* For the most part, in this chapter, we shall concentrate on those decisions that lead to *payoffs* measurable in monetary terms.

Finally, we must have some way of specifying the one or more random factors that will affect the outcome of a decision. In other words, we must be able to identify the various *states of nature* that will influence the outcome

decisions made under certainty

The outcome of each decision is known; nothing is left to chance.

decisions made under uncertainty

The outcome of any particular decision (or choice) is influenced by one or more random factors or events.

decisions

The various choices, alternatives, or actions that confront a decision maker.

utility

When the outcome of a decision is measured in terms of values other than strict monetary gain.

of any particular choice (or *act*) that we make. Further, we need some way of determining the likelihood of occurrence of each *state of nature*. Although Section 9.2 goes into this more fully, a simple illustration may serve to introduce a conceptual understanding of this problem.

Suppose, for example, the two jobs differ in income depending upon such things as, say, low, medium, or high sales. Thus, the outcome for any particular choice (e.g., choose Job 1) is dependent upon which *event* occurs—low, medium, or high sales. Or, the choice of a major of, say, real estate versus data processing may involve assessing the probabilities associated with the growth potential of these fields.

Once we can assign probability values (based either on past records— *empirical probabilities*—or on experience and judgment—*subjective probabilities*), we are on our way to formalizing the problem and taking an action based on *decision analysis*.

To summarize: In order to evaluate a problem within the rigors of decision analysis, we must have:

1. *A choice of alternatives (decisions; actions).* For example: Take job 1 or job 2? Take this apartment or buy this house? Major in real estate or data processing? The decisions must be mutually exclusive and exhaustive.

2. *A way to measure payoff of the various choices.* Payoff may be measured in terms of gain, loss, costs, rewards or penalties. We shall first deal with payoffs that can be expressed in monetary terms.

3. *Varying states of nature and events* that we cannot control but to which we can assign probability values (related to their likelihood of occurrence). For example: Low, medium, or high sales; low growth potential or high growth potential. The events must be mutually exclusive and exhaustive.

4. *An objective.* A decision maker must decide whether his or her objective is to maximize monetary gain, minimize risk, or obtain some combination thereof.

We speak of the objectives as our criteria for decision making. Four commonly used criteria are:

1. Probabilities alone—we take the course of action that has the highest associated probability of success.

2. Economic consequences alone—we pursue an action that either minimizes our losses or maximizes our gains.

3. Probability and economic consequences together—we take an action that has the highest *expected value,* i.e., we select an act in which the long-run average is the highest.

4. Utility—Our choice is made on the basis of some factor or factors other than long-range expected value, i.e., we take out insurance to guard against the possibility of catastrophic loss or we may choose one job over another because of the ambience associated with the work situation.

Initially, we shall concentrate on the *expected value criterion,* in which we choose the act that leads to the maximum expected gain or payoff in the long run, or "on the average."

Let us follow the problem of choosing between jobs, and see whether we can reduce the dilemma to an acceptable conclusion.

payoff

The way in which we measure the outcome of a decision or act. A payoff may be positive or negative; it may be measurable in terms of money or time or lives. If we cannot calculate a payoff, we cannot use decision analysis.

states of nature

Events, not under the control of the decision maker, that will influence the outcome of any particular action.

decision analysis

The evaluation of a decision-making problem using logical and quantitative methods.

expected value criterion

Choosing the alternative that leads to the maximum expected payoff in the long run.

To simplify matters, assume there are only two jobs to choose from and you must select only one. Thus, these jobs are both mutually exclusive and exhaustive. Assume they both pay on a salary-commission-incentive basis.

Let us say that job 1 offers $150 salary plus fringes worth at least another $50. This salary is guaranteed as long as a certain minimal level of sales is achieved (low sales). Suppose the salary (but not the fringes) doubles if you achieve medium sales. For a high level of sales, your salary doubles again (another $150) and this time $100 in fringes is added.

Job 2 offers a minimal salary of $250 with $50 in fringes. An extra $100 is added every time you move into the next category—medium and then high sales.

Summarizing this information into an initial payoff table (Table 9.1) we see:

Table 9.1
Payoff table for decision between two jobs

	Job 1	Job 2
Low sales	$200	$300
Medium sales	350	400
High sales	600	500

Now comes the big question. What do you think the likelihood is of achieving low, medium, and high sales? We will assume that these three events exhaust all possibilities and that they are mutually exclusive. In other words, if you achieve low sales, you cannot *also* achieve medium or high sales. Let us assume you would assign the same probabilities, regardless of the job. Suppose you feel reasonably confident that you have at best a 50–50 chance of achieving medium sales. Since this is all new to you, you're not quite sure whether you have a better chance to be tops (high sales) or simply achieve the minimum required. Let's be a little confident and say the probability is slightly greater that you will do extremely well (i.e., p(high sales) = 0.30; p(low sales) = 0.20); p (medium sales) = 0.50. Once we have these probabilities, we can calculate the expected value of each decision.

The procedures are the same as described in Section 5.2 for obtaining the expected value of a random variable. Recall that the procedure involved multiplying each value of the variable by its associated probability and obtaining the sum over all values of the variable. This sum represented the expected

Table 9.2
Calculation of expected payoffs for decision between two jobs

	Job 1			Job 2		
	Payoff	Prob-ability	Payoff × probability	Payoff	Prob-ability	Payoff × probability
Low sales	$200	0.2	$ 40	$300	0.2	$ 60
Medium sales	350	0.5	175	400	0.5	200
High sales	600	0.3	180	500	0.3	150
Expected payoff			395			410

	Job 1	Job 2
Low sales	$100	$ 0
Medium sales	50	0
High sales	0	100

Table 9.3
Opportunity loss table for job-decision problem

value of that variable. Similarly, the sum of the expected values of each decision represents the expected value of that decision (Table 9.2).

Thus, in terms of maximizing expected monetary gain, you should choose job 2 since, over the long run, you will make more money. This does not mean that this action will *necessarily* be superior. It is, rather, the best monetary decision we can make in the face of uncertainty concerning so many questionable states of nature.

There's one other way to view a problem of this sort. We could look at it in terms of *opportunity loss* or *regret*. That is, if you take one action (e.g., take job 2) you will lose under certain conditions. For example, if your sales are extremely high, and job 2 had been selected, then that decision would be regretted since you could have made more money with job 1 ($600 versus $500).

It should be noted that the calculation of opportunity loss is very straightforward when dealing with payoff tables. However, as we shall see in later examples, entries in a payoff table may represent costs rather than gain or profit. In these cases, the objective would be to *minimize* cost, i.e., the best action would be the one with the lowest cost.

However, in the present example, opportunity loss is merely the numerical value resulting from the comparison of the payoff for each state of nature with the "best" payoff. In the case of two decisions there is only one opportunity loss per state of nature. If you make the best decision for each state of nature, you experience zero opportunity loss (OL). For example, if you selected job 1 under conditions of low sales, you would make $100 less than if you had selected job 2 (see Table 9.1). That $100 constitutes your opportunity loss. Had you selected job 2, there would be zero opportunity loss. Let us look at an opportunity loss table (Table 9.3).

Once again, by viewing this in terms of the probability or likelihood of occurrence, we can come up with an expected value for each decision. In this case, it would be the expected OL [E(OL)] (see Table 9.4).

opportunity loss (OL) (sometimes called regret)

The difference in profit that could be realized under a given state of nature if you had made a different decision. OL is zero when the optimal action is taken for a specific state of nature.

	Job 1			Job 2		
	OL	p	OL $\times$ p	OL	p	OL $\times$ p
Low sales	$100	0.2	20	0	0.2	0
Medium sales	50	0.5	25	0	0.5	0
High sales	0	0.3	0	100	0.3	30
Expected OL			45			30

Table 9.4
Calculation of expected opportunity loss for job decision

It is interesting (but something that always occurs) that a decision made on the basis of *minimizing* expected OL is the same decision made on the basis of *maximizing* expected monetary gain. So, once again we choose job 2 since this leads to the smaller expected value of OL.

You might ask, "If these two techniques yield the same result, why analyze opportunity loss?" There are two primary advantages to this analysis. First, an analysis of OL yields zero values, which will make arithmetic operations easier. But, more importantly, as we shall see, expected OL provides us with the cost of uncertainty. This leads us to the next concept to be explored.

Suppose you were able to get perfect information. That is, suppose you *knew* with absolute certainty which state of nature would occur. One might ask, "How much is it worth to reduce uncertainty?" One way of deciding the value of this information is to calculate the *expected value of perfect information (EVPI)*.

expected value of perfect information (EVPI)

The expected profit with perfect information minus the expected profit under uncertainty; sometimes called the cost of uncertainty. When dealing with costs rather than profit, EVPI = the expected cost under uncertainty minus the expected cost with perfect information.

The expected value of perfect information (EVPI) is the expected profit with perfect information minus the expected profit under uncertainty. However, if the entries in the payoff table are costs rather than profits, the EVPI is the expected *cost* under uncertainty minus the expected *cost* with perfect information. Example B illustrates the calculation of EVPI with a cost table.

Going back to the original payoff table (Table 9.1), we see that if low or medium sales occur, we would be better off selecting job 2. However, if sales are high, job 1 is the better choice. The way we figure the expected profit with perfect information is to assume that we make the correct decision each time.

Thus, the expected profit with perfect information = $60 + $200 + $180 = $440. A decision made under uncertainty yielded an expected gain of $410 (see Table 9.2). Hence, EVPI = $440 − $410 = $30. In this example, perfect information is worth up to $30 a week, or it's worth up to $30 a week to eliminate uncertainty. The reason we say "up to" is if you pay more, you are actually worse off financially than if you made the decision under uncertainty. Note that the value obtained for the EVPI ($30) is the same as obtained in Table 9.4 [E(OL) = $30]. We should recognize that, in the real world, we never obtain perfect information. The EVPI sets the upper limit to the amount we should be willing to pay to move our decision in the direction of certainty. Thus, it provides a basis for deciding the value of such strategies as sampling as a means of improving our information base. We would not be willing to spend $30 for sampling since sampling

Table 9.5

Calculation of expected profit with perfect information

	Job 1			Job 2		
	Payoff of best decision	p	Expected profit	Payoff of best decision	p	Expected profit
Low sales	—	—	—	$300	0.2	$ 60
Medium sales	—	—	—	400	0.5	200
High sales	$600	0.3	$180	—	—	—

information is not perfect. However, we may find it to our advantage to spend, say, $10 or $20 for this information.

Example A: Supply and demand decision problem

Saudi Arabia is one of the leading suppliers of crude oil to the United States. Because of uncertainties in the world situation—including the possibilities of oil cutoffs due to war in the Persian Gulf and boycotts—and because of energy conservation in the United States, the oil companies find themselves repeatedly on the horns of a dilemna. They find it necessary to over-order in the event of a shutoff but they don't want inventories to get so large that prices must be cut. Suppose they must sign contracts with Saudi Arabia for crude oil for a 90-day period. Based on past experiences, there are four possible demands during the 90-day period (we have purposely restricted the number of possibilities to four in order to keep the decision table a manageable size): 180, 225, 270, and 315 million barrels. Imagine that the cost of purchasing the crude and processing it into various products for distribution is $40 per barrel. When sold for $42, a barrel yields a profit of $2. However, if inventory is left after 90 days, the price must be discounted 5 percent to make room for new inventory. Thus, the loss is 10 cents a barrel for any quantity left in inventory (i.e., $42 $\times$ 0.05 = $2.10; $42 − $2.10 = $39.90; $40 − $39.90 = $0.10).

Find:

a. The initial payoff table and the initial opportunity loss table.
b. The expected profits from taking each of four possible contract actions.
c. The optimum monetary action.

Solution

    a.

Demand (million barrels)	Decision or action			
	A_1 180	A_2 225	A_3 270	A_4 315
E_1 180	360.00	355.50	351.00	346.50
E_2 225	360.00	450.00	445.50	441.00
E_3 270	360.00	450.00	540.00	535.50
E_4 315	360.00	450.00	540.00	630.00

Table 9.6
Payoff table

The payoffs, in millions of dollars, are shown in each cell of Table 9.6. Each of these entries is a conditional value; that is, the payoff for a particular action is conditional on which event occurs. Thus, if 180 million barrels are ordered (A_1) and all are sold, the profit is $360 million ($180 million $\times$ $2). In fact, no matter how much is demanded (assuming a constant profit per barrel), the maximum profit on an order of 180 million barrels is $360 million (see column A_1). Note that if the demand is only 180 million barrels, the profit goes down as the size of the order increases. Therefore, if 315 million barrels are ordered and 180 million demanded, the profit is $346.50 million: $360 million profit on 180 barrels sold less $13.50 million loss on 135 million barrels not sold. This is the minimum profit possible. Some decision makers are more conservative than others. In fact, they take the pessimistic view that the *worst* state of nature will prevail—in this case, demand of 180 million barrels. They would choose that action that leads to the *max*imum profit under this *min*imum demand *(maximin).* In other words, they would order 180 million barrels.

On the other side of the ledger, the act of ordering 315 million barrels accompanied by a demand for the like amount will produce the largest profit ($630 million). Some individuals are optimistic and their decisions reflect this optimism. Regardless of the decision they make, they assume the *best* possible state of nature. They would choose the action that leads to the *max*imum profit under the *max*imum demand *(maximax).* They would order 315 million barrels. Thus, the largest possible order promises the greatest profits but it also engenders the risk of the smallest profits.

In calculating the opportunity loss table for this problem, you take each state of nature and pick out the best decision. For example, for E_1 (demand = 180 million barrels), A_1 represents the best decision since it has the highest associated payoff. Therefore, OL = 0 for this decision because you have *no* opportunity loss; you have made the best decision. However, all other decisions yielding values less than the best represent a loss. Since it is not a true loss, we call it an "opportunity loss." You have not actually lost money, just the *opportunity* to make more. That's why opportunity loss is sometimes referred to as "regret." You regret making a decision when, it turns out, you would have done better had you made a different decision. But, in a sense, this is all hindsight. You choose a decision that, for various reasons, seems right at the time. It is only after the event has occurred that we can look back with regret at what might have been. (See Table 9.7.)

One other approach called the *minimax loss* or *regret,* involves choosing the minimum of a set of maximum opportunity losses. In other words, we choose the action in which the worst possible outcome (highest OL) is the least. Using this approach, they would order 315 million barrels since this decision has a maximum OL of $13.50 million as compared with, say, ordering 180 million barrels which has a maximum opportunity loss of $270 million.

b. Taking probabilities into account. Recall that the decision matrix shown in Table 9.6 is based on economic considerations alone. The table does not reveal the number of barrels of crude oil that should be ordered to produce maximum profits. It simply shows the economic consequences

---

maximin criterion

The decision maker pessimistically selects the action providing maximum benefit or gain under the assumption that the "worst will happen."

maximax criterion

The decision maker optimistically selects the action providing maximum gain under the assumption that the "best will happen."

minimax loss or regret criterion

The decision maker selects the action in which the maximum opportunity loss is the smallest.

| | Decisions or actions | | | |
Demand (million barrels)	A_1 180	A_2 225	A_3 270	A_4 315
E_1 180	0	4.50	9.00	13.50
E_2 225	90.00	0	4.50	9.00
E_3 270	180.00	90.00	0	4.50
E_4 315	270.00	180.00	90.00	0

Table 9.7
Opportunity loss table

of each act given that a particular event occurs. Let us suppose that consumer surveys, past knowledge of energy conservation practices, and other sources of information (including informed hunches) permit us to assign probability values to the various events. These are shown in Table 9.8.

Event (demand)	Probability
180	0.40
225	0.30
270	0.20
315	0.10

Table 9.8
Events and their associated probabilities

Combining these probabilities with the values shown in Table 9.6, we are in a position to calculate the expected profit of each possible buying action.

Let's first look at the expected profit from act A_1 (ordering 180 million barrels) when the demand may take on any of four different values (E_1 through E_4). The calculations are shown in Table 9.9.

Demand (million barrels)	Conditional profit ($ millions)	Probability of demand	Expected profit ($ millions)
180	$360.00	0.40	$144.00
225	360.00	0.30	108.00
270	360.00	0.20	72.00
315	360.00	0.10	36.00
		1.00	$360.00

Table 9.9
Expected profit from ordering 180 million barrels of petroleum (based on column of A_1 of Table 9.6)

The sum of the last column is $360 million. This expected profit of $360 million is precisely what we obtain from Table 9.6, since the conditional profit of $360 million would always result from selling 180 million barrels, even if the demand was 180, 225, 270, or 315 million barrels.

We may now calculate the expected profit from each of the three remaining acts, i.e., ordering 225, 270, and 315 million barrels. These are shown in Tables 9.10, 9.11, and 9.12.

Table 9.10

Expected profit from ordering 225 million barrels of petroleum (based on column A_2 of Table 9.6)

Demand (million barrels)	Conditional profit ($ millions)	Probability of demand	Expected profit ($ millions)
180	$355.50	0.40	$142.20
225	450.00	0.30	135.00
270	450.00	0.20	90.00
315	450.00	0.10	45.00
		1.00	$412.20

Table 9.11

Expected profit from ordering 270 million barrels (based on column A_3 of Table 9.6)

Demand (million barrels)	Conditional profit ($ millions)	Probability of demand	Expected profit ($ millions)
180	$351.00	0.40	$146.40
225	445.50	0.30	133.65
270	540.00	0.20	108.00
315	540.00	0.10	54.00
		1.00	$436.05

Table 9.12

Expected profit from ordering 315 million barrels (based on column A_4 of Table 9.6)

Demand (million barrels)	Conditional profit ($ millions)	Probability of demand	Expected profit ($ millions)
180	346.5	.40	$138.60
225	441.00	.30	132.30
270	535.50	.20	107.10
315	630.00	.10	63.00
		1.00	$441.00

We have now calculated the expected profit resulting from taking each of four possible actions: ordering 180, 225, 270, or 315 million barrels of crude oil. To summarize:

If 180 million barrels are ordered, the expected profit is $360 million.

If 225 million barrels are ordered, the expected profit is $412.2 million.

If 270 million barrels are ordered, the expected profit is $436.05 million.

If 315 million barrels are ordered, the expected profit is $441.00 million.

c. The optimum action. Ordering 315 million barrels is the optimum action, since the expected value of exercising that option is the largest. This does not mean that this action will *necessarily* be the most profitable. It is, rather, the best decision we can make, over the long run, in the face of uncertainty.

Example B: Dominant action

Mr. and Mrs. Scott have finally found the home of their dreams. They will purchase the home for $100,000, with a down payment of $20,000, and will finance the remaining $80,000. They have a choice of three options available for the financing:

A_1. They can apply for a variable rate mortgage (VRM) at a savings and loan at an initial rate of 14 percent. The VRM is a new type of mortgage that is tied to a federal index that measures changes in interest rates. Under a VRM, the savings and loan may raise their interest rates up to 1 percent a year, with a maximum increase of 5 percent above the initial rate.

A_2. The owner has agreed to give them a one-year, $80,000 loan at 15.5 percent. At the end of the year, they would then apply for a VRM at the prevailing rates.

A_3. The owner has also offered the option of giving them a five-year loan at a fixed rate of 15.5 percent.

After analyzing the economy and the behavior of interest rates, they conclude that there are three possible events that can occur over the five-year period:

B_1. Interest rates can go up to 1 percent a year.

B_2. Interest rates can go up 0.5 percent a year.

B_3. Interest rates can stay the same.

Since they deem it highly unlikely that interest rates will go down, they don't even consider that possibility. After discussions with real-estate people, their accountant, banker, etc., they assign the following probabilities to the three events:

$$p(B_1) = 0.5$$
$$p(B_2) = 0.4$$
$$p(B_3) = 0.1$$

The following table shows how we obtain the total costs over the five-year period for each decision (A_1, A_2, A_3) under each state of nature (B_1, B_2, B_3).

It should be noted that, where the VRM is involved, payments usually include principal and interest. However, for pedagogical purposes, we have calculated all values on the basis of simple interest.

State of nature	Year	Decision					
		A_1		A_2		A_3	
		Rate	Cost	Rate	Cost	Rate	Cost
B_1	1	14.0%	$11,200	15.5%	$12,400	15.5%	$12,400
	2	15.0	12,000	15.0	12,000	15.5	12,400
	3	16.0	12,800	16.0	12,800	15.5	12,400
	4	17.0	13,600	17.0	13,600	15.5	12,400
	5	18.0	14,400	18.0	14,400	15.5	12,400
			$64,000		$65,200		$62,000
B_2	1	14.0%	$11,200	15.5%	$12,400		
	2	14.5	11,600	14.5	11,600		
	3	15.0	12,000	15.0	12,000		
	4	15.5	12,400	15.5	12,400		
	5	16.0	12,800	16.0	12,800		
			$60,000		$61,200		$62,000
B_3	1	14.0%	$11,200	15.5%	12,400		
	2	14.0	11,200	14.0	11,200		
	3	14.0	11,200	14.0	11,200		
	4	14.0	11,200	14.0	11,200		
	5	14.0	11,200	14.0	11,200		
			$56,000		$57,200		$62,000

Summarizing this information, we obtain the following payoff table (Table 9.13). Note that the entries represent costs rather than gain or profit.

Table 9.13
Payoff table for borrowing decision

	A_1	A_2	A_3
B_1	$64,000	$65,200	$62,000
B_2	60,000	61,200	62,000
B_3	56,000	57,200	62,000

a. Based on the information provided, which action should the Scotts take?

b. What is the maximum they should pay for perfect information?

dominant action

When one decision or action is always preferred over another, regardless of the state of nature, we say that decision is dominant.

Solution

a. As we look at the payoff table we see that, regardless of the state of nature, decision A_1 always costs less than decision A_2. When one action is always preferred over another, no matter what conditions exist, we say it *dominates* that action. Thus, since action A_2 is dominated by action

Table 9.14
Expected costs for
borrowing decision

State of nature	p	Take VRM now (A_1)		Take owner's five-year loan (A_3)	
		Cost	Expected cost	Cost	Expected cost
B_1	0.5	$64,000	$32,000	$62,000	$31,000
B_2	0.4	60,000	24,000	62,000	24,800
B_3	0.1	56,000	5,600	62,000	6,200
			$61,600		$62,000

A_1, we say that A_2 is *inadmissible* as an alternative, and therefore may be eliminated. We can now calculate the expected costs of the two remaining decisions under the three possible states of nature. (See Table 9.14.)

Thus, based on the lower expected cost, the Scotts should choose A_1; that is, they should apply for a VRM now at 14 percent.

b. To determine the maximum they should pay for perfect information, we first calculate the expected cost with perfect information. In other words, we assume that for each state of nature, they make the best decision.

inadmissible action

In general, whenever one action is dominated by another one, it is inadmissible.

State of nature	Expected cost of best decision
B_1	$31,000
B_2	24,000
B_3	5,600
	$60,600

Thus, the expected value of perfect information is equal to the expected cost under uncertainty minus the expected cost with perfect information, or EVPI = $61,600 − $60,600 = $1,000. Therefore, the maximum they should consider paying to reduce uncertainty is $1,000. Note that since we are dealing with costs, we use the "cost" definition of EVPI.

EXERCISES

9.1 In each of the following decision situations, enumerate possible states of nature and possible payoffs.

a. A candidate for political office must visit four different localities. Should he go by plane, auto, or train?

b. A publisher is going to publish a new book in 30 days. How many copies should be included in the first printing?

c. You are trying to decide among three jobs.

    d. A multinational oil company must decide on the amount of crude oil for which it will contract over a 90 day period.

    e. A couple of potential homeowners have a choice among three different methods of financing their dream home.

9.2 State plausible states of nature, possible actions, and types of payoffs in each of the following:

    a. The campaign manager for a gubernatorial candidate must arrange the appearance of the candidate in three different regions of the state. There are three different possible weather conditions (clear, rain, fog) and three modes of transportation (plane, auto, train). The candidate wishes to spend the least possible time in travel.

    b. As the heating season approaches, a heating contractor must decide how many replacement units to order for customers who may need replacements during the winter. Any unit sold provides a gross margin of profit. A unit that is unsold must be returned to the manufacturer at a fixed loss per unit.

    c. An investor is looking over her investment portfolio. She finds that she has a certain amount of money available for investment. She considers investing the money in a passbook account, Treasury bonds, speculative stocks, or gold on the commodities market. She wonders, "Is the rate of inflation likely to decrease, increase, or remain the same?"

9.3 The gubernatorial candidate in Exercise 9.2*(a)* has decided to

    a. Take a train because it is the best under the worst conditions.

    b. Take a plane because it is the best under the best conditions.

What decision criteria was he using?

9.4 The secretary for the gubernatorial candidate learns the following: If the weather is clear, the plane requires 20 hours; auto, 35; and train, 50; if rainy, plane is 25 hours, auto is 40 hours, and train is 53 hours; if foggy, plane is 60 hours, auto is 56, and train is 54.

    a. Construct a payoff table.

    b. Does any action dominate any other action?

    c. Construct a table of opportunity loss.

    d. What mode of transportation provides the least opportunity loss when it is clear, when it rains, or when it is foggy?

9.5 Refer to Exercise 9.4. The secretary contacts the local meteorology department of the U.S. Weather Bureau to learn the likelihood of the various states of nature at the time of the year that the candidate's tour is to be made. She learns that it is clear 70 percent of the time, rainy 25 percent of the time, and foggy 5 percent of the time.

    a. What action provides maximum expected payoff (i.e., lowest expected time)?

    b. What action provides minimum opportunity loss?

    c. Calculate the expected value of pefect information (EVPI).

9.6 After being shown the proposed mode of transportation, the candidate states, "I'm sick and tired of going by plane. For a change, I want the leisurely relaxing pace of a train." What decision-making criterion is he using?

9.2 PROBABILITIES ASSOCIATED WITH STATES OF NATURE

One of the most perplexing, but extremely important, aspects of decision analysis concerns the probabilities associated with the various states of nature. How do we assign these probabilities?

As we indicated earlier, we may have had previous experience with the relative frequency of occurrence of the various events or states of nature. Suppose, for example, we have accurate records on past demand for a particular product. Thus, we may assign empirical probabilities to the various states of nature (e.g., low, medium, or high demand) based on the relative frequency of occurrence in the past.

However, some probabilities are not so easy to determine. There may be no previous data available or conditions may have changed so that past data cannot serve as a reliable guide in assessing future probabilities. For example, suppose you wanted to assess the degree of success associated with a new job application procedure. Since you have never tried this procedure before, you have no previous experience to guide you in assigning probabilities.

It is especially difficult to assign probabilities in situations that cannot be repeated, particularly in situations where changing conditions influence or affect the outcome. To illustrate, suppose a tampon manufacturer introduced a new type of tampon early in 1980. Based on market research at that time, the manufacturer might conclude that there was a high probability of success if he or she marketed the product. However, if the product was introduced into the market several months later, changing conditions (in this case, controversy surrounding the use of tampons) might drastically alter the probabilities associated with success or failure of the product.

Anticipating mortgage rates a year from now, or assigning probabilities to changes in interest rates in general, are examples of events that are extremely difficult to predict because of constantly changing economic conditions.

Unfortunately, there is no single prescription for selecting subjective probabilities. In many of these instances, we simply must rely on judgment, intuition, and any objective information that may be available. These subjective probabilities are a measure of the degree of faith or confidence that the decision maker has concerning the likelihood that a particular event will occur. Since the outcomes may be extremely sensitive to even slight changes in the probabilities associated with the various events, it is important that the assigned probabilities be as meaningful and as precise as possible. Thus, the decision maker must use any and all objective information available. Even so, it may still come down to relying on one's personal judgment and experience in assessing this information to arrive at the probabilities. The best we can hope for is that we have assigned reasonably accurate assessments of the likelihood of occurrence to the various states of nature. Incorrect or inaccurate probabilities may lead to misleading conclusions and consequently to wrong decisions.

The following example illustrates how changing probabilities affects the decision-making process.

Example C: Changing probabilities associated with states of nature

Rosa has to choose between two apartments. Apartment 1 rents for $295 a month. It is located 15 miles from where she works. There is a possibility of joining a car pool that will cost her $30 a month. If she can't, she must use her own car. She calculates that it will cost her approximately $102 a month if she has to use her own car.

Apartment 2 is 30 miles away from her job. It rents for $225 a month. If she can get into a car pool, it will cost her $60 a month. If she can't, she estimates that using her own car will cost her approximately $204 a month.

a. Assuming that the probability of joining a car pool is 0.70 (from either location), which apartment should she choose?
b. Assume that the probability of joining a car pool is 0.30. Does this change her decision?
c. Disregarding the probabilities given in *(a)* and *(b)*, what decision would you make using the maximin criterion? What decision would the maximax criterion dictate?

Solution

a. We set up a payoff table (Table 9.15) to calculate the expected costs of the two apartments under the two different states of nature. In a problem of this sort, we make a decision based on the *lower* expected cost.

Table 9.15

Expected cost for decision between two apartments when p(carpool) = 0.7

State of nature	p	Apartment 1		Apartment 2	
		Cost	Expected cost	Cost	Expected cost
Carpool	0.7	($295 + $30)	$227.50	($225 + $60)	$199.50
Own car	0.3	($295 + $102)	119.10	($225 + $204)	128.70
			$346.60		$328.20

In this case, Rosa would choose apartment 2 since the expected cost is less.

b. *The expected costs will differ because of the change in probabilities (see Table 9.16).*

Table 9.16

Expected cost for decision between two apartments when p(carpool) = 0.3

State of nature	p	Apartment 1		Apartment 2	
		Cost	Expected cost	Cost	Expected cost
Carpool	0.3	$325	$ 97.50	$285	$ 85.50
Own car	0.7	397	277.90	429	300.30
			$375.40		$385.80

State of nature	Apartment 1	Apartment 2
Carpool	$325	$285
Own car	397	429

Table 9.17
Costs of the two apartments under two states of nature

This example illustrates the importance of the probabilities assigned to the various states of nature. If the assigned probabilities are not reliable or based on inadequate information, the obtained results may be incorrect or misleading. As can be seen in Table 9.16, when the probabilities associated with the two states of nature (carpool versus own car) change, the optimal decision shifts. She would now select Apartment 1.

c. **Maximin criterion.** Assuming the *worst* state of nature, i.e., using own car, select Apartment 1 since this provides the maximum benefit (i.e., the lower cost) under the worst possible condition (see Table 9.17).

Maximax criterion. Assuming the *best* state of nature, i.e., carpool, select Apartment 2 since this provides the maximum benefit under the best possible condition (see Table 9.17).

Unless Rosa can come up with a reasonably accurate way to assign the probabilities, she should not attempt to make a choice based on expected costs. She probably should use one of the above criteria to make her decision.

9.3 REVISING PROBABILITIES—BAYES' THEOREM

We have seen the importance of assigning probabilities to the various states of nature that are as accurate as possible. There are many ways that we can improve the accuracy of the prior probabilities. For example, before making a decision, an investor might consult with a financial advisor, a marketing director might conduct a market research survey before deciding to market a new product, or a travel agent might consult a meteorologist before planning a vacation trip for clients.

In each of these instances, the decision maker is utilizing additional information or evidence to revise the *prior probabilities*. If we are able to evaluate the reliability of this new information, we can use Bayes' theorem to revise our prior probabilities. These revised probabilities are sometimes called *posterior probabilities*.

Let us use the problem presented in Example B to illustrate how we may revise probabilities and make decisions based on these posterior probabilities. Recall that the Scotts had to choose between taking the VRM loan (A_1) and taking the owner's five-year loan (A_3). The various states of nature affecting their decision and their associated probabilities were:

B_1. Interest rates go up 1 percent a year; $p(B_1) = 0.50$.

B_2. Interest rates go up 0.5 percent a year; $p(B_2) = 0.40$.

B_3. Interest rates remain the same; $p(B_3) = 0.10$.

prior probability

The probability of an event before it is revised by additional information or evidence.

posterior (revised) probability

The prior probability that is modified in light of new information. The revised probability is usually obtained by employing Bayes' theorem.

Table 9.18

Conditional probabilities $[p(P_i|B_i)]$ of three different predictions made by the forecasting service when the interest rates actually went up 1 percent (B_1), 0.5 percent (B_2), and stayed the same (B_3).

State of nature	Prediction		
	P_1	P_2	P_3
B_1	0.70	0.10	0.20
B_2	0.20	0.75	0.05
B_3	0.05	0.10	0.85

These probabilities (which we shall now refer to as prior probabilities) were based on subjective evaluations. On the basis of these prior probabilities, recall that the Scotts decided to take the VRM loan. However, under certain circumstances (specifically, if interest rates go up 1 percent a year) they would be making the wrong decision. In fact, when we calculated the expected value of perfect information (EVPI) we found the cost of uncertainty was $1,000. This amount is also the expected opportunity loss. (Note: confirm this by calculating the expected OL).

Suppose the Scotts decide to call in a professional forecasting service in order to get expert opinion on the probable changes in interest rates. Let us examine the effect this would have on their decision as well as how much the service's information is worth.

Let

P_1 = Service's prediction that interest rates go up 1 percent a year;

P_2 = Service's prediction that interest rates go up 0.5 percent a year;

P_3 = Service's prediction that interest rates remain the same.

Table 9.18 summarizes the past record of the forecasting service by presenting the conditional probabilities of each prediction, given the three possible events that may occur. For example, in the past, when the service predicted interest rates would go up 1 percent a year (P_1), they were correct 70 percent of the time (i.e., interest rates actually went up 1 percent a year). However, when interest rates actually went up 0.5 percent a year, the service incorrectly predicted a 1 percent increase 20 percent of the time. Similarly, when interest rates actually remained the same, the service's prediction of 1 percent increase was wrong 5 percent of the time.

Our goal is to revise the prior probabilities, i.e., $p(B_i)$, based on the service's prediction. In other words, we want to obtain the conditional probabilities $p(B_i|P_i)$. Let us use as a model the calculation of $p(B_1|P_1)$ to illustrate how these revised probabilities are obtained.

Recall that in our discussion of Bayes' theorem (Section 4.8), we presented the following general formula:

$$p(B_1|P_1) = \frac{p(B_1 \text{ and } P_1)}{p(P_1)}$$

[We have substituted the notations B_1 and P_1 for those presented in Formula (4.4).]

We calculate the joint probability, $p(B_1 \text{ and } P_1)$ in the numerator by the multiplication rule [Section 4.7, Formula (4.5)]:

$$p(B_1 \text{ and } P_1) = p(B_1)p(P_1|B_1)$$

In order to calculate the denominator, we observe that

$$p(P_1) = p(B_1 \text{ and } P_1) \text{ or } p(B_2 \text{ and } P_1) \text{ or } p(B_3 \text{ and } P_1)$$

Since these joint events $(B_1 \text{ and } P_1)$ and $(B_2 \text{ and } P_1)$ and $(B_3 \text{ and } P_1)$ are mutually exclusive,

$$p(P_1) = p(B_1 \text{ and } P_1) + p(B_2 \text{ and } P_1) + p(B_3 \text{ and } P_1)$$

We express each of these joint probabilities according to the multiplication rule:

$$p(B_1 \text{ and } P_1) = p(B_1)p(P_1|B_1)$$
$$p(B_2 \text{ and } P_1) = p(B_2)p(P_1|B_2)$$
$$p(B_3 \text{ and } P_1) = p(B_3)p(P_1|B_3)$$

then

$$p(P_1) = p(B_1)p(P_1|B_1) + p(B_2)p(P_1|B_2) + p(B_3)p(P_1|B_3)$$

or

$$p(B_1|P_1) = \frac{p(B_1)p(P_1|B_1)}{p(B_1)p(P_1|B_1) + p(B_2)p(P_1|B_2) + p(B_3)p(P_1|B_3)}.$$

Substituting the appropriate values, we find

$$p(B_1|P_1) = \frac{(0.50)(0.70)}{(0.50)(0.70) + (0.40)(0.20) + (0.10)(0.05)}$$

$$= \frac{0.35}{0.350 + 0.080 + 0.005}$$

$$= 0.8046.$$

Let us summarize this in Table 9.19.

Table 9.19

Calculation of revised probabilities in the borrowing decision based on service's prediction of interest rate increases of 1 percent a year (P_1).

Event	Prior probability $p(B_i)$	Conditional probability $p(P_1\|B_i)$	Joint probability $p(B_i)p(P_1\|B_i)$	Revised $p(B_i\|P_1)$
B_1	0.05	0.70	0.350	0.8046
B_2	0.40	0.20	0.080	0.1839
B_3	0.10	0.05	0.005	0.0115
	1.00		0.435	1.0000

Each of the revised probabilities in the last column is obtained by dividing the corresponding joint probability by the sum of the joint probabilities (i.e., $p(P_1) = 0.435$).

Similarly, Tables 9.20 and 9.21 present the calculation of the revised probabilities based on the service's prediction of 0.5 percent increases (P_2) and interest rates remaining the same (P_3).

Now, suppose the service predicts interest rates will go up 1 percent a year. The expected cost of the VRM loan is

$$(0.8046)(\$64,000) + (0.1839)(\$60,000) + (0.0115)(\$56,000) = \$63,172.40$$

Table 9.20

Calculation of revised probabilities in the borrowing decision based on service's prediction of interest rate increases of 0.5 percent a year (P_2).

Event	Prior probability $p(B_i)$	Conditional probability $p(P_2\|B_i)$	Joint probability $p(B_i)p(P_2\|B_i)$	Revised $p(B_i\|P_2)$
B_1	0.50	0.10	0.05	0.1389
B_2	0.40	0.75	0.30	0.8333
B_3	0.10	0.10	0.01	0.0278
	1.00		0.36	1.0000

Table 9.21

Calculation of revised probabilities based on service's prediction that interest rates remain the same (P_3).

Event	Prior probability $p(B_i)$	Conditional probability $p(P_3\|B_i)$	Joint probability $p(B_i)p(P_3\|B_i)$	Revised $p(B_i\|P_3)$
B_1	0.05	0.20	0.100	0.4878
B_2	0.40	0.05	0.020	0.0976
B_3	0.10	0.85	0.085	0.4146
	1.00		0.205	1.0000

The expected cost of the VRM loan if the service predicts 0.5 percent increases is

$$(0.1389)(\$64{,}000) + (0.8333)(\$60{,}000) + (0.0278)(\$56{,}000) = \$60{,}444.40$$

Finally, the expected cost of the VRM loan if the service predicts interest rates remain the same is

$$(0.4878)(\$64{,}000) + (0.0976)(\$60{,}000) + (0.4146)(\$56{,}000) = \$60{,}292.80$$

It is not necessary to calculate the expected cost of the owner's loan since it is a fixed-rate loan and, regardless of the prediction, it will always be $62,000.

Table 9.22 summarizes these expected costs.

How do we interpret Table 9.22? If the service predicts 1 percent increases (P_1) the Scotts should take the owner's loan. If the service predicts 0.5 percent increases or no increases, the Scotts should take the VRM loan. Note that the Scott's decision is now made on the basis of the service's prediction. Thus, instead of a single decision (i.e., take VRM loan) we have different decisions corresponding to the outcome or prediction made by the service. In other words, the service will make *one* prediction, which will lead to a specific decision by the Scotts. Which prediction will they make? In order to find this out, the Scotts will have to contract for their services. That is, they will have to pay the service and in return will receive a specific prediction. How much should they pay?

In order to calculate the worth of the service, we compare the previously obtained cost ($61,600) under uncertainty (Table 9.14) to the expected cost using the service. We obtain this value by multiplying the cost associated with the decision dictated by the prediction by the probability of that prediction. In other words,

$$p(P_1)(\$62{,}000.00) + p(P_2)(\$60{,}444.40) + p(P_3)(\$60{,}292.80) =$$
$$(0.435)(\$62{,}000.00) + (0.36)(\$60{,}444.40) + (0.205)(\$60{,}292.80) = \$61{,}090.01$$

The difference between these two values (cost under uncertainty versus cost using the service) is $61,600 − $61,090.01 = $509.99, which is the

Expected costs for borrowing decision for three different predictions made by the forecasting service.   **Table 9.22**

Service's prediction	Take VRM	Take owner's loan
P_1	$63,172.40	$62,000
P_2	60,444.40	62,000
P_3	60,292.80	62,000

expected value of sample information (EVSI)

The expected *payoff* with sample information minus the expected *payoff* with uncertainty; the expected *cost* under uncertainty minus the expected *cost* with sample information.

most the Scotts should pay for the forecasting service. This value is analogous to the EVPI but applies to less reliable information gleaned from a sample, market research, or expert opinion and is known as the *expected value of sample information (EVSI)*. Thus, the most the Scotts should pay for the forecasting service is $509.99. As long as they pay less for the service than this amount, they are better off using the service than they would be if they had not used the service.

EXERCISES

9.7 Bob B. has to decide which of three apartment buildings to buy. Building 1 has a gross income of $20,000 a year, and a net income, after all expenses, of $4,400 a year. Building 2 grosses $30,000 and nets $6,000; Building 3 grosses $45,000 and nets $8,000. However, these net income figures are based on zero vacancy. Bob knows that the vacancy rate can reduce the gross income by 5 percent, 10 percent, 15 percent, or 20 percent. This, of course, will result in a decrease in the net income. Bob has decided he would like to buy the building with the highest net income. When he calculates the net income of each building under the four possible vacancy rates, he finds:

Vacancy rate	Building		
	1	2	3
5%	$3,400	$4,500	$5,750
10	2,400	3,000	3,500
15	1,400	1,500	1,250
20	400	0	−1,000

After researching the area, he figures the probability of each of the four vacancy rates to be the same: i.e., $p(5\%) = 0.25$; $p(10\%) = 0.25$; $p(15\%) = 0.25$; $p(20\%) = 0.25$.

Since he is really not sure that his probabilities are accurate, he considers consulting a real estate firm that specializes in predicting vacancy rates. The past record of this firm is:

Actual vacancy rate	Prediction			
	5 percent	10 percent	15 percent	20 percent
5%	0.70	0.15	0.10	0.05
10	0.15	0.75	0.04	0.06
15	0.10	0.03	0.85	0.02
20	0.10	0.05	0.05	0.80

a. Which building should Bob buy based on his probabilities (prior probabilities)?

b. What is the expected value of perfect information (EVPI)?

c. Calculate the revised probabilities based on the real estate firm's predictions.

  d. What decision should Bob make if the firm predicts a 5 percent vacancy rate? 10 percent? 15 percent? 20 percent?

  e. The real estate firm charges $100 for these predictions. Is the information worth $100? (Calculate the EVSI).

9.8 Refer to the data in Exercise 9.7. Assume that Bob's prior probabilities are $p(5\%) = 0.10$; $p(10\%) = 0.30$; $p(15\%) = 0.30$; $p(20\%) = 0.30$.

  a. Which building should Bob buy based on these prior probabilities? Has his decision changed?

  b. What is the EVPI?

  c. Calculate the revised probabilities using these prior probabilities.

  d. What decision should Bob make now if the firm predicts a 5 percent, 10 percent, 15 percent, or 20 percent vacancy rate? Has the change in the prior probabilities changed his decisions based on the predictions?

  e. Calculate the EVSI and determine whether he should pay $100 for the information.

9.9 Refer to Exercise 9.7. Suppose the real estate firm has a *perfect* record. That is, they are correct in their predictions 100 percent of the time. If they predict, for example, a 5 percent vacancy rate, there is 100 percent chance that the vacancy rate *will* be 5 percent. In other words, their past record looks like this:

Actual vacancy rate	Prediction			
	5 percent	10 percent	15 percent	20 percent
5%	1.00	0.0	0.0	0.0
10	0.0	1.00	0.0	0.0
15	0.0	0.0	1.00	0.0
20	0.0	0.0	0.0	1.00

  a. Calculate the revised probabilities.

  b. Calculate the EVSI and compare this value to the EVPI. Discuss the relationship between these two values.

9.4 POSTERIOR ANALYSIS AND THE VALUE OF SAMPLING INFORMATION AFTER SAMPLING

  In decisions involving acceptance sampling, we have often compiled a considerable amount of data on various vendors. We may know, for example, the percentage of times that a vendor has shipped lots in which various proportions of defectives have been found. These constitute our prior probabilities. These probabilities often provide rather stable estimates of the long-term performance of a vendor. However, they do not allow us to decide whether a particular shipment should be accepted or rejected. For this, we commonly take a random sample from the entire shipment, test the items in the sample, find the number of defectives, and then decide whether or not to accept. The following example illustrates the entire process of obtaining expected costs on the basis of prior probability and expected costs based on the revision of the prior probabilities to take into account sampling information that we have collected.

Example D

We usually purchase 500 components at a time from a vendor. Each component is individually wired into a larger piece of equipment. Previously the vendor has supplied components in which the proportion of defectives has been 0.01, 0.05, 0.10, or 0.15. These proportions of defectives have occurred 40 percent, 30 percent, 20 percent, and 10 percent of the time, respectively. There is a choice of two decisions: either reject or accept the shipment. If we accept a shipment and a defective component is placed into the equipment, it costs $0.50 to remove each mistake. On the other hand, if we return a shipment, we are required to pay $12 in delivery costs. Thus, if we accept a shipment in which the proportion of faults is 0.01, we will later have to remove five items at a cost of $5 \times 0.50 = 2.50. If we reject the shipment, we will incur a cost of $12. In contrast, if the proportion of defectives is actually 0.15, we will incur a cost of $37.50 by mistakenly accepting the shipment (i.e., $500 \times 0.15 \times \$0.50) = \$37.50$.

Now let us say that we obtain a random sample of $n = 10$ from a shipment of 500 components and we find that 2 are defective. What is the expected cost of accepting or rejecting the shipment when we revise our probabilities in the light of a sample that we have already collected? For purposes of simplifying computations, we shall assume that the proportion of defectives can take on only four values: 0.01, 0.05, 0.10, and 0.15.

Solution

Our first step is to construct a conditional table of costs and the expected costs of accepting or rejecting a shipment. These are shown in Table 9.23.

At this point we make a surprising discovery. On the basis of historical records, the expected cost of acceptance from this vendor is *greater* than the expected cost of rejection! Had we done this analysis before, we would have realized that this vendor should have been terminated a long time ago

Table 9.23
Conditional costs and expected costs of accepting or rejecting a shipment

Proportion defective	Conditional cost of Accepting	Conditional cost of Rejecting	Prior probability	Expected cost of Accept	Expected cost of Reject
0.01........	$ 2.50	$12.00	0.40	$ 1.00	$ 4.80
0.05........	12.50	12.00	0.30	3.75	3.60
0.10........	25.00	12.00	0.20	5.00	2.40
0.15........	37.50	12.00	0.10	3.75	1.20
			1.00	$13.50	$12.50

Proportion defective	Prior probability	Conditional probability of sample result	Joint probability	Posterior probability
0.01	0.40	0.0042	0.0017	0.0188
0.05	0.30	0.0746	0.0224	0.2478
0.10	0.20	0.1937	0.0387	0.4281
0.15	0.10	0.2759	0.0276	0.3053
			0.0904	1.0000

Table 9.24
Revision of prior probabilities taking into account the sample result

(and perhaps we should have also been terminated for neglecting to do this analysis). Be that as it may, we are faced with the decision concerning the acceptance or rejection of a shipment that we have just received. We have found 2 components out of 10 to be defective. We ask the question, "Given that 2 out of 10 are defective, what is the probability that the sample was drawn from a population in which the true proportion, P, was 0.01, 0.05, 0.10, 0.15?" This is a situation to which the binomial applies (see Section 5.3) since there are two mutually exclusive and exhaustive possibilities (defective versus non-defective). Thus, we refer to Table I and follow the procedures for finding the probability of $x = 2$, when $n = 10$ and $P = 0.01$, $P = 0.05$, $P = 0.10$, and $P = 0.15$. We find these to be, respectively, 0.0042, 0.0746, 0.1937, and 0.2759. We multiply these probabilities by their associated prior probabilities to obtain the joint probabilities. These are, in turn, used to find the posterior or revised probabilities. The procedures are summarized in Table 9.24.

We now construct a decision table for accepting or rejecting the shipment by taking into account the posterior probabilities. The format of the table is the same as Table 9.23 except that posterior probabilities are substituted for prior probabilities.

Note that the expected cost of acceptance is more than twice the expected cost of rejection. The decision is clear: reject the shipment and, based on prior records, terminate the vendor. Note also that, since the decision has not been changed by the sample data, we must conclude, in this case, that the value of the sample information is zero. The truth of the matter is that we had sufficient prior information to have made the decision. The cost of collecting and testing the sample was without monetary value. However,

Proportion defective	Conditional cost of		Posterior probability	Expected cost of	
	Accepting	Rejecting		Accepting	Rejecting
0.01	$ 2.50	$12.00	0.0188	$ 0.0470	$ 0.2257
0.05	12.50	12.00	0.2478	3.0975	2.9736
0.10	25.00	12.00	0.4281	10.7025	5.1372
0.15	37.50	12.00	0.3053	11.4488	3.6636
				$25.2958	$12.0001

Table 9.25
Table for accept-versus-reject decision based on sample data

its value may have been utilitarian. It made us feel more secure that we had made the correct decision.

EXERCISES

9.10 Your company manufactures a solar heating unit that uses silicon cells. Your major supplier ships the cells in lots of 1,000. The proportion of defectives can be one of five different levels: 0.01; 0.02; 0.03; 0.04; 0.05. Your prior experience with this supplier indicates that the prior probabilities of obtaining the various proportions of defective cells are 0.45, 0.25, 0.15, 0.10, and 0.05, respectively. The costs associated with acceptance are $2 to locate, remove, and replace a defective cell. We will reject a shipment if the proportion of defectives is judged to be greater than 0.03. The supplier tests all components of a rejected shipment and, by contractual agreement, if the true proportion of defectives is found to be 0.03 or less, we must pay both the testing and reshipping costs. This comes to $150.

    a. Set up a decision table based on the use of prior probabilities.

    b. Find the expected cost of acceptance.

    c. Find the expected cost of rejection.

9.11 Refer to Exercise 9.10. Since the expected cost of rejection is considerably greater than the expected cost of acceptance, we wish to avoid rejecting a shipment without additional evidence. In our latest shipment, we select a random sample of 30 components and find that that 3 are defective. Should we accept or reject the shipment?

[Table I (Appendix C) does not include values of $P = 0.02$, 0.03, and 0.04. The conditional probabilities needed for Exercises 9.11 and 9.12 are shown in the following table.]

Conditional probabilities for $n = 30$

P	x		
	0	3	4
0.02	0.5455	0.0188	0.0026
0.03	0.4010	0.0482	0.0101
0.04	0.2939	0.0863	0.0243

9.12 Refer to Exercise 9.11.

    a. If we had found four defectives, should we have accepted or rejected the shipment?

    b. If we had found no defectives, should we have accepted or rejected the shipment?

    c. Draw a graph showing the expected cost of rejection and expected cost of acceptance when $x = 0$, $x = 3$, and $x = 4$ and $n = 30$. Can we use this cutoff point to establish a criterion for acceptance when $n = 30$?

9.5   DECISION TREES

Certain decision-making situations cannot be easily portrayed in the form of a payoff table. This is especially true when different events may affect the outcome of the various decisions. For example, a salesman may have to decide which of two accounts to call on. He may not have enough time to call on both; therefore, he must choose one. The probability of getting an order may differ for the two accounts and there may be a different set of events affecting the ultimate outcome of each decision.

Sometimes, decisions must be made at different times, with uncertain events occurring between actions. Often the outcome of the first decision affects the outcome of the next decision. In this case, it may be difficult to present the decision in the form of a payoff table. For example, in September 1980, Pacific Southwest Airlines (PSA) flights were grounded by a strike. Some carriers made the decision to add flights on some PSA routes. All these carriers had to make another decision at another point in time. Specifically, what do they do when the PSA strike ends? In many cases, this decision was affected by what happened in the interim. For example Pan American added three flights a day for one particular route after the PSA strike started but, based on disappointing results during the strike, they decided to cut back service when the PSA strike ended. On the other hand, based on success during the strike, other carriers decided to stay in the market after the strike.

The outcome of many decisions is not determined completely by the simple interaction of an action or choice with a specific set of events or states of nature. Often, the initial action is merely the first step in a long series of decisions to be made with various uncertain events occurring between actions. Decisions may have to be made at various points. Each of these decisions will not only be influenced by the action preceding it, but will also be affected by uncertainties all along the way. The ultimate outcome may not be known for a long time. However, decisions still have to be made during this period.

A decision tree is a convenient device for portraying decision making. It presents the decision problem as a series of branches or forks. There are two kinds of forks in a decision tree. A *decision fork* (usually represented by a square) represents a choice point where the decision-maker has to choose between actions. A *chance fork* (usually represented by a circle) represents the various states of nature that will affect the outcome of any particular decision or action. Every path starting from the initial decision fork leads to a final position that represents the outcome or consequence of the decision. Every possible combination of decisions and events leads to a specific outcome.

Let us look at an example which involves a series of uncertain events.

It's Tuesday morning and the buyer for a large variety chain has open buying hours from 9:30 A.M. to noon. Based on previous experience with this particular buyer, Mindy figures she has a 75 percent chance of getting an order that will earn her $400 *if* she can be the first person the buyer sees. If she gets there by 8:15 she is assured of being first. However, she got up a little late. Therefore, the only way she can make it by 8:15 is to take the freeway. She guesses that she has a 50–50 chance to encounter traffic enroute. If she encounters traffic, she will be delayed and may not reach the buying office until 8:40. By that time, there will be a lineup of salespeople. She calculates that her chances of getting the order, given this delay, have been reduced to about 10 percent.

On the other hand, if she takes the surface streets she figures she will arrive, with no traffic delays, around 8:30. If she arrives at that time, she figures she has a 60 percent chance of getting the order. The probability of encountering traffic that would delay her as much as 10 minutes (i.e., arrive 8:40 A.M.) is pretty low, around 0.05.

The structure of the decision problem is provided by the decision tree in Figure 9.1. When a decision is to be made, we indicate this by a square. A circle represents the chance events that can occur, e.g., traffic versus no traffic; success or failure in obtaining the order.

Beginning at the left side, the square represents the decision. Mindy can either follow the branch representing the freeway or the branch representing the surface streets. Following either branch, she comes to the first chance fork, representing the uncertainty of whether or not she will encounter traffic. Each of these chance forks leads to another chance fork influencing the success or failure of getting the order.

The probabilities of the various events are shown on the branches. For example, the probability of encountering traffic if she takes the freeway is 0.50.

The probabilities placed in parentheses are *conditional probabilities*. For instance, the probability of success given that Mindy encounters no traffic

Figure 9.1

Decision tree for Mindy's problem of which route to take

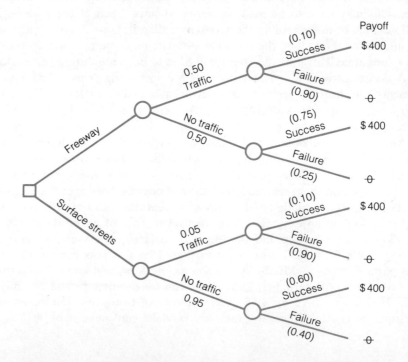

along the freeway is 0.75. However, if she encounters traffic, the probability of success is 0.10.

Finally, we list the payoffs associated with each outcome at the terminal points.

In order to analyze this problem and help Mindy decide whether to take the freeway or the surface streets, we must calculate the expected payoffs for the various outcomes. Figure 9.2 shows the expected payoffs given the various probabilities.

The decision analysis used in decision trees is called *backward induction.* We start at the terminal (or right side of the tree) and work backwards to arrive at the expected monetary payoff of being located at each fork. (To simplify the discussion, we have labeled the various chance forks *a, b, c, d, e,* and *f.*) For example, starting at the top, we calculate the expected monetary payoff for success versus failure in the usual way: $400(0.10) + 0(0.90) = $40. This represents the expected monetary payoff of being located at that fork *(b).*

The expected monetary payoff of being located at fork *(c)* is $400(0.75) + 0(0.25) = $300.

Continuing to work backwards, we now calculate the expected monetary payoff of being at fork *(a):* $40(0.50) + $300(0.50) = $170.

Decision tree diagram for freeway-versus-surface-streets decision.   Figure 9.2
The expected payoffs for the two decisions are obtained through the process of backward induction. The expected monetary payoff of being at a particular fork is entered above that fork.
Since taking the surface streets results in a higher expected payoff ($230 versus $170), Mindy should choose that route.

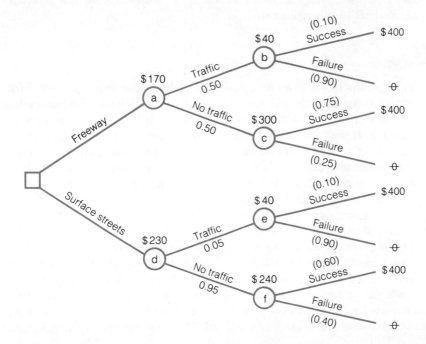

This represents the expected monetary gain if Mindy takes the freeway.

Similarly, we work backwards to obtain the expected monetary gain of the alternate route. Thus, the expected payoff of being at fork *(e)* is $400(0.10) + 0(0.90) = $40; at fork *(f)*, $400(0.60) + 0(0.40) = $240.

Finally, the expected monetary payoff of being at fork *(d)* is $40(0.05) + $240(0.95) = $230.

This represents the expected monetary gain if Mindy takes the surface streets. Since this value exceeds the one for the freeway, Mindy should choose this route.

Example E

Solartronics is considering marketing a do-it-yourself solar hot-water heating unit. The board of directors must decide whether or not to commit $1 million to develop a kit that "even a child can assemble." Based on results of a survey reported by the head of marketing, they decide to pursue further development. If the results of the development effort are favorable ($p = 0.80$), they will market the product. They must then decide on whether to market at a high level, medium level, or low level. If they market at a high level and the demand is high ($p = 0.60$), they will earn $8 million. If the demand is low ($p = 0.40$), a loss of $1.5 million will be sustained. If they market at a medium level and demand is high ($p = 0.60$), they will earn $5 million. If the demand is low ($p = 0.40$), they will earn $1 million. If marketed at a low level and demand is high ($p = 0.60$), the earnings will be $2 million. However, if the demand is low ($p = 0.40$), a loss of $500,000 will be sustained. Conduct a decision tree analysis of this situation.

Solution

The decision tree will involve a sequential decision-making process. First, the directors must decide whether or not to develop the unit. Then, depending on whether or not this is favorable, they will manufacture the unit or abort the mission. If they discontinue, the cost is $1 million. If favorable, a second decision must be made. Should they market at a high, medium, or low level? Having made this decision, the success of the enterprise depends in large part upon whether a high or low level of demand results. Figure 9.3 shows the decision tree for this situation.

We are now ready to roll back our analysis from right to left (see Figure 9.4). To obtain the expected value at chance fork *(b)*, we sum the product of $8 million times 0.60 and −$1.5 million times 0.40 to obtain $4.2 million. Similarly, multiplying $5 million by 0.60 and adding to the product of $1 million times 0.40 we obtain $3.4 million at chance fork *(c)*. Using the same procedures, we find the expected value at fork *(d)* to be $1 million. Since the expected value is greatest at a high level of manufacturing, the two lower branches (medium and low) were pruned. The expected value of development

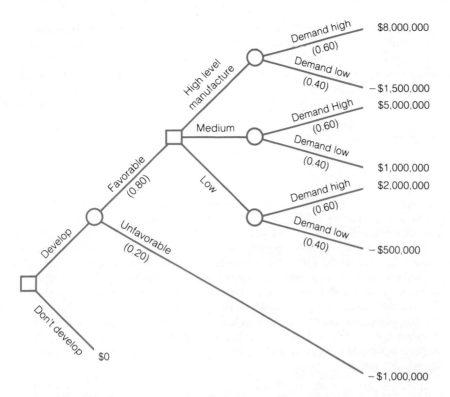

Figure 9.3
Decision tree for
development decisions

Decision tree for development decision. The expected monetary gains are obtained through   **Figure 9.4**
backward induction.

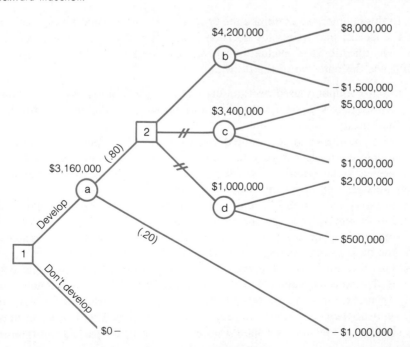

is then obtained by rolling back the sum of the product of $4.2 million times 0.80 and −$1 million times 0.20. This is found to be $3.16 million. Based on this backward induction from the future to the present, the decision is to go ahead with a high level of manufacture if the development is favorable.

EXERCISE

9.13 Following an explosion in the engine room, an oil tanker was left without power and steerage. After a discussion with the boiler room crew, the captain concluded that power was unlikely to be restored within 5 hours (probability of being restored = 0.10). The meteorological service had issued a warning that a tropical storm was headed for the tanker and was due to intercept it in four hours. The probability that the storm would veer off was 0.30. If it veered off and power was restored, the probability that the ship would still be lost was 0.05. If power was restored and the storm struck, the probability that the ship would be lost was 0.10. If power was not restored and the storm veered off, the probability that the ship would be lost was 0.10. If power was not restored and the storm struck, the probability that the ship would be lost was 0.80. The ship and cargo was valued at $80 million. If a tug was called immediately, the tanker would be taken in tow and would claim salvage rights of $10 million. What should the captain do? (For the purposes of arriving at expected monetary values, assume that the lives of the captain and crew are without value).

SUMMARY

In this chapter, we examined the process by which we make daily decisions in the face of uncertainty. A decision is an act that we make as a result of our identification and evaluation of various events or states of nature that influence the outcome of the decision.

1. Four commonly used decision criteria are: probabilities alone, economic consequences alone, probability and economic consequences together, and utility.
2. Most business and economic decisions involve both probability and economic consequences to yield expected monetary values. Using the expected value criterion, we choose an act that leads to the maximum expected gain or payoff in the long run or on the average.
3. We may also evaluate various acts in terms of their expected opportunity loss or regret. However, the decision we make on the basis of minimizing expected opportunity loss (OL) is the same as the decision made on the basis of maximizing expected monetary gain.
4. One criterion for deciding the value of information in reducing uncertainty is the expected value of perfect information (EVPI). We assume that we make the correct decision for each state of nature and obtain the sum of these corresponding expected values. The EVPI sets the upper limit to the amount we should be willing to pay for perfect information.

The expected value of perfect information is equal to the expected cost under uncertainty minus the expected cost with perfect information.

5. When conducting a decision analysis, we may find that a given action is superior to another action, no matter what the state of nature is. We say that the first dominates the second and we should drop the second action from consideration.

6. The assignment of probability values is a crucial aspect of decision making. Our probabilities may be based on prior objective information, sample information, or judgments. When probabilities are little more than guesswork, use of the maximin or maximax criteria for decision making is often warranted.

7. Prior probabilities may or may not be accurate. Bayes' theorem provides the basis for revising prior probabilities based on additional or more up-to-date sources of information. These revised probabilities are referred to as posterior probabilities. Bayes' theorem is regarded by many as the cornerstone of modern decision theory.

8. Posterior analysis may be used to revise prior probabilities based on obtaining samples from a population. The procedures are often used in deciding questions such as lot acceptance.

9. Decision tree analyses are often used in situations that cannot easily be represented in the form of a payoff table, e.g., in sequential analyses where one decision is affected by a prior decision. The procedures involve backward induction or roll-back, in which we work backward from the terminal to arrive at the expected monetary payoff at each of the chance or decision forks.

TERMS TO REMEMBER

decisions made under certainty (258)
decisions made under uncertainty (258)
decisions (258)
utility (258)
payoff (259)
states of nature (259)
decision analysis (259)
expected value criterion (259)
opportunity loss (OL) (sometimes called regret) (261)
expected value of perfect information (EVPI) (262)

maximin criterion (264)
maximax criterion (264)
minimax loss or regret criterion (264)
dominant action (268)
inadmissible action (269)
prior probability (273)
posterior (revised) probability (273)
expected value of sample information (EVSI) (278)

EXERCISES

9.14 Holiday Greeting Cards must decide many months in advance on the number of Chanukah and Christmas greeting cards that will be demanded by retailers for sale prior to the holiday season. Past experience leads them to believe that no more than 11 million nor less than 9 million cards will be ordered by the various jobbers who distribute to the retailers. They formulate three possible decisions: manufacture 9, 10, or 11 million cards. The gross profit on the sale of each greeting card is three cents. However, any stock remaining after the first of the year is sold to a jobber at a loss of 2 cents a card.

The three states of nature (demand) are 9, 10, and 11 million orders. Based on previous experience, the probability of a demand for 9 million is 0.20, 10 million is 0.50, and 11 million is 0.30. Find

    a. The payoff table when probabilities are not taken into account.
    b. The decision using the maximin criterion.
    c. The decision using the maximax criterion.
    d. The expected value of each decision.
    e. The best decision in terms of expected values.
    f. The expected payoff with perfect information.
    g. The expected value of perfect information (EVPI).

9.15 Refer to Exercise 9.14. Imagine that Holiday Greeting Cards decides to hire an independent survey organization to sample the jobbers and obtain an estimate of the number of cards they plan to order. This organization provides you with its prior record of forecasts of this type. The results are summarized below:

Actual order	Prediction (millions)		
	9	10	11
9	0.80	0.15	0.05
10	0.10	0.75	0.15
11	0.05	0.10	0.85

Based on the survey results, suppose that the firm recommended that 11 million cards be manufactured.

    a. Calculate the posterior or revised probabilities.
    b. Find the expected value of sample information (EVSI).

9.16 In October 1980 more than $98 billion in six-month money market certificates came due. Thousands of investors were faced with the question of whether or not to roll over their $10,000 T-bill accounts for another six months. This was not an easy decision because, according to the *Los Angeles Times* business section (October 27, 1980), "no one, including leading economists and financial advisers, seems to know how interest rates are likely to behave in the coming weeks and months."

According to one expert, if interest rates have nearly peaked, "then you probably can't lose by reinvesting in a six-month money market certificate."

Rates on six-month certificates, at this point, were about 11½ percent. This compared, for example, with an average of 10 percent return on money market mutual funds.

But, since interest rates have been so volatile, many experts expect rates to shoot up even higher. Thus, many investment counselors were recommending the lower yielding money market funds over savings certificates. The main reason is that money-market certificates carry heavy penalties for early withdrawals. For example, an investor who pulls out within the first three months not only loses all the earned interest, but some portion of the initial $10,000 deposit, under the latest government rules. On the other hand, although money market funds often yield less, they do not penalize for early withdrawal. Thus, if interest rates jump appreciably during this period, money fund investors have the flexibility of cashing in (at no cost) and then buying a savings certificate or some other fixed-rate investment at that time.

  a. Specify some possible actions these investors have to decide among.
  b. State the possible states of nature that will affect the outcome of their decision.
  c. How will payoff be measured?
  d. If an investor used the maximin criterion to make a decision, what would he or she probably do?
  e. If an investor used the maximax criterion, what are some possible actions he or she might take?

9.17 During the spring and summer of 1980, a prolonged heat wave and drought devastated cotton crops throughout Arkansas, Oklahoma, and Texas. As a result, cotton futures approached $1 a pound. Cotton farmers in parts of the country not affected by the drought were faced with a dilemma: to sell or not to sell on the futures market when the price appeared to have peaked. The decision is complicated by the fact that, if they contract to sell more than they actually raise, they must buy the cotton themselves to make up the difference, no matter what the price. To simplify matters somewhat, let's suppose the following. Mike and Rich are cotton farmers in Parker, Arizona, who have 500 acres under cultivation. If they have an exceptional year, they may obtain four bales per acre; if a poor year, two bales per acre. An average year would yield about three bales per acre. Let us suppose they decide to sell on the futures market at $450 a bale. Everything over the amount they contract for is sold at $380 a bale. Any amount under the number of bales contracted for must be purchased at $500 a bale and sold at $450 a bale. In 1980, it cost approximately $340 to raise a bale of cotton in Parker, Arizona. It should be noted that the cost to raise cotton is zero when it is purchased to make up a deficit in the amount contracted for.

  a. Prepare a conditional income table.
  b. Prepare a payoff or profit table.
  c. What is the decision using the maximin criterion?
  d. What is the decision using the maximax criterion?

9.18 Refer to Exercise 9.17. Suppose that, based on past experience and early indications of growth of the cotton crop, Mike and Rich assign the following probabilities to the various events:

Bales raised	Probability
1,000	0.20
1,500	0.50
2,000	0.30

  a. Calculate the expected payoff (profit) with each decision.
  b. What decision would they make based on expected profits?
  c. Find the expected value of perfect information (EVPI).

9.19 Refer to Exercises 9.17 and 9.18. Mike and Rich decide to hire a forecasting service. Based on its analysis, it is recommended that they contract for 1,500 bales. In the past, this service has been correct 60 percent of the time when predicting 1,500 bales. However, when they have recommended 1,500 bales in the past, 30 percent of the time 1,000 bales were actually raised, and 10 percent of the time 2,000 bales were actually raised.

    a. Calculate the posterior probabilities.

    b. Find the expected value for each decision using these posterior probabilities.

    c. What decision would they make using the maximum expected payoff as their decision criterion?

9.20 A machine produces aluminum extrusions for the building industry at the rate of 500 per hour. Under normal conditions, the proportion of defectives is 0.01; this occurs 90 percent of the time. However, for reasons still under investigation, the machine goes haywire about 10 percent of the time and produces 20 percent defectives. The cost of accepting a defective extrusion is $1. If a lot of 500 is rejected, the cost of transporting and remelting the scrap is 0.15 each.

    a. Set up a conditional hourly cost table.

    b. Find the expected cost of accepting a lot.

    c. Find the expected cost of rejecting a lot.

    d. In the absence of additional information, should a given lot be accepted or rejected?

9.21 Refer to Exercise 9.20. If a random sample of 15 extrusions were selected for testing, what is the cutoff point for acceptance of the lot?

9.22 In the following decision situation, state the events or states of nature, the decision choices, and the probable payoffs.

In October 1980, Ford Motor Co. set a new industry record for losses incurred in a quarter. In a related development, the *Los Angeles Times* (October 29, 1980) reported that Standard & Poor's Corporation downgraded Ford's bonds to "A" from "AA." S&P said it took the action because "it is increasingly uncertain that Ford will return to strong levels of profitability in the near term."

9.23 A distributor has the opportunity to buy a close-out of roller skates. His cost will be $36 and he figures to sell them at $50. He knows he can easily dispose of any unsold stock at $25 each. In order to take advantage of the close-out, he is required to buy in lots of 1,000. Since there are only a total of 3,000 available, his choice is between 1,000, 2,000, or 3,000. On the basis of previous experience, he estimates the probability of selling 1,000 to be 0.5; 2,000, 0.3; and 3,000, 0.2.

    a. Summarize the information in an initial payoff table.

    b. What decision should he make using the maximin criterion? The maximax criterion?

    c. What decision should the distributor make using maximum expected payoff as his decision criterion?

    d. Calculate the expected opportunity loss for each decision. Using minimum expected opportunity loss as his decision criterion, how many should he buy?

    e. Calculate the expected value of perfect information (EVPI).

9.24 Refer to Exercise 9.23. An independent research firm will do a survey for the distributor. On the basis of their market research, they will recommend the quantity he should buy. They charge $2,500 for this service.

Their past record of success is:

Actual sales	Market research prediction		
	1000	2000	3000
1,000	0.60	0.30	0.10
2,000	0.20	0.70	0.10
3,000	0.15	0.05	0.80

a. Calculate the revised probabilities based on the market research.

b. What is the probability that the market research will predict that he will sell 1,000? 2,000? 3,000?

c. What decision should he make if the market research predicts 1,000? 2,000? 3,000?

d. What is the EVSI? Should the distributor pay $2,500 for the market research?

9.25 Refer to Exercise 9.23. The distributor is approached by another independent research firm. They offer to do the market research and make recommendations regarding the quantity he should buy. They only charge $1,500. Their previous record of success is:

Actual sales	Market research prediction		
	1,000	2,000	3,000
1,000	0.50	0.35	0.15
2,000	0.20	0.50	0.30
3,000	0.20	0.30	0.50

a. Calculate the revised probabilities based on this company's record.

b. What is the probability that this firm will predict sales of 1,000? 2,000? 3,000?

c. What decision should he make if the firm predicts 1,000? 2,000? 3,000?

d. Calculate the EVSI. Which firm would you recommend the distributor use to do the market research?

9.26 Back in 1973 when the oil ministers of OPEC were meeting to establish the prices of crude oil for a six-month period, a scenario like the following may well have taken place. The oil ministers considered these possible acts: double the price of crude; increase the price by 50 percent; or permit prices to remain the same. The Minister of Aba Abu must decide which of these actions he will support. He recognizes that the action might have one of three possible consequences. It might trigger war; it might trigger a world-wide recession; or it might be accepted by the West.

If prices were doubled, the minister assessed the probabilities of each of the events as follows: war, 0.2; recession, 0.2; acceptance, 0.6. If war broke out, income would be zero since the production capacity would be destroyed or the country would be overwhelmed by the invader; if a recession ensued, the income could be the same as with no price increase, $1 billion; if the decision were accepted, the income would be $2 billion.

If the price was increased by 50 percent, the probability of war would be 0.15 with zero income; the probability of a recession would be 0.10 with $0.75 billion income; the probability of acceptance would be 0.75 with $1.5 billion income.

Finally, if prices were to remain the same, the probability of war would be 0.02 with zero income; the probability of a recession would be 0.03 with $0.5 billion income; and the probability of acceptance would be 0.95 with $1 billion income.

    a.   Construct a decision tree diagram of the oil minister's decision-making problem.

    b.   What are the expected values associated with each of these decisions?

9.27   A city in the sunbelt of the United States has undergone a continuing population explosion over the past few years. Based on this growth and the failure of a competing newspaper, a daily newspaper has decided to increase production by replacing the present equipment. Three alternatives under consideration are: a new conventional press (CP); an offset press (OP); and a computerized system (CS). The acquisition costs of the three alternatives are about the same. The annual operating costs, however, depend on the demand. The table shows the annual operating costs at each level of demand and the probability that each level of demand will prevail (adapted from Chase and Aquilano, 2d ed, 1977).

		Annual operating costs ($000)		
Demand (000)	*p*	*CP*	*OP*	*CS*
Under 80	0.2	$350	$200	$200
80–100	0.4	370	250	200
101–110	0.3	475	250	400
More than 110	0.1	510	300	600

    a.   Construct a decision tree for this decision-making situation.

    b.   Find the expected value of each act.

    c.   Based on expected costs, what decision should the newspaper make?

9.28   J.J. owns a $266,200 trust deed (TD) that he acquired as a result of selling some residential income units. The TD has six more years to go. In the meantime, he receives 10 percent interest only. Since he first made the deal, interest rates have gone up considerably. Consequently, he finds that the most he can sell the TD for is $175,000.

If he doesn't sell the TD, his total income would be ($26,620 × 6) = $159,720. In addition, the trust deed would be worth $266,200.

If he sells the TD, he sees at least two possible alternative ways he can invest his money.

1.   He can use the money from the sale to buy residential income units. He would only buy these units in certain prime areas. At this point, these areas are under rent-control laws; therefore, the maximum yearly increase he can *count on* is 7 percent. It is uncertain whether this condition will prevail over the entire six-year period. He can buy property worth $275,000 with a net income (after payments on an underlying loan of $100,000 plus taxes and expenses) of $9,000. Figuring a 7 percent increase on the gross income of $27,000 each year, his total net income (over the six years) would be approximately $85,000. Based on the same 7 percent appreciation factor each year, the building would be worth approximately $413,000 at the end of six years. Thus, his equity would be approximately $313,000.

However, there is a 30 percent chance that rent control will last only another two years. Based on previous real estate experience he figures a 15 percent increase both in gross income and on the appreciation of the building for the remaining four years. Based on these projections, his total net income (over the six years) would be approximately $114,000. In addition, the building would be worth about $550,000 at the end of six years. Thus, his equity would be approximately $450,000.

2. He can use the money to go into an investment that will yield 20 percent interest annually if conditions remain the same ($p = 0.20$). However, if interest rates jump ($p = 0.80$), the investment will yield an average of 25 percent return over the six years. Thus, if interest rates remain the same he will earn $210,000 plus his initial $175,000 investment. If interest rates jump, he will accumulate a total of $262,500 in interest plus his $175,000 investment.

Set up a decision tree and use backward induction to help J.J. decide which action to take.

(For purposes of simplification, we have not taken into consideration the tax consequences of the various decisions).

9.29 The Las Vegas Hilton hotel has two restaurants that can be used to serve breakfast. Usually, they open only one of these restaurants. However, if the crowd gets too large or the waiting time too long, they will open the second restaurant to accommodate the overflow.

    a. State the decisions or actions in this situation.
    b. Specify the states of nature that will influence the actions.
    c. What are some of the decision criteria management might use to make this decision?
    d. What are some of the ways payoff would be measured?

9.30 A publisher of college textbooks intends to publish a manuscript by Porter and Bloom. He has to decide whether the first printing should be small or large. The marketing editor, based on comments from the field, estimates a 50% chance of success (selling all copies printed) and a 50 percent chance of failure (selling none of the copies printed).

It will cost $25,000 to run a small printing and $65,000 to run a large printing. A small successful printing will net the publisher $20,000, whereas profits from a large successful printing will net $335,000.

Thus, the initial payoff table is:

	Printing	
	Small	Large
Success	$20,000	$335,000
Fail	−25,000	−65,000

Using decision tree analysis, what action should the publisher take?

9.31 Refer to Exercise 9.30. Before making the printing decision, the publisher decides to ask Professor Tech to review the manuscript. In the past, the professor's record of success has been:

Actual outcome	Review		
	Favorable	Mediocre	Unfavorable
Success	0.70	0.20	0.10
Failure	0.05	0.20	0.75

a. Calculate the revised probabilities based on the professor's record.

b. Calculate the probability that the review will be favorable, mediocre, unfavorable.

c. Summarize these probabilities in a decision tree.

9.32 Refer to Exercises 9.30 and 9.31. Set up another decision tree and place the expected values above the chance and decision forks.

a. Assuming the review costs $1,000, should the publisher use the reviewer?

b. Use backward induction to help the publisher make his decisions.

9.33 Amy G. owned a warehouse valued at $1 million, of which the equity was $300,000. Since her total liquid assets were under $10,000, the loss of the building could bring about her bankruptcy. She was considering one of three possible acts: not insuring the building; insuring it for half value; and insuring it for full value. The respective costs of exercising these options were $0; $1,400; and $2,100. The structure was such that, if fire broke out, the building would be a total loss. The probability of a fire was assessed at 0.001, and the probability of not having a fire was 0.999.

a. Find the expected value of each act (i.e., the cost).

b. Explain why Amy G. might reject the decision with the lowest expected value (recall that, since we're dealing with cost, the lowest expected value is the best monetary decision).

9.34 Sally needs a car and has to decide whether to buy a new car or a used car. A new car will cost her $5,000 with no maintenance costs the first year since it will be under warranty. However, the next two years, she figures to spend $1,000 on maintenance. If the car lasts more than three years, the annual maintenance costs will be $2,000. She figures the probability of a new car lasting a year (with no extra costs) is 0.70; 2–3 years, $p = 0.20$; and $p = 0.10$ that the new car will last more than 3 years.

On the other hand, a used car will cost $3,500 and the probability of no additional costs the first year is 0.65. However, the annual maintenance costs should be $2,000 if the used car lasts between two and three years ($p = 0.25$), and if the car lasts more than three years ($p = 0.10$), it will cost her $3,000 a year to keep it in running condition.

Using backward induction and a decision tree analysis, should she buy a new or used car?

Analysis of Variance

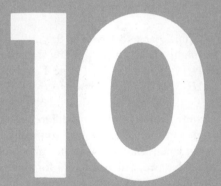

10.1 COMPARISONS OF SEVERAL GROUPS OR CONDITIONS

In Chapter 8, we took our first look at the fascinating process of testing null hypotheses. We examined such questions as: Compared to a gasoline without an additive, what is the effect of a gasoline with an additive on gasoline mileage? How do the mean life expectancies compare between fluorescent bulbs produced at two different factories? Which of two different packaging designs leads to greater sales? Each of these comparisons involved a classic two-group design.

In the real world of decision making, many of the questions we ask simply cannot be fitted into the mold of a choice between two alternatives. More often we raise questions that involve the comparison of several groups or experimental conditions. We collect data or perform experiments to answer such questions as: Which of five different additives leads to greatest gasoline mileage? Which of four different packaging designs leads to greatest sales? Which of eight different combinations of metals produces the strongest alloys? Which of three different recruiting methods leads to the greatest number of qualified applicants?

To answer these questions, the research design must involve more than two groups. But why should this fact have any bearing on the statistical analysis? If we wish to determine which of five packaging designs leads to greatest sales, is there any reason that we should not set up five experimental groups and, after collecting the data, compare condition 1 with 2, 1 with 3, 1 with 4, etc? This would involve 10 comparisons between pairs of obtained statistics.

If you think back to Chapter 8, you'll recall that there are two types of error we might make when applying the criterion for rejecting H_0. We might mistakenly reject a true H_0 (Type I error) or we might mistakenly fail to reject an H_0 that is false (Type II error). Typically, we are more concerned about a Type I error since this involves a positive assertion that we have found a statistically significant difference. To guard against this error, we usually set our α-level quite low (typically 0.05 or 0.01). Now, what if we make 10 independent comparisons in which H_0 is actually true. What is the chance that we will reject H_0 for one or more comparisons when $\alpha = 0.05$? We may answer this question by making use of the binomial table. Here under $n = 10$, $P = 0.05$, we find the probability that $x \geq 1$ is approximately 0.40. Thus, the chances are about 40 percent that we will falsely reject at least one H_0.

analysis of variance (ANOVA)

A statistical technique used for testing null hypotheses concerning two or more means.

The *analysis of variance* (abbreviated ANOVA) permits us to avoid this ambiguity when several comparisons are made. It permits us to test the null hypothesis that the samples were drawn from populations with equal means, i.e., $\mu_1 = \mu_2 = \mu_3 = \ldots \mu_k$. The analysis of variance may also be used in the two-sample case. It yields precisely the same probabilities as the t-ratio. However, ANOVA is typically used when we wish to test null hypotheses involving three or more groups.

10.2   *F*-RATIO AND SUM OF SQUARES

In Section 7.1.1, we noted that the sample variance,

$$s^2 = \frac{\Sigma(X - \overline{X})^2}{n - 1}$$

is an unbiased estimator of the population variance. In the numerator of s^2 is the quantity $\Sigma(X - \overline{X})^2$. This term shows the sum of squared deviations about the mean. When the deviations are large, this sum is also large. When small, the sum is correspondingly small. In analysis of variance, which makes much use of this term, it is convenient to substitute SS *(sum of squares)* for the quantity $\Sigma(X - \overline{X})^2$. Expressed in these terms, the unbiased estimate of the population variance is defined as follows:

$$s^2 = \frac{SS}{n - 1}$$

In a one-way analysis of variance, there are two independent estimates of variability. One involves the variability between means and is called the *between-group variance estimate*. The other is based on the variability within the various groups and is referred to as the *within-group variance estimate*. The within-group variance estimate is the mean of the variances of each group.

The test statistic is the *F-ratio*. It is a ratio between two variance estimates, the between-group variance estimate and the within-group variance estimate. The within-group estimate is based on the dispersion of scores within the various groups or conditions. The between-group estimate, on the other hand, is based on the dispersion of group means. In the one-way analysis of variance, independent group design, the *F*-ratio is

$$F = \frac{\text{Between-group variance estimate}}{\text{Within-group variance estimate}} = \frac{s^2_{\text{bet}}}{s^2_{\text{w}}} \qquad (10.1)$$

When the difference among means is large, so also is the between-group variance estimate. If s^2_{bet} is large relative to the within-group variance estimate (i.e., the differences among scores are relatively small), the *F*-ratio is large. The larger the *F*-ratio, the greater the likelihood of rejecting H_0.

One way of stating the null hypothesis is that the samples were drawn from populations with equal means, i.e., $\mu_1 = \mu_2 = \mu_3 = \cdots \mu_k$. An alternative way is to hypothesize that the between-group and the within-group variances are both estimates of random error (σ^2_e). When H_0 is true, there is no treatment effect. The between-group variance estimate will provide an estimate of σ^2_e. The within-group variance, which contains no treatment effect, is also estimating random error. Thus, when H_0 is true, the *F*-ratio would consist of two independent estimates of random error:

sum of squares (SS)

Deviations from the mean, squared and summed.

between-group variance estimate

An independent estimate of the population variance based on the variability between the groups.

within-group variance estimate

An independent estimate of the population variance based on the variability within the groups.

***F*-ratio**

The between-group variance estimate divided by the within-group variance estimate.

$$F = \frac{\sigma_e^2}{\sigma_e^2}.$$

Allowing for random variations in error estimates from sample data, the expected value of F is 1.00 when H_0 is true.

However, when H_0 is false, the between-group variance estimates contain both treatment effects (σ_a^2) and random error (σ_e^2). Thus, when H_0 is false and there is a treatment effect, the F-ratio is

$$F = \frac{\sigma_e^2 + \sigma_a^2}{\sigma_e^2}$$

The greater the treatment effects (σ_a^2), the greater the F-ratio. The greater the F-ratio, the greater the likelihood of rejecting the null hypothesis.

The significance of the difference between the within- and between-group estimates is provided by the Fisher *F-distributions*. Like the *t*-distributions, they are a family of distributions, expressed in terms of degrees of freedom (see Figure 10.1). Unlike *t*, however, it is tridimensional. To use the *F*-table (Table VIII, Appendix C), we refer to the column showing the number of degrees of freedom of the between-group variance and move down the column until we find the row showing the number of degrees of freedom of the within-group variance estimate. At the junction, we find the critical value required for rejecting the null hypothesis.

The assignment of degrees of freedom is as follows:

F-distributions

A family of distributions distinguished by degrees of freedom associated with two different variance estimates.

$$df_{bet} = k - 1 \qquad (10.2)$$

in which k is the number of groups, and

$$df_w = n - k \qquad (10.3)$$

in which n is the total number of measurements
in all of the groups.

Finally, the total degrees of freedom, df_{tot}, equals $df_{bet} + df_w$.

To illustrate the use of the *F*-table (Table VIII), imagine we obtained a between-group variance estimate of 26.53 with four different conditions. The number of degrees of freedom, df_{bet}, is $k - 1 = 4 - 1 = 3$. If the within-group variance estimate is 4.53 with $n_1 = 5$, $n_2 = 5$, $n_3 = 5$, $n_4 = 5$, $df_w = n - 4 = 20 - 4 = 16$.

Our *F*-ratio is $F = 26.53/4.53 = 5.856$; df $= 3,16$.

Referring to Table VIII and using $\alpha = 0.05$, we find the critical value under 3 and 16 to be 3.24. Since the obtained F exceeds the critical value, we may reject H_0 and assert that the four samples were not all drawn from populations with equal means.

As noted earlier, the sum of the squared deviations about the mean is

F-distributions with varying combinations of degrees of freedom. Note that as df increases, the distribution approaches normality. Note also that as df increases, the critical value of F that marks off 0.05 of the area decreases.   **Figure 10.1**

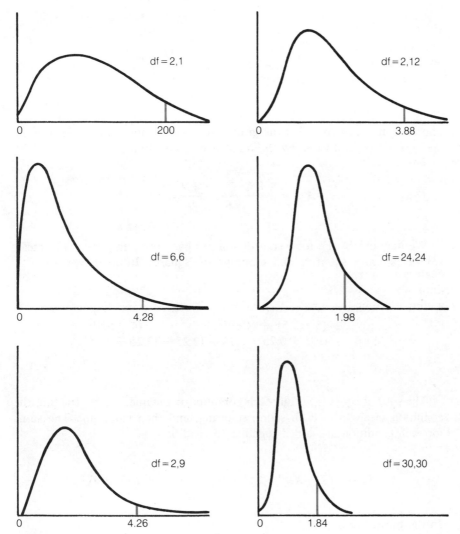

total sum of squares (SS$_{tot}$)

Deviations from the total (or grand) mean squared and summed.

within-group sum of squares (SS$_w$)

Sum of squares within each group pooled or added together.

between-group sum of squares (SS$_{bet}$)

Deviations of each group mean from the grand mean, squared and summed.

known as the sum of squares, and symbolized as SS. A convenient raw-score formula for calculating the sum of squares is

$$SS = \Sigma X^2 - \frac{(\Sigma X)^2}{n}$$

The analysis of variance might better be termed analysis of sums of squares. As we shall see, we partition the *total sum of squares* (SS$_{tot}$) into two components, the *within-group sum of squares* (SS$_w$) and the *between-group sum of squares* (SS$_{bet}$).

This can be shown very simply. Imagine the following three conditions with two observations in each condition.

	X_1	X_2	X_3
	5	7	2
	6	9	4
Sum	11	16	6

The total mean, often called the grand mean and denoted $\overline{X}_{tot}$ is sum of all the measurements divided by the number of observations, 6. Thus,

$$\overline{X}_{tot} = \frac{5 + 6 + \cdots + 4}{6} = 5.5$$

We may obtain the total sum of squares by subtracting the grand mean from each score, squaring and summing the squares. In other words,

$$\begin{aligned} SS_{tot} &= \Sigma(X - \overline{X}_{tot})^2 \\ &= (5 - 5.5)^2 + (6 - 5.5)^2 + \cdots + (4 - 5.5)^2 \\ &= 0.25 + 0.25 + 2.25 + 12.25 + 12.25 + 2.25 \\ &= 29.5 \end{aligned}$$

The sum of squares within each condition is obtained by subtracting the condition mean from each score, squaring, and then summing. The sums for each condition are added together to yield SS_w:

$$SS_w = \Sigma(X - \overline{X}_1)^2 + \Sigma(X - \overline{X}_2)^2 + \cdots + \Sigma(X - \overline{X}_k)^2$$

In the present example,

$$\begin{aligned} SS_w &= [(5 - 5.5)^2 + (6 - 5.5)^2] + [(7 - 8)^2 + (9 - 8)^2] + [(2 - 3)^2 + (4 - 3)^2] \\ &= [0.25 + 0.25] + [1 + 1] + [1 + 1] \\ &= 4.50. \end{aligned}$$

Finally, we subtract the grand mean from each group or condition mean, square, multiply by the number of observations in each condition, and then sum. This yields the between-group sum of squares, SS_{bet}:

$$SS_{bet} = \sum_{i}^{k} [n_i(\overline{X}_i - \overline{X}_{tot})^2]$$

in which

$n_i =$ The number of observations in
      the ith condition

$\overline{X}_i =$ The mean of the ith condition

$k =$ The number of conditions

Thus, in the present illustration,

$$SS_{bet} = 2(5.5 - 5.5)^2 + 2(8 - 5.5)^2 + 2(3 - 5.5)^2$$
$$= 0 + 12.50 + 12.50$$
$$= 25$$

Note that we have now partitioned the total sum of squares into its two components:

$$SS_{tot} = SS_w + SS_{bet}$$
$$29.5 = 4.50 + 25.00$$
$$29.5 = 29.5$$

This example was presented primarily to provide an intuitive grasp of the term *partitioning the sum of squares.* In practice, the mean deviation formulas are impractical. Throughout the remainder of the chapter, we will use raw score formulas because of their greater computational ease.

To expose the basic characteristics of ANOVA, we'll confine our initial illustrative material to three independent groups. The procedures may then be readily generalized to any number of independent conditions.

10.3 AN EXAMPLE INVOLVING THREE GROUPS

Noxie Automotive has developed two compounds, each of which has shown promise of improving gasoline mileage. Eighteen late model cars are randomly assigned to three experimental conditions—no additive (condition 0), Additive 1 (condition 1), and Additive 2 (condition 2). The cars are run over a measured course of 100 miles. The gasoline mileage achieved by each car was then calculated. The results are shown in Table 10.1.

Let's set up the problem in formal statistical terms:

1. *Null hypothesis (H₀):* $\mu_1 = \mu_2 = \mu_3$.

2. *Alternative hypothesis (H₂):* At least one population mean is different from the others.

3. *Statistical test:* Since we are using more than two independent groups, analysis of variance using the F-ratio is appropriate.

4. *Significance level:* $\alpha = 0.01$.

Table 10.1
Miles-per-gallon ratings of
automobiles run with no
gasoline additive, Additive
1, and Additive 2

	Condition				
	1		2		3
X_0	X_0^2	X_1	X_1^2	X_2	X_2^2
18.3	334.89	21.3	453.69	21.4	457.96
15.7	246.49	16.8	282.24	25.3	640.09
19.2	368.64	17.4	302.75	22.1	488.41
14.6	213.16	15.2	231.04	20.6	424.36
13.5	182.25	16.3	265.69	26.4	696.96
20.4	416.16	18.1	327.61	24.2	585.64
101.7	1,761.59	105.1	1,863.02	140.0	3,293.42

$$n_0 = 6; \; n_1 = 6; \; n_2 = 6; \; n = 18.$$
$$\Sigma X_{tot} = 101.7 + 105.1 + 140.0 = 346.8$$
$$\overline{X}_{tot} = \Sigma X_{tot}/n = 346.8/18 = 19.27$$
$$\Sigma X_{tot}^2 = 1{,}761.59 + 1{,}863.02 + 3{,}293.42 = 6918.03$$

5. *Sampling distribution:* F with $df_{tot} = 17$, $df_{bet} = 2$, and $df_w = 15$.
6. *Critical value:* At $\alpha = 0.01$ and df = 2 and 15, the critical value
for rejecting H_0 is 6.36.

The mean for X_0 is 16.95. for X_1 it is 17.52, and for X_2 it is 23.33. If
we were to subtract the overall mean ($\overline{X}_{tot} = 19.27$) from each score, square
the differences, and sum the squares, we would obtain the total sum of squares,

$$SS_{tot} = \Sigma(X - \overline{X}_{tot})^2$$

The more convenient raw score formula is

$$SS_{tot} = \Sigma X_{tot}^2 - \frac{(\Sigma X_{tot})^2}{n} \qquad (10.4)$$

$$= 6{,}918.03 - \frac{(346.8)^2}{18}$$
$$= 236.35$$

The within-group sum of squares pools, or adds together, the sum of
squares within each group. Thus,

$$SS_w = SS_0 + SS_1 + SS_2$$

$$SS_0 = \Sigma X_0^2 - \frac{(\Sigma X_0)^2}{n_0}$$

$$= 1761.59 - \frac{(101.7)^2}{6}$$

$$= 37.78$$

$$SS_1 = 1863.02 - \frac{(105.1)^2}{6}$$

$$= 22.02$$

$$SS_2 = 3293.42 - \frac{(140.0)^2}{6}$$

$$= 26.75$$

Thus,

$$SS_w = 37.78 + 22.02 + 26.75$$
$$= 86.55$$

The SS_{bet} may be obtained by subtracting the overall mean from each condition mean, squaring, multiplying each by its corresponding n, and then summing across all groups.

The raw score formula for SS_{bet} is

$$SS_{bet} = \sum \frac{(\sum X_i)^2}{n_i} - \frac{(\sum X_{tot})^2}{n} \tag{10.5}$$

$$= \frac{(101.7)^2}{6} + \frac{(105.1)^2}{6} + \frac{(140)^2}{6} - 6681.68$$

$$= \frac{10342.89 + 11046.01 + 19600}{6} - 6681.68$$

$$= 6831.48 - 6681.68$$

$$= 149.80$$

As a check on the calculations, $SS_{tot} = SS_w + SS_{bet}$. In this example, $SS_{tot} = 236.35$, $SS_w = 86.55$, and $SS_{bet} = 149.80$. The calculations check: $236.35 = 86.55 + 149.80$.

10.3.1 Obtaining estimates of variance and calculating F

Now that we have calculated the between-group and the within-group sum of squares, we may obtain variance estimates by dividing each by the appropriate number of degrees of freedom. Since $k = 3$ and $n = 18$, $df_{bet} = 3 - 1 = 2$ and $df_w = 18 - 3 = 15$. Thus,

$$s_{bet}^2 = \frac{SS_{bet}}{df_{bet}} = \frac{149.80}{2} = 74.90$$

and

$$s_w^2 = \frac{SS_w}{df_w} = \frac{86.55}{15} = 5.77$$

Table 10.2
ANOVA table

Source of variation	df	Sum of squares	Variance estimates	F
Between-group	2	149.80	74.90	12.98
Within-group	15	86.55	5.77	
Total	17	236.35		

Now we calculate the F-ratio and test H_0 that the variance estimates were selected from the same population. Rejecting H_0 implies that the significantly larger between-group variance estimates were due to the experimental treatments, i.e., the additives.

$$F = \frac{s^2_{bet}}{s^2_w} = \frac{74.9}{5.77} = 12.98;\ df = 2,15$$

Table 10.2 summarizes the results of the analysis.

Looking up the critical value of F required for significance at $\alpha = 0.01$ and df = 2 and 15, we find 6.36. Since obtained F exceeds the critical value, we reject H_0 at $\alpha = 0.01$ (see Figure 10.2). The different additives lead to different mileage ratings. But does condition 0 differ significantly from condition 1 and 2? Or does condition 1 differ significantly from condition 2? The analysis of variance does not answer these questions. To find answers, we must turn our attention to a multicomparison test.

10.3.2 Tukey's HSD test

A number of multicomparison tests are available that permit us to test specific hypotheses concerning differences between and among means. In this text, we shall demonstrate the use of the Tukey HSD (honestly significant differences) test. It is used when the overall F-ratio is significant and we wish to make pairwise comparisons of means. The number of such pairwise

Figure 10.2
Distribution of F at 2 and 15 df showing region of rejection at $\alpha = 0.01$

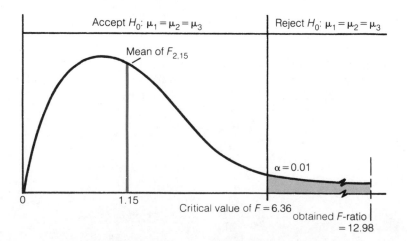

comparisons is $k(k - 1)/2$. Thus, if there are six groups, the number of differences between means is $6(5)/2 = 15$.

A difference between two means is significant, at a given α-level, if it equals or exceeds HSD which is defined as

$$\text{HSD} = q_\alpha \sqrt{\frac{s_w^2}{n}} \qquad (10.6)$$

where

$s_w^2 = $ The within-group variance estimate
$n = $ The number of subjects in *each condition*
$q_\alpha = $ Tabled value at a given α-level for df_w
and k(number of means)

We'll use the data in Section 10.3 to illustrate the calculation of the HSD test statistic. Using $\alpha = 0.01$, we shall test the significance of the difference between each pair of means.

The first step is to prepare a table of the differences among the means (Table 10.3).

Table 10.3
Differences among means

	$\bar{X}_0$ 16.95	$\bar{X}_1$ 17.52	$\bar{X}_2$ 23.33
$\bar{X}_0$ 16.95	—	0.57	6.38
$\bar{X}_1$ 17.52	—	—	5.81
$\bar{X}_2$ 23.33	—	—	—

We next refer to Table IX, Appendix C. Under error $df = 15$ and $k = 3$ at $\alpha = 0.01$, we find $q_\alpha = 4.84$. We find HSD by multiplying q_α by $\sqrt{\frac{s_w^2}{n}}$. We previously found that $s_w^2 = 5.77$. Thus,

$$\text{HSD} = 4.84 \sqrt{\frac{5.77}{6}}$$

$$= 4.75$$

This is the difference between means required for significance at $\alpha = 0.01$. Referring back to Table 10.3, we find that the differences between $\bar{X}_0$ and $\bar{X}_2$ and between $\bar{X}_1$ and $\bar{X}_2$ are statistically significant. Therefore, we conclude that Additive 2 produces significantly better gasoline mileage than Additive 1 or no additive.

10.3.3 Assumptions underlying analysis of variance

Like the t-ratio, the analysis of variance is known as a robust test. This means that, even though some of the assumptions may be violated to a degree, the statistical decisions will not often be significantly altered. However, we should be aware of the basic assumptions so that we do not incorrectly use the test when the assumptions are seriously compromised. The three most important assumptions are that the populations being compared are normally distributed, that the population variances are equal, and that the observations are independent.

In Chapter 11 we demonstrate a test for normality of a distribution. If you have serious reservations that your sample comes from a population that is not approximately normal, this test should be applied to the data.

EXERCISES

10.1 The enormous increases in the price of gold that occurred in the late 1970s made it feasible to consider extracting gold from previously discarded tailings. Using three different techniques of reclaiming the gold, the following amounts (in troy ounces) were extracted from 21 randomly selected samples of ore.

X_1	X_2	X_3
8.2	4.4	7.1
5.3	3.1	4.2
2.1	2.1	5.3
4.6	2.8	3.5
4.4	4.5	4.4
3.2	6.1	1.3
6.3	1.1	2.8

a. Set up and test H_0, using $\alpha = 0.05$.

b. If warranted, test for significance of differences among paired means.

10.2 Refer to Exercise 10.1. Another way of looking at the within-group variance estimate is that it is the average or mean of the sample variances obtained from each of the experimental groups, i.e.,

$$s_w^2 = \frac{s_1^2 + s_2^2 + \cdots + s_k^2}{n_k}$$

Confirm that this is true for Exercise 10.1.

10.3 In a foundry that stamps chassis for the electronics industry, there are four different machines in operation. Each day a random sample of a fixed size is selected from the output of each machine. The number of defective chassis is recorded. The following are the results obtained over a 15-day period.

Number of defective chassis

	Machine		
1	2	3	4
6	8	5	6
2	15	3	8
0	12	0	4
4	7	4	2
5	11	0	3
9	10	3	5
3	6	1	9
7	8	6	2
4	13	1	1
0	9	2	6
1	8	0	8
4	6	4	7
8	9	7	4
6	12	3	3
5	10	5	1

a. Set up and test H_0, using $\alpha = 0.01$.

b. If warranted, test for significance of differences among paired means.

10.4 Referring to Exercise 10.3, confirm that

$$s_w^2 = \frac{s_1^2 + s_2^2 + s_3^2 + s_4^2}{4}$$

10.4 TWO-WAY ANALYSIS OF VARIANCE: FACTORIAL DESIGN

So far, we have looked at a *one-way analysis of variance*. Using this technique we have compared means resulting from several levels or categories of a *single* treatment variable. However, one of the advantages of the analysis of variance technique is that it permits us to extend our analysis to situations in which the measure of interest is affected by two or more variables. For example, consider the following:

> We are interested in comparing gasoline mileage for four different makes of cars using two different types of additives.

> We wish to assess the effect on productivity of two types of training programs and frequency of training sessions (daily, weekly, monthly).

> We want to know if there are price differences among three different supermarkets according to type of item (meat, grocery, bakery, dairy).

In each of these examples, our interest goes beyond the question of whether the variability in the measure of interest is due either to variability between the groups or variability within the groups (differences due to "chance").

one-way analysis of variance

Statistical test of various categories or levels of a *single* treatment variable.

310

two-way analysis of variance

Statistical test of various categories or levels of two treatment variables.

interaction

The joint effect of two treatment variables on the measure of interest, separate from the individual effects of either variable (or factor).

These examples illustrate the *two-way analysis of variance* in which our interest extends to the *joint* effects of the two treatment variables as well as their *separate* effects. In other words, we are interested in whether the treatment variables work together or interact in some way to produce differences in the dependent measure. This concept of *interaction* is crucial to an understanding of two-way analysis of variance. Let us look at some examples.

Suppose, for instance, that one type of additive produces higher miles-per-gallon ratings for city but not highway driving. In contrast, a second additive improves miles-per-gallon ratings only for highway driving. If the effect of the two additives did not change with city versus highway driving conditions, there would be no interaction. But as long as the effect of one variable (type of additive) depends on the level or type of a second variable (conditions of driving), we say these two variables interact. A possible interaction between types of additive and conditions of driving is shown in Figure 10.3.

Figure 10.3

Interaction of two variables. Note that performance of one variable depends on the status of a second variable. Additive 2 is more effective than additive 1 in city driving. Conversely, additive 1 produces better mileage ratings during highway driving.

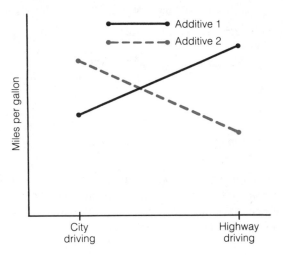

Let's look at another example. Suppose one of two training methods improved productivity only when administered daily, whereas the other training method improved productivity when administered either weekly or monthly. If the effects of the training methods were the same regardless of frequency of training session, there would be no interaction. Thus, the variability in the dependent measure (productivity) can best be understood by looking at the *joint* effect of the two independent variables, i.e., the *interaction* between these factors.

In the one-way analysis of variance, we saw that total variation may be partitioned into between-group variability and within-group variability. In the two-way analysis of variance, between-group variability may be further partitioned:

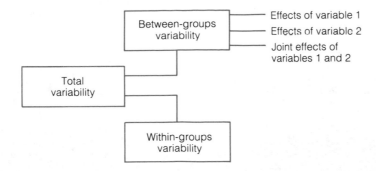

Although there are only two variables in a two-way analysis of variance, there may be several levels of each variable, for example, four types of cars and two types of additives. We call this a 4 × 2 design; two types of training programs and three different frequencies of training sessions require a 2 × 3 design; three different supermarkets and four types of items involve a 3 × 4 design.

All of these examples illustrate what we call a *factorial design*. In this type of design, the number of treatment combinations is dictated by the type of design. For example, in a 4 × 2 design, there are eight treatment combinations. Thus, one treatment combination might be the second type of car with the first type of additive.

Let's look at the various treatment combinations in these examples:

1. Treatment combinations in a 4 × 2 factorial design:

Make of car	A_1		A_2		A_3		A_4	
Type of additive	B_1	B_2	B_1	B_2	B_1	B_2	B_1	B_2
Treatment combinations	A_1B_1	A_1B_2	A_2B_1	A_2B_2	A_3B_1	A_3B_2	A_4B_1	A_4B_2

2. Treatment combinations in a 2 × 3 factorial design:

Training program	A_1			A_2		
Frequency of session	B_1	B_2	B_3	B_1	B_2	B_3
Treatment combinations	A_1B_1	A_1B_2	A_1B_3	A_2B_1	A_2B_2	A_2B_3

Let us look at an example to illustrate the calculations involved in a two-way analysis of variance.

Suppose we set up two different types of training programs (A_1 and A_2) and three different frequencies of training sessions (B_1, B_2, and B_3). We randomly select 24 workers from the total labor pool and randomly assign them to the different treatment combinations. Since there are eight different

factorial design

A research design in which independent observations are obtained for each level of the treatment variables. We call each specific combination of experimental conditions the treatment combination.

Table 10.4

Productivity scores for subjects receiving two different types of training over three different frequencies of training sessions

Training program	A_1			A_2		
Frequency of training session	B_1	B_2	B_3	B_1	B_2	B_3
Treatment combinations	A_1B_1	A_1B_2	A_1B_3	A_2B_1	A_2B_2	A_2B_3
	7	4	5	2	6	8
	8	3	4	3	8	7
	9	3	4	3	9	7
	7	5	3	4	9	9
Sums	31	15	16	12	32	31
Means	7.75	3.75	4.00	3.00	8.00	7.75

$$\Sigma X = 137$$

treatment combinations, three subjects are randomly assigned to each treatment combination. We wish to determine whether level of productivity is affected by training program and/or the frequency of training sessions. We shall use $\alpha = 0.01$.

Table 10.4 presents the productivity measures (number of units produced in a given time period) for the 24 randomly selected workers according to the appropriate treatment combinations.

In a two-way analysis of variance, the total sum of squares is partitioned into two broad components:

$$SS_{tot} = SS_w + SS_{bet}$$

The between-group sum of squares is further partitioned into three components—the sum of squares for each variable plus the sum of squares for the interaction between these two variables:

$$SS_{bet} = SS_A + SS_B + SS_{A \times B} \tag{10.7}$$

Each of the treatment effects provides an independent estimate of the population variance when divided by the appropriate degrees of freedom. Thus,

$$s_A^2 = \frac{SS_A}{df_A}; \qquad df_A = A - 1 \tag{10.8}$$

in which A is the number of levels of the A-variable. Likewise,

$$s_B^2 = \frac{SS_B}{df_B}; \qquad df_B = B - 1 \tag{10.9}$$

in which B is the number of levels of the B-variables. And finally,

$$s_{A \times B}^2 = \frac{SS_{A \times B}}{df_{A \times B}} \,; \qquad df_{A \times B} = df_A \times df_B \qquad\qquad (10.10)$$

The within-group sum of squares provides an independent estimate of the population variance when divided by the appropriate degrees of freedom:

$$s_w^2 \,(\text{error}) = \frac{SS_w}{df_w} \,; \qquad df_w = n - AB$$

in which n is the total number of observations and AB is the number of treatment combinations.

When we have completed all the necessary calculations, we will summarize them in an analysis of variance table as shown below (Table 10.5).

Source of variation	Sum of squares		df		Variance estimate
Between-group	SS_{bet}			$AB - 1$	
A-variable	SS_A	$A - 1$			$SS_A/(A - 1)$
B-variable	SS_B	$B - 1$			$SS_B/(B - 1)$
A-B interaction	$SS_{A \times B}$	$(A - 1)(B - 1)$			$SS_{A \times B}/(A - 1)(B - 1)$
Within-group		SS_w		$n - AB$	$SS_w/(n - AB)$
Total		SS_{tot}		$n - 1$	

Table 10.5
ANOVA table

Let us now proceed with the calculations for this example. We shall summarize our calculations in Table 10.6.

1. $$SS_{tot} = \Sigma X^2 - \frac{(\Sigma X)^2}{n}$$

$$= 7^2 + 8^2 + 9^2 + \ldots + 7^2 + 7^2 + 9^2 - \frac{(137)^2}{24}$$

$$= 911 - 782.04$$

$$= 128.96.$$

$$df_{tot} = n - 1 = 23.$$

Source of variation	Sum of squares		df	Variance estimate	F
Between-group		110.71	5		
A-variable	7.04		1	7.04	6.97
B-variable	1.34		2	0.67	0.66
A-B interaction	102.33		2	51.16	50.65
Within-group (error)		18.25	18	1.01	
Total		128.96	23		

Table 10.6
ANOVA table

2. $$SS_{bet} = \frac{(\Sigma A_1 B_1)^2}{n_{A_1 B_1}} + \frac{(\Sigma A_1 B_2)^2}{n_{A_1 B_2}} + \ldots + \frac{(\Sigma A_2 B_3)^2}{n_{A_2 B_3}} - \frac{(\Sigma X)^2}{n}$$

$$= \frac{(31)^2}{4} + \frac{(15)^2}{4} + \ldots + \frac{(31)^2}{4} - 782.04$$

$$= 892.75 - 782.04$$

$$= 110.71.$$

$$df_{bet} = AB - 1 = 5.$$

3. $$SS_w = SS_{tot} - SS_{bet}$$
$$= 18.25$$
$$df_w = n - AB = 24 - 6 = 18$$

4. $$SS_A = \frac{(\Sigma X_{A_1})^2}{n_{A_1}} + \frac{(\Sigma X_{A_2})^2}{n_{A_2}} - \frac{(\Sigma X)^2}{n}$$

$$= \frac{(62)^2}{12} + \frac{(75)^2}{12} - 782.04$$

$$= 789.08 - 782.04$$

$$= 7.04$$

$$df_A = A - 1 = 1$$

5. $$SS_B = \frac{(\Sigma X_{B_1})^2}{n_{B_1}} + \frac{(\Sigma X_{B_2})^2}{n_{B_2}} + \frac{(\Sigma X_{B_3})^2}{n_{B_3}} - \frac{(\Sigma X)^2}{n}$$

$$= \frac{(43)^2}{8} + \frac{(47)^2}{8} + \frac{(47)^2}{8} - 782.04$$

$$= 783.38 - 782.04$$

$$= 1.34$$

$$df_B = B - 1 = 2$$

6. $$SS_{A \times B} = SS_{bet} - SS_A - SS_B$$
$$= 110.71 - 7.04 - 1.34$$
$$= 102.33$$
$$df_{A \times B} = (A - 1)(B - 1) = 2$$

All of these results are shown in Table 10.6. Let us now find the *variance estimates:*

7. $$s_A^2 = \frac{SS_A}{df_A} = \frac{7.04}{1} = 7.04$$

$$s_B^2 = \frac{SS_B}{df_B} = \frac{1.34}{2} = 0.67$$

$$s_{A \times B}^2 = \frac{SS_{A \times B}}{df_{A \times B}} = \frac{102.33}{2} = 51.16$$

$$s_w^2 = \frac{SS_w}{df_w} = \frac{18.25}{18} = 1.01$$

We are now ready to find the F-ratios and test the null hypotheses. It should be noted that, in a two-way analysis of variance, we actually test three separate null hypotheses (see 8, 9, and 10):

8. H_0: There is no interaction; i.e., the effects of the treatment variables are independent of each other.

H_1: There is an interaction; i.e., the effects of the treatment variables are not independent of each other.

The interaction F-ratio is obtained by dividing the interaction variance estimate by the within-group variance estimate:

$$F = \frac{s_{A \times B}^2}{s_w^2} = \frac{51.16}{1.01} = 50.65; \quad df = 2, 18$$

Looking in Table VIII under 2 and 18 df, we find the critical value of F required to reject H_0 at $\alpha = 0.01$ is 6.01. Therefore, we may reject H_0 and assert that the interaction is statistically significant.

9. H_0: There is no difference in the training programs, i.e.,

$$\mu_{A_1} = \mu_{A_2}$$

H_1: There is a difference in the training program, i.e.,

$$\mu_{A_1} \neq \mu_{A_2}$$

The F-ratio for the A variable:

$$F = \frac{s_A^2}{s_w^2} = \frac{7.04}{1.01} = 6.97; \quad df = 1, 18$$

Looking in Table VIII under 1 and 18 df we find the critical value of F at $\alpha = 0.01$ is 8.28. Therefore, we may not reject H_0.

10. H_0: There is no difference in the level of productivity among subjects undergoing different frequencies of training sessions, i.e.,

$$\mu_{B_1} = \mu_{B_2} = \mu_{B_3}$$

H_1: The population mean productivity level of at least one labor group is different from the others.

The F-ratio for the B-variable:

$$F = \frac{s_B^2}{s_w^2} = \frac{0.67}{1.01} = 0.66; \quad df = 2, 18$$

Since obtained F is less than the critical value, we fail to reject H_o.

All of these calculations are summarized in Table 10.6.

Conclusion: Of the three effects evaluated, only the interaction was found to be statistically significant. It is now appropriate to employ a multicomparison test in order to determine which of the various combinations are statistically significant. The Tukey HSD test shown in Section 10.3.2 may be used to make these comparisons.

The first step is to find the mean of the scores obtained from each treatment combination. These are as follows:

$$\overline{X}_{A_1B_1} = 7.75 \qquad \overline{X}_{A_2B_1} = 3.00$$
$$\overline{X}_{A_1B_2} = 3.75 \qquad \overline{X}_{A_2B_2} = 8.00$$
$$\overline{X}_{A_1B_3} = 4.00 \qquad \overline{X}_{A_2B_3} = 7.75$$

(Figure 10.4 summarizes these findings.)

Next arrange them in ascending order (from low to high means) and tabulate the differences between means as shown in Table 10.7. Referring to Table IX under df = 18 and $k = 6$ at $\alpha = 0.05$, we find $q_\alpha = 4.49$. Thus,

$$\text{HSD} = 4.49 \sqrt{\frac{1.01}{4}}$$
$$= 2.26$$

Figure 10.4

Mean productivity associated with two training programs and three different frequencies of training sessions. A two-way analysis of variance of these data yielded a statistically significant interaction.

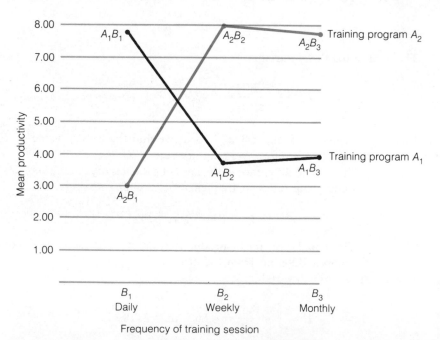

	A_2B_1 3.00	A_1B_2 3.75	A_1B_3 4.00	A_2B_3 7.75	A_1B_1 7.75	A_2B_2 8.00
A_2B_1　3.00	—	0.75	1.00	4.75*	4.75*	5.00*
A_1B_2　3.75	—	—	0.25	4.00*	4.00*	4.25*
A_1B_3　4.00	—	—	—	3.75*	3.75*	4.00*
A_2B_3　7.75	—	—	—	—	0.00	0.25
A_1B_1　7.75	—	—	—	—	—	0.25
A_2B_2　8.00	—	—	—	—	—	—

Table 10.7
Pairwise comparisons among six different treatment combinations

Thus, any difference between means shown in Table 10.7 that equals or exceeds 2.26 is significant at $\alpha = 0.05$. These differences are marked with an asterisk on Table 10.7.

EXERCISES

10.5 A large alfalfa farm was divided into 63 plots. Prior to planting, each plot was treated with one of three levels of a root stimulating hormone. After planting, each plot received either 1, 2, or 3 acre-feet of water per irrigation. The results, in terms of number of bales per plot, are shown in the following table.

Amount of hormone Amount of irrigation	A_1 2 gal/plot			A_2 4 gal/plot			A_3 6 gal/plot		
	B_1	B_2	B_3	B_1	B_2	B_3	B_1	B_2	B_3
	2	3	5	3	4	5	4	5	6
	3	3	4	2	3	7	2	4	8
	1	3	3	4	4	6	3	3	7
	2	4	6	5	5	7	5	5	9
	3	5	4	4	3	5	5	6	7
	4	4	7	3	5	8	3	7	9
	2	2	4	4	6	9	4	6	10

  a. Draw a graph of the means and judge whether an interaction is likely.
  b. Set up and test H_0, using $\alpha = 0.05$.
  c. If warranted, conduct multicomparison tests at $\alpha = 0.05$.

10.6 Refer to Exercise 10.5. Another way of regarding s_w^2 is that it is the mean of the variances for all the separate treatment combinations. Confirm that this is true for the data in Exercise 10.5.

10.4.1 Randomized block design

A variation of the two-way analysis of variance is a design called the *randomized block design.* In this design we are primarily interested in the differences between or among one treatment variable. Our second variable consists of blocks of relatively homogeneous units. The second variable (blocks) is included to reduce the variation in the error term (SS_w). Although many different components contribute to the magnitude of the error term, we can reduce the variation caused by differences among relatively homogeneous units (e.g., same cars, same subjects) by using blocks. Thus, the total variation can be distributed among three components:

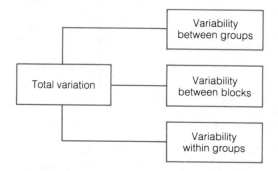

As we shall see, the randomized block design is especially useful for analyzing business and economic data.

For example, in Section 10.3 we used a one-way analysis of variance to analyze the effects of additives (or no additive) on gasoline mileage. We randomly assigned 18 cars to the three conditions. Thus, the design looked like this:

	Additive		
	None	1	2
Automobile	2,3,18,6,9,13	1,5,8,12,15,16	4,5,10,11,14,17

In this particular analysis, we obtained a significant difference among the treatments. However, some of the variation in the error term can be attributed to differences among the various automobiles. We can eliminate this variation by administering each treatment condition to each of the 18 cars. That is, each automobile would receive each of the treatments (no additive, Additive 1, Additive 2) in a random sequence. Table 10.8 illustrates the randomized block design using the various automobiles as blocks. Each additive in a single automobile is a unit in our experiment.

In Table 10.8, each of the three additive conditions are assigned to each auto. We assume that the additives are administered in the sequence shown

Automobiles (blocks)				
1	2	3	4	18
⓪	②	①	①	⓪
②	①	②	⓪	②
①	⓪	⓪	②	①

Table 10.8
A randomized block design using 18 automobiles as blocks and three types of additives (0,1,2) as treatment (circled)

from top to bottom. Each treatment condition occurs only once in each block. For example, auto 1 receives the treatments in the following order: no additive (0), Additive 2 (2), Additive 1 (1).

We obtain 18 measurements for each of the three treatments using the randomized block design. In the one-way analysis of variance (or completely randomized design), we would need 54 autos to obtain 18 measurements for each treatment.

In a randomized block design, we randomly assign the treatments within each block. In this particular experiment, it means that each car is tested under the three conditions, with the order varied across blocks (or autos). Just in case using Additive 1 affects subsequent performance with, say, no additive, randomly assigning the treatments will eliminate this source that usually contributes to error. In a sense, the *t*-test for matched samples is a simple example of a randomized block design.

Blocking is intended to remove variation between homogeneous units, such as the same auto, same subjects, same stores, etc. In the randomized block design, we are primarily interested in the treatment differences. The differences among blocks are usually only of secondary importance. The primary purpose of the blocks is to identify a source of variation and remove it from error.

Unlike the treatment variable, which is free to vary and is manipulated by the experimenter, the blocks are not free to vary. They represent preexisting conditions or events. Thus, although we can randomly assign different types of additives to different cars, we cannot randomly assign "type of car" to each automobile. A Ford is a Ford and a Chevy a Chevy.

In a randomized block design the total sum of squares is partitioned into three components:

$$SS_{tot} = SS_{treat} + SS_{blks} + SS_w$$

The degrees of freedom associated with each component are

$$df_{tot} = n - 1$$

where n is the total number of observations;

$$df_{treat} = k - 1$$

where k is the number of treatment groups;

$$\mathrm{df_{blks}} = b - 1$$

where b is the number of blocks; and

$$\mathrm{df_w} = (n - 1) - [(k - 1) + (b - 1)] = n - k - b + 1$$

Table 10.9 presents the ANOVA table for a randomized block design.

Table 10.9
ANOVA table

Source of variation	df	SS	Variance estimate
Blocks	$b - 1$	SS_b	$\dfrac{SS_b}{b - 1}$
Treatments	$k - 1$	SS_t	$\dfrac{SS_t}{k - 1}$
Error	$n - k - b + 1$	SS_w	$\dfrac{SS_w}{n - k - b + 1}$
Total	$n - 1$	SS_{tot}	

Let us look at a worked example utilizing the randomized block design.

Suppose we want to know if there are price differences among three different gasoline stations. We are aware of the fact that premium usually costs more than regular or nonleaded. However, we are primarily interested in whether differences exist among the three stations (treatments). In order to eliminate the variation caused by differences among types of gasoline, we will use them as blocks. Table 10.10 illustrates the randomized block design using type of gasoline as blocks. The price of each of these items is obtained from each of the three stations in the order shown. Since prices may vary across time, we obtain the prices over a three-week period.

Table 10.10
Randomized block design using type of gasoline as blocks and three different gasoline stations as treatments (circled)

Type of gasoline (blocks)		
Regular	Unleaded	Premium
①	②	③
③	①	②
②	③	①

Thus, during week 1 we obtain the price of regular gas at station 1, unleaded at station 2, and premium at station 3. Therefore, in case prices for items are higher in any one week, this gives each station an equal chance of representation during that week.

Table 10.11 summarizes the data collected over the three-week period from each of the three gasoline stations. Let us use these data to illustrate the calculations involved in a randomized block design.

Table 10.11
Price (in cents) at each of
three stations for each type
of gasoline

	Type of gasoline	
Regular	Unleaded	Premium
1	2	3
124.8	129.9	135.9
3	1	2
129.7	129.9	129.9
2	3	1
119.9	133.7	132.9

The various sums necessary for the calculations are the following.

Total:

$$\Sigma X_{\text{tot}} = 1{,}166.6 \qquad \Sigma X_{\text{tot}}^2 = 151{,}402.08 \qquad \frac{(\Sigma X_{\text{tot}})^2}{9} = 151{,}217.28$$

Blocks:

$$\Sigma X_{b_1} = 374.4 \qquad \Sigma X_{b_2} = 393.5 \qquad \Sigma X_{b_3} = 398.7$$

Treatments:

$$\Sigma X_{t_1} = 387.6 \qquad \Sigma X_{t_2} = 379.7 \qquad \Sigma X_{t_3} = 399.3$$

The sums of squares are the following.

$$SS_{\text{tot}} = \Sigma X_{\text{tot}}^2 - \frac{(\Sigma X_{\text{tot}})^2}{n}$$

$$= 184.8.$$

$$
\begin{aligned}
SS_{\text{blks}} &= \frac{(\Sigma X_{b_1})^2 + (\Sigma X_{b_2})^2 + (\Sigma X_{b_3})^2}{n_{\text{blk}}} \\
&= \frac{(374.4)^2 + (393.5)^2 + (398.7)^2}{3} = 151{,}217.28 \\
&= 151{,}326.43 - 151{,}217.28 \\
&= 109.15
\end{aligned}
$$

$$
\begin{aligned}
SS_{\text{treat}} &= \frac{(\Sigma X_{t_1})^2 + (\Sigma X_{t_2})^2 + (\Sigma X_{t_3})^2}{n_{\text{treat}}} \\
&= \frac{(387.6)^2 + (379.7)^2 + (399.2)^2}{3} - 151{,}217.28 \\
&= 151{,}282.11 - 151{,}217.28 \\
&= 64.83
\end{aligned}
$$

Table 10.12
ANOVA table

Source	df	SS	Variance estimate	F
Blocks............	2	109.15	54.575	20.18
Treatments	2	64.83	32.415	11.98
Error	4	10.82	2.705	—
Total	8	184.80	—	—

$$SS_w = SS_{tot} - SS_{blks} - SS_{treat}$$
$$= 184.8 - 109.15 - 64.83$$
$$= 10.82$$

All of these calculations are entered in Table 10.12. By dividing each sum of squares by its appropriate df, we obtain the variance estimates.

Finally, we obtain the F-statistic related to treatments:

$$F = \frac{s^2_{treat}}{s^2_w} = \frac{32.415}{2.705} = 11.98; \; df = 2,4$$

The critical value of F ($\alpha = 0.05$) for 2 and 4 df is 6.94. Thus we may reject the null hypotheses and assert that there is a difference in gasoline prices among the three stations.

We may also test the null hypothesis that no difference exists among the three types of gasoline. Although this difference is not of primary importance, it does suggest that variability exists among the three types of gasoline and that blocking is desirable.

$$F = \frac{s^2_{blks}}{s^2_w} = \frac{54.575}{2.705} = 20.18; \quad df = 2, 4$$

Since our obtained F exceeds the critical value ($\alpha = 0.05$), we may conclude that a difference exists among the three types of gasoline. This result supports our decision to block with respect to type of gasoline.

EXERCISE

10.7 Suppose we want to compare the price-earnings ratios (high calendar year market prices divided by net earnings per common share) of three different stock groups (industrials; chemicals; fertilizers) with regard to stability over time.

The following data are based on Standard & Poor's industry Group Stock Price Indexes.

	Year						
	1973	1974	1975	1976	1977	1978	Sums
Industrials	15.13	11.62	12.52	11.31	10.31	9.06	69.95
Chemicals	14.30	10.13	13.54	13.61	11.76	8.33	71.67
Fertilizers	43.86	6.07	5.63	8.63	8.90	10.46	54.12
Sums	43.86	27.82	31.69	33.55	10.97	27.85	—

Source: Adapted from *Standard & Poor's Industry Surveys,* January 1980, p. C41.

    a.   What type of design should be employed?
    b.   Perform the appropriate analysis of variance on these data.
    c.   Using $\alpha = 0.05$, summarize and state your findings.

SUMMARY

Many of the questions we ask involve a choice between more than two alternatives. We may wish to compare several different investment strategies, several different dosage levels of a drug, a variety of training approaches. There are also many occasions when we wish to test several different categories or levels of two different variables. Our interest may focus on whether or not the variables interact in this more complex type of design. The analysis of variance technique is ideally suited for testing null hypotheses when complex experimental designs are used.

1. In this chapter, we covered both the one-way and the two-way ANOVA.
2. In the one-way ANOVA, we wish to test H_0 that the groups or conditions were drawn from populations with the same means. If we reject H_0, we know that the treatment variable had a significant effect.
3. The test statistic in ANOVA is the F-ratio. The F table shows critical values at $\alpha = 0.05$ and 0.01 for two different variance estimates: s_{bet}^2 and s_w^2. When we reject H_0, we may make paired comparisons among the group means. We demonstrated the use of one multicomparison test, the Tukey HSD (honestly significant difference) test. The test allows us to determine which pairs of means differ significantly from one another.
4. In the two-way analysis of variance, there are three separate null hypotheses. One relates to the effect of the A variable, a second to the B variable, and the third to the interaction of the two variables.
5. There may be any number of levels of each variable in a two-way ANOVA. Three levels of one variable and five of another is known as a 3×5 design.
6. The number of treatment combinations is dictated by the type of design. A 3×5 design involves 15 different treatment combinations.
7. If H_0 is rejected, we may use Tukey's HSD to evaluate differences among the paired means for the condition or conditions that are statistically significant.

324

8. Finally, we examined the randomized block design in which treatments are randomly assigned within a set of blocks in order to identify and remove a source of error. This design is similar to the t-test for matched samples.

TERMS TO REMEMBER

analysis of variance (ANOVA) (298)
sum of squares (SS) (299)
between-group variance estimate (299)
within-group variance estimate (299)
F-ratio (299)
F-distributions (300)
total sum of squares (SS_{tot}) (301)
within-group sum of squares (SS_w) (301)

between-group sum of squares
(SS_{bet}) (301)
one-way analysis of variance (309)
two-way analysis of variance (310)
interaction (310)
factorial design (311)
randomized block design (318)

EXERCISES

10.8 In an effort to match the availability of police protection to the incidence of crime, a city was subdivided into four geographical areas, each including 25,000 residences. Five equal-size samples of residences were randomly selected in each area. The number of crimes reported per area was as follows.

X_1	X_2	X_3	X_4
7	12	9	12
3	6	7	14
9	9	6	9
8	6	8	16
7	3	6	13

a. Set up and test H_o, using $\alpha = 0.05$.
b. If warranted, conduct a multicomparison test.

10.9 ABC Corporation is planning to market a new product. It is considering two different types of packaging. It test markets the product in three different geographical regions using two different packaging styles. Gross sales, in hundreds of thousands of dollars, are shown in the table.

a. Graph the means and judge whether there might be an effect of packaging, an effect of geography, or an interaction.
b. Set up and test the appropriate null hypotheses using $\alpha = 0.01$.
c. If warranted, conduct a multicomparison test.

Geographical region	A_1		A_2		A_3	
Packaging	B_1	B_2	B_1	B_2	B_1	B_2
	22	11	15	9	12	9
	10	7	12	11	9	11
	12	9	14	5	7	14
	14	4	27	8	13	10
	9	2	24	3	8	6

10.10 Following a leak of radioactive gas at a nuclear power plant, samples of air were tested at three distances from the stack. The following table shows the number of radioactive particles captured at each distance.

a. Set up and test H_o at $\alpha = 0.01$.

b. Conduct a multicomparison test, if warranted.

100 ft	500 ft	2,000 ft
X_1	X_2	X_3
26	21	14
22	12	8
14	8	3
18	14	5
17	16	9
19	11	4
30	7	2

10.11 The following data represent dividends (as a percentage of earnings) of three different stock groups over a five-year period.

These data are based on Standard & Poor's Industry Group Stock Price Indexes.

	1974	1975	1976	1977	1978
Industrials	38.6	43.4	39.5	42.9	41.0
Chemicals	32.6	39.6	37.6	45.1	43.3
Fertilizers	7.3	11.2	33.9	38.9	52.5

Source: Adapted from *Standard & Poor's Industry Survey,* January 1980, p. C39.

Perform the appropriate analysis of variance and, using $\alpha = 0.01$, summarize and state your findings.

10.12 As canned food moves along a conveyor belt following processing, a random sample is selected for analysis by a laboratory technician. Using five alternative testing procedures, the following testing times (in seconds) were recorded. We wish to determine if the various testing procedures produce the same results.

X_1	X_2	X_3	X_4	X_5
21	21	16	20	19
16	22	20	19	29
20	20	18	19	23
15	18	20	19	26
13	16	14	16	24

a. Set up and test H_o, using $\alpha = 0.05$.

b. If warranted, conduct a multicomparison test, using $\alpha = 0.05$.

10.13 A large insurance company decides it wants to expand its program of bonus incentives. The personnel manager decides to study employee preferences before implementing a new incentive program. She suspects that employees differ in terms of their preference for certain rewards. She randomly selects five single and five married employees from each of three age groups and administers an incentive preference test to each individual. The test is designed to determine preference among two broad categories: incentives related to monetary increases versus incentives related to less working time. The test was constructed so that it yielded normally distributed interval measurements. A positive score indicates a preference for monetary increases. A negative score indicates a preference for shorter working days or weeks. She obtains the following results:

Age group	A_1 (18–34)		A_2 (35–49)		A_3 (50–65)	
Marital status	B_1 (S)	B_2 (M)	B_1 (S)	B_2 (M)	B_1 (S)	B_2 (M)
	−0.68	1.30	1.40	2.00	−1.80	1.20
	−1.00	1.20	1.10	2.50	−2.00	1.25
	−0.75	1.10	1.25	1.90	−0.03	0.50
	0.02	2.00	−0.02	2.60	0.01	1.00
	−1.05	0.05	1.80	2.40	−3.20	0.80
Sums	−3.46	5.65	5.53	11.40	−7.02	4.75

a. Draw a graph of the means and judge whether an interaction is likely, and/or whether there are age and marital differences.

b. Perform the appropriate ANOVA, using $\alpha = 0.05$. Summarize your findings.

10.14 The following data represent factory shipments of selected electrical appliances during the first six months of 1979. Perform the appropriate analysis of variance on these data. Using $\alpha = 0.05$, summarize and state your results.

	Shipments (in 000 units)					
	Jan	Feb	Mar	Apr	May	June
Refrigerators	375	382	514	412	581	562
Washing machines ...	416	397	476	354	455	436
Dryers	306	291	327	233	298	273
Dishwashers	300	260	334	275	308	268
Sums	1,397	1,330	1,651	1,274	1,642	1,539

Source: Adapted from *Standard & Poor's Industry Survey,* January 1980 p. E3.

10.15 A plant manager wishes to evaluate the effects on the hardness of an alloy of adding three different amounts of a second metal and two different methods of tempering the resulting alloy. The following hardness measurements were obtained:

Amounts of a second metal Tempering method	A_1		A_2		A_3	
	B_1	B_2	B_1	B_2	B_1	B_2
	3.46	3.04	3.01	2.44	3.49	2.58
	3.57	2.91	2.76	2.17	3.60	2.85
	3.39	2.90	2.73	2.16	3.74	3.02
	2.87	2.87	2.83	2.26	3.61	2.82
Sums	13.29	11.72	11.33	9.03	14.44	11.27

a. Graph the means and judge whether there might be an interaction, a difference among the effects of the amounts of the second metal, and/or a difference between the effects of the tempering methods.

b. Perform the appropriate analysis of variance and, using $\alpha = 0.05$, summarize and state your findings.

10.16 Keypunch operators at a large computer facility were trained by one of three different methods. Following training, 200 punch cards were randomly selected from each trainee's punch box and were checked for errors. The results are as follows:

	Training method	
X_1	X_2	X_3
4	12	11
8	10	14
9	14	11
8	13	11
7	10	12

a. Set up and test H_0, using $\alpha = 0.01$.

b. If warranted, conduct a multicomparison test.

Chi-Square and Nonparametric Tests
of Significance

11.1 MULTINOMIAL DISTRIBUTION

Priscilla F. was responsible for setting up a field test, collecting and analyzing data, and recommending the type of packaging for a new type of exotic ground coffee. So far marketing had narrowed the selection down to four different types of packaging. It was now Priscilla's task to find the one with the greatest eye appeal. Her plan of attack was simple and straightforward. She would conduct a controlled aroma test. Each subject, seen individually, would be asked to select the brand of coffee that produced the most pleasant aroma. In actuality, all four brands were identical. The only feature that differentiated the brands was the packaging. This difference in packaging was explained to the subjects as merely a means of keeping track of the different brands. Priscilla was really interested in ascertaining whether the packaging influenced the choice of brand.

What test of significance is appropriate in a situation in which there are several mutually exclusive and exhaustive qualitative categories? Let's pursue this question in greater detail. If the probability of selecting each is P_1, P_2, P_3, and P_4 and the trials are independent, the expected number of subjects selecting each brand is nP_1, nP_2, nP_3, and $nP_4 = n(0.25)$. To illustrate, if $n = 12$, the expected frequencies would be: $nP_1 = (12)(0.25) = 3$, $nP_2 = 3$, $nP_3 = 3$, and $nP_4 = 3$. In other words, on the presumption that the aromas are actually the same, we would expect an equal likelihood of selecting each brand. It would be reasonable, then, to use the expected frequencies against which to evaluate the outcome actually obtained in the experiment. Stated another way, we are trying to find out how well the obtained data fit the hypothetical model specified under the null hypothesis. Tests of this sort are known as *goodness of fit*. If the obtained data do not fit or conform to the expectancies based on the null hypothesis (i.e., we reject H_0), we may look to our experimental manipulation as the source of the observed disparity. In other words, we would conclude that the difference in the packaging of the coffee influenced the preference.

One possible test involves the use of the *multinomial distributions*, which are merely extensions of the rationale of the binomial to more than two categories.

Let us assume k mutually exclusive and exhaustive categories with associated probabilities of P_1, P_2, . . . , P_k. If n independent and random observations are made, the probability that exactly n_1 will be of type 1, n_2 of type 2, ... , and n_k of type k is

$$\frac{n!}{n_1!n_2! \ldots \ldots n_k!}(P_1)^{n_1}(P_2)^{n_2} \ldots (P_k)^{n_k}$$

Imagine Priscilla had 12 people select the coffee they felt had the superior aroma and she had obtained 6, 4, 2, and 0 respectively. The probability of obtaining *exactly* this result, when $P_1 = P_2 = P_3 = P_4 = 0.25$, is

$$\frac{12!}{6!4!2!0!}(0.25)^6(0.25)^4(0.25)^2(0.25)^0 = 0.000826$$

goodness of fit

A statistical test that ascertains how well obtained data conform to a theoretical distribution.

multinomial distributions

Distributions that involve more than two mutually exclusive and exhaustive categories.

We could construct a complete sampling distribution of the multinomial for $n = 12$ and the hypothesized values for P_1, P_2, P_3, and P_4. From this we could obtain the probabilities of more unusual outcomes (e.g., 6, 5, 1, 0; 6, 6, 0, 0; 7, 5, 0, 0; etc.). As you may imagine, the computations of these probabilities would represent a *tour de force*. Indeed, with a large n, the computational task would be staggering.

Fortunately, at the turn of the century, Karl Pearson, whom many regard as the father of modern statistical analysis, demonstrated that a family of distributions is eminently suited for approximating multinomial values when n is large. He started out with the recognition that we could describe goodness of fit in terms of the sum of deviations of obtained frequencies from the frequencies expected under the null hypothesis, i.e., $\Sigma(f_o - f_e)$, in which f_o is the obtained frequency and f_e the expected frequency. However, it can be shown that this sum is always equal to zero. Indeed, the fact that $\Sigma(f_o - f_e) = 0$ provides a valuable computational check. Pearson then showed that by squaring each deviation, dividing by its expected frequency, and then summing, he would have a valuable test statistic appropriate for use in the multicategory case:

$$\chi^2 = \sum_{i=1}^{k} \frac{(f_o - f_e)^2}{f_e} \tag{11.1}$$

If the null hypothesis is true and n is sufficiently large, the sampling distribution approximates that of chi-square (χ^2).

11.2 THE CHI-SQUARE DISTRIBUTIONS

Like the *t*-distributions discussed earlier (Chapter 7), χ^2 is a family of distributions with each based on the number of degrees of freedom. To conceptualize a chi-square distribution, imagine that we have a normally distributed variable. Now we select values of the variable, one at a time, express as a z score, and square, i.e., $z^2 = \dfrac{(X - \mu)^2}{\sigma^2}$. The squared standardized score is called $\chi^2_{(df = 1)}$. Thus, $\chi^2_{(df = 1)} = z^2$. The sampling distribution of this variable when $df = 1$ is shown in Figure 11.1.

Recall that the normal distribution of the original scores can range over

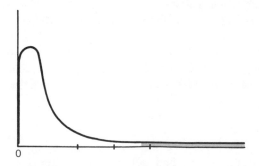

Figure 11.1
The distribution of χ^2
when $df = 1$

Figure 11.2   Sampling distributions of chi-square for various degrees of freedom. In each distribution, we show the critical value of chi-square at the 0.05 level of significance. Note that as df increases, the distributions start to approach the form of the normal distribution.

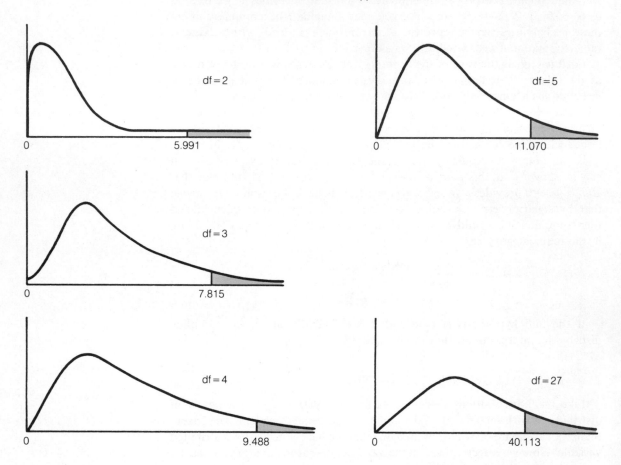

all real numbers. Since $\chi^2_{(df\,=\,1)}$ is a squared quantity, its range is all nonnegative numbers from zero to positive infinity. Recall further that, with normally distributed variables, 68 percent of the area lies between standard scores of -1 and 1. Similarly, when df $= 1$, 68 percent of the χ^2 values are between 0 and 1. Thus, as Figure 11.1 shows, the χ^2 distribution is skewed to the right.

Now let us imagine that two values of the variable are randomly and independently selected from a normally distributed variable. We obtain the squared z scores of each observation and add them together to obtain

$$\chi^2_{(df\,=\,2)} = z_1^2 + z_2^2$$

The sampling distribution of the random variable $\chi^2_{(df\,=\,2)}$ ranges over all nonnegative real numbers and is less skewed than when df $= 1$. Thus, the probability is less that the sum of two squared standard scores will fall between 0 and 1. Similarly, we can imagine extending this reasoning to n independent observations from a normal population. The sum of the squared standard scores will be distributed as χ^2 with n degrees of freedom. Figure 11.2 presents graphs of the sampling distributions of χ^2 for selected degrees of freedom. It should be noted that the expected value of χ^2 for a given number of degrees of freedom is equal to the degrees of freedom, i.e., $E[\chi^2_{df = k} = k]$ and the standard deviation is equal to the square root of twice the number of degrees of freedom (i.e., $\sigma[\chi^2_{df = k}] = \sqrt{2k}$).

> The general rule for assigning the number of degrees of freedom that is it is equal to the number of cells minus one degree of freedom for each independent linear restriction placed on the observed cell counts. If there is one linear restriction, df $= k - 1$, if two, df $= k - 2$, etc. One restriction is always present: $n_1 + n_2 + \cdots \cdots + n_k = n$.

11.3   CHI-SQUARE AND GOODNESS OF FIT

Let us return to Priscilla F.'s packaging problem. To illustrate the multinomial, we used a small sample size. In real life, however, a much larger n would have been used in research of this sort. The χ^2 distribution provides a reasonable approximation to multinomial probabilities only when n is large. Further, Pearson's χ^2 statistic is distributed exactly as chi square only when n is infinitely large. What size n is sufficient? Various rules of thumb abound in the literature and have been the source of much controversy. A widely used rule is that each expected frequency should be 5 or more.

Note that this rule applies *only to expected frequencies* and not to obtained frequencies.

Let us suppose that Priscilla's sample consisted of 360 independent judgments. The null hypothesis is a uniform distribution of probabilities, i.e., $P_1 = P_2 = P_3 = P_4 = 0.25$. The alternative hypothesis is that the population proportions are not equal. Her obtained frequencies were: $f_1 = 115$, $f_2 = 72$, $f_3 = 85$, $f_4 = 88$. The corresponding expected frequencies were: $nP_1 = 90$, $nP_2 = 90$, $nP_3 = 90$, and $nP_4 = 90$. These are summarized in Table 11.1.

	\multicolumn{4}{c}{Type of packaging}
	A_1	A_2	A_3	A_4
Obtained frequency	115	72	85	88
Expected frequency	90	90	90	90
$(f_o - f_e)$	25	−18	−5	−2
$(f_o - f_e)^2$	225	324	25	4

Table 11.1
Summary table of obtained frequencies (f_o) in four different categories, expected frequencies (f_e), $(f_o - f_e)$, and $(f_o - f_e)^2$

334

Figure 11.3
Region of acceptance and
region of rejection in
sampling distribution of χ^2
when df = 3.

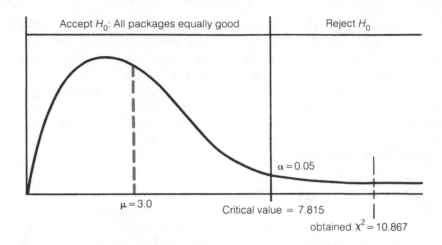

Let us state this problem in formal statistical terms:

1. *Null hypothesis* (H_0):   All packages are equally preferred.
2. *Alternative hypothesis* (H_1):   All packages are not equally preferred.
3. *Statistical test:*   Since we are dealing with multinomial data in which the number of observations is large, we use the χ^2 test.
4. *Significance level:* $\alpha = 0.05$.
5. *Sampling distribution:* χ^2, in which df equals the number of categories minus 1. Thus, df $= k - 1 = 3$.
6. *Critical region:*   Table X, Appendix C, is like the table of t in that it shows the probability that selected values of χ^2 will exceed a specified amount. Hence, for df $= 3$, the probability is 0.05 that χ^2 will exceed 7.815. The obtained value of χ^2 is

$$\chi^2 = \frac{(115 - 90)^2}{90} + \frac{(72 - 90)^2}{90} + \frac{(85 - 90)^2}{90} + \frac{(88 - 90)^2}{90} = 10.867$$

Since the obtained value of χ^2 exceeds the critical value at $\alpha = 0.05$, we reject H_0. Not all packages are equally preferred. Figure 11.3 diagrams the region of rejection for χ^2 when df $= 3$.

Example A: Goodness of fit to the binomial distribution

When exploring for oil in many geologically promising offshore sites, three exploratory wells are drilled at a cost of about $6 million each (1978 dollars). Past experience has shown that the probability is about 0.20 that a given hole will yield sufficient indications of oil to justify further exploration of a potential field. It has been hypothesized that the number of promising wells at a given test site follows a binomial distribution with $P = 0.20$. Let's suppose that, at 640 test sites, 320 found no oil whatsoever, 253 found one promising well, 60 found two, and 7 found three promising wells. Do the observed results fit a binomial model?

Solution

First we refer to Table I (Appendix C) under $P = 0.20$ and $n = 3$. Using the procedures for finding the exact probabilities associated with each value of x, we find: $p_{(x=0)} = 0.512$, $p_{(x=1)} = 0.384$; $p_{(x=2)} = 0.096$; and $p_{(x=3)} = 0.008$. These constitute our expected probabilities under the null hypothesis. Multiplying n by each of these probabilities yields the corresponding expected frequencies. We enter the obtained frequencies, the expected probabilities, and the expected frequencies in Table 11.2 and calculate χ^2. In the present example, this is the only linear restriction. Therefore, df $= 4 - 1 = 3$.

Table 11.2
χ^2 test of goodness of fit of binomial model to oil drilling operations

Number of promising wells	Obtained frequency	Expected probability	Expected frequency	$\frac{(f_o - f_e)^2}{f_e}$
0	320	0.512	327.68	0.1800
1	253	0.384	245.76	0.2133
2	60	0.096	61.44	0.0346
3	7	0.008	5.12	0.6903
	640	1.000	640.00	1.1182

As we can see, the calculated value of χ^2 (1.1182) is quite small. It does not even approach the value required to reject H_0 at $\alpha = 0.05$ (i.e., $\chi^2 = 7.815$). Consequently, we fail to reject H_0. Although, as we have previously noted, we cannot prove H_0 true, we may say that the obtained results are consistent with a binomial model.

Example B: Goodness of fit to the Poisson distribution

In Section 5.4, we saw that the Poisson distribution is ideally suited for estimating probabilities when the variable of interest is discrete but it is distributed over space or an interval of time. In many scheduling problems, we may be interested in learning whether the load or demand follows a Poisson distribution. To illustrate, a telephone order house receives, on the average, 80 orders per eight-hour shift. During 100 randomly selected one-hour periods (Table 11.3), the following frequency distribution of orders was observed.

Using $\alpha = 0.05$, determine whether the obtained distribution is consistent with the null hypothesis that it conforms to the distribution of a Poisson variable.

Solution

First we ascertain the mean number of calls, μ, per one-hour period. We find that $\mu = (80)(\frac{1}{8}) = 10.0$. Referring to Table IV (Appendix C) under $x = 10$, we find the probabilities of various values of x (calls per hour). These are shown in Table 11.4, column C.

The expected frequencies (column D) are obtained by multiplying each

Number of orders	f
0	0
1	0
2	1
3	1
4	1
5	4
6	6
7	12
8	10
9	15
10	16
11	9
12	8
13	4
14	6
15	3
16	2
17	1
18	0
19	0
20	1
21	0
22	0
23	0
24	0
	$n = 100$

probability by n. Recall, however, our rule of thumb that the *expected* frequency in each category should be at least five to achieve a reasonably good approximation to the chi-square distribution. Thus, we must combine categories so that the expected frequency in each one equals five or more. These combined categories are shown in brackets in Table 11.4. Table 11.5 shows the obtained and expected frequency distributions after combining the categories.

Since there are two linear restrictions placed on these data (i.e., $n_1 + n_2 + \cdots + n_k = 100$, and we estimated μ from the sample), df $= k - 1 - 1$. The number of categories after combining is 11. Thus, df $= 9$. Reference to Table X (Appendix C) reveals that the critical value for rejecting H_0 at $\alpha = 0.05$ is 16.919. Since obtained χ^2 is less than this critical value, we fail to reject H_0. Therefore, the obtained distribution is consistent with a Poisson distribution.

Example C: Goodness of fit to a normal distribution

Many of the test procedures discussed in this text make the assumption that the population measurements are normally distributed. However, as we have frequently noted, population distributions are rarely known. We

Obtained and expected frequency distributions of telephone orders during 100 randomly   **Table 11.4**
selected one-hour periods. Expected probabilities and expected frequencies are based
on H_0 that the variable follows a Poisson distribution in which $\mu = 10$.

(A) Number of orders	(B) f	(C) p	(D) $np = f_e$
0	0 ⎫	0.0000	0.00 ⎫
1	0 ⎪	0.0005	0.05 ⎪
2	1 ⎪	0.0023	0.23 ⎪
3	1 ⎬ 7	0.0076	0.76 ⎬ 6.71
4	1 ⎪	0.0189	1.89 ⎪
5	4 ⎭	0.0378	3.78 ⎭
6	6	0.0631	6.31
7	12	0.0901	9.01
8	10	0.1126	11.26
9	15	0.1251	12.51
10	16	0.1251	12.51
11	9	0.1137	11.37
12	8	0.0948	9.48
13	4	0.0729	7.29
14	6	0.0521	5.21
15	3 ⎫	0.0347	3.47 ⎫
16	2 ⎪	0.0217	2.17 ⎪
17	1 ⎪	0.0128	1.28 ⎪
18	0 ⎪	0.0071	0.71 ⎪
19	0 ⎬ 7	0.0037	0.37 ⎬ 8.35
20	1 ⎪	0.0019	0.19 ⎪
21	0 ⎪	0.0009	0.09 ⎪
22	0 ⎪	0.0004	0.04 ⎪
23	0 ⎪	0.0002	0.02 ⎪
24	0 ⎭	0.0001	0.01 ⎭

(A) Number of orders	(B) f	(C) p	(D) $np = f_e$	(E) $\dfrac{(f_o - f_e)^2}{f_e}$
0– 5	7	0.0671	6.71	0.0125
6	6	0.0631	6.31	0.0152
7	12	0.0901	9.01	0.9922
8	10	0.1126	11.26	0.1410
9	15	0.1251	12.51	0.4956
10	16	0.1251	12.51	0.9736
11	9	0.1137	11.37	0.4940
12	8	0.0948	9.48	0.2311
13	4	0.0729	7.29	1.4848
14	6	0.0521	5.21	0.1198
15–24	7	0.0835	8.35	0.2183
	100	1.0001	100.01	$\chi^2 = 5.1781$

Table 11.5
Obtained and expected
frequency distributions of
telephone orders with
categories combined and
the calculation of the test
statistic

have learned to use statistics calculated from samples to estimate parameters. Similarly, we may not know the true *form* of the population distribution. Just as we used statistics to estimate parameters, we may use sample data to ascertain the goodness of fit to the normal curve model. Let us look at an example.

In the three months preceding the 1980 presidential election, over 3 billion shares were traded on the New York Stock Exchange (NYSE). Table 11.6 shows a frequency distribution of the number of shares traded on each of the 65 business days of this period.

Table 11.6
Obtained frequency distribution of number of shares traded over a three-month period, prior to the 1980 presidential election

Number of shares (in millions)	f
30—under 35	3
35—under 40	6
40—under 45	20
45—under 50	16
50—under 55	10
55—under 60	5
60—under 65	4
65—under 70	1
	65

Let us see whether this distribution conforms to a normal distribution.

Solution

First, we must estimate the population mean and standard deviation from the sample data:

$$\overline{X} = \frac{\Sigma fX}{n} = \frac{3062.5}{65} = 47.12$$

$$s = \sqrt{\frac{3815.39}{64}} = 7.72$$

In order to find the expected frequencies for a normal distribution, we must first find the proportion of area associated with each class. We then multiply this proportion by $n = 65$ to determine the frequency expected for each class. Table 11.7 shows these calculations.

Table 11.8 shows the calculations necessary to determine the χ^2 value.

Before we can determine whether our obtained $\chi^2 = 3.95$ is statistically significant, we first must calculate the degrees of freedom. The basic formula

Number of shares traded	z-scores of class limits*	Proportion of area†	f_e
30–under 35	−2.22 to −1.57	0.0582	3.78
35–under 40	−1.57 to −0.92	0.1206	7.84
40–under 45	−0.92 to −0.27	0.2148	13.96
45–under 50	−0.27 to 0.37	0.2507	16.30
50–under 55	0.37 to 1.02	0.2018	13.12
55–under 60	1.02 to 1.67	0.1064	6.92
60–under 65	1.67 to 2.32	0.0373	2.42
65–under 70	2.32 to 2.96	0.0102	0.66
		1.0000	65.00

Table 11.7
Calculation of expected frequencies for volume on the NYSE over a 65-day period

* The z-scores are based on $\bar{X} = 47.12$, $s = 7.72$; for example, for $X = 30$,

$$z = \frac{X - \bar{X}}{s} = \frac{30 - 47.12}{7.72} = -2.22.$$

† The first proportion of area (0.0582) is the proportion of area in the entire left tail beyond a z-score of −1.57. Likewise, the proportion of area corresponding to the last class (0.0102) is the proportion of area in the entire right tail beyond a z-score of 2.32. In order to allocate the entire area under the normal curve, we must follow this procedure for the first and last classes.

Number of shares traded	f_o	f_e	$\dfrac{(f_o - f_e)^2}{f_e}$
30–under 35	3 ⎫ 9	3.78 ⎫ 11.62	0.59
35–under 40	6 ⎭	7.84 ⎭	
40–under 45	20	13.96	2.61
45–under 50	16	16.30	0.01
50–under 55	10	13.12	0.74
55–under 60	5 ⎫	6.92 ⎫	0.00
60–under 65	4 ⎬ 10	2.42 ⎬ 10.00	
65–under 70	1 ⎭	0.66 ⎭	
			$\chi^2 = 3.95$

Table 11.8
Observed and expected frequencies for NYSE volume data and calculation of chi-square

for this purpose is that the appropriate df equals the number of classes minus 1 df for each parameter estimated. In order to follow the rule that the *expected* frequencies equal at least five, we were forced to collapse to five classes. In addition, we used the sample mean and standard deviation to estimate these two population parameters. Therefore, df $= 5 - 1 - 2 = 2$.

Using the 0.05 level of significance, we see that when df $= 2$, the critical value of $\chi^2 = 5.991$. Since our obtained $\chi^2 = 3.95$ is less than the critical value, we fail to reject the null hypothesis that the distribution of shares traded follows the normal distribution.

EXERCISES

11.1 Gail L. has collected much data on the number of female children born to mothers who have had four normal-term children. She hypothesized that the probability is 0.50 that any given child is female. The table below summarizes her findings on 1,200 mothers with four children.

Number of female children	Observed frequency
0	84
1	317
2	449
3	283
4	67
	1,200

Test H_0 that the observed distribution conforms to a binomial distribution in which $P = 0.50$. Use $\alpha = 0.05$.

11.2 During a two-hour period between 7 and 9 A.M., 600 cars pass through a toll booth. During 65 randomly selected one-minute observation periods, R & J research associates obtained the following distribution of cars passing through the toll booth.

Using $\alpha = 0.05$, test H_0 that the distribution of traffic flow through the toll booth follows a Poisson distribution.

x	f
0	4
1	1
2	0
3	9
4	12
5	17
6	8
7	5
8	4
9	3
10	0
11	0
12	2
13	0
14	0
15	0
	65

11.3 One of the most important indicators of New York Stock Exchange (NYSE) activity is the average for the Dow Jones Industrials. For example, on the day after the 1980 presidential election, the Dow Jones Industrials (DJ Ind.) closed at 953.16, up 15.96 points. Changes in the DJ Ind. are reported daily. The following frequency distribution show the changes in the DJ Ind. for the three-month period preceding the election.

Changes in DJ Ind.	f
−20 to under −15	4
−15 to under −10	9
−10 to under − 5	6
− 5 to under 0	12
0 to under 5	15
5 to under 10	11
10 to under 15	6
15 to under 20	2
	65

Using the mean and standard deviation calculated from these sample data ($\overline{X} = -0.42$, $s = 9.10$), test the hypothesis that this distribution conforms to a normal distribution. Use $\alpha = -0.05$ and indicate the df.

11.4 CHI-SQUARE TEST OF INDEPENDENCE

Decision makers are often called upon to evaluate data in which there are two or more categories (or continuous variables arranged into mutually exclusive classes). The question raised is whether or not the two variables are independent of one another. Consider the following:

> Does the support of a piece of legislation depend on political affiliation or are the two variables (attitude toward legislation and political party) independent of one another? In other words, are Democrats more or less likely than Republicans to favor a piece of legislation?

> Does the packaging of a product depend on the demographics of the intended market? For example, does our preference for the packaging of a deodorant depend on gender, age, income status, region of the country, etc?

> Does the tendency for equipment to break down depend on the age of the equipment or are the two variables (tendency to break down and age) independent of one another?

> Do the qualities cited by chief executives as important for advancement depend on the size of the firm?

> Does the tendency to favor a strike call depend on the number of years of membership in a union?

Each of these examples involve the chi-square two-variable case. The frequency data on each variable are presented in a tabular form known as a *contingency table*. If there are two classes of each variable, we refer to the tabular presentation as a 2 × 2 contingency table. If there are two classes of one variable and three of another, the table is known as a 2 × 3 contingency table; if three classes of each, we have a 3 × 3 contingency table, etc.

To illustrate, in The Wall Street Journal–Gallup Poll reported November 14, 1980, chief executive officers at large and small firms were asked to cite traits they considered failings in subordinates. A total of 59 out of 282 at large firms cited "inability to understand others." At small firms, 22 out of 200 cited this characteristic. These findings are summarized in the 2 × 2 contingency table shown as Table 11.9.

We see that the cells show the numbers of executives at large and small firms who cited or failed to cite "inability to understand others" as an important trait. Thus, at large firms, 223 did not cite this characteristic. At small firms, 178 did not cite this trait. The sum of the columns yields the total

contingency table

A two-way table for classifying members of a group according to two or more classes of each variable.

	Size of firm		Totals
	Large	Small	
Not cited	223	178	401
Cited	59	22	81
Totals	282	200	482

Table 11.9
Contingency table: Frequency with which executives cited "inability to understand others" as a failing in subordinates

number of respondents at each type of firm. These are shown on the bottom marginal total (282 and 200 respectively). The marginal totals at the right show the total number of respondents who failed to cite or cited the trait of interest (401 and 81 respectively).

The number of degrees of freedom in a contingency table is equal to the number of row classes minus 1 ($r - 1$) times the number of column classes minus 1 ($c - 1$). Thus, df = ($r - 1$) ($c - 1$). In the present example, df = $(2 - 1) (2 - 1) = 1$.

Since we have no prior or hypothetical basis for obtaining expected frequencies, they must be estimated from the obtained data. The null hypothesis is that the citing of inability to understand others is independent of the size of the firm. Stated another way, there is no difference in the population proportion of executive officers in large and small firms who cited "inability to understand others." On the basis of H_0, we would expect (401/482) × 282 = 234.61 executives at large firms not to cite this trait; the corresponding value for small firms would be (401/482) × 200 = 166.39. A more direct way to obtain any expected cell frequency involves multiplying together the marginal frequencies common to a cell and dividing by n. Thus, the expected frequency for the cell large firm–citing is (81 × 282)/482 = 47.39. Box 11.1 ties the calculation of expected cell frequencies to our prior discussion of joint probabilities (Section 4.6).

You should note that, once one expected cell frequency is obtained, the remainder may be found by subtraction since there is only one degree of freedom. Thus, once we know that the expected frequency for large firm–citing is 47.39,

f_e for large firm–not citing is 282 − 47.39 = 234.61
f_e for small firm–not citing is 401 − 234.61 = 166.39
f_e for small firm–citing is 200 − 166.39 = 33.61

Note that the sum of the expected frequencies must equal n. Thus, 47.39 + 234.61 + 166.39 + 33.61 = 482.

Table 11.10 summarizes the obtained and expected cell frequencies[1] and the calculation of χ^2. The formula for calculating χ^2 in the two-variable case is

$$\sum_{r=1}^{k} \sum_{c=1}^{k} \frac{(f_o - f_e)^2}{f_e} \tag{11.2}$$

in which the double summation sign indicates that we must sum over both rows and columns.

---

[1] Many texts recommend Yates correction for continuity when df = 1. This consists of subtracting 0.5 from the absolute value of $f_o - f_e$ prior to squaring. We have dropped this recommendation based on convincing evidence (Dunlap, 1974; Camilli and Hopkins, 1978).

Box 11.1

CONTINGENCY TABLES
AND JOINT PROBABILITY

You may have recognized the similarity between a contingency table and the joint probability table encountered earlier (see Section 4.6). Recall that the marginal totals may be converted into probabilities by dividing each total by n. The cell probabilities are the product of their marginal probabilities. We previously referred to these cell probabilities as joint probabilities. These marginal and cell probabilities are shown in the table.

| | Size of firm | | | | Marginal | |
| | Large | | Small | | | |
	f_o	p	f_o	p	Totals	Probabilities
Not cited	223	0.48674	178	0.34521	401	0.83195
Cited	59	0.09832	22	0.06973	81	0.16805
Marginal totals	282	0.58506	200	0.41494	482	1.00000

Under the null hypothesis that the two variables are independent, the expected frequencies may be obtained by multiplying each cell probability by n (see table below). If the expected and obtained cell frequencies are not very different from one another, the value of χ^2 is not large and we are not likely to reject the null hypothesis of independence. The greater the disparity between the obtained and the expected cell frequencies, the larger the value of χ^2 and the more likely are we to conclude that the variables of interest are related.

| | Expected frequencies, f_e | | Totals |
| | Size of firm | | |
	Large	Small	
Not cited	234.61	166.39	401
Cited	47.39	33.61	81
Totals	282	200	482

The critical value of χ^2 at $\alpha = 0.01$ and df = 1 is 6.635. Since obtained χ^2 exceeds this value, we may reject H_0. The two variables do not appear to be independent. More precisely, executives of small firms do not cite inability to understand others as a failing of subordinates as often as executives at large firms.

Table 11.10
Observed and expected frequencies and calculation of χ^2

	Size of firm	
	Large	Small
Not cited	$f_o = 223$ $f_e = 234.61$	$f_o = 178$ $f_e = 166.39$
Cited	$f_o = 59$ $f_e = 47.39$	$f_o = 22$ $f_e = 33.61$

$$\chi^2 = \frac{(223 - 234.61)^2}{234.61} + \frac{(178 - 166.39)^2}{166.39} + \frac{(59 - 47.39)^2}{47.39} + \frac{(22 - 33.61)^2}{33.61}$$
$$= 0.5745 + 0.8101 + 2.8443 + 4.0105$$
$$= 8.2394$$

Example D: Tests of differences among proportions

There are numerous occasions when we are interested in ascertaining whether or not a number of sample proportions may reasonably have been drawn from a common population. For example, the χ^2 test finds wide application in monitoring various industrial processes to determine when they are in or out of control. Let's look at an example. Extrudalum Corporation produces aluminum extrusions for the construction industry. Each day a sample of 40 extrusions is selected from the day's production and is tested for defects. Table 11.11 shows the record compiled over a 22-day period.

Table 11.11
Numbers of failures and successes in daily samples of 40 extrusions in an industrial process

Day	Number passed	Number failed	Day	Number passed	Number failed
1	36	4	12	33	7
2	38	2	13	36	4
3	35	5	14	37	3
4	31	9	15	28	12
5	37	3	16	38	2
6	29	11	17	32	8
7	39	1	18	30	10
8	32	8	19	35	5
9	34	6	20	36	4
10	37	3	21	31	9
11	30	10	22	34	6

In this example, we have neither a prior nor a hypothetical basis for estimating the probability of pass and fail. We must use the obtained results to estimate the population proportion of passes and fails. Since there were 132 failures out of 880 items tested, the probability of a failure is $132/880 = 0.15$. The corresponding probability of a success is $1.00 - 0.15 = 0.85$. The null hypothesis is that the proportion of failures in each day's output is the same. Thus, $P_1 = P_2 = P_3. \ldots = P_{22}$. If H_0 is rejected, it means that the proportion of failures from day to day is variable. Thus, the production process is out of control and warrants a careful look by management.

1. *Null hypothesis* (H_0): $P_1 = P_2 = \ldots\ldots = P_{22}$. The proportion of failure in each day's output is the same.

2. *Alternative hypothesis* (H_1): $P_1 \neq P_2 \ldots\ldots \neq P_{22}$. The proportion of failure in each day's output is not the same.

3. *Statistical test:* Since we are dealing with multinomial data in which the number of observations is large, we use the χ^2 test.

4. *Significance level:* $\alpha = 0.01$.

5. *Sampling distribution:* χ^2. Since there are 22 categories in which there is one linear restriction $(n_1 + n_2 + \ldots\ldots n_k = n)$, the number of degrees of freedom is $k - 1 = 21$.

6. *Critical region:* Reference to Table X under 21 df and $\alpha = 0.01$ shows the critical value of χ^2 to be 38.932.

Since the probability of failure is 0.15, the expected frequency of failures

Table 11.12
Summary of calculation of χ^2 involving defective aluminum extrusions

	Success			Failure		
Day	Obtained	Expected	$\frac{(f_o - f_e)^2}{f_e}$	Obtained	Expected	$\frac{(f_o - f_e)^2}{f_e}$
1	36	34	0.1176	4	6	0.6667
2	38	34	0.4706	2	6	2.6667
3	35	34	0.0294	5	6	0.1667
4	31	34	0.2647	9	6	1.5000
5	37	34	0.2647	3	6	1.5000
6	29	34	0.7353	11	6	4.1667
7	39	34	0.7353	1	6	4.1667
8	32	34	0.1176	8	6	0.6667
9	34	34	0.0000	6	6	0.0000
10	37	34	0.2647	3	6	1.5000
11	30	34	0.4706	10	6	2.6667
12	33	34	0.0294	7	6	0.1667
13	36	34	0.1176	4	6	0.6667
14	37	34	0.2647	3	6	1.5000
15	28	34	1.0588	12	6	6.0000
16	38	34	0.4706	2	6	2.6667
17	32	34	0.1176	8	6	0.6667
18	30	34	0.4706	10	6	2.6667
19	35	34	0.0294	5	6	0.1667
20	36	34	0.1176	4	6	0.6667
21	31	34	0.2647	9	6	1.5000
22	34	34	0.0000	6	6	0.0000
			6.4115			36.3338

$$\chi^2 = \Sigma \frac{(f_o - f_e)^2}{f_e} = 42.745$$

for each day is obtained by multiplying the sample size, 40, by 0.15. Multiplying 40 by 0.85 yields the expected number of successes. Table 11.12 summarizes the calculation of χ^2 for this example. Since obtained $\chi^2 = 42.745$ exceeds the critical value at $\alpha = 0.01$, we reject H_0 and assert that the process appears to be out of control. Management intervention is warranted in making an effort to discover and correct the cause or causes of the variable quality of extrusions produced at the plant.

EXERCISES

11.4 In The Wall Street Journal–Gallup Poll study cited above, the following results were obtained. The 2×3 contingency table shows integrity as a quality cited by chief executive officers as important for advancement.

    a. What is the number of degrees of freedom?

    b. Are the two categories independent? Use $\alpha = 0.05$.

	Size of firm		
Integrity	Large	Medium	Small
Not cited	180	219	152
Cited	102	81	48

11.5 Management wished to learn if the tendency to favor a strike call depends on the number of years of membership in a union. An independent research firm divided the employees into four categories with respect to years of membership. A random selection of employees in each group was selected in the survey. The number in each group was selected to be proportional to their representation in the entire employee pool. The results are shown below. Using $\alpha = 0.01$,

    a. State the null hypothesis.

    b. Determine df.

    c. Test the null hypothesis.

	Strike vote	
Years of membership	Favor	Oppose
0–under 5	25	10
5–under 10	40	55
10–under 15	30	28
15 and over	15	25

11.5 OTHER NONPARAMETRIC TECHNIQUES

Although we have not labeled them as such, the various statistical tests that make assumptions of normality about the population distribution are referred to as *parametric tests* of significance. However, many types of data do not meet these assumptions. For example, frequency data such as we encountered in our chi-square examples, or data that consist of ranks do not lend themselves to analysis through the use of parametric tests. However, there are a wide variety of *nonparametric* techniques that have been developed for dealing with such data.

Nonparametric techniques are sometimes called distribution-free statistical procedures. We prefer not to use the term *distribution-free* as it tends to be misleading. True, no assumption regarding the underlying population distribution restricts the use of nonparametric techniques. Be aware, however, that this freedom of distribution refers only to the *underlying population distribution(s)* that the sample(s) is drawn from. We are not given such freedom with the *test statistic. Each and every one of the nonparametric techniques requires that we make certain assumptions about the sampling distribution of the test statistic.*

For example, when we are dealing with a chi-square test, we make very specific assumptions about the form of the sampling distribution of the test statistic. If our outcomes are independent, and the null hypothesis is true, we assume that the test statistic comes from a particular chi-square distribution distinguished by the degrees of freedom. Our conclusions are based on whether or not this test statistic could reasonably have occurred by chance under these circumstances (i.e., H_0 true).

In the following sections, we shall describe some of the more important nonparametric tests.[2]

parametric tests

Statistical tests that make the assumption that the parent or population distribution is normal.

nonparametric tests

Statistical tests that do not make assumptions about the shape of the parent or population distribution.

11.6 THE SIGN TEST

The sign test is one of the simplest of the nonparametric techniques. It is most commonly used to compare two samples that are matched or related in some way. We are usually interested in testing the null hypothesis that the two samples come from the same population. Thus, if H_0 is true, and we can designate the difference between each measure by either a plus or a minus, there should be approximately the same number of pluses as minuses. You may recognize that we are dealing with two mutually exclusive and exhaustive categories, which is the characteristic of Bernoulli trials. A difference is either a plus *or* minus. If it is zero, we eliminate it from our calculations. Since we are interested in finding the probability of x equal to or greater than obtained x, the appropriate sampling distribution of the test statistic is the binomial.

We let

---

[2] See Workbook for a test involving the analysis of variance for ranked data.

348

P = Proportion of pluses
Q = Proportion of minuses
x = Number of pluses.

If H_0 is true, $P = Q = 0.50$. In other words, if the two groups were drawn from the same population, the probability of obtaining a plus or a minus is 0.50. Thus, we would expect the proportion of pluses and minuses obtained from the sample to divide themselves approximately equally.

If n is small ($n \leq 30$), we may use Table I to determine the probability of obtaining a value of x or greater by chance in the distribution under H_0. If $n > 30$, we use the normal approximation to the binomial and z as our test statistic. Let's look at a few examples.

Example E: $n \leq 30$

The National Research Council conducts an annual survey of doctorate recipients. The analyses pertain to scientists, engineers, and humanists who earned their doctorate between 1936 and 1978 and who were residing in the United States in 1979.

Table 11.13 lists the median annual salaries for men and women with varying numbers of years of professional experience. We shall use the sign test to determine whether there is a statistically significant difference in median income between men and women.

Table 11.13
Median annual salaries of full-time employed male and female Ph.D.s by years of professional experience

Years of professional experience	Median annual salary ($000)		Sign of the difference, (M − F)
	Male	Female	
0– 1	19.7	17.4	+
2– 5	22.3	19.7	+
6–10	26.0	22.0	+
11–15	30.1	25.1	+
16–20	32.2	27.2	+
21–25	35.0	28.9	+
26–30	37.4	29.3	+
Over 30	39.1	30.5	+

Source: Adapted from *Science, Engineering, and Humanities Doctorates in the United States, 1979 Profile.* p. 21, Table 1.7.

1. *Null hypothesis (H_0):* There is no difference in the median salary of men and women; therefore, the proportion of pluses *(P)* equals the proportion of minuses *(Q)*, or $P = Q = 0.50$.

2. *Alternative hypothesis (H_1):* There is a difference in the median salary between men and women, or $P \neq Q \neq 0.50$. Note that since we have no

prior reason to specify the *direction* of the difference, we use a nondirectional hypothesis and a two-tailed test.

3. *Statistical test:* Since we are dealing with a Bernoulli variable and a small n ($n = 8$), we use x (the number of pluses) as our test statistic.

4. *Significance level:* $\alpha = 0.05$.

5. *Sampling distribution:* The binomial distribution with $P = 0.50$ and $n = 8$.

6. *Critical region:* Since we are using a two-tailed test, we shall reject H_0 if $p(x \geq 8)$ plus $p[(n - x) \leq 0]$ is less than or equal to 0.05.

Looking in Table I for $n = 8$, $P = 0.50$, we find that $p(x \geq 8) = 0.0039$. Therefore, the two-tailed probability is 0.0078. Thus, we reject H_0. There is a statically significant difference between men and women in terms of median salaries.

Example F: $n > 30$

As a service to the accounting and finance professions, a survey is conducted each year to determine the median salary for five position categories within both public and industrial accounting. The results are further subdivided according to the size of the organization. Table 11.14 presents the results of the 1980 survey for two geographical areas—Minneapolis/ St. Paul and the San Francisco Bay Area. Let us use the sign test to determine whether these two areas differ with respect to salary.

In this example, we are dealing with $n > 30$ ($n = 40$); therefore, we may use the normal approximation to the binomial and z as our test statistic (see Section 5.6).

1. *Null hypothesis (H_0):* There is no difference between the two locations; therefore, the proportion of pluses (P) equals the proportion of minuses (Q), or $P = Q = 0.50$.

2. *Alternative hypothesis (H_1):* There is a difference between the two locations, or $P \neq Q \neq 0.50$. We have no basis to use a directional hypothesis, therefore, we employ a two-tailed test.

3. *Statistical test:* Since we are dealing with a Bernoulli variable and $n > 30$ ($n = 40$), we may use z as our test statistic.

4. *Significance level:* $\alpha = 0.01$.

5. *Sampling distribution:* The normal probability distribution.

6. *Critical region:* Since H_1 is nondirectional, any value of $z \geq 2.58$ or $z \leq -2.58$ will lead to rejection of H_0.

In order to calculate the z-statistic, we must first find the mean and standard deviation of the distribution under the null hypothesis.

Recall that the mean (μ) of a binomial distribution equals nP. In this case, $\mu = 40(0.50) = 20$.

The standard deviation (σ) of a binomial distribution is $\sqrt{nPQ}$. Thus, $\sigma = \sqrt{(40)(0.50)(0.50)} = 3.16$.

Table 11.14
Median annual salaries for
various position categories
in different size
organizations within public
and industrial accounting
for two geographical areas

	Firm size*	Median salary ($000)		Sign of the difference (M/Stp — SF)
		Minn/St. Paul	San Francisco	
Public accounting				
Staff	S	$ 15.7	$ 15.0	+
	M	17.2	16.3	+
	L	18.0	18.1	−
Senior	S	19.8	20.3	−
	M	21.7	21.9	−
	L	25.5	23.5	+
Manager	S	28.5	26.1	+
	M	34.0	31.5	+
	L	40.2	36.1	+
Partner	S	38.1	43.4	−
	M	56.6	63.3	−
	L	89.8	73.1	+
Senior partner	S	65.7	75.5	−
	M	95.3	86.1	+
	L	196.6	191.5	+
Industrial accounting				
Junior	S	14.2	12.5	+
	M	15.6	14.2	+
	ML	16.4	15.9	+
	L	17.8	16.8	+
	VL	18.7	18.0	+
Senior	S	19.2	19.3	−
	M	20.1	21.7	−
	ML	21.8	23.0	−
	L	23.1	25.5	−
	VL	24.8	28.3	−
Manager	S	23.4	23.2	+
	M	25.2	25.4	−
	ML	30.5	28.5	+
	L	36.9	36.5	+
	VL	48.1	43.5	+
Controller, Treasurer	S	28.8	28.7	+
	M	34.1	36.0	−
	ML	41.1	39.3	+
	L	57.8	55.7	+
	VL	98.7	80.6	+
Chief financial officer	S	40.7	46.3	−
	M	51.1	57.9	−
	ML	71.6	82.8	−
	L	97.1	88.2	+
	VL	179.2	173.2	+

* L = large; M = medium; S = small; V = very large.
Source: Reprinted with permission from the *1980 Local Metropolitan Accounting and Finance Salary Survey*, published by Source Finance.

Since $x = 24$ (number of pluses),

$$z = \frac{24 - 20}{3.16} = 1.27$$

Since our obtained z does not fall in the critical region, we may not reject H_0. There is no statistically significant basis for asserting a difference between the two geographical areas.

Example G: The median

An extremely useful application of the sign test is to determine whether a particular sample comes from a population in which the median is equal to some specified value. For example, suppose you know that the median price of a single-family home nationwide is $65,000. You might wish to know whether a sample of single-family homes from a particular locale are typical of the rest of the country. If they are, they should have the same median. If they have the same median, then the number above $(+)$ and below $(-)$ the median should, by definition, be the same. Again, we are dealing with two mutually exclusive and exhaustive categories (a value is either above the median *or* below—we don't count those that fall at the median).

Thus, this is merely a variation of the sign test. Instead of comparing two samples, we are comparing one sample against a hypothesized population median. If the null hypothesis is true, $P = Q = 0.50$.

Median annual salary for staff members of medium-sized public accounting firms for various geographical areas. The national median salary for this position is $16,900.   Table 11.15

Geographical area	Median ($000)	Above (+) or below (−) national median
Boston	$16.1	−
Chicago	17.1	+
Dallas/Ft. Worth	17.6	+
Denver	16.9	0
Detroit	17.3	+
Houston	18.5	+
Los Angeles	17.2	+
Milwaukee	17.1	+
Minneapolis/St. Paul	17.2	+
New Jersey....................	19.3	+
New York	19.8	+
San Francisco	16.3	−

Source: Reprinted with permission from the *1980 Local Metropolitan Accounting and Finance Salary Survey*, published by Source Finance.

Let us return to the accounting and finance salary survey. The national median salary for a member of the staff of a medium sized public accounting firm is $16,900. Table 11.15 lists the median salary for this position at 12 different geographical locations. We shall test to see whether this sample comes from a population in which the median equals $16,900 ($\alpha = 0.05$).

In this case, one of the values (Denver) falls at the median; therefore, we exclude Denver from our calculations and $n = 11$.

Since we are testing the null hypothesis that this sample comes from a population in which the median $= 16.9$, P(the proportion of values *above* the median, or pluses) should equal Q (the proportion of values *below* the median, or minuses).

$$H_0: P = Q = 0.50$$
$$H_1: P \neq Q \neq 0.50$$

Looking up $n = 11$ for $P = 0.50$ in Table I, we find the $p(x \geq 9) = 0.0328$. However, our alternative hypothesis is nondirectional, therefore, we double the probability value to obtain the two-tailed probability of 0.0656. Since $\alpha = 0.05$, we cannot reject H_0.

EXERCISES

11.6 Based on the survey cited in Example E, the following data were obtained on the percentage of science/engineering Ph. D.s who were unemployed and seeking employment according to the field of their doctorate. Group 1 consists of those individuals who received their Ph.D.s between 1936 and 1978. Group 2 consists of those who received their Ph.D.s between 1973 and 1978. Use the sign test, $\alpha = 0.05$, to determine whether there is a statistically significant difference in the percentage of "older" versus "younger" Ph.D.s unemployed and seeking employment.

Field	Group 1 1936–78 graduates	Group 2 1973–78 graduates
Mathematics	0.4%	0.5%
Computer science	0.0	0.0
Physics/Astronomy	0.9	1.5
Chemistry	1.0	1.4
Environmental science	0.5	0.7
Engineering	0.5	1.3
Agriculture	0.8	1.0
Medicine	1.0	1.4
Biology	1.1	1.3
Psychology	1.1	1.8
Social sciences	1.1	1.4

11.7 Based on the survey cited in Example E, the median annual salary of all full-time employed doctoral scientists and engineers was $29,200 in 1979.

The following data show the median annual salary by field of doctorates working in business or industry and of those working in educational institutions.

Field	Median annual salary ($000)	
	Business/industry	Educational institutions
Mathematics	31.3	25.3
Computer science	32.1	24.0
Physics/Astronomy	33.1	27.7
Chemistry	33.4	26.1
Environmental science	33.6	26.0
Engineering	35.0	29.7
Agriculture	29.6	26.9
Medicine	35.7	29.1
Biology	33.0	25.7
Psychology	36.7	25.4
Social sciences	33.6	25.6

Use the sign test and $\alpha = 0.05$ to answer the following questions:
a. Is there a difference between the two groups in median salary?
b. Are those working in business and industry typical of the total population of Ph.D.s with respect to median salary?
c. Are those working in educational institutions typical of the total population of Ph.D.s with respect to median salary?

11.8 It has often been said that changes in the prime rate (the charge by large U.S. money-center banks to their best business borrowers) leads to changes in other money rates. The following data show changes in the prime rate and changes in the typical rates paid by major banks on new issues of six-month certificates of deposit randomly selected over a four-month period. Test the null hypothesis that changes in the prime rate are no different from changes in the six-month certificate of deposit rate. Use the sign test with $\alpha = 0.05$.

Exercise 11.8

Prime rate	6-mo. C.D.
+0.25%	+0.10%
0	+0.10
+0.25	+0.15
+0.25	+0.50
0	−0.50
+0.25	0
+0.25	+0.50
0	−0.10
+0.25	+0.50
0	+0.12
0	−0.37
+0.25	+0.37
0	+0.05
+0.50	+0.37
0	+0.75
0	−0.25
+0.25	−0.05
+0.50	+0.05
0	−0.45
0	+0.15
+0.25	+0.30
0	−0.25
0	+0.12
0	+0.37
+0.50	+0.50
0	−0.25
0	+0.40
+0.25	+1.10
+0.75	+0.12
0	−0.57
0	−0.10
+0.50	+0.25
0	−0.10
0	−0.25

11.7 THE MANN-WHITNEY *U*-TEST

The Mann-Whitney *U*-test is a nonparametric technique that enables us to compare two independent samples by utilizing ranking information. Unlike the sign test, which employs only pluses or minuses, the Mann-Whitney *U*-test uses more of the information inherent in the data. We use this procedure to test whether two independent samples come from a population with the same mean. If this H_0 is true, then any particular value is as likely to occur in one sample as the other. Therefore, if we were to merge all values of the variable from both samples together, and rank them from low to high, the ranks should be randomly distributed between the two groups. Hence, the obtained value of the test statistic, *U*, should be one that would occur relatively frequently in the *U*-distribution. Suppose, however, the null hypothesis is false. We would then expect a preponderance of high ranks in one group and low ranks in the other. If this is the case, then the obtained *U*-statistic would have a low probability of occurring by chance in the distribution specified under H_0 and we would reject H_0.

11.7.1 The *U*-test with small samples

Suppose you are a personnel director and training supervisor for a vacuum cleaner manufacturer that uses a direct sales approach. You have just hired 22 new sales personnel and you would like to evaluate the effectiveness of a new training program. You randomly assign 11 to the new training program and 11 to the traditional program. Neither group is informed that it is participating in an experiment. Following the training, each trainee is evaluated in a variety of simulated sales situations. A group of trained observers, unaware of the type of training each subject had received, rate each trainee on overall effectiveness of presentation. The results of their combined ratings are presented in Table 11.16. The higher the rating, the better the performance.

Table 11.16
Ratings of effectiveness
of two groups of sales
trainees

(A) New training	(B) Traditional training
86	80
84	74
80	72
78	69
73	64
71	62
68	59
67	57
65	54
63	53
60	50

in either ascending or descending order. You should maintain the identity of the experimental condition associated with each score. Then rank each

You are reluctant to assume that these ratings come from a normally distributed population. Therefore, you apply the Mann-Whitney *U*-test since it makes no assumptions concerning the form of the population distribution.

The first step is to combine the two groups into one, arrange the scores rating in either ascending or descending order. Table 11.17 summarizes these procedures.

The personnel director is interested only in ascertaining whether or not the new training procedures lead to improved ratings of sales performance. Thus, a one-tailed test of significance is used. Since a shift to new procedures would entail considerable expense, a conservative α-level is chosen, i.e., $\alpha = 0.01$.

The test statistic, U, is

$$U_1 = n_1 n_2 + \frac{n_1(n_1 + 1)}{2} - R_1 \qquad (11.3)$$

or

$$U_2 = n_1 n_2 + \frac{n_2(n_1 + 1)}{2} - R_2 \qquad (11.4)$$

in which
$R_1 = $ The sum of ranks assigned to the group with sample size n_1
$R_2 = $ The sum of ranks assigned to the group with sample size n_2

In this example, we'll call R_1 the sum of the ranks of Group A and R_2 the sum of the ranks of Group B. Thus, we find $R_1 = 95.5$ and $R_2 = 157.5$. As a check for computational error, $R_1 + R_2 = n/2$ (lowest rank + highest rank) $= n/2 (1 + 22) = 253$.

We now find

$$U_1 = (11)(11) + \frac{(11)(12)}{2} - 95.5$$

$$= 91.5;$$
$$U_2 = (11)(11) + \frac{(11)(12)}{2} - 157.5$$

$$= 29.5$$

Rating	Condition	Rank	Rating	Condition	Rank
86	A	1	67	A	12
84	A	2	65	A	13
80	A	3.5*	64	B	14
80	B	3.5*	63	A	15
78	A	5	62	B	16
74	B	6	60	A	17
73	A	7	59	B	18
72	B	8	57	B	19
71	A	9	54	B	20
69	B	10	53	B	21
68	A	11	50	B	22

Table 11.17
Trainees' ratings and their associated ranks

* Note that when there are tied ranks, each rating receives the mean of the shared rank. Thus, ratings of 80 shared 3d and 4th ranks. Therefore, each is assigned $(3 + 4)/2 = 3.5$. The next rating (i.e., 78) is assigned the rank it would have received had there been no ties.

As an additional check, $U_1 + U_2 = (n_1)(n_2) = 91.5 + 29.5 = 121$.

Referring to Table XI(d) (Appendix C) under $n_1 = 11$ and $n_2 = 11$, we find that a U equal to or greater than 87 or equal to or less than 34 leads to a rejection of the null hypothesis. Since $U_1 = 91.5$ exceeds the critical value of 87, we reject H_0. The new training program appears to produce improved performance in the simulated sales situation.

11.7.2 The U-test with large samples

As with many of the test statistics we have examined so far, as the sample size increases, the sampling distribution of U begins to approximate the normal distribution. The larger the sample size, the better the approximation. Since we may use Tables XI(a)–XI(d) when both n_1 and n_2 are 20 or less, we shall reserve the use of the normal approximation only for cases in which either n_1 or n_2 exceeds 20.

The mean of the U-distribution is

$$\mu_U = \frac{n_1 n_2}{2} \tag{11.5}$$

The standard deviation of the U-distribution is

$$\sigma_U = \sqrt{\frac{n_1 n_2 (n_1 + n_2 + 1)}{12}} \tag{11.6}$$

Therefore, with a large n, the test statistic is

$$z = \frac{U - \mu_U}{\sigma_U}$$

and the sampling distribution is the normal probability distribution.

Let's look at an example using this procedure.

Example H

L & J, an independent research company, has developed a new small business index that measures changes in total sales, industrial sales, consumer sales, and accounts receivable turnover for small businesses. The higher the index, the more successful the company. Table 11.18 presents the L & J index for 32 randomly selected small firms (those with monthly sales less than $100,000) and for 32 randomly selected large firms (those with sales greater than $500,000 monthly). Use the Mann-Whitney U-test to determine whether there is a difference between large and small firms in terms of the L & J index (use $\alpha = 0.05$).

Small firms	Rank	Large firms	Rank
3.2	10.5	3.6	15.5
5.4	48.5	5.3	46
4.7	36	5.1	41.5
2.5	6	4.0	23.5
2.4	4	3.2	10.5
2.5	6	2.9	8
4.9	40	5.5	51.5
3.8	19.5	4.4	30.5
3.4	13	4.2	28
3.6	15.5	4.1	25.5
6.2	59	6.1	58
4.7	36	5.2	44
4.7	36	5.1	41.5
5.7	54	6.3	60
4.6	34	5.4	48.5
4.5	32.5	5.4	48.5
3.9	22	5.2	44
3.2	10.5	4.5	32.5
2.5	6	3.8	19.5
2.3	3	3.7	17
2.2	2	4.8	38.5
2.1	1	5.7	54
3.5	14	4.2	28
4.4	30.5	3.8	19.5
4.0	23.5	5.4	48.5
3.2	10.5	4.8	38.5
3.8	19.5	5.5	51.5
6.7	61	8.0	64
5.9	56	7.4	63
5.2	44	7.0	62
4.2	28	6.0	57
4.1	25.5	5.7	54
	807.5		1272.5

Table 11.18
L & J business index for small and large firms and their associated ranks

In this example,

$$\mu = \frac{n_1 n_2}{2} = \frac{(32)(32)}{2}$$

$$= 512;$$

$$\sigma = \sqrt{\frac{(32)(32)(32 + 32 + 1)}{12}}$$

$$= 74.48;$$

$$U = (32)(932) + \frac{(32)(33)}{2} - 807.5$$

$$= 744.5$$

Therefore,

$$z = \frac{744.5 - 512}{74.48} = 3.12.$$

Thus, since $3.12 > 1.96$, we may reject H_0. Small firms do not fare as well according to the L & J index as large firms.

EXERCISES

11.9 Many financial analysts have speculated that the price of gold is a barometer of investor anxiety. Thinking that the behavior of gold on one day might serve as a harbinger of common stock performance on the following day, J, R and S Associates conducted the following analysis. They recorded the direction of the change in gold prices on 26 randomly selected trading days. Then they recorded the change in the closing price of the Dow Jones Industrials on the following day. The results are shown below.

Change in DJI on the day after the gold price	
Rose	Fell
−8.28	+10.44
+9.22	−15.61
+2.30	+0.85
−13.74	−11.43
−2.56	+15.96
+4.70	−17.55
−6.40	+10.24
+0.68	+17.49
+4.09	+3.93
−11.86	−5.03
−3.41	
+6.74	
+12.71	
−2.99	
+1.37	
+20.90	

Test for the significance of the difference in the change in the closing prices following the price of gold going up versus the price of gold going down. Calculate Mann-Whitney U, using $\alpha = 0.05$, two-tailed test.

11.10 J, R and S Associates speculate that the typical rates paid by major banks on six-month certificates of deposit are related to changes in the Dow Jones Industrials. They randomly select 24 days on which the DJI went up and 24 days on which it went down. They then recorded the rate on six-month C.D.s for those days. The results are shown below:

Rates on 6-month CDs on days when the DJI was:	
Up	Down
9.30%	9.20%
9.40	9.50
9.50	9.50
9.95	9.25
10.00	9.75
11.50	9.80
11.37	10.25
10.75	10.50
11.15	10.75
11.75	11.00
11.87	11.12
11.70	11.25
12.75	11.50
12.70	12.25
12.12	13.00
12.30	12.45
12.50	12.37
12.75	12.45
13.12	13.00
14.00	13.50
14.50	13.75
14.80	14.15
14.50	15.25
14.40	15.37

Use the Mann-Whitney U-test to determine if there is any difference in 6-month C.D. rates when the DJI goes up versus when it goes down. Use $\alpha = 0.05$, two-tailed test.

11.8 ONE-SAMPLE RUNS TEST

As we have seen throughout the text, the assumption of randomness is basic to our efforts to estimate parameters from sample statistics and for making statistical decisions concerning the rejection of the null hypothesis. We assume that observed data have been selected randomly from a population or, in experimental settings, that the subjects are randomly assigned to the different conditions. We have even used tables of random numbers to accomplish this assignment. But how do we know that the numbers are random? The runs test provides a procedure for judging the randomness of a sequence of numbers. Let's look at a simple example.

Let us suppose that you are given the following list of numbers that are alleged to be random: 0, 1, 0, 1, 2, 3, 4, 4, 3, 2, 5, 8, 7, 9, 6, 9, 8, 7, 5, 6. Casual examination reveals that there is an equal number of each digit. However, you also notice that there appear to be many low numbers at the beginning of the sequence and high numbers at the end. You decide to arbitrarily assign a plus sign to a digit between 0 and 4 and a minus sign to a digit between 5 and 9. You obtain the sequence $+++++++++----------$. Intuitively, you would doubt the randomness of the sequence. You see two *runs*—one of 10 pluses and another of 10 minuses. It would be like winning gin rummy ten times in a row and then losing ten in a row—a very unlikely event. Similarly, if you had found $+-+-+-+-+-+-+-+-+-+-$ you might also have been suspicious. The alternations of pluses and minuses appear to be too regular and frequent to be a random process. Indeed, there would be 20 runs, each with a sequence of one. In both cases, too many runs or too few runs, you would question the randomness of the process that generated the runs.

Runs tests have broad applications in business and economics. An investor might take a careful look at a stock that, over a 16-day period, compiled a record like the following: $------+-+++++++$. Cyclic trends in business or seasonal variations might produce runs that raise serious doubts that random processes are in operation.

Let us look at an example of the use of the runs test with small samples (n_1 and n_2 less than 20). Table 11.19 shows the change in the Dow Jones Industrials over a 19-day period prior to the national election held on November 4, 1980.

We might wish to ascertain if the uncertainty concerning the upcoming elections was reflected in the behavior of stock market prices. The following runs were observed between October 8 and November 3: $\underline{+}$ $\underline{--}$ $\underline{+++}$ $\underline{--}$ $\underline{+}$ $\underline{-}$ $\underline{+}$ $\underline{-}$ $\underline{+}$ $\underline{-}$ $\underline{+}$ $\underline{--}$ $\underline{++}$. Each run is underlined. There are a total of 13 runs, i.e., $R = 13$. Let's let n_1 equal the number of days during which the Dow Jones posted gains and n_2 the number of times it posted losses. Thus, $n_1 = 10$ and $n_2 = 9$.

Table XII shows the critical values for rejecting H_0 at $\alpha = 0.05$, two-tailed test, and $\alpha = 0.025$, one-tailed test. When $n_1 = 10$ and $n_2 = 9$, we find two critical values, 6 and 16. If R is equal to or less than 6, there are significantly fewer runs than expected under a random process. Similarly, if $R \geq 16$, there are a significantly greater number of runs. Since $R = 13$, we cannot reject H_0. There is, therefore, no statistically significant basis for inferring unusual behavior of the closing prices in the 19-day period prior to the election.

Table 11.19

Changes in Dow Jones Industrials during a 19-day period prior to national elections on November 4, 1980

Nov. 3	+12.71	Oct. 20	+4.70
Oct. 31	+6.74	17	−2.56
30	−11.43	16	−13.74
29	−3.41	15	+10.24
28	+0.85	14	+2.30
27	−11.86	13	+9.22
24	+4.09	10	−8.28
23	−15.61	9	−5.03
22	+0.68	8	+3.32
21	−6.40		

Runs test when n is large

Table XII (Appendix C) provides critical values if both n_1 and n_2 are equal to or less than 20. For larger values of n, the sampling distribution of R begins to approximate the normal distribution. Thus, the normal curve may be used to evaluate the statistical significance of the runs.

The expected value of R,

$$\mu_R = \frac{2n_1 n_2}{n_1 + n_2} + 1 \tag{11.7}$$

and the standard deviation of R,

$$\sigma_R = \sqrt{\frac{2n_1 n_2 (2n_1 n_2 - n_1 - n_2)}{(n_1 + n_2)^2 (n_1 + n_2 + 1)}} \tag{11.8}$$

We may establish the region for rejecting H_0 as follows:

Critical region at $\alpha = 0.05$, two-tailed test $= \mu_R \pm 1.96 \ \sigma_R$
Critical region at $\alpha = 0.05$, one-tailed test $= \mu_r \pm 1.645 \ \sigma_R$
Critical region at $\alpha = 0.01$, two-tailed test $= \mu_R \pm 2.58 \ \sigma_R$
Critical region at $\alpha = 0.01$, one-tailed test $= \mu_R \pm 2.33 \ \sigma_R$

Example I

Table 11.20 shows the seasonally adjusted unemployment rates for civilian workers between the years of 1947 through 1979. Using the mean (5.1) as the cutoff point, every rate less than 5.1 is assigned a negative sign and every rate above 5.1 receives a positive sign. Thus, n_1 (negative sign) $= 15$, and $n_2 = 18$. Note that the number of runs, R, equals 10. The null hypothesis is that the number of runs conforms to a random process.

Year	Unemployment rate X	Sign of $X - \bar{X}$	Year	Unemployment rate X	Sign of $X - \bar{X}$
1947	3.9	−	1964	5.2	+
1948	3.8	−	1965	4.5	−
1949	5.9	+	1966	3.8	−
1950	5.3	+	1967	3.8	−
1951	3.3	−	1968	3.6	−
1952	3.0	−	1969	3.5	−
1953	2.9	−	1970	4.9	−
1954	5.5	+	1971	5.9	+
1955	4.4	−	1972	5.6	+
1956	4.1	−	1973	4.9	−
1957	4.3	−	1974	5.6	+

Table 11.20
Seasonably adjusted unemployment rates over a 33-year period and the calculation of the two-tailed critical values for rejecting H_0 at $\alpha = 0.05$

Year	Unemployment rate X	Sign of $X - \bar{X}$	Year	Unemployment rate X	Sign of $X - \bar{X}$
1958	6.8	+	1975	8.5	+
1959	5.5	+	1976	7.7	+
1960	5.5	+	1977	7.0	+
1961	6.7	+	1978	6.0	+
1962	5.5	+	1979	5.8	+
1963	5.7	+			

$$\mu_R = \frac{2n_1 n_1}{n_1 + n_2} + 1$$

$$= \frac{2(15)(18)}{33} + 1$$

$$= 17.36.$$

$$\sigma_R = \sqrt{\frac{2n_1 n_2 (2n_1 n_2 - n_1 - n_2)}{(n_1 + n_2)^2 (n_1 + n_2 - 1)}}$$

$$= \sqrt{\frac{2(15)(18)[2(15)(18) - 15 - 18]}{33^2(32)}}$$

$$= 2.80$$

The critical region at $\alpha = 0.05$ is $17.36 \pm 1.96(2.80)$, or 11.87 to 22.85. We reject H_0 if the number of runs is outside this interval; $R = 10$, therefore we reject H_0 and assert that this sequence is not random.

Note that we reject H_0. The data suggest that periods of high unemployment alternate with periods of low unemployment in a cyclical fashion.

EXERCISES

11.11 In the period 1947–78, the median percentage of total working time lost due to work stoppages was 0.185. The following data represent the percentage of total working time lost during this period. Determine whether these data represent a random sequence, using the median as the cutoff point.

Percent of estimated total working time idle during month
0.30
0.28
0.44
0.33
0.18
0.48
0.22

Percent of estimated total working time idle during month
0.18
0.22
0.24
0.12
0.18
0.50
0.14
0.11
0.13
0.11
0.15
0.15
0.15
0.25
0.28
0.24
0.37
0.26
0.15
0.14
0.24
0.16
0.19
0.17
0.18

Source: Adapted from T. A. Kochan, *Collective Bargaining and Industrial Relations* (Homewood, Ill. Richard D. Irwin, 1980), p. 250, Table 8–1. © 1980 by Richard D. Irwin, Inc.

11.12 One measure of stock market activity is the behavior of New York Stock Exchange bonds. The Dow Jones average of 20 bonds is recorded daily and changes are thought to reflect bond activity in general. Between August 1, 1980, and October 31, 1980, the DJ average of 20 bonds went from 72.12 to 65.80. The following data represent the daily changes in this measure over this three-month period. A minus indicates the average went down and a plus indicates an increase from the preceding trading day. Determine if changes in bond activity follow a random sequence. Use $\alpha = 0.05$. (Read from left to right.)

−0.57,	−0.28,	−0.34,	−0.42,	−0.01,	−0.13,	−0.01,	−0.16,	−0.18,	
−0.05,	−0.10,	−0.20,	−0.10,	−0.71,	−0.22,	+0.22,	−0.16,	+0.04,	−0.21
−0.23,	−0.05,	+0.07,	+0.10,	+0.16,	+0.25,	+0.10,	−0.26,	+0.27,	−0.03,
−0.16,	−0.17,	−0.17,	−0.84,	+0.08,	+0.04,	−0.05,	−0.20,	−0.21,	−0.46,
−0.53,	−0.51,	−0.23,	−0.20,	+0.68,	+0.39,	+0.09,	−0.11,	−0.06,	+0.19,
+0.25,	+0.04,	+0.08,	+0.04,	−0.09,	−0.10,	+0.01,	−0.34,	−0.35,	+0.02,
−0.40,	−0.12,	−0.01,	−0.28						

SUMMARY

In this chapter we looked at χ^2 and several nonparametric tests of significance.

1. An extension of the rationale of the binomial provides a test of significance when there are more than two mutually exclusive and exhaustive classes of a variable. The multinomial test may be used to test the goodness of fit of obtained data to expectations based on the null hypothesis. However, computational complexities virtually exclude its use for all applications involving a large n.

2. Pearson's χ^2 statistic has been shown to approximate the chi-square distribution, at a given number of degrees of freedom, when n is sufficiently large. The χ^2 statistic has broad applications to goodness-of-fit problems.

3. We demonstrated the application of Pearson's χ^2 statistic to situations involving goodness of fit to *(a)* a binomial distribution, *(b)* a Poisson distribution, and *(c)* a normal distribution.

4. We also examined the use of the χ^2 two-variable case when frequency data are available on two or more classes of two discrete variables. Cast in the form of a contingency table, the null hypothesis is that the categories are independent of one another.

5. Tests making assumptions about normality are known as parametric tests of significance. Nonparametric tests of significance make no assumptions concerning the distribution of the populations from which the samples are drawn. However, the nonparametric tests do involve assumptions about the sampling distribution of the test statistics.

6. The sign test is used with matched or related samples. With a small n ($n \leq 30$) the binomial is used to determine if the signs of the differences between paired measures conform to $H_0: P = 0.50$. For $n > 30$, the normal approximation to binomial values is used.

7. The Mann-Whitney U-test enables us to test the significance of the difference between two independent samples when the values of the variable are ranked. When H_0 is false, the Mann-Whitney U-test is more likely than the sign test to lead to rejection of H_0 since it utilizes more of the quantitative information inherent in the data.

8. Finally, the one-sample runs test provides a basis for assessing the randomness of a sequence of events. The runs test has broad applications in business and economics, including the evaluation of the behavior of financial indexes and cyclical phenomena.

TERMS TO REMEMBER

goodness of fit   (330)

multinomial distributions   (330)

contingency table   (341)

parametric tests   (347)

nonparametric tests   (347)

run   (360)

EXERCISES

11.13   There is a geological hypothesis that, as you go from higher grades of ore to lower grades, the available amount is double the preceding amount. Thus, if the amount of a high-grade ore is 1, the amount of the next lower grade is 2, the next lower is 4, the next lower 8, etc.

The following table describes the results of 300 core samples of copper-bearing ore.

Grade of ore	f
High grade	10
Medium high	30
Medium	50
Medium low	80
Low	130
	300

Using $\alpha = 0.05$, test H_0 that the grade of ore follows the hypothesized function.

11.14   Ozone concentrations in the air are reported in parts per million (ppm). The following table gives the maximum single highest one-hour average of the year (1979) for 14 stations in Los Angeles County and 9 stations in San Bernardino County.[3]

Determine whether there is a significant difference between the two counties, using $\alpha = 0.05$.

Los Angeles	San Bernardino
0.34	0.34
0.45	0.16
0.39	0.34
0.21	0.37
0.33	0.42
0.35	0.40
0.19	0.21
0.32	0.13
0.32	0.12
0.20	
0.44	
0.29	
0.39	
0.26	

11.15   When the ozone concentration exceeds 0.20 ppm, a stage-one episode alert is declared. The following table shows the numbers of days such levels were reached in 1979 and 1978 for 14 locations in Los Angeles County.

---

[3] Source: Adapted from *Summary of Air Quality in the South Coast Basin of California 1979* (El Monte, Calif.: South Coast Air Quality Management District), p. 27, Table VI.

Location	1978	1979
Los Angeles	16	14
Azusa	76	71
Burbank	30	26
Long Beach	0	1
Reseda	16	24
Pomona	72	57
Lennox	2	0
Whittier	18	16
Newhall	45	59
Lancaster	5	1
Pasadena	85	78
Lynwood	0	6
Pico Rivera	48	38
West Los Angeles	10	7

Source: Adapted from *Summary of Air Quality in the South Coast Air Basin of California 1979*, p. 2, Table VI.

Using $\alpha = 0.01$, test H_0 that there has been an improvement in 1979 over 1978.

11.16 A matter of extreme concern among financial backers of films is their appeal to audiences of various ages. In a survey of audiences attending showings of films dealing with mature adult relationships, data were collected on the age of the respondents and the ratings of the films. The sample polled consisted of movie-goers in various parts of the country who were attracted to and paid to see the film during the first three days of its release.[4]

	Age	
Ratings	Under 25	25 and over
A	432	518
A—	418	626
C	309	550

Using $\alpha = 0.01$, test H_0 that the proportion of viewers under 25 and 25 and over is independent of the ratings of the films.

11.17 In a survey of attendance at 112 U.S.-made films opening in 1979, it was found that approximately 46 percent of the total sample (described in Exercise 11.16) were under 25. Using the data in Exercise 11.16, test H_0 that the proportion of the audience at the films dealing with mature adult relationships is equal to 0.46. Use $\alpha = 0.01$.[5]

11.18 The median rate for conventional first mortgages on new home purchases from January 1973 to December 1976 was 8.68 percent. Using the median as the cutoff, determine if the following data follow a random sequence. Use $\alpha = 0.05$.

---

[4] Source: *Cinemascore* 1980.
[5] Ibid.

	Month	Rate		Month	Rate
1973:	January	7.52	1975:	January	9.09
	February	7.52		February	8.88
	March	7.51		March	8.79
	April	7.53		April	8.71
	May	7.55		May	8.63
	June	7.62		June	8.73
	July	7.69		July	8.66
	August	7.77		August	8.63
	September ..	7.98		September ..	8.70
	October	8.12		October	8.75
	November ..	8.22		November ..	8.74
	December ...	8.31		December ...	8.74
1974:	January	8.33	1976:	January	8.71
	February	8.40		February	8.67
	March	8.43		March	8.67
	April	8.47		April	8.67
	May	8.55		May	8.75
	June	8.65		June	8.69
	July	8.75		July	8.76
	August	8.87		August	8.79
	September ..	8.97		September ..	8.85
	October	8.95		October	8.85
	November ..	9.04		November ..	8.83
	December ...	9.13		December ...	8.87

Home mortgage rates (conventional first mortgages on new home purchases)

11.19 The median rate for conventional first mortgages on new home purchases for the 14 months preceding the 1980 recession was 10.29 percent. Using the median as the cutoff point, determine whether these data follow a random sequence. Use $\alpha = 0.05$.

Nov 1978	9.63%
Dec	9.76
Jan 1979	9.92
Feb	9.94
Mar	10.02
Apr	10.06
May	10.20
June	10.39
July	10.49
Aug	10.73
Sept	10.72
Oct	10.91
Nov	11.04
Dec	11.30

11.20 Among those attending the movie (during the first three days of its release) *North Dallas Forty* (a story about a professional football player), 342 were males and 140 were females.

Among those attending *Promises in the Dark* (a story about a terminally ill 17-year-old), 111 out of 347 were male. It is likely that both audiences were drawn from the same population of movie-goers? Use $\alpha = 0.01$.[6]

11.21 Refer to Exercise 11.20.

    a. Test the null hypothesis that an equal proportion of males and females were attracted to *North Dallas Forty*.

    b. Test the null hypothesis that an equal proportion of males and females were attracted to *Promises in the Dark*.

11.22 On the average, 36 ships arrive at a given port during a 24-hour period. During 24 randomly selected one-hour periods, the following distribution of arrivals was obtained. Test the goodness of fit of the obtained distribution to expectations based on the Poisson model, using $\alpha = 0.01$.

Number of ships	f
0	6
1	7
2	5
3	2
4	2
5	0
6	1
7	1
8	0
9	0

11.23 The following table presents the average hourly gross earnings per production worker on private manufacturing payrolls from 1947 to 1979 for two types of nondurable goods industries. Using $\alpha = 0.01$, determine if there is a difference between these two groups.

	Tobacco manufactures	Textile mill products
1947	$0.905	$1.035
8	0.956	1.155
9	0.999	1.181
50	1.076	1.228
1	1.14	1.32
2	1.18	1.34
3	1.25	1.36
4	1.30	1.36
5	1.34	1.38
6	1.45	1.44
7	1.53	1.49
8	1.59	1.49
9	1.64	1.56
60	1.70	1.61

[6] Source: *Cinemascore* 1980.

	Tobacco manufactures	Textile mill products
1	1.78	1.63
2	1.85	1.68
3	1.91	1.71
4	1.95	1.79
5	2.09	1.87
6	2.19	1.96
7	2.27	2.06
8	2.48	2.21
9	2.62	2.34
70	2.91	2.45
1	3.16	2.57
2	3.47	2.74
3	3.74	2.95
4	4.10	3.19
5	4.51	3.40
6	4.91	3.67
7	5.54	3.99
8	6.13	4.30
9	6.65	4.66

11.24 The median prime rate for the 10 months preceding the recession beginning at the end of 1973 was 7.56 percent. The following data present the prime rate during this period: 5.6, 6.14, 6.82, 6.97, 7.15, 7.98, 9.19, 10.18, 10.19, 9.07. Using the median as the cutoff point, determine whether these data follow a random sequence. Use $\alpha = 0.05$.

11.25 The median prime rate for the 10 months preceding the recession that began in January 1980 was 10.30 percent. Using the median as the cutoff, determine whether these data follow a random sequence. Use $\alpha = 0.05$.

9.94, 9.90, 9.98, 9.79, 9.99, 10.62, 11.70, 13.44, 13.53, 13.31

11.26 During the recession from December 1973 to March 1975, the median prime rate was 9.11 percent. The following data present the prime rate during this period. Using the median as the cutoff point, determine whether these data follow a random sequence. Use $\alpha = 0.05$.

Dec. 1973	8.94%
Jan. 1974	8.72
Feb.	7.83
Mar.	8.43
Apr.	9.61
May	10.68
June	10.79
July	11.88
Aug.	12.08
Sept.	11.06
Oct.	9.34
Nov.	9.03
Dec.	9.19

Jan. 1975	7.54
Feb.	6.35
Mar.	6.22

11.27 In the 20 months following the recession ending March 1975, the median prime rate was 5.61 percent. The following data present the prime rate for this period. Determine whether these data represent a random sequence, using the median as the cutoff point. Use $\alpha = 0.05$.

April 1975	6.15%	Jan 1976	5.08%
May	5.76	Feb	4.99
June	5.70	Mar	5.18
July	6.40	Apr	5.03
Aug	6.74	May	5.53
Sept	6.83	June	5.77
Oct	6.28	July	5.50
Nov	5.79	Aug	5.32
Dec	5.72	Sept	5.28
		Oct	5.06
		Nov	4.90

11.28 Based on past experience, the probability that a given computer chip will be defective is 0.05. In 30 samples of $n = 100$, the following distribution of defective components was observed. Test H_0 that the obtained results conform to a binomial model. Use $\alpha = 0.05$.

x	f
0	8
1	9
2	10
3	3
4	0
5	0
6	0
7	0
8	0

11.29 Do changes in the prime rate affect the behavior of the stock market? In an attempt to answer this question, 43 days were randomly selected. The prime rate went up on 18 of these days and stayed the same on the remaining 25. The change in the DJI was recorded for each of these days. Using $\alpha = 0.05$, determine if changes in the DJ Ind. are different when the prime rate goes up versus when the prime rate stays the same.

Changes in DJI on days that the prime rate	
Went up	Stayed the same
+12.71	−3.84
−8.78	−0.42
+9.72	−1.28
−2.82	+8.45

Changes in DJI on days that the prime	
Went up	Stayed the same
−12.71	+5.46
−7.85	+3.16
−12.38	−1.96
−4.78	+2.21
+1.11	+8.19
+7.26	−4.35
−12.54	+6.15
−15.87	+3.75
+7.00	+2.82
+2.82	+2.73
+4.70	−18.17
+0.85	+8.44
−17.75	−5.03
−2.99	−8.28
	−13.74
	+0.68
	−15.61
	+4.09
	−3.41
	+12.71
	+1.37

11.30 During the eight-hour period between 9 A.M. and 5 P.M., an airlines reservation switchboard receives, on the average, 2,496 calls. During 50 randomly selected one-minute periods, the following distribution of calls was received. Test the goodness of fit to a Poisson distribution, using $\alpha = 0.05$.

In the 1979 survey of science/engineering and humanities doctorates, data were collected to determine whether differences between respondents and nonrespondents were independent of gender.

Exercises 11.31 and 11.32 present these data for science/engineering and for humanities Ph.D.s. Using $\alpha = 0.01$, determine whether these variables are independent.

Number of calls	f
0	0
1	0
2	0
3	5
4	6
5	12
6	14
7	4
8	3
9	1
10	4
11	0
12	1
13	0
14	0
15	0
16	0

11.31

	Science/Engineering Ph.D.s.	
	Male	Female
Respondents	200,652	25,818
Nonrespondents	101,947	12,323

11.32

	Humanities Ph.D.s	
	Male	Female
Respondents	38,243	13,437
Nonrespondents	17,745	5,665

11.33 The following data present the number of men and women who received their doctorates between 1936 and 1978 by field. Test the hypothesis that the proportion of men and women is independent of field. Use $\alpha = 0.01$.

Field	Males	Females
Mathematics	15,657	1,343
Computer science	1,679	121
Physics/Astronomy	26,897	803
Chemistry	43,478	3,122
Environment	9,922	478
Engineering	48,211	389
Agriculture	14,386	414
Medicine	7,578	1,422
Business	48,475	10,425
Psychology	29,497	10,203
Social science	42,031	7,769

Source: Adapted from *Science, Engineering, and Humanities Doctorates in the United States 1979 Profile.*

11.34 The same survey (Exercise 11.33) reported that not all individuals who receive their doctorate in a particular field actually work in that field. Use the data in the table below to test the hypothesis that the proportion of individuals working in the field of their doctorate is independent of Ph.D. fields. Use $\alpha = 0.01$.

Field of doctorate	Number working in field	Number working in other field	n
Mathematics	11,810	4,390	16,200
Computer science	1,346	454	1,800
Physics/Astronomy	16,013	10,587	26,600
Chemistry	31,161	12,239	43,400
Environment	8,420	1,580	10,000
Engineering	37,982	9,318	47,300
Agriculture	10,261	3,439	13,700
Medicine	6,783	1,717	8,500
Business	38,368	16,132	54,500
Psychology	31,902	5,498	37,400
Social science	33,992	12,508	46,500
	228,038	77,862	305,900

11.35 The following data show the range of safety and health provisions found in collective bargaining agreements for contracts sampled by the Bureau of National Affairs. Using $\alpha = 0.01$, test the following:
Is there a difference in the percentage of these items included in
    a. Manufacturing industry contracts as opposed to mining industry contracts?
    b. Manufacturing industry contracts as opposed to all contracts?
    c. Mining industry contracts as opposed to all contracts?

	Percent in all contracts	Percent in mfg. industry contracts	Percent in mining industry contracts
Some provision on safety	82	87	100
General statement of responsibility	50	58	75
Company to comply with laws	29	29	50
Company to provide safety equipment	42	46	92
Company to provide first aid	21	26	50
Physical examinations	30	30	75
Hazardous work provisions	22	19	67
Accident investigations	18	24	58
Safety committees	43	55	92
Dissemination of safety information to employees	16	18	38
Dissemination of safety issues to union	19	21	44
Employees to comply with safety rules..............	47	50	67
Right of inspection by union or employees safety committees	20	30	56
Wage differentials for hazardous work	15	6	6

Source: Adapted from T. A. Kochan, *Collective Bargaining and Industrial Relations* (Homewood, Ill.: Richard D. Irwin, 1980), p. 363, Table 11–2. © 1980 by Richard D. Irwin, Inc.

11.36 In 1947 the Taft-Hartley amendments to the National Labor Relations Act provided procedures whereby members could vote to decertify a union. Although the number of decertification elections is small relative to the number of union representation elections, unions lose members whenever they lose a decertification election.

The following is the percentage of decertification elections lost by the union from 1948 (the first year after the passage of the Taft-Hartley) through 1977.

Percentage of elections lost	f
60–under 63	2
63–under 66	6
66–under 69	7
69–under 72	10
72–under 75	4
75–under 78	1

Test the hypothesis that this distribution conforms to a normal distribution. Use the sample mean ($\overline{X} = 68.6$) and sample standard deviation ($s = 3.74$) calculated from these data. Employ $\alpha = 0.05$ and state the number of df.

11.37 Total outstanding consumer installment credit has steadily increased from almost $13 billion in 1947 to over $322 billion in 1979. However, the proportion of credit liquidated relative to the amount outstanding has fluctuated over the years. The following table presents the proportion of credit liquidated to credit outstanding for total consumer installment credit and for automobile paper from 1947 to 1979. (A proportion greater than 1.00 means more credit was liquidated (or paid back) than borrowed for that year.) Determine if this proportion differs for automobile credit versus total, using $\alpha = 0.05$.

Year	Total	Auto	Year	Total	Auto
1947	0.802	0.745	1964	0.898	0.889
1948	0.852	0.790	1965	0.896	0.871
1949	0.857	0.779	1966	0.935	0.942
1950	0.856	0.822	1967	0.963	1.008
1951	0.975	1.011	1968	0.917	0.899
1952	0.861	0.850	1969	0.914	0.921
1953	0.886	0.838	1970	0.947	1.011
1954	0.982	1.002	1971	0.919	0.883
1955	0.863	0.782	1972	0.889	0.871
1956	0.930	0.938	1973	0.875	0.881
1957	0.949	0.944	1974	0.941	0.989
1958	1.006	1.084	1975	0.954	0.942
1959	0.887	0.876	1976	0.894	0.837
1960	0.925	0.930	1977	0.861	0.817
1961	0.981	1.033	1978	0.850	0.773
1962	0.931	0.886	1979	0.890	0.861
1963	0.894	0.870			

11.38 The closing prices of stocks listed on the New York Stock Exchange are published daily in The Wall Street Journal. The Dow Jones Average for transportation stocks is based on 20 representative stocks. The behavior of the DJ 20 transportation stocks is thought to represent transportation stocks in general. The DJ average for utilities is based on 15 representative utility stocks. Finally, the DJ Futures is a commodity index thought to reflect activity in commodities in general.

The following table presents the daily changes of these three indexes from August 1, 1980, through October 31, 1980. A minus indicates the average went down and a plus reflects an increase over the previous trading day.

Use $\alpha = 0.05$ and test the following:

    a. Changes in the DJ Transportation averages reflect random activity.

    b. Changes in the DJ Utilities averages reflect random activity.

    c. Changes in the DJ Futures index reflect random activity.

Day	DJ Transportation	DJ Utilities	DJ Futures Index
1	−2.46	−0.62	+4.39
2	−0.93	−0.94	+2.06
3	−0.77	−0.99	−3.33
4	+0.16	−0.55	+4.61
5	+1.04	+0.08	+4.40
6	+2.80	+0.69	−5.73
7	+1.86	+0.73	+4.68
8	+0.88	−0.15	−1.08
9	−0.93	+0.19	+5.58
10	−2.36	−0.26	+2.68
11	+6.52	+0.37	+0.66
12	+1.09	+0.47	−0.05
13	−3.99	−0.73	−0.95
14	−0.33	−0.07	+3.38
15	+3.34	+0.47	+0.71
16	+4.54	+0.91	−1.30
17	+2.80	−0.32	−7.45

Exhibit 11.38 table *(continued)*

Day	DJ Transportation	DJ Utilities	DJ Futures Index
18	−3.12	−0.26	+7.05
19	−1.26	−0.26	+7.74
20	−3.18	−0.32	−8.31
21	−2.19	−0.19	+1.99
22	+2.13	+0.70	+0.83
23	+4.71	+0.54	+1.67
24	+4.33	+1.86	+2.47
25	−1.37	−0.40	−3.94
26	−2.35	−0.18	+2.55
27	−4.38	−1.06	+3.38
28	−0.28	+0.04	+0.81
29	−0.11	+0.62	+1.10
30	+0.55	+0.04	+3.30
31	+0.05	−0.26	+1.94
31	+8.22	−0.36	−0.47
33	+9.53	+0.91	+0.73
34	+6.90	−0.11	−5.11
35	−0.82	−0.22	+2.14
36	+1.43	−0.04	+2.70
37	+3.50	−0.29	+2.89
38	−0.82	−1.20	+0.70
39	−0.22	−0.26	+0.50
40	−3.39	−0.73	−2.86
41	−8.06	−1.05	−2.93
42	−8.76	−1.43	+3.83
43	+5.09	+0.44	+5.48
44	+3.73	+0.88	−0.56
45	+3.45	+0.73	−6.34
46	+4.38	+1.78	+6.16
47	+3.18	+2.63	+4.46
48	−0.55	−0.11	+1.16
49	+0.27	+0.22	+1.51
50	+0.88	−0.55	+3.82
51	+0.88	−0.29	+1.55
52	+2.02	−0.11	+3.74
53	+0.93	+0.04	−1.52
54	+1.76	+0.76	−0.05
55	+1.42	−0.84	−1.47
56	+0.11	−0.03	+1.59
57	+6.57	−0.19	−1.93
58	+6.52	+0.19	+3.70
59	+11.67	+0.14	+1.84
60	−7.23	−0.14	−0.17
61	−0.38	−0.19	+1.43
62	−8.33	−0.73	−4.35
63	−1.20	−0.18	+1.54
64	+1.80	−0.40	−1.96
65	−4.87	−1.28	−1.25

11.39 Using the data presented in Exercise 11.38, test the following ($\alpha = 0.05$): Is there a difference in changes in

a. The DJ Transportation averages versus the DJ Utilities?
b. The DJ Transportation averages versus the DJ Futures index?
c. The DJ Utilities versus the DJ Futures?

11.40 Using the data in Exercise 11.38, find the number of days each of the three measures posted a gain and the number of days each posted a loss. Use $\alpha = 0.05$ to test whether the proportion of gains versus losses is independent of the index selected.

11.41 The following table presents two measures of the propensity to strike in U.S. manufacturing industries between 1954 and 1975. Measure I represents the percentage of contract negotiations involving a strike. Measure II represents the percentage of workers in strikes. Using $\alpha = 0.05$, determine whether there is a difference between these two measures.

Two measures of the propensity to strike in U.S. manufacturing

Year	Propensity to strike (I)	Propensity to strike (II)
1954	0.078	0.094
1955	0.138	0.265
1956	0.119	0.242
1957	0.133	0.119
1958	0.130	0.304
1959	0.114	0.307
1960	0.060	0.094
1961	0.074	0.063
1962	0.073	0.075
1963	0.053	0.062
1964	0.082	0.092
1965	0.068	0.127
1966	0.168	0.181
1967	0.222	0.256
1968	0.190	0.164
1969	0.268	0.256
1970	0.181	0.113
1971	0.189	0.163
1972	0.136	0.118
1973	0.160	0,150
1974	0.297	0.261
1975	0.153	0.132

Source: Adapted from Bruce E. Kaufman, "The Propensity to Strike in American Manufacturing," (Madison, Wis.: Industrial Relations Research Associations Proceedings, 1978), p. 423.

Linear Regression and Correlation

12.1 REGRESSION AND CORRELATION DISTINGUISHED

There is hardly a field of human endeavor in which correlations between or among variables fail to play a significant role. Consider the following:

> The presence of such variables as high fever, aching muscles, chest rattles, nasal mucus, and a known outbreak in the general population leads the medical diagnostician to say, "The disorder appears to be the flu."

> A stock broker may look at a number of indexes such as price to earnings ratios, consumer price index, prime rate, etc., before predicting the behavior of a given stock. Each of these indexes is a predictor variable used to predict the behavior of the stock.

> A farmer distributes varying amounts of fertilizer, pesticide, and irrigation water to different plots and notes the variations in crop growth. To what extent do these procedural variations permit the farmer to predict yield?

> A personnel director administers a battery of tests to a group of prospective employees. Will one or more of the test scores increase the prospect of better predicting the subsequent performance of the employees? Can the test results be used with advantage as a selection device?

All of these situations involve the use of one or more variables to predict a single *dependent variable* or measure. Each or any one of these predictor variables is called an *independent variable*. Use of a single variable as a predictor involves *simple regression analysis*. If more than one predictor variable is used, we refer to the procedures as *multiple regression analysis*. In the following sections, we shall restrict our sights to regression and correlation for the case in which there is a dependent variable and a single independent variable. In the next chapter we examine multiple regression.

Before proceeding further, it is helpful to distinguish between relationships that are completely determined and those that are not. Many of the relationships that attract the interest of mathematicians are completely determined. For example, when a mathematician describes a relationship between the length of a side of a square (the predictor or X variable) and the perimeter of a square (the dependent or Y variable), the value of Y is completely determined by its associated X value. Thus, if we know that one side of a square is 3 inches ($X = 3$), it follows with certainty that the perimeter is 12 inches ($Y = 12$). A diagram of this mathematical relationship is shown in Figure 12.1.

The business executive, the manager, or the economist rarely, if ever, deals with relationships that are completely determined. The reason is that the variables that interest these individuals are typically influenced by a multitude of factors, many of which may not even have been identified. Is inflation produced *only* by high demand for and low availability of goods and services? If it were, we could presumably find an equation that completely determines inflationary rate (Y variable) from the ratio of demand to availability (X variable). When the behavior of a given Y variable is influenced by a wide assortment of different variables, the relationship between any single variable and the Y variable is unlikely to produce data points that fall precisely on

dependent variable

The variable we are trying to predict based on its relationship with one or more other variables.

independent variable

The variable that we use to predict values of a dependent variable.

simple regression analysis

The process by which we predict one variable from another. The purpose of simple regression analysis is to describe the nature of the relationship between two variables in terms of a mathematical equation.

multiple regression analysis

The process by which two or more independent variables are used to predict values of a dependent variable.

Diagram of a completely determined relationship. The relationship shown is a straight line or linear relationship. When a given value of *X* is plotted, its associated *Y* value *must* fall on the line that describes the relationship.   **Figure 12.1**

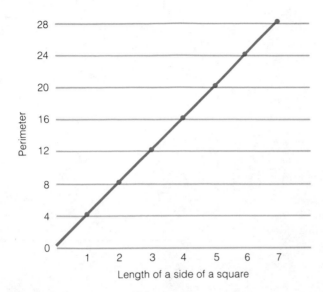

a line, such as in Figure 12.1. More frequently, we find the data points scattered around some straight line or curve that represents an overall relationship (see Figure 12.2). The purpose of regression analysis is to find an equation that describes that straight line or curve. Once we have done so, we may use values of the *X* or independent variable to predict values of the dependent or *Y* variable. *Correlational analysis* results in a single number that describes the extent of the relationship between the two variables. It provides answers to questions such as "How good is the relationship? Is it weak? Moderate? Strong?".

Regression and correlational analysis involves the relationship between variables that are matched in some way. When dealing with a two-variable problem, each observation provides a value for the *X*-variable and a matching value for the *Y*-variable. For example, suppose we were interested in the relationship between various interest rates. Our focus might be on two measures of interest rates such as federal discount rate and prime rate. Each observation would consist of two values, one for each variable. These two variables would be matched, across time, for example.

correlational analysis

The process by which we obtain a single number that summarizes the *degree* of the relationship between two or more variables.

12.2 LINEAR REGRESSION

There are any number of ways that two variables may be related. Several possibilities are shown in Figure 12.2. In this text, we shall limit our discussion to finding linear regression equations (Figure 12.2*c* and 12.2*d*).

It should be pointed out, however, that fitting a straight line to data does not prove that a relationship is linear. A straight line is merely one of several

Figure 12.2 Different types of possible relationships between a predictor and a dependent variable. The dots represent the actual measurements obtained on the two variables. Note that many of these dots do not fall on the curve or line representing the relationship between the variables.

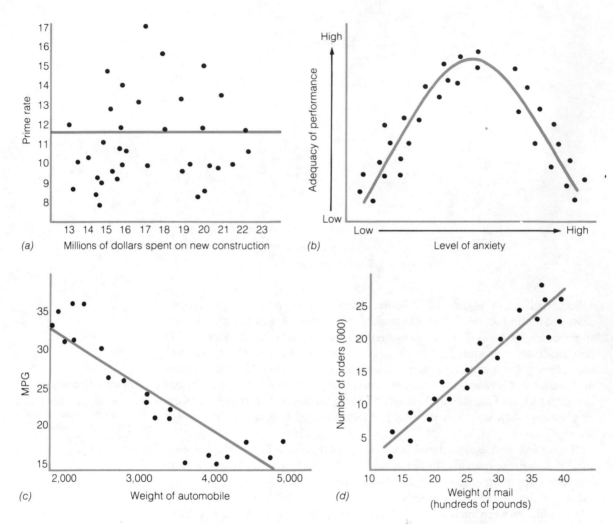

different possible lines that can be used to describe the form of a relationship. It is true, however, that many of the relationships within the real world approach the form of a straight line. Thus, the linear model is often appropriate.

What is the value of finding a regression equation that relates a predictor variable to a dependent variable? A use that quickly comes to mind is the prediction of future values of a variable. If an index of economic health, such as orders for new model automobiles, is related in a determinable fashion to total demand for steel in the economy, the steel industry would be better able to establish future steel production goals in the light of present information.

At other times obtaining the dependent measure may be difficult, expensive, and/or time-consuming. If a more readily available predictor variable can be shown to accurately predict the dependent measure, much expense, time, and, perhaps, lives may be saved. For example, there is a characteristic change in the patterns of electrical activity of the brain immediately prior to entering one of the early stages of sleep. With the advent of inexpensive microcomputers, it is possible that future truckers will have electrical recording equipment attached to their scalps. As soon as changes signalling approaching sleep are detected, an alarm is sounded, thereby arousing the driver before a catastrophe can occur.

Finally, when the predictor variable temporally precedes the dependent measure, it is possible that a causal relationship is at work. In other words, it is possible that variations in the X variable are causing variations in the Y variable. However, finding a relationship does not *prove* that one causes the other. For example, a third variable may cause both X and Y and therefore be responsible for the relationship between X and Y. Nevertheless, finding a relationship is the first step in establishing a causal link. Thus, the possibility of a cause-effect relationship should not be overlooked. Box 12.1 shows an amusing example that illustrates all three values of regression analysis.

12.2.1   The equation for a straight line

Recall from algebra that the equation for a straight line is

$$Y = a + bX$$

in which

Y is the dependent variable;
a is the Y intercept (i.e., the value of Y when $X = 0$);
b is the slope of the line; and
X is any value of the predictor.

The slope of the line indicates how many units Y changes with each change of one unit in X. To illustrate, look at Figure 12.3 (Panel A). It shows a straight line with Y intercept of 5 (i.e., $a = 5$) and a slope of about 0.64. This means that for every change of one unit in X there is a corresponding change of 0.64 unit in Y. This slope may be approximated by the triangle shown in the figure. Note that an increase of 14 units in X (i.e., $\Delta X = 14$) is accompanied by an increase of 9 units in Y (i.e., $\Delta Y = 9$). The ratio of the two,

$$\frac{\Delta Y}{\Delta X} = \frac{9}{14} = 0.64,$$

is the slope of the line. Note that substituting any value of X in the equation for the straight line yields the approximate value of Y shown in the graph.

Figure 12.3

Graphs of a straight line when $b = 0.64$ (Panel A), $b = -0.92$ (Panel B), and $b = 0$ (Panel C). Note that a is the value of Y when $X = 0$. The slope of the line may be approximated by constructing a triangle and finding out how much Y changes for each unit of change in X. Thus, $b = \Delta Y/\Delta X$, in which $\Delta = $ "change in." Note that when the slope is zero (Panel C), no triangle can be constructed since each change in X is accompanied by zero change in Y.

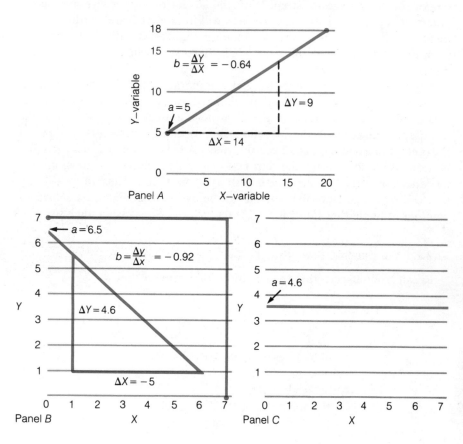

Thus, if $X = 15$, $Y = 5 + 0.64\ (15) = 14$ (rounded to the nearest whole number).

Box 12.1

FLUSH ME A COMMERCIAL

If you were to conduct a house-to-house survey of television viewing, the educational channels would seem to dominate nighttime viewing. And how many homes do you imagine would admit to subscribing to such magazines as *Playboy, Playgirl, Penthouse,* or *Oui?* Now if we could find indirect measures—mea-sures correlated with what we want to assess but sufficiently different that the respondent's defenses are not aroused—we'd be able to get at the truth without offending anyone. One of my favorite examples of such an indirect measure is the *flushometer*. What's the flushometer?

Box 12.1 (*continued*)

Some years ago a water district on Long Island puzzled over the fact that its demand for water during the evening hours was punctuated by short bursts of enormous activity, followed by long periods of quiescence. At first the water company was at a complete loss as to what forces were orchestrating and synchronizing these visits to the family john. The picture that these nightly activities conjured up was eerie. Was some UFO out there in space exercising mind control and directing people to march robotlike to the throne? A possibility, however remote.

Then, some bright-eyed individual got the idea that the toilet flushes might coincide with the commercial breaks on television programs. He followed up his hunch by recording the exact amount of water demanded during specific time periods in the evening and subsequently determining when the commercial breaks took place on the television programs being viewed during these hours. The results were most astounding. It was found that the greatest number of flushes occurred during the commercial breaks of those programs that ranked highest on the Nielsen ratings. In fact, over the first ten ratings, the correlation was absolutely perfect. The number one program, which was "I Love Lucy," received the greatest surge of flush power during its commercial break, and the next nine in order showed decreasing amounts of water consumed during their corresponding commercial breaks. In this one instance, at least, reasoning from correlation to causation would seem to be fairly straightforward and direct. Few would argue that the flushing of toilets was causing the commercials to go on, but one would probably be reasonably correct in hypothesizing that the commercials were sending people scurrying to the bathroom.

Imagine the flush power if all TV programs synchronized their commercial breaks! End of the energy crisis.

Source: Richard P. Runyon, *How Numbers Lie: A Consumer's Guide to the Fine Art of Numerical Deception* (Lexington, Mass.: Lewis Publishing Co., 1981) pp. 125–26. Copyright © Lewis Publishing Co., Lexington, Mass.

EXERCISE

12.1 Given the following graphs of straight lines, estimate both *a* and *b*.

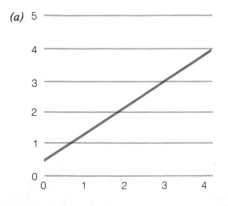

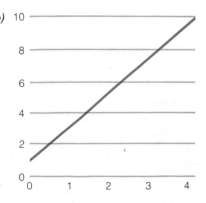

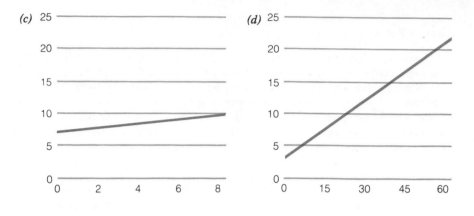

12.2.2 A model of simple linear regression

We have noted throughout this text that we rarely know populations. It is for this reason that we resort to collecting samples and using the statistics calculated from these samples to estimate parameters. The same considerations hold true for simple linear regression. In the populations of paired measures of two variables, there is an equation that best describes the relationship between these two variables. In the case of linearly related variables, this equation is

$$\mu_{y \cdot x} = A + BX + e$$

where

$\mu_{y \cdot x}$ is a specific, individual value in the population model;
A is the population Y-intercept;
B is the population slope of the regression line; and
e represents the dispersion or scatter in the population associated with omitted, unobserved and unobservable variables.

In regression analysis, the dependent variable Y is considered a random variable. The independent variable, X, is fixed. Thus, for any given value of X, we assume there is a distribution of Y-values in the population. Only if the relationship between X and Y were perfect, would there be one Y value associated with any given X value. However, in real life, where many random variables affect the dependent measure, perfect relationships simple do not occur. Therefore, there will always be some scatter around the regression line.

One reason for collecting samples on paired variables is, as we have seen, to develop a prediction formula relating dependent measures to a predictor variable. Another goal is to use data derived from samples to obtain the estimating equation or mathematical formula that describes the regression line. We take a sample of measurements on an independent variable *(X)* and a matching sample of measurements on a dependent variable *(Y)*. However, data from samples represent only a part of the total population. As such, calculations based on these sample data are regarded as estimates of the true population values.

In other words, the obtained regression line and its corresponding equation are really *estimates* of the true relationship that exists in the population. Therefore, the obtained values of *a* and *b* may be considered point estimates of the true population values, *A* and *B*. Further, the linear equation for estimating the true, but unknown, relationship that exists in the population is

$$Y_c = a + bX \qquad (12.1)$$

where

Y_c is the point estimate of *E(Y|X)* (i.e., the expected value of *Y* for a given *X*);
a estimates *A;* and
b estimates *B*.

Like any statistics, these estimates are subject to sampling error. The true regression equation may have a slope of 1.00, but repeated random samples from the population will yield a distribution of *b* coefficients that distribute themselves about the true population value. The same is true of *a*. Thus, as with other sampling data, we may establish confidence intervals of the corresponding population parameters and test hypotheses concerning point estimates of the parameters.

12.2.3   Constructing a scatter diagram

It is not always clear which variable is the predictor *(X)* and which is the dependent variable *(Y)*. Generally, if one precedes the other in time, the earlier is the predictor variable. Thus, if we are using an economic index to predict future trends, the economic index would be the *X* variable. Occasionally, however, we may wish to predict "backwards," in which case the economic index would be the *Y* variable. Such an occasion might arise when, for one reason or another, we are missing information on the economic index and wish to estimate what it might have been. Such backward "predicting" is commonly done in psychology where the intelligence of past historical figures is estimated from knowledge of their accomplishments.

There are other occasions when the two measures are simultaneous. The decision about which to use as a predictor variable depends on our goals or purposes. Thus, we may have two measures on a metal alloy: one involving hardness and the other resistance to stress. If we wish to predict resistance to stress from hardness, then the predictor variable would be hardness.

In any event, after we obtain a pair of measures on each person, object, or event, it is desirable to present these paired measures in a form of a chart known as a scatter diagram. The measures on the predictor or independent variable are represented along the X or horizontal axis. The dependent measure is represented along the Y axis.

As Figure 12.2 illustrated, not all variables are related in a linear fashion. Therefore, before attempting to estimate the relationship by a linear (or straight line) equation, it is often helpful to construct a diagram that allows us to determine visually whether a linear relationship seems appropriate.

Table 12.1 presents data on two variables. The X-variable is the federal discount rate. The Y-variable is the prime rate matched over the same time period.

Let us suppose we wish to know whether knowledge of the federal discount rate at any given time will allow us to predict the prime rate. First, we will construct a scatter diagram so that we may visually inspect the nature of this relationship.

In plotting a scatter diagram, we take a value of the X-variable, for example $X = 10.0$. We locate this value on the X-axis. Imagine a perpendicular line erected at this point. The corresponding value on the Y-variable is 9.85. Now imagine a line perpendicular to the Y-axis drawn at $Y = 9.85$. The point representing these paired values is where the two lines intersect. This is done with every set of paired values until all are represented on the chart.

Figure 12.4 presents a scatter diagram of these data. Although it would not be possible, in this case, to find a straight line that goes through every single point, it appears that a linear relationship between these two variables possibly exists.

Note that the points representing the paired scores appear to be arranged in a roughly diagonal fashion from the lower left-hand corner to the upper right-hand corner. Moreover, if we took a transparent ruler, we could readily

Table 12.1

Values of the federal discount rate *(X)* and the prime rate *(Y)* over an 18-month period

Observation	X Federal discount rate	Y Prime rate
1	9.50%	9.79%
2	9.69	9.99
3	10.24	10.62
4	10.70	11.70
5	11.77	13.44
6	12.00	13.53
7	12.00	13.31
8	12.00	13.15
9	12.52	14.01
10	13.00	17.10
11	13.00	15.63
12	12.94	9.60
13	11.40	8.31
14	10.87	8.58
15	10.00	9.85
16	10.10	12.15
17	11.00	13.86
18	11.53	15.94

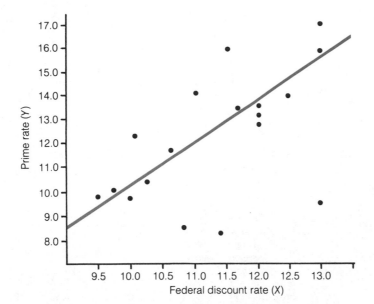

Figure 12.4
Scatter diagram showing the relationship between the federal discount rate *(X)* and the prime rate *(Y)*

"eyeball" a straight line that "fits" these paired data. This line would minimize the squared deviations between the line and the various data points. The line shown in Figure 12.4 was fitted to the data in precisely this fashion. Admittedly, the procedure of fitting a straight line by eye is crude. However, it illustrates one technique widely used by mathematicians and statisticians, namely, the method of least squares.

12.2.4   Constructing a regression line

There are a number of different methods for fitting a straight line to a set of points. The technique most widely used is known as the *least squares method*.

You may recall that, when discussing the properties of the mean (Section 3.2.2), we demonstrated that the mean is that point in a distribution that makes the summed deviations around it equal to zero $[\Sigma(X - \bar{X}) = 0]$. Moreover, the mean is also the point that makes the sum of the squared deviations around it minimal (least sum of squares). The line of "best fit," or the *regression line,* is analogous to the mean. It is the one line that makes the squared deviations around it minimal. And, as with the mean, the sum of the deviations about this line is zero.

Expressed symbolically, $\Sigma(Y - Y_c) = 0$ and $\Sigma(Y - Y_c)^2$ is minimal, where Y is the obtained score on the Y-variable for a given X and Y_c is the predicted value on the Y-variable for a given X.

We shall use the data presented in Table 12.1 to illustrate the use of the least squares method for obtaining the regression line.

We may use Formula 12.1 $(Y_c = a + bX)$ as our estimating equation to describe the relationship between these two variables.

The formulas for a and b may be derived algebraically. However, for

least squares method

A method of fitting a line to a set of points so that the squared deviations are minimal.

regression line

The line that "best fits" a set of data points so that the squared deviations around it are minimal.

our purposes, we shall present formulas that may be used to calculate these values:

$$b = \frac{n(\Sigma XY) - (\Sigma X)(\Sigma Y)}{n(\Sigma X^2) - (\Sigma X)^2}$$ (12.2)

$$a = \frac{\Sigma Y - b\Sigma X}{n}$$ (12.3)

Table 12.2 presents the values necessary to obtain the linear equation: ΣX, ΣY, ΣXY, and ΣX^2. In addition we have calculated Y^2 for later use.

Table 12.2
Calculations for data in Table 12.1

Observation	Federal discount rate (percent) X	Prime rate (percent) Y	XY	X²	Y²
1	9.50	9.79	93.0050	90.2500	95.8441
2	9.69	9.99	96.8031	93.8961	99.8001
3	10.24	10.62	108.7488	104.8576	112.7844
4	10.70	11.70	125.1900	114.4900	136.8900
5	11.77	13.44	158.1888	138.5329	180.6336
6	12.00	13.53	162.3600	144.0000	183.0609
7	12.00	13.31	159.7200	144.0000	177.1561
8	12.00	13.15	157.8000	144.0000	172.9225
9	12.52	14.01	175.4052	156.7504	196.2801
10	13.00	17.10	222.3000	169.0000	292.4100
11	13.00	15.63	203.1900	169.0000	244.2969
12	12.94	9.60	124.2240	167.4436	92.1600
13	11.40	8.31	94.7340	129.9600	69.0561
14	10.87	8.58	93.2646	118.1569	73.6164
15	10.00	9.85	98.5000	100.0000	97.0225
16	10.10	12.15	122.7150	102.0100	147.6225
17	11.00	13.86	152.4600	121.0000	192.0996
18	11.53	15.94	183.7882	132.9409	254.0836
Sums (ΣX, etc.)	204.26	220.56	2,532.3967	2,340.2884	2,817.7394

$$\bar{X} = 11.35 \qquad \bar{Y} = 12.25$$

The resulting regression equation $Y_c = a + bX$ is $Y_c = -2.73 + 1.32X$.

We may interpret this equation in the following way: for any given federal discount rate (X), the predicted prime rate equals 1.32 times the discount rate minus 2.73. Thus, for a federal discount rate equal to 12.50, the equation estimates a prime rate of $-2.73 + 1.32 (12.50) = 13.77$.

There are certain points that are important to understand in regression analysis. As we said previously, the regression line is analogous to the mean. Therefore, for a given value of X, the estimating equation predicts an *average* value of Y, *not* an *exact* value. In addition, we are restricted to values of X and Y that lie within the range of the sample data. That is, the same relation-

Substituting the values from Table 12.2,

$$b = \frac{18(2,532.3967) - (204.26)(220.56)}{18(2,340.2884) - (204.26)^2}$$

$$= \frac{45,583.1406 - 45,051.5856}{42,125.1912 - 41,722.1476}$$

$$= \frac{531.5550}{403.0436}$$

$$= 1.32$$

$$a = \frac{220.56 - (1.32)(204.26)}{18}$$

$$= \frac{-49.0632}{18}$$

$$= -2.73$$

ship may not hold if we attempt to predict values that lie outside this range. Moreover, the Y-intercept frequently has little practical meaning in many business and economic applications since $X = 0$ may simply never occur, e.g., prime rate equal to zero.

Scatter diagram showing the regression line derived from the estimating equation ($Y_c = -2.73 + 1.32X$). Also shown are the means of the X and Y variables. Note that the regression line passes through the intersection of the two means. Note also that the Y-intercept ($a = -2.73$) is well outside of the range of values shown in this diagram. Here is one of many instances in which a has no real practical significance.

Figure 12.5

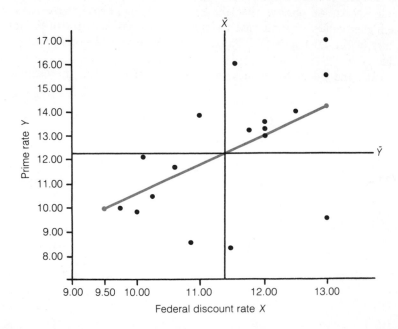

To construct the best fitting regression line to these data, we take two extreme values of X and find Y_c.

Thus, for $X = 9.50$, $Y_c = 9.81$ and for $X = 13.00$, $Y_c = 14.43$. We locate these two points and join them with a straight line. This line represents the regression line. Figure 12.5 shows this regression line. Note that the regression line passes through the intersection of $\overline{X}$ and $\overline{Y}$. This is one of the properties of the regression line.

Example A: Negative Slope

direct relationship

If a direct relationship exists between two variables, then increases (higher values) of one are associated with higher values of the other. This is represented by a positive value of b (the slope of the regression line).

inverse relationship

When high values of one variable are associated with low values of the other variable, and vice versa. This is represented by a negative value of b.

As we indicated, the value of b represents the slope of the regression line. If b is a positive number then the relationship between the two variables is *direct*. That is, as X increases, Y increases. However, if the slope is negative, then the relationship is *inverse*. Let us look at an example of an inverse relationship.

Suppose you are thinking of investing in bonds. One measure of bond activity is the daily average of 20 Dow Jones bonds. You would like to find a regression equation that will estimate values of the DJ bonds based on knowing the values of another variable.

Table 12.3 presents 15 values of the prime rate (X) and the corresponding value of the DJ bonds randomly selected over a four-month period. Also shown are ΣX, ΣY, ΣX^2, ΣY^2, and ΣXY.

First, let us construct a scatter diagram so that we can determine whether a linear relationship appears reasonable. Figure 12.6, which presents this relationship, tells us several things. First, it certainly appears that the relationship between these two variables is consistent with a linear model. Next, we observe that the relationship appears inverse—that is, the higher values of prime rate (X) appear to be associated with the lower values of the DJ bonds (Y). Thus, we would expect to obtain a negative value of b.

Using the calculated values in Table 12.3, let us find the estimating equation.

$$b = \frac{n(\Sigma XY) - (\Sigma X)(\Sigma Y)}{n(\Sigma X^2) - (\Sigma Y)^2}$$

$$= \frac{15\,(13{,}702.70625) - (205)(1008.61)}{15\,(2{,}884.28125) - (205)^2}$$

$$= -0.99$$

$$a = \frac{\Sigma Y - b((\Sigma X)}{n}$$

$$= \frac{1{,}008.61 - (-0.99)(205)}{15}$$

$$= 80.77$$

Therefore, the resulting regression equation is $Y_c = 80.77 - 0.99X$.

This equation tells us that as the prime rate increases, the value of DJ bonds decreases. Thus, if you were thinking of investing in bonds, you probably would hesitate if increases in the prime rate were expected.

Table 12.3
Fifteen randomly selected observations of the prime rate (X) and the DJ bonds average (Y) with the corresponding calculations

Observation	Prime rate X	DJ Bonds Y
1	10.875%	71.50
2	11.000	70.59
3	11.125	69.21
4	11.500	68.82
5	12.000	69.24
6	12.250	68.14
7	12.500	67.80
8	13.750	67.03
9	13.750	67.42
10	14.000	66.94
11	14.500	65.80
12	15.500	65.14
13	16.250	64.04
14	17.000	63.98
15	19.000	62.96
	$\Sigma X = 205.000$	$\Sigma Y = 1{,}008.61$

$$\Sigma X^2 = 2{,}884.28125$$
$$\Sigma Y^2 = 67{,}906.3159$$
$$\Sigma XY = 13{,}702.70625$$

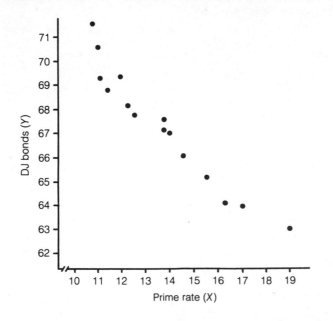

Figure 12.6
Scatter diagram showing relationship between prime rate and DJ bonds for 15 randomly selected days over a four-month period

12.2.5 Assumptions underlying linear regression

conditional distributions

The distribution of Y-values
for any given value of X.

We have noted that, for any given value of X, there is a distribution of different possible values of Y. These are known as *conditional distributions* since the value of the random variable, Y, depends on or is conditional upon its corresponding X value. The scatter or variability within each conditional distribution represents random error. It is assumed that the standard deviations of each conditional distribution are equal. In other words, there are equal degrees of dispersion or error in each conditional distribution. This condition is known as *homoscedasticity.* Moreover, the sums of the error terms are assumed to equal zero and to be independent of one another.

homoscedasticity

The assumption that
all of the conditional
Y-distributions have the
same standard deviation.

Regression analysis also assumes that the values of the Y variable are independent of one another. In practice, this assumption may not always be valid since there are times when some extraneous variable, operating over a period of time, may exert a similar influence on several different Y values. Thus, if we are measuring regression of price-earnings ratios (X) on stock prices (Y), a general downturn in the economy may depress stock prices over an extended period of time. Thus, at least some of the stock prices might not be independent since they may be affected by the external economic climate.

Finally, the assumption of normality is not required to derive the least squares line. However, the normality assumption is made in order to obtain interval estimates and to test hypotheses (see Figure 12.7). Thus, for any given value of X, we assume a normal distribution of Y-values. The mean

Figure 12.7

To test hypotheses, we assume that, for any given value of X, there is a conditional distribution of Y values that is normally distributed. The mean of each of these distributions is $\mu_{Y \cdot X}$ for the corresponding X value. Further, we assume that all the conditional Y distributions have the same standard deviation.

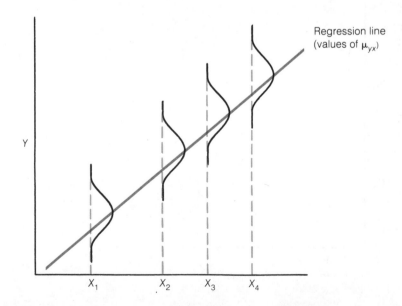

of each distribution ($\mu_{Y \cdot X} = A + BX$) is estimated by $Y_c = a + bX$. The standard deviation of each distribution is called the *standard error of estimate*. Since one of the assumptions we make is that all these conditional Y-distributions have the same standard deviation, there is only one standard error of estimate. It should be noted that the normality assumption can often be justified by the central limit theorem, i.e., if the error term represents a large number of independent variables, then e is approximately normally distributed regardless of the underlying distributions of the individual variables.

standard error of estimate

The standard deviation of each of the conditional Y-distributions; a measure of the variability, or scatter, of the observed values around the regression line.

EXERCISES

12.2 Find and plot the regression line for the data presented in Example A.

12.3 During 20 monthly periods, the following were the amounts spent on new construction (in $ millions) and the average prime rate for the same month.

Plot a scatter diagram of the relation between new construction (X) and prime rate (Y). If a linear model appears appropriate, find the regression equation.

New construction ($ millions)	Prime rate
$14.0	10.29%
13.4	10.01
15.8	9.94
17.1	9.90
19.3	9.98
20.7	9.79
21.5	9.99
22.3	10.62
22.2	11.70
22.5	13.44
20.9	13.53
18.9	13.31
16.7	13.15
15.8	14.01
17.0	17.10
17.9	15.63
18.9	9.60
19.7	8.31
20.0	8.58
20.3	9.85

12.4 The table below presents data on the importance of various plant location factors for British and German manufacturing investors in the United States. Assuming we wish to use British values (X) to predict German values (Y):

  a. Construct a scatter diagram of these data.
  b. Obtain an eyeball estimate of the regression line.
  c. Calculate both a and b.
  d. Draw the regression line from the equation for a straight line.
  e. Find the regression equation and find Y_c if $X = 1.5$; 2.0; 3.0.

Importance of plant location factors for British and German manufacturing investors in the United States*

Factors	Average importance	
	U.K. X	Germany X
Labor attitudes	3.71	3.94
Availability of transportation services	3.68	3.51
Ample space for future expansion	3.63	3.89
Availability of suitable plant sites	3.61	3.62
Availability of utilities	3.56	3.30
Nearness to markets within the U.S.	3.54	3.74
Attitudes of local citizens	3.46	3.57
Labor laws	3.27	3.55
Availability of managerial and technical personnel	3.22	2.96
Availability of skilled labor	3.17	3.45
Salary and wage rates	3.17	3.45
Cost of transportation services	3.10	2.92
Education facilities	3.07	3.13
Cost of construction	3.05	3.45
Cost of suitable land	3.02	3.53
Availability of unskilled labor	3.02	2.76
Cost of utilities	3.00	3.17
State tax rates	2.93	3.08
Local tax rates	2.90	2.91
Police and fire protection	2.90	3.06
Attitudes of government officials	2.85	3.00
Proximity to suppliers	2.76	2.83
Housing facilities	2.68	2.98
Facilities for importing and exporting	2.63	2.79
Availability of local capital	2.29	2.17
Climate	2.24	2.64
Cost of local capital	2.24	2.26
Proximity to raw material sources	2.12	2.34
Government incentives	2.07	2.34
Nearness to home operation	1.73	1.68
Proximity to export markets (outside the U.S.)	1.42	1.96
Nearness to operations in a third country	1.17	1.30

* The sample includes 41 British firms and 53 German firms.
Source: Hsin-Min Tong, ''An Analysis of Plant Location Decisions at English and German Manufacturing Investors in the United States,'' AIDS, 1980.

12.5 Over the course of the past year, RLC Sales Associates has run advertisements in the local newspaper to recruit salespeople, trying different size ads and recording the number of responses each ad produced. Shown below is a randomly selected sample showing the number of lines in the ad and the corresponding number of responses to that ad.

    a. Construct a scatter diagram.

    b. If a linear model seems appropriate, find the regression equation and plot the regression line.

    c. How many responses would you estimate they would obtain to ads with 7, 11, or 15 lines?

Number of lines	Number of responses
3	3
3	7
4	3
5	5
6	7
8	7
8	16
10	10
10	12
12	10
12	8
14	14
14	16
16	14
125	132
X	Y

12.3 THE STANDARD ERROR OF ESTIMATE

When we use the regression equation to estimate a value of Y for any given X value, we are naturally concerned with the reliability of our estimate. The more scatter or dispersion around the regression line, the less reliable is our estimate. We can estimate the amount of scatter in the population from the dispersion of the sample observations around the estimated regression line. This measure is known as the standard error of estimate and symbolized $s_{Y \cdot X}$. The following formula defines the standard error of estimate:

$$s_{Y \cdot X} = \sqrt{\frac{\Sigma (Y - Y_c)^2}{n - 2}}$$

where

$Y =$ Observed values of the dependent variable
$Y_c =$ Predicted values of the dependent variable estimated from the estimating regression equation
$n =$ Number of observations.

You may recognize the similarity between this formula and Formula (3.5), the formula for the standard deviation. However, in this case, we divide the sum of the squared deviations by $(n - 2)$. This is because we lose two degrees of freedom when we use sample data to estimate A and B in the regression equation. Note also that the deviations we are interested in are those around the regression line (i.e., $Y - Y_c$).

For computational purposes, we do not usually use the defining formula. Rather, we use the following formula, which has been algebraically derived from the definition:

$$s_{Y \cdot X} = \sqrt{\frac{\Sigma Y^2 - a \Sigma Y - b \Sigma XY}{n - 2}} \qquad (12.4)$$

The standard error of estimate may be interpreted in the same way as the standard deviation. The greater the dispersion, the larger the standard error of estimate.

As stated in the previous section, the computation and interpretation of the standard error of estimate are based on the assumption that the conditional Y-distributions have equal standard deviations (homoscedasticity). The standard error of estimate, $s_{Y \cdot X}$, obtained from sample data is an estimate of the standard deviation, $(\sigma_{Y \cdot X})$, around the true, but unknown, regression line.

Table 12.4 presents the data from Exercise 12.5 with all the calculations necessary to obtain the standard error of estimate.

Table 12.4
Calculation of $s_{Y \cdot X}$ for relationship between number of lines in ad *(X)* and number of responses *(Y)*

Observation	X	Y
1	3	3
2	3	7
3	4	3
4	5	5
5	6	7
6	8	7
7	8	16
8	10	10
9	10	12
10	12	10
11	12	8
12	14	14
13	14	16
14	16	14
	$\Sigma X = 125$	$\Sigma Y = 132$

$$\Sigma X^2 = 1359 \qquad \Sigma Y^2 = 1502$$

$$\bar{X} = 8.93 \qquad \bar{Y} = 9.43$$

$$\Sigma XY = 1373$$

$$b = 0.80; \ a = 2.29$$

$$Y_c = 2.29 + 0.80 \ X$$

$$s_{Y \cdot X} = \sqrt{\frac{1502 - 2.29(132) - 0.80(1373)}{12}}$$

$$= 2.91$$

12.4 CONFIDENCE INTERVALS IN ESTIMATING PARAMETERS

Recall that the linear regression equation provides point estimates of $E(Y|X)$, i.e., the mean value of Y for a particular value of X in the sample data. The obtained value of Y_c is *not* a point estimate for $\mu_{Y \cdot X}$ (a specific, individual value in the population model). In this section, we consider two types of interval estimators of the parameters:

1. The interval estimator of the mean value of Y for a given value of X. Since this estimator leads to an interval presumed to bracket the population mean for a given value of X, it is termed a confidence interval.

2. The interval estimate for a specific value of Y for a given X. This interval is customarily referred to as the *predictive interval* of Y since it predicts an individual value rather than a mean.

For example, in Table 12.5, we used the number of lines in an ad *(X)* to estimate the number of responses *(Y)*. If we are interested in predicting the *average* number of responses for a six-line ad, we are concerned with the *conditional mean*. In other words, we want to estimate the mean of the conditional Y-distribution for $X = 6$. If, however, our interest lies in predicting the number of responses for an *individual* six-line ad, we are concerned with a *single value* of Y rather than a conditional average value.

When we wish to estimate the parameters of the population, we should recall that the obtained values of a and b will generally differ from the population values of A and B. However, repeated random samples from the population will yield a distribution of sample values that are scattered about the parameters and have A and B as their respective mean values. Thus, in a randomly selected sample, a provides an unbiased estimate of A and b provides an unbiased estimate of B. Whether we are predicting the conditional mean or the single value of Y, the regression equation will yield the same point estimate. The difference lies in the interval estimates. Predictions for single or individual values cannot be as precise as estimates for mean values. Since the width of the interval indicates the precision of the estimate, we would expect wider confidence intervals when predicting individual values of Y. Thus, the standard error of estimate must be modified to take into account the fact that we are trying to estimate population values from sample data and that the predictive interval is wider than the confidence interval.

Let us look at the two types of interval estimates.

predictive interval

Interval estimate for a specific value of Y for a given X.

conditional mean

The mean of the distribution of Y-values for any given X.

12.4.1 Interval estimation of a conditional mean

Let us use the data in Table 12.5 to illustrate the calculation of an interval estimate for a conditional mean. Suppose we wanted to estimate the mean number of responses to a six-line ad. Since the estimated regression equation is $Y_c = 2.29 + 0.80\ (X)$, substituting $X = 6$, we find $Y_c = 7.09$.

The confidence interval for the regression line, also called the confidence interval estimate for the conditional mean is

$$Y_c \pm t s_{Y_c} \qquad (12.5)$$

where

s_{Y_c} is the estimated standard error of the conditional mean, and t is the t-value for $(n-2)$ df for the desired confidence interval.

As we mentioned earlier, we lose two degrees of freedom because A and B are estimated from the sample data.

The estimated standard error of the conditional mean may be obtained through the use of the following formula:

$$s_{Y_c} = s_{Y \cdot X} \sqrt{\frac{1}{n} + \frac{(X - \bar{X})^2}{\Sigma X^2 - \frac{(\Sigma X)^2}{n}}} \qquad (12.6)$$

Let us continue our illustration of obtaining an interval estimate of the conditional mean number of responses to a six-line ad. We shall find the 95 percent confidence interval for the conditional mean.

$$s_{Y_c} = 2.91 \sqrt{\frac{1}{14} + \frac{(6 - 8.93)^2}{1359 - \frac{(125)^2}{14}}}$$

$$= 0.95.$$

Thus, the 95 percent confidence interval $Y_c \pm t s_{Y_c}$ is

$$7.09 \pm (2.179)(0.95) = 5.02 \text{ to } 9.16$$

where the t-value for. df $= n - 2 = 12$ for a 0.95 confidence coefficient was obtained from Table VII. Therefore, the 95 percent confidence limits are approximately five to nine mean responses to a six-line ad.

12.4.2 Predictive interval for individual values

As we indicated earlier, we would obtain the same predicted Y value for a given X whether we were predicting the conditional mean or an individual Y value. That is, the best point estimate for a given X is obtained from the regression equation $Y_c = a + bX$.

However, we would not expect the same precision when predicting an individual value of Y for a given X as when we estimate a conditional mean. In a sense, when dealing with individual values, we are involved with the dispersion of variability in the population. On the other hand, mean values are analogous to a sampling distribution of means from that population. We would always expect a smaller standard deviation for a distribution of means than for a distribution of individual scores (see Chapter 6).

Thus, the predictive interval estimate for an individual score, sometimes called an individual forecast, is

$$Y_c \pm t s_{Y_{\text{ind}}},\qquad (12.7)$$

where $s_{Y_{\text{ind}}}$ is the estimated standard error of the individual scores (sometimes called the standard error of forecast) and t is the value for $(n-2)$ df for the desired predictive interval.

A convenient computation formula for the standard error of the individual scores is

$$s_{Y_{\text{ind}}} = s_{Y \cdot X} \sqrt{1 + \frac{1}{n} + \frac{(X - \bar{X})^2}{\Sigma X^2 - \frac{(\Sigma X)^2}{n}}}\qquad (12.8)$$

Note the similarity between Formulas (12.8) and (12.7). The only difference is the addition of 1 to the expression under the square root sign. We previously noted that we are unable to predict an individual value of Y as precisely as we can estimate a conditional mean of Y. This is due to the fact that a predicted individual value has two components: the sampling error when we use Y_c to estimate $\mu_{Y \cdot X}$ and the variability in the conditional probability distributions of Y values. These are shown below:

$$s_{Y_{\text{ind}}}^2 = s_{Y \cdot X}^2 \left(\frac{1}{n} + \frac{(X - \bar{X})^2}{\Sigma (X - \bar{X})^2} \right) + s_{Y \cdot X}^2$$

where the first term on the right estimates the sampling error in estimating a single mean, and the second term estimates the conditional variance of individual Y observations.

Factoring and taking the square root of $s_{Y_{\text{ind}}}^2$, we obtain

$$s_{Y_{\text{ind}}} = s_{Y \cdot X} \sqrt{\frac{1}{n} + \frac{(X - \bar{X})^2}{\Sigma (X - \bar{X})^2} + 1}$$

or

$$= s_{Y \cdot X} \sqrt{1 + \frac{1}{n} + \frac{(X - \bar{X})^2}{\Sigma (X - \bar{X})^2}}$$

Returning to our ad-response illustration, we find

$$s_{Y_{\text{ind}}} = 2.91 \sqrt{1 + \frac{1}{14} + \frac{(6 - 8.93)^2}{1359 - \frac{(12.5)^2}{14}}}$$

$$= 3.06$$

Thus, the 95 percent predictive interval for the individual forecast of responses to a six-line ad is $7.09 \pm (2.179)(3.06)$, or 0.42 to 13.76.

Note, as we indicated earlier, the limits for the individual forecast are wider than those obtained for the conditional mean.

If we were to calculate the confidence and predictive intervals for all possible values of X, we would obtain two confidence bands around the regression line, as shown in Figure 12.8.

There are a number of factors that affect the width of the estimated intervals for both the conditional mean and the individual forecast.

First, the larger the sample size *(n)*, the smaller will be the standard errors. Thus, the larger the sample size, the narrower will be the widths of the intervals.

Second, the more dispersion around the regression line, i.e., larger $s_{Y \cdot X}$,

Figure 12.8 Confidence bands for conditional means and individual forecasts. Note that the farther from the mean, the wider the interval. Note also that the intervals for conditional means are narrower than for individual scores.

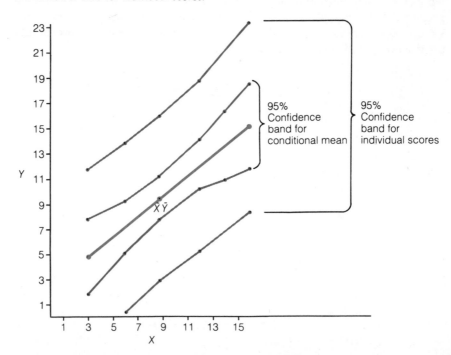

the wider will be the intervals. Thus, the more variable the data, the less precise the predictions.

Finally, X scores closest to the mean will yield narrower intervals than those far from the mean. In other words, very low or high values of X (as compared to the mean of X) will yield wider intervals, or less precision in prediction.

EXERCISES

12.6 The manager of the planning department of a large retail store chain with divisions all over the country decides to systematically study the relationship between advertising dollars and sales volume. He wants to be able to predict sales volume from a predetermined advertising budget. However, on reviewing past records, he finds that the amount budgeted for advertising is often determined by the previous month's sales volume. Therefore, he decides to preset the advertising budget for the entire year in advance. He randomly assigns amounts of advertising money monthly from $2,000 to $20,000 to the various divisions. He then records the sales volume associated with each specific amount. Finally, he randomly selects 12 observations and obtains the following data:

Observation	X Amount spent on advertising ($000)	Y Sales (100 units)
1	17	10
2	8	4
3	6	6
4	10	7
5	2	6
6	13	13
7	15	10
8	3	14
9	18	15
10	5	3
11	11	9
12	20	13

   a. Plot a scatter diagram of these data.

   b. Find the regression equation.

   c. Find the 95 percent confidence interval for the mean sales volume for the company as a whole for a $10,000 advertising budget.

   d. Find the 95 percent predictive interval for sales volume for an individual division for a $10,000 advertising budget.

   e. Compare and explain the difference in interval estimates between *(c)* and *(d)*.

12.7 Referring to Exercise 12.6,

   a. Find the 95 percent confidence interval for the company as a whole if the advertising budget is $15,000.

   b. Find the 95 percent predictive interval for an individual division with a $15,000 advertising budget.

     c. Compare and explain the difference in the interval estimates for Exercise 12.6 *(c)* versus 12.7 *(a)* and for Exercise 12.6 *(d)* versus 12.7 *(b).*

12.8 The personnel manager of a large industrial concern notes that many employees do not work a full eight-hour day. Extended lunch hours, coffee breaks, tardiness in the morning, and leaving a little bit earlier all tend to reduce the total work week. She randomly selects 10 employees who have been with the company a varying number of years and records the total number of hours worked during a "40-hour" week. Find the 95 percent interval for the mean number of hours worked by

    a. Employees who have been with the company for 10 years.
    b. A specific employee who has been with the company 10 years.
    c. Employees who have been with the company for six years.
    d. A specific employee who has been with the company for six years.

Observation	Number of years with company	Number of hours worked per week
1	1	39.5
2	7	36
3	5	37
4	3	40
5	0.5	32
6	15	35
7	12	31
8	13	30
9	4	38
10	2	37.5

12.5 CORRELATIONAL ANALYSIS

So far we have concentrated on regression analysis, which provides the basis for predicting values of one variable from known values of another variable. We have developed a mathematical equation that describes this relationship. Thus, we have used regression analysis as a tool for describing the nature of the relationship between two variables in terms of an estimating equation. We now turn our sights to correlational analysis, in which we concentrate on the degree or strength of the relationship between two variables.

12.5.1 Coefficient of determination

We have seen that through the use of regression analysis, we are able to estimate (or predict) values of Y from known values of X. In the absence of any information on X, our best prediction for Y would be the mean of Y $(\overline{Y})$. Thus, one legitimate question that may be raised is, How can we determine whether our predictions based on the regression equation are superior to predictions based on $\overline{Y}$? If predictions based on the regression equation are no better than those based on the mean value of Y, then the regression analysis is of no value. Therefore, we need a way to measure the amount of variability in Y that can be attributed to, or explained by, the relationship between X and Y. Fortunately, such a measure exists. The *coefficient of determination (r²)* measures the proportion of variation that has been accounted for or explained by the relationship between X and Y. In fact, we define the coefficient of determination in terms of the following proportion:

coefficient of determination (r²)

The ratio of explained variation to total variation.

$$r^2 = \frac{\text{Explained variation}}{\text{Total variation}}$$

Let us return to the example presented in Section 12.4. Table 12.5 reproduces the data from Table 12.4 with one additional column, Y_c.

X	Y	Y_c*
3	3	4.69
3	7	4.69
4	3	5.49
5	5	6.29
6	7	7.09
8	7	8.69
8	16	8.69
10	10	10.29
10	12	10.29
12	10	11.89
12	8	11.89
14	14	13.49
14	16	13.49
16	14	15.09
$\Sigma X = 125$	$\Sigma Y = 132$	

Table 12.5
Relationship between number of lines in ad *(X)* and number responses *(Y)*

* Calculated through use of the obtained regression equation $Y_c = 2.29 + 0.80\ X$.

$$\Sigma X^2 = 1359 \qquad \Sigma Y^2 = 1502$$
$$\bar{X} = 8.93 \qquad \bar{Y} = 9.43$$
$$\Sigma XY = 1373$$

Figure 12.9 shows three scatter diagrams based on the data in Table 12.5. Each diagram shows the regression line and the mean of Y. Let us examine each diagram separately as each represents a different source of variation.

a. Variation of observed scores around the sample mean $(Y - \bar{Y})$. The sum of squares of these deviations represents the *total variation* $\Sigma(Y - \bar{Y})^2$.

b. Variation of predicted scores around the sample mean $(Y_c - \bar{Y})$. The sum of squares of these deviations represents the *explained variation* $\Sigma(Y_c - \bar{Y})^2$. If the relationship between X and Y were perfect—that is, all the variation in Y could be accounted for by the relationship between X and Y—then all the obtained Y scores would fall on the regression line. If this occurred then, for each X, Y would equal Y_c. Thus, $\Sigma(Y_c - \bar{Y})^2$ would equal $\Sigma(Y - \bar{Y})^2$ or the total variation would be completely explained by the relationship between X and Y.

c. Variation of observed scores from the regression line $(Y - Y_c)$. The

total variation

Variation of observed scores from the sample mean $\Sigma(Y - \bar{Y})^2$.

explained variation

Variation of predicted scores from the sample mean $\Sigma(Y_c - \bar{Y})^2$.

404

Figure 12.9 Scatter diagram of paired scores on two variables (see Table 12.5), regression line, and the mean of the Y-scores $(\bar{Y})$.
A. Deviations of scores from the mean of Y $(Y - \bar{Y})$: total variation.
B. Deviations of predicted scores from the mean of Y $(Y_c - \bar{Y})$: explained variation.
C. Deviations of obtained scores from the regression line $(Y - Y_c)$: unexplained variation.

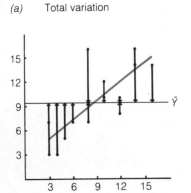

(a)   Total variation

Total variation
Deviation of Y scores from $\bar{Y}$

$\Sigma(Y - \bar{Y}) = 0$

$\Sigma(Y - \bar{Y})^2 = 257.05$

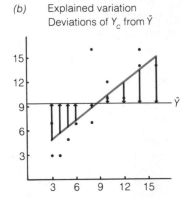

(b)   Explained variation
Deviations of Y_c from $\bar{Y}$

$\Sigma(Y_c - \bar{Y}) = 0$

$\Sigma(Y_c - \bar{Y})^2 = 155.73$

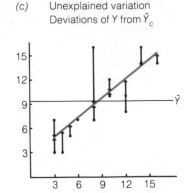

(c)   Unexplained variation
Deviations of Y from $\bar{Y}_c$

$\Sigma(Y - Y_c) = 0$

$\Sigma(Y - Y_c)^2 = 101.32$

unexplained variation

Variation of observed scores from the regression line $\Sigma(Y - Y_c)^2$.

sum of squares of these deviations represents the *unexplained variation,* $(Y - Y_c)^2$. This is the difference between the obtained scores and the predicted scores. If the relationship were perfect, all the obtained scores would fall on the regression line and $\Sigma(Y - Y_c)^2$ would be zero. However, when the relationship is not perfect, many of the scores do not fall on the regression line. Thus, the deviations of the scores around the regression line is the variation that is *not* explained in terms of the relationship between X and Y.

It can be demonstrated mathematically that the total variation is composed of two parts: explained and unexplained variation. In other words,

$$\Sigma(Y - \bar{Y})^2 \quad = \quad \Sigma(Y_c - \bar{Y})^2 \quad + \quad \Sigma(Y - Y_c)^2$$

| Total variation | Explained variation | Unexplained variation |

Figure 12.9 shows these calculations. Note that the sum of the explained variation (155.73) plus the sum of the unexplained variation (101.32) is equal to the total variation (257.05).

Recall that we defined the coefficient of determination as the proportion of explained variation to total variation. Thus,

$$r^2 = \frac{\text{Explained variation}}{\text{Total variation}} = \frac{\Sigma(Y_c - \bar{Y})^2}{\Sigma(Y - \bar{Y})^2}$$

$$= \frac{155.73}{257.05} = 0.61.$$

Therefore, approximately 61 percent of the variation in Y can be accounted for or explained by variations in X. In terms of this example, it means that approximately 61 percent of the variability in the number of responses to an ad can be accounted for by the number of lines in the ad.

Or, looked at another way, 39 percent of the variation in Y is *not* explained by the relationship between X and Y. That is, since r^2 represents the proportion of variation explained, $(1 - r^2)$ represents the proportion that is not explained. We call this the *coefficient of nondetermination* and use k^2 to symbolize this concept. Thus, k^2 represents the proportion of variation in Y that must be explained by variables other than X.

coefficient of nondetermination (k^2)
Proportion of variation not accounted for in terms of the relationship between X and Y.

As with many of the other statistics we have encountered so far, the calculation of r^2 can be expressed in terms of a shortcut formula that utilizes quantities already calculated:

$$r^2 = \frac{a \Sigma Y + b \Sigma XY - n\overline{Y}^2}{\Sigma Y^2 - n\overline{Y}^2} \qquad (12.9)$$

Substituting the values from Table 12.5,

$$r^2 = \frac{2.29(132) + 0.80(1373) - 14(9.43)^2}{1502 - 14(9.43)^2}$$

$$= \frac{155.73}{257.05}$$

$$= 0.61$$

12.5.2 Coefficient of correlation

The most commonly used measure of the degree or strength of linear relationship between two variables is the *Pearson r correlation coefficient.* This is merely the square root of the coefficient of determination; i.e., $r = \sqrt{r^2}$.

Pearson r correlation coefficient
A number that summarizes the degree of relationship between two variables; the square root of the coefficient of determination.

Pearson r has the following properties:

1. Two sets of measurements are obtained on the same individuals, objects, or events that are matched in some way.

2. The values of r vary between -1.00 through $+1.00$. A value of r equal to $+1.00$ or -1.00 indicates a perfect linear relationship. A value of r equal to zero indicates no linear relationship. Therefore, values of r close to zero indicate weak linear relationships, whereas values of r approaching $+1.00$ or -1.00 indicate a strong degree of linear relationship between the two variables.

3. The algebraic sign of r is the same as that of the slope of the regression line, b. Thus, the positive square root of r^2 is taken when $b > 0$ (a positive or direct linear relationship) and the negative square root of r^2 is taken when $b < 0$ (a negative linear relationship).

4. The value of r does not depend on either the units of measurement for X and Y or which variable is selected as the independent variable. Thus,

if we're correlating the retail price of an item with the availability of that item, price is one variable and quantity of the item is a second variable. Either can be considered an independent variable.

Figure 12.10 presents scatter diagrams that illustrate various values of the correlation coefficient.

Although we indicated that $r = \sqrt{r^2}$, it is not always convenient to calculate r in this way. Formula (12.9) for r^2 requires that we calculate values of a and b. There are alternative computational formulas for r that do not require

Figure 12.10
Various scatter diagrams and the corresponding values of the correlation coefficient

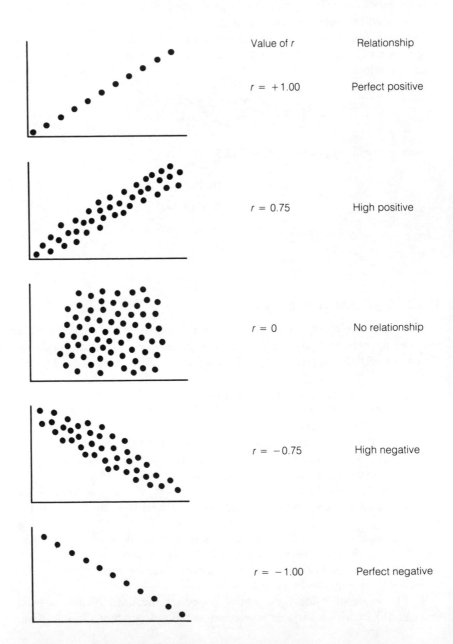

Value of r	Relationship
$r = +1.00$	Perfect positive
$r = 0.75$	High positive
$r = 0$	No relationship
$r = -0.75$	High negative
$r = -1.00$	Perfect negative

calculation of the regression coefficients. We shall use the following formula to calculate r directly from the raw data:

$$r = \frac{\Sigma XY - \dfrac{(\Sigma X)(\Sigma Y)}{n}}{\sqrt{\left[\Sigma X^2 - \dfrac{(\Sigma X)^2}{n}\right]\left[\Sigma Y^2 - \dfrac{(\Sigma Y)^2}{n}\right]}} \qquad (12.10)$$

Let us use the data in Table 12.5 to calculate r. (Recall that we found $r^2 = 0.61$ for these data. Thus, $r = \sqrt{0.61} = 0.78$.) Substituting into Formula (12.10):

$$r = \frac{1373 - \dfrac{(125)(132)}{14}}{\sqrt{\left[1359 - \dfrac{(125)^2}{14}\right]\left[1502 - \dfrac{(132)^2}{14}\right]}}$$

$$= \frac{194.43}{250.07}$$

$$= 0.78$$

EXERCISES

12.9 A 1975 study found that employee preferences for certain rewards differ according to variables such as age, marital status, or years of service. The following data present employee mean preferences for eight different options (e.g., extra vacation, pay increase, etc.) for two types of employees, clerical and operating. Is there a relationship between these two types of employees in terms of their preferences for certain types of rewards?

| Reward | Job title | |
option	Clerical	Operating
1	5.00	4.81
2	4.70	4.37
3	4.15	3.97
4	2.70	3.86
5	3.09	3.38
6	3.48	2.93
7	2.48	3.03
8	1.63	1.17

Source: Adapted from J. Brad Chapman and Robert Otteman, "Employee Preference for Various Compensation and Fringe Benefit Options," *The Personnel Administrator*, November 1975, p. 34. Reprinted from the November 1975 issue of *Personnel Administrator*, 30 Park Drive, Berea, OH 44017, $26 per year.

12.10   A stock market analyst wants to determine whether there is any relationship between the volume of shares traded on the NYSE and changes in the Dow Jones Industrials. He randomly selects 40 trading days and records the number of shares traded *(X)* and the corresponding change in the DJI, both the direction and amount *(Y)*. From the data he calculates the following quantities: $\Sigma X = 9{,}985.84$; $\Sigma Y = 61.01$; $\Sigma X^2 = 103{,}885.46$; $\Sigma Y^2 = 3759.30$; $\Sigma XY = 3{,}945.92$; $n = 40$.

    a.   Calculate the correlation coefficient for these data.

    b.   What proportion of variation in the DJI can be explained by variations in volume?

12.11   Refer to Exercise 12.10. Not satisfied with the results obtained in Exercise 12.10, the analyst decides to ignore the direction of change in the DJI. Thus, for the same 40 randomly selected days, he uses the absolute value of the change and obtains the following results: $\Sigma X = 1{,}985.84$; $\Sigma Y = 326.03$; $\Sigma X^2 = 103{,}885.46$; $\Sigma Y^2 = 3759.30$; $\Sigma XY = 17{,}032.07$; $n = 40$.

    a.   Find the coefficient of correlation.

    b.   Find the coefficient of determination.

12.6   TESTS OF SIGNIFICANCE FOR CORRELATION AND REGRESSION

Thus far we have dealt with measurements derived from samples. However, our interest ordinarily extends beyond the obtained samples. We wish to make inferences from these results to the population parameters. Therefore, before we can make inferences from relationships obtained on samples, we need to apply tests of significance of the relationship. We shall use the hypothesis testing techniques from Chapter 8 to determine the significance of obtained correlational and regression relationships.

12.6.1   Inferences about the population correlation coefficient

Suppose we take a random sample of *n* pairs of observations and calculate the sample correlation coefficient as demonstrated in Section 12.5.2. We wish to test the hypothesis that the obtained coefficient, *r*, comes from a population in which the true relationship (ρ) is zero. Stated in terms of hypotheses: The null hypothesis (H_0): $\rho = 0$ versus the alternative hypothesis (H_1): $\rho \neq 0$, where ρ is the population correlation coefficient.

We use the *t*-ratio to test H_0: $\rho = 0$. However, the following assumptions underlie the use of this test statistic:

1. The relationship between the two variables is linear.
2. Both variables are random variables.
3. The condition of homoscedasticity is satisfied.
4. The conditional distributions of one variable given values of a second variable are normally distributed.

The standard error of *r* may be estimated by

$$s_r = \sqrt{\frac{(1 - r^2)}{(n - 2)}} \qquad (12.11)$$

and the test statistic for H_0: $\rho = 0$ is

$$t = \frac{r - \rho}{s_r} = \frac{r}{\sqrt{\frac{(1 - r^2)}{n - 2}}} \qquad (12.12)$$

where the test statistic, t, comes from the t-distribution with df $= n - 2$.

For example, suppose we wish to test the null hypothesis that $\rho = 0$ for our ad-response problem from Section 12.5.2. Recall that we obtained $r = 0.78$ for $n = 14$. Thus, substituting into Formula (12.12):

$$t = \frac{0.78}{\sqrt{\frac{(1 - 0.78^2)}{14 - 2}}} = 4.318$$

Referring to Table VII (Appendix C) for df $= 12$, we see that, for $\alpha = 0.01$, two-tailed test, the critical value of t is 3.055. Thus, since our obtained t exceeds this value, we may reject H_0 and assert that there is a positive relationship between size of ad and number of responses in the population from which our sample was drawn.

This procedure is valid only when testing hypotheses of $\rho = 0$. We cannot use the same procedures for testing population correlation coefficients other than zero.

The sample size is a key factor here. First, as n increases, the t-distribution approaches the form of the normal curve. Thus, for large sample sizes, we may use z as our test statistic and the normal distribution as the sampling distribution. Second, as n increases, lower values of a sample r will lead to rejection of the null hypothesis. This means that a sample r may be statistically, but not practically, significant. For example, suppose we obtained a sample r of 0.30 for a sufficiently large n. Although we may find this differs significantly from 0, the relationship between the two variables may not be particularly useful. Recall that r^2 represents the proportion of variability accounted for by the relationship between X and Y. Thus, an obtained r of 0.30 accounts for less than 10 percent ($r^2 = 0.09$) of the variability in Y.

12.6.2 Inferences about the slope of the regression line

If there is no relationship between two variables, the slope of the regression line equals zero. Figure 12.11 illustrates this graphically.

However, as we know, through chance or random sampling errors, it is possible to obtain sample data that appear to show a relationship between

Figure 12.11

A population in which no relationship exists between the two variables. The regression line has a slope of zero.

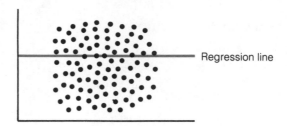

the variables. This can occur even when no relationship exists in the population.

Thus, we need to test the hypothesis that the true population slope is zero against the alternative that it has some value other than zero, or: H_0: $B = 0$; H_1: $B \neq 0$.

We may use the following as our test statistic:

$$t = \frac{b}{s_b}$$

where b is the sample value and

$$s_b = s_{Y \cdot X} \sqrt{\frac{1}{\Sigma X^2 - \frac{(\Sigma X)^2}{n}}} \qquad (12.13)$$

Thus,

$$t = \frac{b}{s_{Y \cdot X} \sqrt{\dfrac{1}{\Sigma X^2 - \dfrac{(\Sigma X)^2}{n}}}} \qquad (12.14)$$

where the sampling distribution is the
t-distribution with df $= n - 2$ for a given α-level.

Once again, using our ad-response data, the following values are reproduced from Table 12.4: $\Sigma X = 125$, $\Sigma X^2 = 1359$, $s_{Y \cdot X} = 2.91$, and $b = 0.80$ for $n = 14$. Thus,

$$t = \frac{0.80}{2.91 \sqrt{\dfrac{1}{1359 - \dfrac{(125)^2}{14}}}} = 4.285$$

As in the previous section, using $\alpha = 0.01$, two-tailed test for df $= 12$, the critical value of t is 3.055. Thus, we may reject the hypothesis that the slope is zero.

12.7 LIMITATIONS AND CAUTIONS

Regression and correlational analysis provide powerful tools for understanding and interpreting data. However, we should be aware of certain limitations so that we may interpret our results with caution.

First, our discussion of regression analysis in this chapter has been restricted to the two-variable linear regression model. Using this model, we assume that Y, the dependent variable, is a random variable and the values of Y are independent. This means that the value of Y on the first observation, for example, does not affect the Y value on the next observation. This assumption may not be valid in time-series data where values of Y may be related from one time period to the next. In addition, we assume that for any given X, there is a normal distribution of Y values that all have the same standard deviation. Our predictions for Y are restricted to values of X that lie within the range of the sample data.

If we do not have sufficient evidence to reject a hypothesis of zero slope, this does not necessarily mean that *no* relationship exists between the two variables. This is true for the significance of the correlation coefficient as well. We must remember that we are testing for *linear* relationships only. Thus, two variables may be related, but in a nonlinear way.

One other situation that may lead us to assume no relationship between two variables occurs when the range of values of one or both of the variables has been restricted. Figure 12.12 illustrates a situation in which there is a high positive relationship between two variables over the entire range of the X and Y variables. However, if a sample is taken over a restricted range, we may mistakenly conclude that no relationship exists.

On the other hand, we must be equally careful in interpreting statistically significant results. When a significant correlation is found, we are sometimes tempted to conclude that one variable has *caused* the effect in the other. Although, at times this assumption *may* be true, a significant relationship between two variables does not prove a cause-and-effect relationship. For example, suppose we found that two variables (price of meat and price of bathing suits) were related. That is, they behaved in a similar fashion: as one goes up or down, so does the other.

Given this information and a corresponding significant correlation, could we conclude that one *caused* the other, that is, do increases in meat prices cause the price of bathing suits to go up? No; the best we could do would be to measure the strength of the relationship. In this situation, we would probably attribute the rise and fall of prices to other variables such as the general state of the economy. In other words, a third variable (or variables) is producing variations in the variables we are studying. When the general state of the economy is down, we would expect both meat and shoe prices to be lower than during times of inflation.

Figure 12.12

Although a high relationship exists between these two variables over the entire range of their values, a sample taken over the restricted range would very likely result in a conclusion of no relationship.

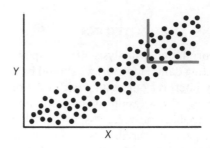

EXERCISES

12.12 Refer to Exercise 12.9. Is the obtained correlation coefficient significant at $\alpha = 0.01$?

12.13 Refer to Exercises 12.10 and 12.11. Are either of the obtained correlation coefficients significant at $\alpha = 0.05$? If so, can we conclude that the volume of shares traded caused changes in the DJI?

12.14 Refer to the data presented in Table 12.2. Test H_0: $B = 0$, using $\alpha = 0.05$.

SUMMARY

In this chapter we have looked at two-variable linear regression and correlation.

1. We have seen that the purpose of regression analysis is to find a mathematical equation that describes the nature of the relationship between two variables.

2. Correlational analysis is used to describe the strength of the relationship between two variables.

3. Both types of analyses involve related or matched observations. The scatter diagram is a useful graphic device to portray the relationship between these variables.

4. To find the straight line that best fits the set of points, we use the method of least squares. This involves finding the equation for the one straight line that makes the squared deviations around it minimal.

5. Using estimation formulas for b (the slope of the line) and a (the Y-intercept), we obtained the following estimating equation for the regression line: $Y_c = a + bX$.

6. If the slope of the line, b, is positive, we say the relationship between the two variables is direct. A negative slope reflects an inverse relationship. The regression line always passes through the intersection of the mean of X and the mean of Y.

7. To use the linear regression model we make certain assumptions. We

assume that the dependent variable, Y, is a random variable. Values of the independent variable, X, are fixed. For any given X, there is a distribution of Y values with a mean equal to Y_c. The standard deviation of all conditional distributions of Y is the same. We call this standard deviation the standard error of estimate. The assumption of normality is not required for deriving estimating equations. However, it is made when estimating intervals and testing hypotheses.

8. We developed two types of interval estimators of the parameters: the interval estimator of the mean value of Y for a given value of X (confidence interval) and the interval estimate for a specific value of Y for a given X (predictive interval). Each has its own associated standard error. The interval for an individual value will always be wider than the corresponding interval for a conditional mean. Moreover, values of X closer to the mean of X yield narrower intervals than those for values of X far from the mean.

9. In correlational analysis, we are primarily concerned with the *degree* of relationship between the two variables. Here we assume *both* X and Y are random variables.

10. The coefficient of determination, r^2, tells us the proportion of variation in Y that is explained, or accounted for, by variations in X.

11. The coefficient of correlation, r, tells us the strength of the relationship between X and Y. Values of r vary between -1.00 to $+1.00$. The closer to ± 1.00, the greater the relationship. Values of r close to zero indicate little or no relationship between the variables. The sign of r corresponds with the slope of the regression line. Thus, a positive r indicates a direct relationship; a negative r, an inverse relationship.

12. We used the procedures of hypothesis testing to determine the significance of the obtained correlation as well as the slope of the regression line.

13. Finally, we discussed certain cautions. Absence of a significant relationship means either the relationship is nonlinear, the range of data on the variables has been restricted, or there is no relationship in the population. When we find a significant relationship, we must exercise care in interpreting these results. A significant relationship does *not* imply a cause and effect relationship.

TERMS TO REMEMBER

dependent variable (378)

independent variable (378)

simple regression analysis (378)

multiple regression analysis (378)

correlational analysis (379)

least squares method (387)

regression line (387)

direct relationship (390)

inverse relationship (390)

conditional distributions (392)

homoscedasticity (392)

standard error of estimate (393)

predictive interval (397)

conditional mean (397)

coefficient of determination (r^2) (403)

total variation (403)

explained variation (403)

unexplained variation (404)

coefficient of nondetermination (k^2) (405)

Pearson r correlation coefficient (405)

EXERCISES

12.15 A professor at The Johns Hopkins School of Hygiene who has studied how the economy affects people claimed that the relationship between unemployment and suicide is so clear that "unemployment can be used as a constant for predicting the mortality rate."[1] The following data present the unemployment rate (as a percentage of the civilian labor force) and suicide rates (per 100,000) for 14 different years.

Year	Unemployment rate	Suicide rate
1950	5.3	11.3
1955	4.4	10.2
1960	5.5	10.6
1965	4.5	11.1
1968	3.6	10.7
1969	3.5	11.1
1970	4.9	11.5
1971	5.9	11.7
1972	5.6	12.0
1973	4.9	12.0
1974	5.6	12.1
1975	8.5	12.7
1976	7.7	12.5
1977	7.0	13.3

a. Construct a scatter diagram of the data.
b. Find the regression equation and plot the regression line.
c. Find the coefficient of determination.
d. Find the coefficient of correlation.
e. Is the obtained r significantly different from zero? Use $\alpha = 0.01$, two-tailed test.
f. Find the standard error of estimate.
g. Find the 95 percent confidence interval for the mean suicide rate when the unemployment rate equals 8.0.
h. Is the obtained slope significantly different from zero? Use $\alpha = 0.01$, two-tailed test.

12.16 Many economists believe that the money supply is a valid indicator of prime rates. That is, as the supply of money goes up, interest rates tend to go up accordingly. *M1-A* is a government index that measures the cash in circulation plus checking deposits at commercial banks. In an attempt to determine whether *M1-A* is, in fact, a predictor of prime rate, the authors randomly selected 15 months and for each of those months, recorded *M1-A* the average prime rate three months later. The following data show our findings:

---

[1] *Los Angeles Times,* December 13, 1980, p 1–A, p. 21.

M1-A ($ billions)	Prime rate (three months later)
$358.4	11.70%
364.0	13.44
361.9	13.53
365.4	13.31
368.2	13.15
370.6	14.01
379.2	17.10
375.6	15.63
365.5	9.60
366.3	8.31
370.9	8.58
362.2	9.85
370.1	12.15
375.6	13.86
377.6	15.94

a. Draw a scatter diagram of these data.
b. Find and plot the regression line.
c. Find the standard error of estimate.
d. Find the coefficient of determination.
e. Find the coefficient of correlation.
f. Test the hypothesis: H_0: $\rho = 0$, using $\alpha = 0.05$
g. Test the hypothesis: H_0: $B = 0$, using $\alpha = 0.05$
h. One economist suggests that the regression equation does not seem plausible. If *M1-A* equals $325 billion, we would have a negative prime rate. Explain the fallacy in the reasoning.

12.17 Shown below are the weights of 12 randomly selected automobiles with manual transmissions and their EPA rating of highway fuel economy.
a. Construct a scatter diagram.
b. Decide whether the data are consistent with a linear model; if so, calculate the regression equation and construct the regression line.
c. Find the standard error of estimate.
d. Test H_0: $B = 0$, using $\alpha = 0.05$.

X (weight, lbs.)	Y (MPG)
2,831	26
2,250	36
3,004	26
1,924	33
2,250	41
2,020	31
3,067	23
4,263	17
3,142	23
2,552	30
2,905	22
3,370	25

12.18 Tommy R. has a Chevrolet van that weighs 6,400 pounds. According to the results obtained in Exercise 12.17, he should expect to get -5.86 miles per gallon. How would you explain this?

12.19 Refer to Exercise 12.17. Find the 95 percent intervals for:
    a. The mean MPG for cars that weigh 2,500 pounds.
    b. Miles per gallon for Sharon's Datsun 710, which weighs 2,500 pounds.

12.20 Refer to Exercise 12.17.
    a. Calculate the coefficient of correlation.
    b. Test H_0: $\rho = 0$, using $\alpha = 0.05$.
    c. Find the coefficient of determination.

12.21 Many of the large mail-order houses hire mailers on a work-available basis. If permanent employees were hired, they would be underutilized during slack times. Conversely, during peak periods, many orders might go unfilled in a timely fashion. Lauren J., in charge of hiring mailers on a daily basis, conducts the following study. When the mail arrives each day, it is weighed; the next day, when opened, the total number of orders is tabulated. Lauren wishes to learn whether the weight of the mail is a predictor of number of orders. If found to be so, mailers can be contacted one day in advance, thereby reducing the disparity between demand and available help. From existing files, she randomly selects 11 records of mail weights and orders and finds the following:

X (weight, lbs.)	Y (000 orders)
10	4.1
35	6.5
13	3.6
34	6.7
21	5.2
24	5.3
18	4.7
29	6.3
16	4.1
25	5.8
20	5.4

    a. Construct a scatter diagram and decide whether the data appear to conform to a linear model.
    b. If so, find the regression equation and plot the regression line.
    c. The owner of the mail-order house criticizes Lauren's results since they suggest that if no mail arrives (i.e., weight is zero), there will still be over 2,500 orders to fill. How should Lauren respond?
    d. Test H_0: $B = 0$, using $\alpha = 0.01$.

12.22 Refer to Exercise 12.21. Find the 99 percent intervals for:
    a. The mean number of orders for mail that weighs 30 pounds.
    b. The number of orders for tomorrow if today's mail weighs 30 pounds.

12.23 Refer to Exercise 12.21.
    a. Calculate the coefficient of correlation.
    b. Test H_0: $\rho = 0$, using $\alpha = 0.01$.
    c. Find the coefficient of determination.

12.24 Although the prime rate is not tied directly to the federal discount rate, they tend to move in the same direction. Since these measures tend to be stable over longer periods of time, the authors have compiled the data for these two variables on an annual basis over a 33-year period. The following represent the results of their investigation:

Federal discount rate	Prime rate
$\Sigma X = 137.75$	$\Sigma Y = 139.13$
$\Sigma X^2 = 770.5625$	$\Sigma Y^2 = 813.9621$
$\overline{X} = 4.17$	$\overline{Y} = 4.22$

$$\Sigma XY = 783.52$$
$$n = 33$$

a. Find and plot the regression line.
b. Find the standard error of estimate.
c. Find the coefficient of correlation.
d. Find the coefficient of determination.
e. Test H_0: $\rho = 0$, using $\alpha = 0.05$.
f. Test H_0: $B = 0$, using $\alpha = 0.05$.

12.25 One of the purposes of regression analysis is to predict future values of a variable from present values of another variable. We have been exploring relationships between the federal discount rate and prime rate (Section 12.2.4 and Exercise 12.17). The following data present the federal discount and the corresponding prime rate one month later for 17 randomly selected months.

Federal discount rate	Prime rate (one month later)
9.50%	9.99%
9.69	10.62
10.24	11.70
10.70	13.44
11.77	13.53
12.00	13.31
12.00	13.15
12.00	14.01
12.52	17.10
13.00	15.63
13.00	9.60
12.94	8.31
11.40	8.58
10.87	9.85
10.00	12.15
10.10	13.86
11.00	15.94

a. Find the regression line.
b. Find the standard error of estimate.
c. Find the coefficient of determination.

d. Find the coefficient of correlation.
e. Test H_0: $\rho = 0$.
f. Based on the result you obtained for *(e)*, do you think you would accept or reject H_0: $B = 0$?

Multiple Regression Analysis

13.1 INTRODUCTION

After completing a simple regression analysis and learning that knowledge of one variable often improves our ability to predict a second variable, it is natural to wonder whether the procedures may be extended to more than one independent variable. After all, in the real world, several factors are usually operating simultaneously on a dependent variable to produce change. Thus, the price of a particular stock on any given day may be affected by corporate earnings, the general state of the economy, a dramatic news development, etc. Multiple regression and multiple correlational analysis are concerned with just such problems as these. In *multiple regression* we use two or more independent variables as predictors of the dependent variable. *Multiple correlation* describes the strength of the overall relationship.

13.2 A WORKED EXAMPLE

To illustrate, let us imagine that we wish to predict the effect on gasoline mileage of mixing varying amounts of two chemical additives to automotive gasoline. We wish to derive a linear equation that allows us to predict the effects of varying amounts of both X_1 and X_2 on the dependent variable. Table 13.1 presents hypothetical data showing the values of Y (gasoline mileage) when various amounts of additives 1 and 2 were mixed with the gasoline.

multiple regression

Two or more variables are used as predictor variables in regression analysis.

multiple correlation

A single value that describes the overall strength of the correlation between two or more independent variables and the dependent variables.

Table 13.1

Amounts of additives 1 and 2 mixed with gasoline and the resulting miles-per-gallon performance of the test car

X_1 Units of additive 1	X_2 Units of additive 2	Y Gasoline mileage
0	0	15.8
1	0	16.0
0	1	15.9
1	1	16.2
2	0	16.5
0	2	16.3
2	2	16.8
3	1	17.4
1	3	17.2

Unlike simple regression analysis, where a single line describes a relationship, this type of problem involves three-dimensional space (for two predictor variables) and multidimensional space (for more than two independent variables). Thus, the estimating equation for two variables specifies a plane, called the *regression plane,* in three-dimensional space. In simple regression, we described a straight line that minimizes the squared deviations of data points around it. With multiple regression, the plane that minimizes squared deviations around it describes a best fit. Figure 13.1 presents a three-dimensional graph of the data in Table 13.1. We shall analyze these data with a view to finding the best fitting plane.

Figure 13.2 shows a regression plane that minimizes the sum of the squared

regression plane

A three dimensional plane that minimizes the sum of the squared deviations in regression analysis.

A three dimensional diagram showing the data points on three variable, X_1, X_2, and Y.   **Figure 13.1**
Multiple regression analysis involves finding a plane that minimizes the squared deviations
of the data points from the plane.

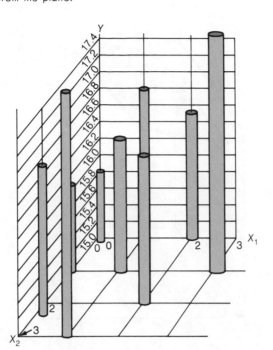

deviations around the plane. Using the least sum-of-squares criterion, it is
the best fitting plane for the data shown.

Recall that $Y_c = a + bX$ is the estimating equation for simple linear
regression. As we increase the number of independent variables, the equation
adds one term for each additional variable. For two independent variables,
the estimating equation is

$$Y_c = a + b_1X_1 + b_2X_2 \qquad (13.1)$$

where

$$Y_c = \text{The predicted dependent measure}$$
$$a = \text{The } Y\text{-intercept}$$
$$X_1 \text{ and } X_2 = \text{Values of two independent variables}$$
$$b_1 \text{ and } b_2 = \text{Slopes associated with the two}$$
$$\text{independent variables}$$

In this text we shall restrict our attention to multiple regression involving
two independent variables. A simple example has been chosen for expository
purposes. In real life, most analysts will avail themselves of existing computer
programs when conducting multiple regression analyses. Moreover, the n
used in this example renders many of the estimating equations crude approxi-

Figure 13.2   A regression plane that minimizes the squared deviations of data points around it. Note: The dashed lines indicate that the data points fall below the plane.

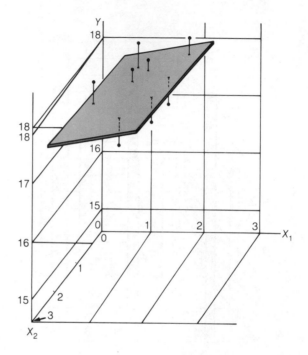

mations, at best. In actual business applications, the sample sizes are sufficiently large that the estimating equations are more than adequate.

It can be shown mathematically that, by simultaneously solving the following three equations, we can find the values of a, b_1, and b_2 for our data.

$$\Sigma Y = na + b_1 \Sigma X_1 + b_2 \Sigma X_2 \qquad (13.2)$$

$$\Sigma X_1 Y = a\Sigma X_1 + b_1 \Sigma X_1{}^2 + b_2 \Sigma X_1 X_2 \qquad (13.3)$$

$$\Sigma X_2 Y = a\Sigma X_2 + b_1 \Sigma X_1 X_2 + b_2 \Sigma X_2^2 \qquad (13.4)$$

The first thing we must do is to arrange our data in a form suitable for obtaining the values required to solve for the three unknown constants. This is shown in Table 13.2.

Step 1. Substituting the appropriate values in Formulas 13.2–13.4, we obtain

$$148.1 = \ 9a + 10b_1 + 10b_2 \qquad (A)$$
$$168.2 = 10a + 20b_1 + 11b_2 \qquad (B)$$
$$167.3 = 10a + 11b_1 + 20b_2 \qquad (C)$$

Step 2. If we multiply Equation (A) by (-10) and Equation (B) by 9 and add, we eliminate a. Thus:

Table 13.2
Values for fitting plane of least squares

Y	X_1	X_2	X_1Y	X_2Y	X_1X_2	X_1^2	X_2^2	Y^2
15.8	0	0	0	0	0	0	0	249.64
16.0	1	0	16.0	0	0	1	0	256.00
15.9	0	1	0	15.9	0	0	1	252.81
16.2	1	1	16.2	16.2	1	1	1	262.44
16.5	2	0	33.0	0	0	4	0	272.25
16.3	0	2	0	32.6	0	0	4	265.69
16.8	2	2	33.6	33.6	4	4	4	282.24
17.4	3	1	52.2	17.4	3	9	1	302.76
17.2	1	3	17.2	51.6	3	1	9	295.84
Sums (ΣY, etc) 148.1	10	10	168.2	167.3	11	20	20	2,439.67

$$n = 9; \quad \bar{Y} = 16.46; \quad \bar{X}_1 = 1.11; \quad \bar{X}_2 = 1.11$$

$$
\begin{array}{lll}
-1481 &= -90a + 100b_1 - 100b_2 & [-10 \times (A)] \\
1513.8 &= 90a + 180b_1 + 99b_2 & [9 \times (B)] \\
\hline
32.8 &= 0 + 80b_1 - b_2 & [(D) = -10(A) + 9(B)]
\end{array}
$$

Step 3. Next we multiply Equation *(A)* by (−10) and Equation *(C)* by 9 and add. We again eliminate *a*.

$$
\begin{array}{lll}
-1481 &= 90a - 100b_1 - 100b_2 & [-10 \times (A)] \\
1505.7 &= 90a + 99b_1 + 180b_2 & [9 \times (C)] \\
\hline
24.7 &= 0 -b_1 + 80b_2 & [(E) = -10(A) + 9(C)]
\end{array}
$$

Step 4. Next we want to eliminate either b_1 or b_2. If we multiply Equation *(D)* by 80 and add the result to Equation *(E)*, we eliminate b_2.

$$
\begin{array}{lll}
2624 &= 6400b_1 - 80b_2 & [80 \times (D)] \\
24.7 &= -b_1 + 80b_2 & (E) \\
\hline
2648.7 &= 6399b_1 & [80(D) + (E)]
\end{array}
$$

$$b_1 = 0.414$$

This value shows the net change in the dependent variable Y for one unit of change in X_1 when X_2 is held at a constant value.

Step 5. We may now find the value of b_2 by substituting the value of b_1 into Equation *(D)*. Thus:

$$
\begin{array}{l}
2624 = (6400)(0.414) - 80b_2 \\
-25.6 = -80b_2 \\
b_2 = 0.320
\end{array}
$$

The interpretation of b_2 is similar to that of b_1. It shows the net change in Y for one unit of change of X_2 when X_1 is held at a constant value.

Step 6. To find the value of *a*, we substitute the values of b_1 and b_2 into Equation *(A)*. Thus:

$$
\begin{array}{l}
148.1 = 9a + (10)(0.414) + (10)(0.320) \\
9a = 140.76 \\
a = 15.64
\end{array}
$$

Step 7. As a check on the accuracy of your work, you should now substitute the calculated values of a, b_1, and b_2 in Equation *(B)* or *(C)*. The sums on both sides of the equation should be equal, allowing a small margin for rounding error. Thus:

$$168.2 = 10a + 20b_1 + 11b_2 \qquad (B)$$
$$168.2 = 156.4 + 8.28 + 3.52$$
$$168.2 = 168.2$$

Step 8. Substituting the values of a, b_1, and b_2 in the multiple regression equation (Formula 13.1), we obtain:

$$Y_c = a + b_1 X_1 + b_2 X_2$$
$$Y_c = 15.64 + 0.414 X_1 + 0.320 X_2 \qquad (F)$$

From this equation, we may estimate the effects on gasoline mileage resulting from the addition of varying amounts of the two additives to our gasoline mixture. However, as with simple regression, we should restrict the range of values of X_1 and X_2 to the amounts used in the sample. We cannot be certain that, beyond this range, the multiple relationship holds. Thus, we might be interested in predicting the effects of adding 2.5 parts of additive 1 and 2.5 parts of additive 2. Substituting in Equation *(F)*, we find:

$$Y_c = 15.64 + (0.414)(2.5) + (0.320)(2.5)$$
$$= 17.475$$

Note that we might systematically explore the effects of holding one variable constant while varying the second. To illustrate, we could hold X_1, the amount of additive 1, at 1.5 and explore the effects of adding 0.5, 1, 1.5, and 2.5 parts of additive 2. Thus:

When $X_1 = 0.5$ and $X_2 = 0.5$, $Y_c = 15.64 + 0.62 + 0.16 = 16.42$.
When $X_1 = 0.5$ and $X_2 = 1$, $Y_c = 16.58$.
When $X_1 = 0.5$ and $X_2 = 1.5$, $Y_c = 16.74$.
When $X_1 = 0.5$ and $X_2 = 2.5$, $Y_c = 17.06$.

Similarly, we could hold X_2 at some constant value and explore the effects of adding various amounts of additive 1.

Example A: The interpretation of a, b_1, and b_2

One of the most effective applications of regression analysis involves situations in which two or more components (independent variables) are contributing to a dependent variable of interest and we wish to develop an estimating equation that weighs the relative importance of these components on the final outcome (the dependent measure). Let's look at a hypothetical example.

Mr. L. is owner of a small chain of pharmacies that are scattered throughout a region of the country. The pharmacies derive their income from nondrug as well as drug items. Mr. L. wishes to find the relative contribution of each type of item to the profit picture. The results obtained at seven pharmacies are shown in Table 13.3.

Drug store	X_1 Sales of drug items ($000)	X_2 Sales of nondrug items ($000)	Y Profits ($000)
1	$10	$150	$20
2	9	90	42
3	8	50	46
4	4	90	30
5	11	140	15
6	12	100	45
7	5	80	38

Table 13.3
Hypothetical data showing sales of drug items, nondrug items, and profits at seven pharmacies

a. Develop the estimating multiple regression equation.
b. Do drug or nondrug items appear to make a greater contribution to the profit picture?

Solution

After solving the simultaneous equations, the following multiple regression equation was developed: $Y_c = 57.28 + 1.628X_1 + (-0.373)X_2$.

Now, if we hold X_1 constant at any given value and increase X_2 by $10,000, we may see what happens to profits as sales of nondrug items increase. Let's set X_1 at $8,000: We find that, when $X_1 = $8,000$ and $X_2 = $50,000, $60,000,$ and $70,000, respectively, the expected profit is $51,660, $47,930, and $44,200. In other words, for each increase in $10,000 in sales, there is an associated loss of $3,730 ($47,930 − $51,660 = −$3,730). Thus, for each increase in sales of $1,000, the loss is $373.

If we hold X_2 constant at, say $60,000 and increase X_1 by $1,000, the predicted profit is:

X_1	X_2	Y_c
$6,000	$60,000	$44,675
7,000	60,000	46,303
8,000	60,000	47,931

Thus, for each increase of $1,000 in sales of drug items, there is an associated increase in profit of $1,628. Thus, drug items are associated with profits whereas nondrug items are associated with losses.

Had we used simple regression analysis, the picture would have been quite different. Figure 13.3 shows the scatter diagrams of X_1 versus X_2, X_1 versus Y, and X_2 versus Y with their regression lines superimposed. For each $1,000

Figure 13.3   Scatter diagrams, regression lines, a, b, r, and r^2 for X_1 against X_2, X_1 against Y, and X_2 against Y

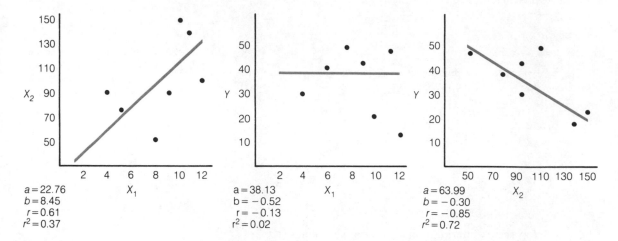

increase in drug sales, there is an associated loss of $520. For each $1,000 increase in nondrug sales, there is an associated loss of $300.

We can see, then, that simple regression analyses may give a misleading picture of the business activities of the drug chain. On the other hand, the multiple regression analysis provides a clear picture of the relationship among profits and sales of drug and nondrug items.

EXERCISES

13.1 Does the multiple regression analysis in Example A suggest that Mr. L. should consider cancelling sales of nondrug items because of their negative contribution to profit?

13.2 A personnel director has hypothesized that employee productivity is a function of at least two variables—hourly wages and working conditions. In a large manufacturing company, eight employees were hired. Each was randomly assigned a different starting hourly wage and different conditions of work. The employees were unaware of these differences. Three months later, records of their productivity levels were obtained. The results were as follows.

Hourly wage X_1	Work conditions X_2	Productivity levels Y
4.6	3	5
5.3	1	4
4.2	5	10
6.4	4	8
4.8	2	7
3.9	1	3
5.0	3	6
4.1	2	4

a. Find a, b_1, and b_2.
b. Holding X_1 constant at 5, predict Y when $X_2 = 1, 2, 3, 4, 5, 6$.
c. Confirm that Y_c increases by 1.44 with each unit increase in X_2 when X_1 is held constant.

13.3 Refer back to Exercise 13.1. Which variable, hourly wages or work conditions, appears to be a better predictor of productivity levels?

13.4 The EPA collects data on various types of pollutants and their sources. Shown below are estimated emissions per automobile mile traveled during seven different time periods.

Carbon monoxide X_1	Hydrocarbons X_2	Nitrogen oxides Y
69.7	8.7	3.8
67.7	8.3	3.6
64.8	7.7	3.4
61.5	7.2	3.1
58.0	6.6	2.9
54.5	6.0	2.7
49.4	5.3	2.6

a. Can carbon monoxide and hydrocarbon levels be used as predictors of nitrogen oxides? Calculate a, b_1, and b_2.
b. Holding X_1 at 55, find Y_c when $X_2 = 6, 7$, and 8.
c. Confirm that Y_c increases by 1.066 when X_1 is held constant and X_2 increases by one unit.
d. Holding X_2 at 7, find Y_c when $X_1 = 52, 53$, and 54.
e. Confirm that Y_c decreases by 0.118 when X_2 is held constant and X_1 increases by one unit.

13.5 Refer back to Exercise 13.4. Predict Y from the following values of X_1 and X_2.

a. $X_1 = 62.4$; $X_2 = 8.1$.
b. $X_1 = 50.3$; $X_2 = 5.8$.
c. $X_1 = 64.3$; $X_2 = 7.7$.
d. $X_1 = 55.4$; $X_2 = 6.9$.

13.3 THE STANDARD ERROR OF ESTIMATE

As with simple regression analysis, the Y-intercept and the slope of the multiple regression line are estimated from sample data. The Y-intercept estimates the value of Y when X_1 and X_2 equal zero. Similarly, b_1 and b_2 describe how changes in the values of X_1 and X_2 affect the dependent variable. Unless the relationship is perfect, there will be scattering or dispersion of data points around the regression plane. To measure this dispersion, we again calculate a standard error of estimate:

$$s_{Y \cdot X_1 X_2} = \sqrt{\frac{\Sigma (Y - Y_c)^2}{n - 3}}$$

where

$Y =$ Obtained sample values of Y

$Y_c =$ Values of the dependent variable estimated from corresponding values of Y

$n - 3 =$ The number of degrees of freedom

Note that the number of degrees of freedom is $(n - 3)$ for the standard error of multiple regression. This is due to the fact that we are estimating three parameters $(A, B_1,$ and $B_2)$. We lose one degree of freedom for each parameter estimated.

The calculation of $s_{Y \cdot X_1 X_2}$ from the defining formula is time-consuming and subject to large rounding errors. Fortunately, there is an alternative formula that permits us to use various values that we calculate in the course of obtaining $a, b_1,$ and b_2:

$$s_{Y \cdot X_1 X_2} = \sqrt{\frac{\Sigma Y^2 - a\Sigma Y - b_1\Sigma X_1 Y - b_2\Sigma X_2 Y}{n - 3}} \qquad (13.5)$$

We have previously found $\Sigma Y^2 = 2{,}439.67$, $a = 15.64$, $\Sigma Y = 148.1$, $b_1 = 0.414$, $\Sigma X_1 Y = 168.2$, $b_2 = 0.320$, $\Sigma X_2 Y = 167.3$, and $n = 9$.

Substituting in Formula (13.5), we find:

$$s_{Y \cdot X_1 X_2} = \sqrt{\frac{2{,}439.67 - (15.64)(148.1) - (0.414)(169.2) - (0.320)(167.3)}{6}}$$

$$= \sqrt{\frac{0.2152}{6}}$$

$$= 0.19.$$

Using this estimate and the t-distribution, we may approximate the confidence interval around a predicted Y. We find the confidence interval of Y_c is: $Y_c \pm t(s_{Y \cdot X_1 X_2})$.

To illustrate, we previously found that when $X_1 = 0.5$ and $X_2 = 1.5$, $Y_c = 16.74$. Using the 95 percent confidence coefficient, we find $t = 2.477$ at $\alpha = 0.05$ and df = 6. Thus, confidence interval of predicted $Y = 16.74 \pm (2.447)(0.19) = 16.28$ to 17.20.

The use of t and the standard error of estimate assumes that e (scatter or random error in the population) is normally distributed and that homoscedasticity prevails, i.e., for any values of X_1 and X_2, the standard deviations of the error are the same. Moreover, the predictive interval for Y is good as an approximation when the sample size is large. In the preceding problem, where n is small for expository purposes, the predictive interval is quite approximate.

EXERCISES

13.6 Refer back to Exercise 13.1.
    a. Calculate the standard error of estimate.
    b. If $X_1 = 3.7$ and $X_2 = 2.4$, find the 95 percent confidence interval for Y_c.

13.7 Refer back to Exercise 13.4
    a. Calculate the standard error of estimate.
    b. If $X_1 = 50$ and $X_2 = 5.4$, find the 99 percent confidence interval for Y_c.

13.8 What assumptions underly the use of the *t*-distributions and the standard error of estimate for establishing confidence intervals of Y_c?

13.4 THE COEFFICIENT OF MULTIPLE DETERMINATION (R^2)

Recall that r^2, the coefficient of determination, shows the proportion of total variation in Y that is accounted for by the regression equation. Similarly, the *coefficient of multiple determination,* multiple R^2, shows the proportion of total variation in the dependent variable that is accounted for by the regression plane. The larger the multiple R^2, the greater the variation explained by the multiple regression. Since calculation of multiple R^2 from the defining formula is cumbersome, we present a useful formula that makes use of previously calculated values:

coefficient of multiple determination

Called multiple R^2, it shows the proportion of the total variation in the dependent variable that is accounted for by the regression plane.

$$R^2 = \frac{a\Sigma Y + b_1\Sigma X_1 Y + b_2\Sigma X_2 Y - \dfrac{(\Sigma Y)^2}{n}}{\Sigma Y^2 - \dfrac{(\Sigma Y)^2}{n}} \qquad (13.6)$$

$$= \frac{(15.64)(148.1) + (0.414)(168.2) + (0.320)(167.3) - \dfrac{(148.1)^2}{9}}{2439.67 - \dfrac{(148.1)^2}{9}}$$

$$= \frac{2.387}{2.602}$$

$$= 0.92$$

Thus, 92 percent of the variation in Y is accounted for in terms of the variations in the values of X_1 and X_2. This means that only 8 percent (100 percent − 92 percent) of the variation is left unexplained. Although we may consider adding a third independent variable to the multiple regression equation, we might question its advisability. With only 8 percent of the variation not accounted for, the cost and time of including a third independent variable might be too high a price to pay for a modest gain in precision.

EXERCISES

13.9 Refer back to Exercise 13.1.
  a. Calculate multiple R^2.
  b. What does this value tell us?
  c. What percentage of variance is not accounted for by the regression plane?

13.10 Refer back to Exercise 13.4.
  a. Calculate multiple R^2.
  b. What does this value tell us?
  c. What percentage of variance is not accounted for by the regression plane?

13.11 Refer back to Example A.
  a. Calculate multiple R^2.
  b. What percentage of the variability in profits is accounted for by the regression plane?

13.5 COMPUTER READOUTS

Thus far in this chapter we have looked at a multiple regression analysis of a very simple data set. This was done primarily to acquaint you with some of the key concepts and to provide some experience with the calculations necessary to find b_1, b_2, $s_{Y \cdot X_1 X_2}$, and R^2, to name a few. As we indicated earlier, the advent of high-speed computers has rendered obsolete the hand calculation of multiple regression analysis. There are presently available a large number of "canned" programs that permit us to use the capacity of the computer even though we may be inexperienced with programming. Although the printouts from the various available programs vary in particulars, they are all quite similar in the essentials. We'll illustrate the printout from one widely used program, the SPSS (Statistical Package for the Social Sciences).

Table 13.4 presents a computer readout of the multiple regression analysis of the gasoline additives that we covered earlier in the chapter.

Let's spend a few moments interpreting the output shown in Table 13.4.

1. The first column identifies the two variables—additive 1 and additive 2 and the Y-intercept (CONSTANT).

2. The second column shows the values of the coefficients. Here we see, as we found earlier, that b_1 is somewhat larger than b_2. Thus, b_1 appears to exert a relatively greater weight in the final outcome.

3. The third column shows the standard error associated with each b coefficient. By dividing each b coefficient by its corresponding standard error, we obtain the test statistic, Student t-ratio, shown in the final column.

VARIABLE	COEFFICIENT	STD. ERROR	T-VALUE
ADTVE 1	0.41392	0.06352	6.51637
ADTVE 2	0.32000	0.06352	5.03778
(CONSTANT)	15.74000		

VARIABLES IN THE EQUATION

Table 13.4
Computer output for
amounts of additive (X_1, X_2)

ANALYSIS OF VARIANCE	DF	SUM OF SQUARES	MEAN SQUARE	F
REGRESSION	2	2.38702	1.19351	33.27321
RESIDUAL	6	0.21520	0.03587	

MULTIPLE R	0.95779
R SQUARE	0.91737
STD. ERROR OF EST.	0.18938

4. The t-ratio tests the nondirectional null hypothesis that $B = 0$, or the directional null hypothesis that $B \leq 0$ or $B \geq 0$.

Since we lose one degree of freedom for each B coefficient that is estimated, the number of degrees of freedom is n minus the number of independent variables. In the present example, df $= 9 - 2 = 7$. Using $\alpha = 0.01$ for a two-tailed hypothesis, we find the critical value of t to be 3.499. Since both b coefficients exceed the critical value, we may reject H_0 and assert that the b coefficients are significantly greater than zero.

5. The analysis of variance shows the total sums of squares partitioned as the two components, regression and residual. The regression is the explained variation (R^2) and the residual is the error term. At $\alpha = 0.01$ and df $= 2,6$, an F-ratio equal to or greater than 10.92 is required to reject the null hypothesis that the population R^2 equals zero. Since obtained $F = 33.27$ exceeds the critical value, we may reject H_0 and assert that the population R^2 is greater than zero.

6. Finally, the bottom figures show the value of the multiple R, R^2, and the standard error of estimate. The coefficient of determination, R^2, tells us that 92 percent of the variance in the sample is accounted for in terms of the regression of X_1 and X_2 on the dependent measure. Thus, only about 8 percent of the total variance is not accounted for.

One note of caution. Recall that we set up a sample problem for expository purposes. Although the t-ratio assumes normally distributed populations of the independent and dependent variables, most business applications of multiple regression use a large n. Thus, the central limit theorem comes onto play so that the sampling distribution of the relevant statistics approach normality.

EXERCISE

13.12 A national department store chain wishes to learn the relationship between size of store, density of surrounding population, and gross sales. Randomly selecting 100 different sites, the following computer output was obtained:

VARIABLES IN THE EQUATION			
VARIABLE	COEFFICIENT	STD. ERROR	T-VALUE
SIZE 1	0.12442	0.01359	9.15526
POP 2	−0.05913	0.08655	−0.68319
(CONSTANT)	5.81206		

ANALYSIS OF VARIANCE	DF	SUM OF SQUARES	MEAN SQUARE	F
REGRESSION	2	25.81468	12.90734	1229.07740
RESIDUAL	97	1.01866	0.01050	

MULTIPLE R	0.98084
R SQUARE	0.96204
STD. ERROR OF EST.	0.58271

a. Write the multiple regression equation for predicting gross sales from the two independent variables.

b. Using $\alpha = 0.01$, test the null hypothesis that the population coefficients are zero.

c. Using $\alpha = 0.01$, test the null hypothesis that population $R^2 = 0$.

d. Interpret the results.

13.6 A FEW WORDS OF CAUTION

Regression and multiple regression analysis can be an important tool in the hands of the decision maker. Under the right circumstances, it can increase the precision of estimates to a degree that substantially reduces the element of risk in the decision making process. But no statistical procedure, even if planned to the nth degree, can remove uncertainty. Some rare or unexpected event can occur (e.g., war, a natural calamity, etc.) that can make a shambles of the best laid plans. Moreover, like any weapon, it can blow up in your face, particularly if not used properly.

When two or more independent variables are used to predict a dependent variable, it is assumed that the predictor variables are not highly correlated with one another. If they are, the problem of multicollinearity arises. *Multicollinearity* consists of a high correlation between and among the predictor variables. When this occurs, it is difficult to independently assess the regression coefficient of each independent variable. For example, in the limiting case, where $r_{X_1 X_2} = 1.00$, it is impossible to distinguish the separate effects of variables X_1 and X_2 on the dependent variable. All we can do is describe the joint effect of both variables. If at some future time, the relationship between the independent variable should change, we might find our prediction equation of precious little use.

Although perfect correlations in the real world are rare, there are many occasions when high correlations exist among variables. For example, daily volumes of activity in the American and New York stock exchanges are highly correlated. This is a fact of economic reality. One would not be wise to use these two measures as predictor variables in a multiple regression analysis. Since such correlations may be regarded as "states of nature," there

multicollinearity

A condition in which independent variables are highly correlated with one another.

may be little we can do to alter these circumstances. However, there are occasions when we may intervene. If it is possible for us to manipulate each independent variable so that they are not correlated, we can remove the problem of multicollinearity. This is what was done in the gasoline problem. Various amounts of additives 1 and 2 were randomly paired so that, on the average, we would expect their correlation to be zero (in the actual example, $r_{X_1 X_2}$ was -0.01). Under these circumstances, the regression equations provide independent predictors of the dependent measure.

An excellent example of multicollinearity is found in Exercise 13.4. You may recall that this exercise used EPA data on three types of automobile pollutants: carbon monoxide, hydrocarbons, and nitrogen oxides. Carbon monoxide and hydrocarbon levels were highly correlated ($r^2 = 0.989$). Either variable, taken alone, accounted for as much of the variation in Y (nitrogen oxides) as the multiple regression plane. Thus, $r_{X_1 Y}^2 = 0.958$, $r_{X_2 Y}^2 = 0.975$, whereas multiple $R^2 = 0.988$. In other words, either X_1 or X_2 would serve as well to predict nitrogen oxide levels as both variables taken together.

As already indicated, there is little we can do when the natural order of things has arranged high intercorrelations between and among independent variables. However, there are times when we may modify the independent measures so as to reduce their intercorrelations. For example, if we are using employment levels and price as predictors of demand for a product, both are correlated merely because both are usually increasing, i.e., the number of employed often goes up merely because the total labor market increases with population increase, even though the rates of unemployment may remain the same. Since inflation often seems endemic, numbers of people employed and inflation may be correlated. However, if *rates* of employment were used as a predictor variable, the correlation with inflation might be reduced.

This example leads to another problem that can arise with regression analysis. One assumption underlying regression analysis is that the errors of the independent and dependent variables are uncorrelated or independent. For example, in business and economics, various time-related indexes are often used in regression analyses. Frequently they are *serially correlated,* i.e., they are related because they are occurring over the same time span. Take the closing value of the Dow Jones Industrials, as an example. If it is high today, it was probably high yesterday and will probably be high tomorrow. The variability or error in stock prices is, therefore, largely a function of the time frame during which the measures were collected.

serial correlation

If one value in a time series is correlated with its value at some other time period, it exhibits serial correlation.

The effect of serial correlation of the error terms is to impair the precision of estimating the B parameters in multiple regression. When serial correlation of error is present, the standard error of estimate may so underestimate the true error that findings may be reported as statistically significant when in fact they are not.

Another effect of serial correlation is to present an appearance of cycles that may not, in fact, exist. To return to the example of closing value of the DJI, if it is above the mean today, it is likely to be above the mean tomorrow and the next day. Thus, it may appear that we are in a high price phase of a cycle of stock prices.

To illustrate, Table 13.5 presents the closing values of the DJI for 27 consecutive trading days. When the mean is calculated and then subtracted from each score, the deviation values (rounded to one decimal place) are obtained.

Table 13.5

Closing values of Dow Jones Industrials for 27 consecutive trading days and the deviation measures obtained by subtracting the mean value from each day's value

Closing value	$(X - \bar{X})$	Closing value	$(X - \bar{X})$	Closing value	$(X - \bar{X})$
986.35	36.7	924.49	−25.2	960.84	11.1
982.42	32.7	917.75	−32.0	956.14	6.4
964.93	15.3	929.18	−20.5	958.70	9.0
944.03	−5.0	932.59	−17.1	972.44	22.7
933.79	−15.9	931.74	−18.0	962.20	12.5
932.42	−17.3	943.60	−6.1	959.90	10.2
935.41	−14.3	939.51	−10.2	950.68	1.0
953.16	3.5	955.12	5.4	958.96	9.3
937.20	−12.5	954.44	4.7	963.99	14.3

When these deviations are plotted on a line graph, the picture shown in Figure 13.4 emerges. Needless to say, it does not require an unusually active imagination to divine some sort of cyclic phenomenon during the 27-day period.

Figure 13.4

Line graph of deviations of daily closing values of Dow Jones Industrials from the mean during 27 consecutive trading days. This graph illustrates both serial correlation and the fact that serial correlation can produce the appearance of a cyclic phenomenon.

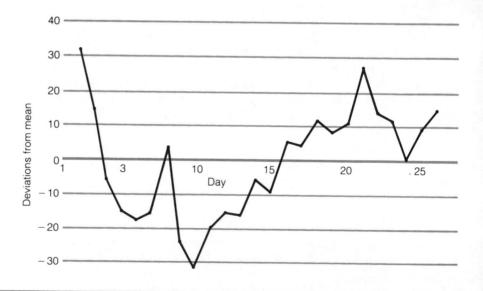

EXERCISES

13.13 In which of the following situations is multicollinearity likely to constitute a problem?

a. Janice B. has collected absentee, lateness, and productivity records of 50 employees. She wishes to predict productivity using absentee and lateness records as the independent variables.

b. The Civil Aeronautics Board wishes to determine whether age and years of experience of airline pilots can be used as predictor variables in a simulated test of reactions to critical incidents.

c. A mail order house wishes to ascertain whether the number of employees and expenditures for advertising are related to sales dollar volume.

d. Gold and platinum closing prices are used as predictor variables of closing values of the Dow Jones Industrials.

13.14   In which of the following three sets of data is multicollinearity likely to constitute a problem?

(a)			(b)			(c)		
X_1	X_2	Y	X_1	X_2	Y	X_1	X_2	Y
1	8	5	2	3	5	6	4	10
4	6	9	5	6	8	5	3	14
5	7	6	9	11	14	4	6	12
6	8	8	12	13	12	7	7	11

13.15   In which of the following is serial correlation likely to constitute a problem?

a. Daily gold prices are recorded for 20 consecutive trading days. A mean is calculated and deviations from the mean are determined. The daily changes are plotted along a time line.

b. The number of guns manufactured domestically and the number imported are used as independent variables to predict the number of murders in each ensuing year.

c. Over a period of 30 consecutive trading sessions, prime rate and closing prices on the precious metal futures market are used as independent variables for predicting the Amsterdam spot market prices.

SUMMARY

In this chapter, we examined an extension of simple linear regression to a situation in which there are two or more independent or predictor variables.

1. Rather than a single line representing best fit, a plane is fitted to the data points that minimizes the sum of their squared deviations about the plane.

2. A different slope is calculated for each independent variable. Thus, when two predictor variables are used, there are two linear slopes, b_1 and b_2.

3. The value b_1 shows the net change in Y for each unit of change in X_1 when X_2 is held constant. Conversely, b_2 shows the change in Y for each unit of change in X_2 when X_1 is held constant.

4. Finding a, b_1, and b_2 involves the simultaneous solution of three equations when two predictor variables are used.

5. When the values of a, b_1, and b_2 are substituted in the equation $Y_c = a + b_1X_1 + b_2X_2$ we are able to predict Y from various values of X_1 and X_2.

6. The standard error of estimate describes the scattering or dispersion of data points around the regression plane. Together with the t-distributions, it permits us to establish confidence intervals around the regression plane.

7. The coefficient of multiple determination, multiple R^2, shows the proportion of variation that is accounted for by the regression plane. The larger the multiple R^2, the larger is the variation explained by multiple regression.

8. When the predictor variables are highly related to one another, multicollinearity is said to exist. The higher the relationship, the more difficult it is to distinguish the independent effects of each predictor variable.

9. Time-dependent measures are frequently serially correlated, thus violating the assumption that the random errors are unrelated. Serial correlation may also lead to an apparent, but spurious, cyclic phenomenon.

TERMS TO REMEMBER

multiple regression (420)
multiple correlation (420)
regression plane (420)

coefficient of multiple
determination (429)
multicollinearity (432)
serial correlation (433)

EXERCISES

13.16 The table below shows the amount of money spent on new home construction, conventional first mortgage rates on new home purchases, and prime rates during nine randomly selected time periods.

    a. Using prime rate as the dependent variable, calculate a, b_1, and b_2.

    b. Find R^2.

    c. Do the two predictor variables permit precise prediction of prime rates?

New construction ($ millions)	Home mortgage rates (percent)	Prime rate (percent)
13.4	9.94%	10.01%
17.1	10.06	9.90
20.7	10.39	9.79
22.3	10.73	10.62
22.5	10.91	13.44
18.9	11.30	13.31
15.8	11.60	14.01
17.9	12.64	15.63
19.7	12.24	8.31

13.17 The owner of a machine shop wishes to estimate weekly production by using number of employees and total wages as predictor variables. Is there likely to be any special problem?

13.18 Explain the advantage of randomly assigning the values of the predictor variable vis-à-vis selecting the values as they exist in nature.

13.19 Explain why EPA figures showing annual averages of carbon monoxide, hydrocarbon, and nitrogen oxide emissions might be highly correlated.

13.20 The director of research for a large conglomerate is interested in determining the effect on sales volume of advertising expenditures, price of the item, and the number of substitute or competing products available. She randomly selects 150 locations, collects the data, and obtains the following computer output:

VARIABLES IN THE EQUATION

VARIABLE	COEFFICIENT	STD. ERROR	T-VALUE
ADV 1	2.04382	0.23035	8.87267
PRICE 2	−3.89734	0.15614	−24.91184
COMPET 3	−0.95316	0.13046	− 7.30615
(CONSTANT)	6.30745		

ANALYSIS OF VARIANCE	DF	SUM OF SQUARES	MEAN SQUARE	F
REGRESSION	3	38.31462	12.77154	129.17508
RESIDUAL	146	14.43538	0.09887	

MULTIPLE R	0.85226
R SQUARE	0.72634
STD. ERROR OF EST.	0.60069

   a. Write the multiple regression equation for predicting sales volume from the three independent variables.
   b. Using $\alpha = 0.01$, test the null hypothesis that the population coefficients are zero.
   c. Using $\alpha = 0.01$, test the null hypothesis that population $R^2 = 0$.
   d. Interpret the results.

13.21 A college admissions officer uses College Board (X_2) and high school grades (X_1) to predict college grade point averages for applying students. The regression equation developed by the admissions department is

$$Y_c = 0.6 + 0.7X_1 + 0.0006X_2$$

Using this equation, predict the college GPA of the following five students.

	A	B	C	D	E
HSGP	2.6	3.5	1.8	3.3	2.4
CEEB	520	630	450	540	550

Time Series

14.1   FORECASTING

How does one predict the future? Will interest rates go up, go down, stay the same? Is the economy headed for a downturn or do the signs indicate an upcoming period of prosperity? Should the head of production plan for increased demand? Should the local automobile dealer decrease inventory to anticipate decreased demand?

Planning for the future—forecasting what will occur—is a complex and often frustrating task. Yet intelligent and successful business entrepreneurs must make decisions about the future. The accuracy of these predictions may well determine future success or failure.

Forecasting is done at all levels. In our everyday life we make informal forecasts about future events. Education and job decisions are made on personal predictions of future happenings. At a more formal level, economists are continually forecasting economic trends. Many different forecasting techniques are used with varying degrees of success. In addition, forecasters rarely agree. Table 14.1 shows a sampling of prime rate forecasts. It clearly illustrates the wide diversity of opinion among bankers and economists.

Although there are many different forecasting techniques, we shall focus

Table 14.1
Forecasts of the prime rate for a single year by several leading bankers and economists.

| | Percent | | | |
	Average	High	Low	Year-end
Irwin Kellner, Manufacturers Hanover	14½	20	11	11
Donald Maude, Merrill Lynch	15	22	12–13	13–14
Francis Schott, Equitable Life	14–15	21½	13–14	15–16
George McKinney, Irving Trust	12	21½	9–10	9–10
Leif Olsen, Citibank	11½–12½	21½	11½	—
David Jones, Aubrey G. Lanston	14	21½	11	14
Alan Greenspan, Townsend-Greenspan	15	21½	10–12	10–12
John O. Wilson, Bank of America	18	22	14	14
Timothy Howard, Wells Fargo	13½	21½	11	12–14
Norman Robertson, Mellon Bank	16	21½	14	16
Allen Sinai, Data Resources	15–36	21½	13	15
Albert Sindlinger, Sindlinger and Co.	18–20	28–30	18½	—

our attention on the most commonly used—the time series model. In this model, past data are analyzed over a given time period and patterns are estimated. A key assumption in the time series model is that patterns that have occurred in the past will continue in the future.

14.2 TIME SERIES COMPONENTS

A *time series* analysis involves data that are arranged in chronological order. Sometimes the interest is simply in describing the historical pattern of the data. However, time series analyses may also provide the basis for forecasting and planning.

The classical or traditional time series model breaks down the variations into the following components:

1. Trend.
2. Cyclical fluctuations.
3. Seasonal variations.
4. Irregular fluctuations.

The *trend* pattern of data is useful for fairly long-range forecasts. Many types of business and economic data show a trend pattern.

Figures 14.1, 14.2, and 14.3 illustrate a number of different trend patterns. The straight line through the data represents the trend. When a trend exists, the trend line will have either a positive or negative slope. A positive slope reflects an increasing series; a negative slope, a decreasing series. If the slope is zero, we say there is no trend pattern (Figure 14.3). Although the values

time series

A sequence of observations that are arranged in chronological order. In graphing, time is usually plotted on the horizontal axis and the value of the variable on the vertical axis.

trend

A relatively smooth upward or downward long-term movement of a time series.

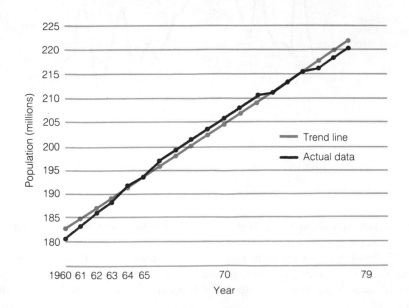

Figure 14.1

Example of an upward trend pattern

442

Figure 14.2
Example of a downward
trend pattern

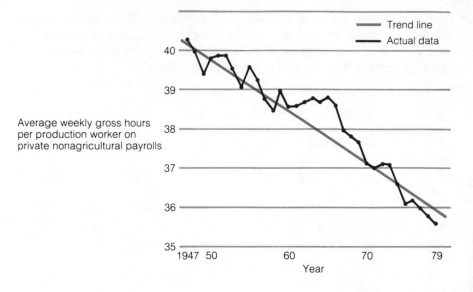

Figure 14.3

Example of no trend pattern. Note that the data fluctuate above and below the trend
line, showing no overall growth or decline.

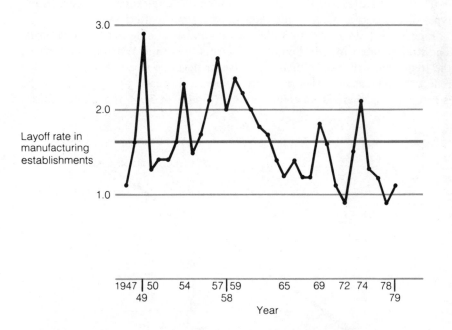

of the data vary above and below the trend line, they show no overall increase
or decrease.

If the only component in the time series were the trend component, forecasting would be a relatively simple and straightforward procedure. Unfortunately, most economic variables differ from the value expected on the basis
of trends.

Cyclical fluctuations, sometimes called business cycle movements, refer to the ups and downs around trend lines observed over long periods of time (greater than one year). Unfortunately, these fluctuations vary in duration as well as size. Moreover, different economic variables do not go up and down at the same time. Figure 14.4 presents a simplified example of a cyclical pattern.

Seasonal variations are cycles that occur during a one-year period and tend to repeat themselves each year. Certain variables are influenced by such factors as weather, holidays, and vacation times that occur during the course of a year. For example, we would expect retail sales to be higher around the Christmas season. Thus, regardless of overall trends, certain variables will show ups and downs at various times of the year. Figure 14.5 shows the seasonal pattern in retail sales in department stores. In this case, the high point is quite clearly end-of-year sales. Notice that the pattern recurs over the years.

cyclical fluctuations

Periodic upward or downward movements in a time series that tend to recur over long periods of time. Cyclical patterns tend to vary in length and size. The irregularity of these cycles precludes a single simple explanation of the so-called business cycle.

seasonal variations

Patterns in the data that tend to recur during a period of one year or less.

Example of cyclical variations in data around a baseline in which the trend component has been removed.

Figure 14.4

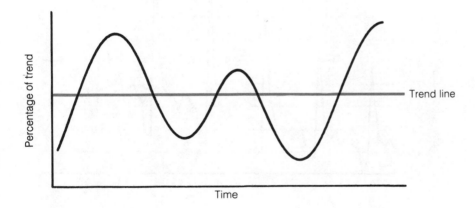

Trend line

Irregular fluctuations in the time series are generally erratic and of short duration. They may occur as a result of unexpected events such as strikes, droughts, or political situations. Since these movements are, by definition, erratic, there is no formal model that may be used to describe or predict them.

The four components of a time series do not usually occur alone. In most cases, all factors operate together in some combination. In the following sections, we will describe how each component may be separately analyzed. Finally, we shall combine all of the components to yield a forecast of the future.

The most common time series model, combining all components, is represented by the following formula:

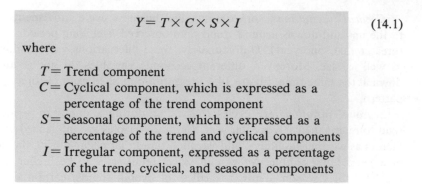

$$Y = T \times C \times S \times I \qquad (14.1)$$

where

$T =$ Trend component

$C =$ Cyclical component, which is expressed as a percentage of the trend component

$S =$ Seasonal component, which is expressed as a percentage of the trend and cyclical components

$I =$ Irregular component, expressed as a percentage of the trend, cyclical, and seasonal components

Figure 14.5   Retail sales in department stores monthly. Note the recurrent seasonal pattern: Sales are highest at the end of the year.

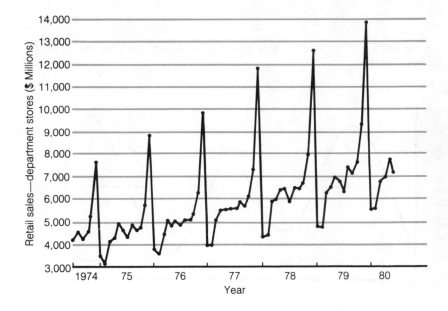

14.3   TREND

The trend line in a time series is obtained in much the same way as the regression line in a correlational analysis (Chapters 12 and 13). There are several different purposes for obtaining trend lines.

We may be interested strictly in the historical description of a trend. We may wish to use the trend line (or its equation) to predict future data. Or, we may wish to eliminate trend movements from a time series so that we can study the other components.

Unfortunately, all economic variables do not follow a straight-line or linear pattern. Some trend movements show increasing or decreasing movements over time. In these instances, a linear model would not be an appropriate description of the trend. In the following examples, we shall describe three models commonly used to describe trend movements: linear, parabolic, and exponential.

Example A: Fitting a straight-line trend

We have already seen how a straight line may be fitted to data (Chapter 12). The procedures we shall use here are essentially the same. We will take our raw data and, using the method of least squares, derive a prediction equation. However, a word of caution is in order. Although the procedures are similar, the underlying theoretical assumptions differ. In regression analysis, we assumed that the predicted Y values were the means of conditional probability distributions for a given X value. Deviations from these predicted values were assumed to be random errors that could be described by a probability distribution. Moreover, the observations of the dependent variable (Y) were assumed to be independent.

However, in fitting a linear trend line to time series data, these assumptions are not met. Deviations from the predicted values are not considered random. If we are dealing with annual data, the deviations represent cyclical and irregular fluctuations. (Seasonal variations are discounted with annual data since they occur *within* a year.) In addition, the condition of independence is not met in time series data. The value of Y for one year is *not* independent of the value for the preceding or following year. For example, if sales were the dependent variable, we would expect the value for, say, 1975 to be related to the corresponding value for 1974 or 1976.

The formula for a linear trend line in a time series is

$$Y_t = a + bx \tag{14.2}$$

where

$$Y_t = \text{Predicted trend value for dependent variable } (Y)$$
$$x = \text{Coded time value}$$

In a time series analysis, we usually simplify the time variable by transforming the year (X) into a coded value, x, which represents the deviation of the year (X) from the mean year $(\overline{X})$, or

$$x = X - \overline{X} \tag{14.3}$$

When we are dealing with an odd number of years, the middle year is the mean, i.e., $x = 0$. Thus, in Table 14.1, $x = 0$ for 1974, the mean year. The x values for years before and after 1974 are expressed as deviations from 1974.

We interpret the constants a and b in the equation in the same way as in regression analysis. Thus a is the trend value for Y when $x = 0$ — (1974) and b is the slope of the trend line. Using these coded values for X, the equations for a and b are

$$a = \overline{Y} \tag{14.4}$$

$$b = \frac{\Sigma xY}{\Sigma x^2} \tag{14.5}$$

Table 14.2 presents the calculations necessary to determine the linear trend line.

Table 14.2

Straight-line trend, using the least squares method, of the average residential cost of 100 kilowatt hours of electricity

Year X	Coded year $x = X - \overline{X}$	Average cost of 100 kwh Y	xY	x^2	Y_t
1970	−4	4.09	−16.36	16	3.83
1971	−3	4.25	−12.75	9	4.20
1972	−2	4.51	−9.02	4	4.58
1973	−1	4,65	−4.65	1	4.95
1974	0	4.99	0	0	5.33
1975	1	5.89	5.89	1	5.70
1976	2	6.15	12.30	4	6.08
1977	3	6.54	19.62	9	6.45
1978	4	6.87	27.48	16	6.83
		$\Sigma Y = 47.94$	$\Sigma xY = 22.51$	$\Sigma x^2 = 60$	

Source: U.S. Bureau of the Census, *Statistical Abstract of the United States: 1979*, 100th ed. (Washington, D.C.: 1979) U.S. Government Printing Office, p. 493.

$$a = \frac{47.94}{9} = 5.327$$

$$b = \frac{22.51}{60} = 0.375$$

$$Y_t = 5.327 + 0.375x$$

Figure 14.6 shows the data and the fitted trend line.

The use of coded time values, as you can see, greatly simplifies the calculations. This is particularly true when there is an odd number of years. However, when the number of years is even, we modify the technique slightly. With an even number, the mean time, $x = 0$, falls midway between two years. For example, suppose we analyzed data for 1970, 1971, 1972, 1973. The mean year, $x = 0$, is 1971½. In this case, 1971 would be $x = -½$ and 1972, $x = +½$. A technique commonly employed to avoid the use of fractions is to view the deviations in half-year intervals. Thus, the

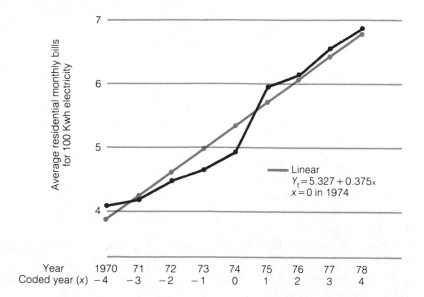

Figure 14.6
Straight-line trend line
fitted to average
residential cost of 100
kilowatt hours of
electricity, 1970–1978

x value for 1971 would be −1; for 1972, $x = 1$; for 1970, $x = −3$; and for 1973, $x = +3$. The constants a and b would be calculated in the same way but we would interpret a as the value of Y for the time period midway between the two middle years. The slope, b, would represent the amount of change per half-year.

Example B: Fitting an exponential trend line

When we use a linear model to fit a trend line, we assume that the values of the dependent variable, Y, are increasing (or decreasing) by a constant *absolute* amount each year. Many business and economic variables do not follow a straight-line trend. They may show periods of greater or lesser *relative* increases over the year. One such nonlinear trend line is represented by an exponential curve, which increases (or decreases) at a constant rate. The equation for the exponential trend line is:

$$Y_t = ab^x \qquad (14.6)$$

The usual procedure in fitting an exponential curve is to take the common logarithms of both sides of the equation:

$$\log Y_t = \log a + x \log b$$

We then use the following formulas to calculate the coefficients a and b:

$$\log a = \frac{\Sigma \log Y}{n} \qquad (14.7)$$

$$\log b = \frac{\Sigma x \log Y}{\Sigma x^2} \qquad (14.8)$$

We obtain the logarithm values from Table XIII (Appendix C). The antilogarithms of the constants from Formulas (14.7) and (14.8) yield the values of a and b for Formula (14.6), which is used to calculate the trend values of the exponential curve. If the trend is increasing, the value of b will be greater than 1. If the trend is decreasing, the value of b will be less than 1.

Table 14.3 presents the calculations necessary to obtain the exponential trend line for the Consumer Price Index (CPI) from 1967 to 1977.

Calculating the trend values for $x \geq 0$ is a straightforward procedure. For example, the trend value for 1974, $x = 2$, is

$$Y_t = (131.2)(1.06)^2 = 147.4$$

However, when $x \leq -1$, we use a slightly different procedure. For example, the trend value for 1970, $x = -2$, is

$$Y_t = (131.2)(1.06)^{-2} = \frac{(131.2)}{(1.06)^2} = 116.8$$

Figure 14.7 shows the data and the fitted exponential trend line.

Table 14.3
Exponential trend line fitted to Consumer Price Index

Year	x	CPI Y	$\log Y$	$x \log Y$	x^2	Y_t
1967	−5	100.0	2.0000	−10.0000	25	98.0
1968	−4	104.2	2.0178	− 8.0712	16	103.9
1969	−3	109.8	2.0407	− 6.1221	9	110.2
1970	−2	116.3	2.0656	− 4.1312	4	116.8
1971	−1	121.3	2.0838	− 2.0838	1	123.8
1972	0	125.3	2.0979	0	0	131.2
1973	1	133.1	2.1242	2.1242	1	139.1
1974	2	147.7	2.1694	4.3388	4	147.4
1975	3	161.2	2.2073	6.6219	9	156.3
1976	4	170.5	2.2316	8.9264	16	165.6
1977	5	181.5	2.2589	11.2945	25	175.6
Sums			23.2972	2.8975	110	

$$\log a = \frac{23.2972}{11} = 2.1179; \quad \text{antilog } 2.1179 = 131.2 = a$$

$$\log b = \frac{2.8975}{110} = 0.0263; \quad \text{antilog } 0.0263 = 1.06 = b$$

$$Y_t = (131.2)(1.06)^x$$

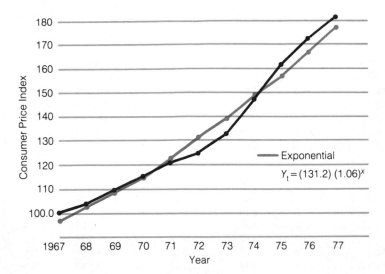

Figure 14.7
Exponential trend line
fitted to Consumer Price
Index

Example C: Fitting a parabolic trend line

Linear and exponential trend lines are useful for describing trends that increase or decrease in the same direction throughout the series. However, certain variables change direction, that is, they may increase for a while and then start to decrease (or vice versa). One type of nonlinear trend line that takes these slope changes into account is represented by a parabolic trend equation:

$$Y_t = a + bx + cx^2 \tag{14.9}$$

A parabolic trend is one that may change direction once and then continue in the opposite direction throughout the rest of the series. The following equations, which use the least squares principle, are employed to calculate the values of the parabolic regression coefficients:

$$\Sigma Y = na + c\,\Sigma x^2 \tag{14.10}$$

$$\Sigma x^2 Y = a\,\Sigma x^2 + c\,\Sigma x^4 \tag{14.11}$$

and $b = \dfrac{\Sigma xy}{\Sigma x^2}$ which we introduced earlier as Formula (4.5).

Table 14.4 presents the calculations necessary to fit a parabolic trend line to new orders for durable goods from 1961 to 1979. The mean year is 1970.

Substituting the appropriate sums in Formulas (14.10) and (14.11), we obtain

$$8{,}252 = 19a + 570c \tag{A}$$
$$279{,}248 = 570a + 30{,}666c \tag{B}$$

Table 14.4
Parabolic trend fitted to
new orders for durable
goods

(1)	(2)	(3) Orders ($ billions)	(4)	(5)	(6)	(7)	(8)
Year	x	Y	xY	x^2	x^2Y	x^4	Y_t
1961	−9	189	− 1,701	81	15,309	6,561	234
1962	−8	208	− 1,664	64	13,312	4,096	230
1963	−7	225	− 1,575	49	11,025	2,401	231
1964	−6	248	− 1,488	36	8,928	1,296	236
1965	−5	279	− 1,395	25	6,975	625	245
1966	−4	314	− 1,256	16	5,024	256	260
1967	−3	310	− 930	9	2,790	81	279
1968	−2	338	− 676	4	1,352	16	303
1969	−1	359	− 359	1	359	1	331
1970	0	329	0	0	0	0	364
1971	1	359	359	1	359	1	402
1972	2	422	844	4	1,688	16	444
1973	3	515	1,545	9	4,635	81	492
1974	4	559	2,236	16	8,944	256	543
1975	5	506	2,530	25	12,650	625	600
1976	6	608	3,648	36	21,888	1,296	661
1977	7	715	5,005	49	35,035	2,401	727
1978	8	842	6,736	64	53,888	4,096	797
1979	9	927	8,343	81	75,087	6,561	872
Sums		8,252	20,202	570	279,248	30,666	

Source: U.S. Bureau of Economic Analysis, *Business Statistics 1979* (Washington, D.C.; U.S.
Government Printing Office, 1980), p. 37.

Multiplying Equation (A) by 570, Equation (B) by (−19), and adding,
we get

$$570(8,252) = \quad 570(19a) + \quad\quad (570)^2c$$
$$-19(279,248) = -19(570a) + -19(30,666)c$$
$$4,703,640 - 5,305,712 = \quad (324,900 - 582,654)c$$
$$-602,072 = \quad\quad\quad -257,754c$$
$$2.3358 = c$$

Substituting $c = 2.3358$ in Equation (A) gives

$$8,252 = 19a + 570(2.3358)$$
$$8,252 - 1,331.406 = 19a$$
$$\frac{6,920.594}{19} = a$$
$$364.2418 = a$$

Substituting the appropriate sums in Formula (14.5), we obtain

$$\frac{20,202}{570} = b$$
$$35.4421 = b$$

Finally, substituting these values for *a, b,* and *c* in Formula (14.9) gives us the parabolic regression equation:

$$Y_t = 364.2418 + 35.4421x + 2.3358x^2$$

Figure 14.8 shows the data and the fitted parabolic trend line.

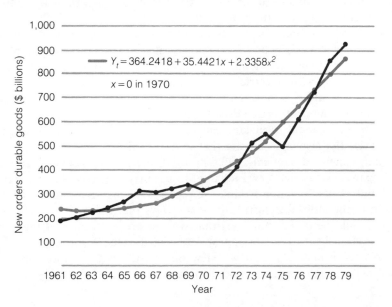

Figure 14.8
Parabolic trend line fitted to new orders for durable goods from 1961–1979

A WORD OF CAUTION

When confronted with raw time series data, how do we decide which trend equation to use? Drawing a line or curve that *seems* to fit the data may be helpful in making this decision. We may employ a goodness-of-fit test (see Chapter 11) to determine whether the selected trend equation adequately describes the data.

Often, the *purpose* of the analysis dictates the choice of the appropriate trend equation. For example, if we are merely interested in describing the historical pattern, any line that fits well will do. Similarly, if we are primarily interested in describing and eliminating the trend component in order to study nontrend movements, any equation that fits reasonably well will be appropriate. However, if we are interested in long-term projection of trend, we must analyze the implications of the type of trend line we have selected. For example, a linear trend implies a *constant* amount of growth over time. This may not be a logical assumption in many situations. Thus, if we are interested in long-range predictions, we must select a model that appears to reasonably describe the phenomenon in question.

When we find a trend that adequately describes past data, we cannot count on the same trend continuing into the future. Thus, future projections based strictly on trend predictions could lead to serious errors in planning. For example, until the early 1970s utility planning was simple. It was based on statistical tables that forecast electricity demand. For most of the century, the tables were "almost always dead right." In planning for the 1980s and beyond, most utilities assumed they could still depend on such forecasts.

They assumed wrong.

One utility company vice president for corporate planning explained that the projections "told us to shoot for 7 percent annual growth"—the industry's historical average growth in demand.

While most power companies busily planned for more fast growth, the chairman of New England Electric hesitated about construction projects. Sensing something "was going to go wrong" with oil supplies, he decided early in the 1970s that electricity demand wouldn't meet industry expectations.

Now most utility companies have too much energy from too many expensive power plants authorized decades ago. In contrast, New England Electric has adequate capital to make a quick, profitable switch in its fuel strategy. However, the chairman admits that it, too, still may have planned wrong. If energy demand suddenly picks up or if alternative energy sources don't live up to expectations, New England Electric may someday "have to scramble to build plants." "I don't think we planned wrong. But, in the uncertain world of the utilities industry, there is always that risk."[1]

It should be noted, however, that we can project values of the variable by simply substituting the appropriate coded value for the year into the trend equation. As long as we view our results with caution, we can at least make rough predictions of the future.

EXERCISES

14.1 The following table presents the quantity of money in the United States (currency, checking accounts, and savings accounts) from 1969 to 1978 in billions of dollars.

Year	Quantity ($ billions)
1969	$392
1970	425
1971	473
1972	526
1973	572
1974	612
1975	665
1976	741
1977	810
1978	872

Source: W. L. Peterson, *Principles of Economics: Macro,* 4th ed. (Homewood, Ill.: Richard D. Irwin, 1980), p. 170. © 1980 by Richard D. Irwin, Inc.

[1] Source: *The Wall Street Journal,* October 9, 1980.

  a. Fit a linear trend line to these data.
  b. Graph the trend line against the actual data.
  c. Calculate the projected trend value for 1979.

14.2 The following table presents the amount of disposable personal income (personal income less personal tax and nontax payments) in billions of dollars for 1969 to 1979.

Year	Disposable income ($ billions)
1969	$ 630
1970	686
1971	743
1972	801
1973	902
1974	985
1975	1,084
1976	1,186
1977	1,305
1978	1,458
1979	1,624

Source: *Business Statistics 1977,* Monthly S–2, p. 8.

  a. Fit an exponential trend line to these data.
  b. Graph the trend line against the actual data.
  c. Calculate the projected value for 1982.

14.3 The following table presents the annual rate (per 10,000) for business failures for every three years from 1949 to 1979.

Year	Failure rate
1949	34.4
1952	28.7
1955	41.6
1958	55.9
1961	64.4
1964	53.2
1967	49.0
1970	43.8
1973	36.4
1976	34.8
1979	27.8

Source: *Business Statistics 1977,* Monthly S–6, p. 41.

  a. Fit a parabolic trend line to these data.
  b. Graph the trend line against the actual data.
  c. Calculate the projected value for 1982.

14.4 CYCLICAL FLUCTUATIONS

Some of the most difficult movements to analyze and predict in time series data are the cyclical fluctuations. These variations are generally erratic and may vary in duration. Practically every aspect of the economy is affected by the ups and downs of the business cycle. However, despite the erratic nature of cyclical fluctuations, we can identify the typical pattern. Any analysis of the history of our economy shows frequent ups and downs. Every few years or so, national output, unemployment, and other related variables go up, level off and then the movement reverses itself. We may identify four phases in a business cycle:

trough

In a business cycle, the lowest point.

expansion

The rising phase of a business cycle.

peak

The highest point in a business cycle.

recession

The falling phase of a business cycle.

1. A *trough* is the lowest point relative to the rest of the cycle.
2. After a downswing has run its course, it reverses direction and starts rising. This is the *expansion* phase.
3. The upswing eventually levels off and reaches its *peak*. This is the highest point relative to a particular cycle.
4. Finally, the upswing starts to turn downward. We refer to this falling phase as a *recession*.

Figure 14.9 identifies these four phases.

As we indicated, one of the difficulties inherent in measuring cyclical variations in a time series is their irregularity. Therefore, it is almost impossible to find a reliable average duration for a cycle. Unfortunately, expansions and recessions do not follow each other at regular intervals. Thus, forecasting cycles is a formidable, if not impossible, task. However, for any particular time series, we can isolate the cyclical variations by removing the trend variations. Recall that we represented the time series model by the following formula:

$$Y = T \times C \times S \times I$$

Removing the trend variations:

$$\frac{Y}{T} = \frac{T \times C \times S \times I}{T} = C \times S \times I$$

Figure 14.9
Four phases of a business cycle

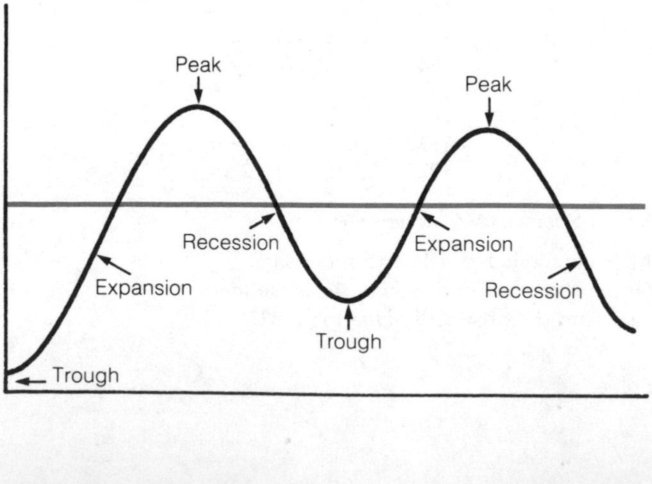

Table 14.5
Isolation of cyclical
fluctuations from the
trend of orders for
durable goods

Year	Orders ($ billions) Y	Trend value Y_t	Actual orders as percentages of trend values
1961	$189	$234	80.8%
1962	208	230	90.4
1963	225	231	97.4
1964	248	236	105.1
1965	279	245	113.9
1966	314	260	120.8
1967	310	279	111.1
1968	338	303	111.6
1969	359	331	108.5
1970	329	364	90.4
1971	359	402	89.3
1972	422	444	95.0
1973	515	492	104.7
1974	559	543	102.9
1975	506	600	84.3
1976	608	661	92.0
1977	715	727	98.3
1978	842	797	105.6
1979	927	872	106.3

New orders for durable goods as percentage of trend. Since we have removed trend variations from the data, the resulting fluctuations represent the cyclical and irregular components.

Figure 14.10

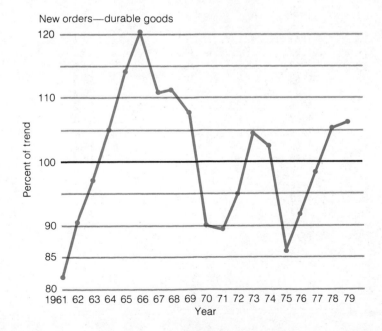

New orders—durable goods

If we are looking at annual data, we need not concern ourselves with seasonal variations since these occur only *within* a year. Thus, we are left with $Y = C \times I$. In other words, any fluctuations left can be explained in terms of cyclical and irregular variations.

The most commonly used method to separate cyclical variations from trend is to express the data as a percentage of the expected trend value.

Table 14.5 presents the relevant data from Table 14.4 (Columns 1, 3, and 8) with an additional column for percentage of trend. Each of these percentages is obtained by dividing the original data *(Y)* by the trend value *(Y_t)* and multiplying by 100. Figure 14.10 presents these results graphically.

EXERCISES

14.4 Refer to Table 14.1, Example A. Calculate the percentage of trend and graph your results.

14.5 Refer to Table 14.2, Example B. Calculate the percentage of trend and graph your results.

14.6 The following table presents the average weekly gross hours worked by production workers on private, nonagricultural payrolls.

The data were fitted to the following linear trend equation: $Y_t = 37.38 - 0.1949x$. The trend values were calculated for each year.
   a.   Graph the trend line against the actual data.
   b.   Calculate the percentage of trend for each year and graph your results.

Year	Average weekly hours	Y_t
1961	38.6	39.1
1962	38.7	38.9
1963	38.8	38.7
1964	38.7	38.5
1965	38.8	38.4
1966	38.6	38.2
1967	38.0	38.0
1968	37.8	37.8
1969	37.7	37.6
1970	37.1	37.4
1971	37.0	37.2
1972	37.1	37.0
1973	37.1	36.8
1974	36.6	36.6
1975	36.1	36.4
1976	36.2	36.2
1977	36.0	36.0
1978	35.8	35.8
1979	35.6	35.6

14.5 SEASONAL VARIATIONS

So far we have looked at data on an annual basis. Our analyses have concentrated on figures that represent an average or total over a complete calendar year. However, we are all aware of the fact that variations exist *within* a year. Some of these are predictable and relatively stable. For example, we would expect household heating bills to be higher during the winter. On the other hand, electricity bills will be higher during summer for businesses and residences that are air conditioned. A retail store specializing in gift items is likely to show higher volume around December than in, say, January.

These seasonal variations are cycles that complete themselves within a calendar year and tend to recur in the same basic pattern over the years. These variations are particularly important in short-range business planning. For example, suppose you have just bought an ice cream concession. Looking over past sales figures, you see that approximately 36,000 units are sold on an annual basis. If you ignored seasonal buying habits, you might plan for 3,000 units a month. In so doing, you could find yourself "eating" much of your stock during the winter months and running short during the summer. In this case, studying past seasonal variations would enable you to project or forecast future patterns. Your planning would then be adjusted accordingly.

Thus, when our interest is primarily in the seasonal variations that occur within a year, production, personnel, inventory, and budget can be planned and scheduled on an informed short-term basis.

On the other hand, we may wish to eliminate seasonal variations from our data so that we may study underlying business fluctuations. Many government statistics are reported as "seasonally adjusted at an annual rate." By eliminating seasonal variations (seasonally adjusting the data), we can compare such things as unemployment rates, housing starts, or personal income without concerning ourselves with the fact that the absolute (unadjusted) values of the variables differ because of the time of the year. For example, suppose a retail gift store reported $8,000 in December volume and $4,000 in January. Does this mean business has dropped? Or is this merely a reflection of what one would expect on the basis of normal or stable seasonal variations?

In this section we shall learn how to calculate seasonal indexes. Although variations occur in some businesses daily or weekly, we shall concentrate on monthly and quarterly data. Thus, our goal will be to obtain 12 monthly or 4 quarterly seasonal indexes that will show us the relative importance of each month or quarter. These indexes may then be used for short-term planning. In other words, if the seasonal variations are stable, we can use these indexes to plan on a monthly or quarterly basis. We will then see how these indexes can be used to "deseasonalize" the data, that is, remove these variations so that we can study movements across time unaffected by seasonal factors.

First, we shall summarize the general procedures. Then, we will look at worked examples that will illustrate the specific method in a step-by-step procedure.

The first step in measuring seasonal variations is to eliminate the nonseasonal elements. The rationale here is the same as we used in the previous section when we removed the trend component so as to isolate the cyclical effects. However, we shall use a slightly different procedure called the *moving average method*. We shall obtain a series of moving averages that include the trend and cyclical components. This will enable us to obtain a base line that, presumably, contains all the nonseasonal elements. We will then divide the original data by these moving averages (to find what is called the ratio-to-moving average) to eliminate the trend and cyclical elements. These remaining figures (or ratios) contain the seasonal and irregular elements.

Then, we will average these ratios by months or quarters to obtain the seasonal indexes. We shall use a technique that will minimize, if not eliminate, the irregular variations.

Finally, we can use these seasonal indexes to remove the seasonal elements from the original data, that is, we can deseasonalize the data.

Now that we have an overview of the procedure, it is time to demonstrate the procedure through examples.

Example D: Seasonal variation by quarters

Table 14.6 contains five columns. We shall use a step-by-step procedure to illustrate how the figures are obtained for each column.

1. Column 1 contains the actual data: quarterly figures for new plant and equipment expenditures in billions of dollars.

2. Column 2 represents a four-quarter moving total. These figures are obtained by summing four quarters and then successively removing the first quarter and adding in the next quarter. To illustrate, $38.01 = 7.80 + 9.16 + 9.62 + 11.43$. The next figure $39.70 = 9.16 + 9.62 + 11.43 + 9.49$. Notice we have dropped the first quarter, 7.80, and added in the next quarter instead, 9.49. By successively advancing one quarter at a time we cancel out the seasonal variations by obtaining annual totals.

3. Column 3 represents an eight-quarter moving total. These figures are obtained by successively adding two four-quarter totals at a time. Thus, $77.71 = 38.01 + 39.70$. Dropping 38.01 and adding the next total, 41.81, we arrive at $81.51 = 39.70 + 41.81$.

The purpose of this step is to center the total at the third quarter so that the moving average (Column 4) corresponds to the third quarter.

4. The moving average, (MA) figures are obtained by dividing the eight-quarter moving totals (Column 3) by 8. Since these moving averages are based on annual totals, they represent the nonseasonal elements in the data.

Figure 14.11 presents the actual data and the moving average.

5. The figures in Column 5, sometimes called the ratio-to-moving-average, are obtained by dividing the actual data by the corresponding moving average and multiplying by 100. Since we are multiplying by 100, we prefer to call these percent-of-moving-average (% MA). These percentages represent the seasonal and irregular variations only, since trend and cyclical fluctuations have been canceled out by dividing by the moving average.

		(1)	(2)	(3)	(4)	(5)
			$ billions			
			Moving total			Expenditures as
		Expenditures			Moving	percentage of
Year	Quarter	Y	4-quarter	8-quarter	average	moving average
1973	I	7.80				
	II	9.16				
	III	9.62	38.01	77.71	9.71	99.07
	IV	11.43	39.70	81.51	10.19	112.17
1974	I	9.49	41.81	85.62	10.70	88.69
	II	11.27	43.81	89.82	11.23	100.36
	III	11.62	46.01	93.37	11.67	99.57
	IV	13.63	47.36	95.60	11.95	114.06
1975	I	10.84	48.24	96.53	12.07	89.81
	II	12.15	48.29	96.25	12.03	101.00
	III	11.67	47.96	96.04	12.00	97.25
	IV	13.30	48.08	96.67	12.08	110.10
1976	I	10.96	48.59	98.99	12.37	88.60
	II	12.66	50.40	102.88	12.86	98.44
	III	13.48	52.48	106.52	13.32	101.20
	IV	15.38	54.04	110.26	13.78	111.61
1977	I	12.52	56.22	114.56	14.32	87.43
	II	14.84	58.34	118.49	14.81	100.20
	III	15.60	60.15	121.45	15.18	102.77
	IV	17.19	61.30	124.52	15.56	110.48
1978	I	13.67	63.22	127.73	15.97	85.60
	II	16.76	64.51	132.13	16.52	101.45
	III	16.89	67.62	137.45	17.18	98.31
	IV	20.30	69.83	141.98	17.75	114.37
1979	I	15.88	72.15	147.52	18.44	86.12
	II	19.08	75.37	154.28	19.28	98.96
	III	20.11	78.91	160.94	20.12	99.95
	IV	23.84	82.03	167.12	20.89	114.12
1980	I	19.00	85.09			
	II	22.14				

Table 14.6
Calculation of percent of-moving-average figures for new plant and equipment expenditures by quarters

Figure 14.12 presents these percentages against a base line of 100. The seasonal variations can be seen clearly.

Seasonal indexes. We are now ready to calculate the four seasonal indexes. Our goal is to end up with four indexes (one for each quarter) that are relatively free of irregular variations. The mean of these four indexes will be 100.0.

First we go to the percent-of-moving-averages (Column 5, Table 14.5) and arrange these figures by quarters for each year. After listing each of these figures in its appropriate column, we eliminate the highest and lowest score for each quarter. By so doing, we hope to eliminate, or at least reduce, the irregular variation in the data. We then obtain a mean of the remaining

Figure 14.11   Actual data and moving average for new plant and equipment expenditures by quarters. The moving average represents the nonseasonal components.

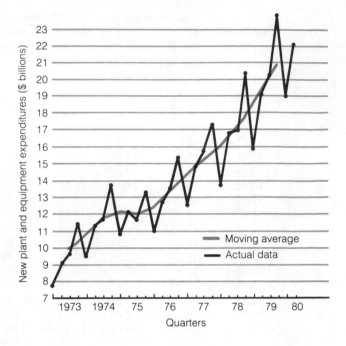

Figure 14.12   Percent-of-moving-average figures for new plant and equipment expenditures. Note the stable seasonal variations present in the data.

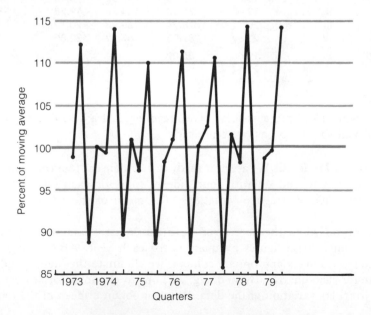

figures for each quarter. We call this a "modified mean" since we have removed the exteme scores. Table 14.7 illustrates these procedures.

In order to obtain the seasonal indexes, the four values must have a mean of 100 (or a sum of 400). Often this requires a slight adjustment to the data. As can be seen from Table 14.6, the sum of the modified means equals 399.95. In order to adjust these indexes we calculate the adjustment factor by dividing the sum into 400, or 400/399.95 = 1.0001. We then multiply each modified mean by this adjustment factor (e.g., 87.71 × 1.0001 = 87.7) to obtain the seasonal indexes. For simplicity, we shall round the seasonal indexes to one place. Thus we should not be surprised if rounding errors result in a mean not exactly equal to 100.0.

Year	Quarter			
	I	II	III	IV
1973			99.07	112.17
1974	88.69	100.36	99.57	114.06
1975	89.81	101.00	97.25	110.10
1976	88.60	98.44	101.20	111.61
1977	87.43	100.20	102.77	110.48
1978	85.60	101.45	98.31	114.37
1979	86.12	98.96	99.95	114.12
Modified means	87.71	100.13	99.62	112.49
Seasonal indexes*	87.7	100.1	99.6	112.5

Table 14.7
Calculation of quarterly seasonal indexes from percent-of-moving-average figures

* Each obtained by multiplying the corresponding modified mean by an adjustment factor (400 divided by sum of modified means, i.e., 400/399.95 = 1.0001), then rounding.

If our interest is solely in the pattern of seasonal variation, our analysis stops once we have calculated the seasonal indexes. However, if we desire to eliminate seasonal variations from the data to examine, for example, trend or cyclical effects, we must deseasonalize the data. In other words, by dividing the actual data by the corresponding seasonal index and multiplying by 100, we obtain figures free of seasonal factors. If we then wish, we can find the trend and cyclical components of the deseasonalized data. In other words, by removing the seasonal effects, we can isolate the trend and cyclical components.

Table 14.8 presents the deseasonalized figures for each quarter. Figure 14.13 plots these deseasonalized values.

Example E: Seasonal variation by months

Many business and economic variables show monthly variations. It is sometimes necessary to anticipate and plan for these fluctuations by the month rather than quarterly. Retail sales in department stores provide a good example of data that show month-to-month fluctuations.

The calculations involved are essentially the same as with quarterly data except that we will obtain 12 seasonal indexes instead of 4. Let us use Table 14.9 to illustrate the step-by-step procedure to obtain moving averages, percent-of-moving-averages, seasonal indexes, and deseasonalized data.

1. Column 1 contains the actual monthly data in millions of dollars.

2. Column 2 represents a 12-month moving total. These figures are obtained in essentially the same way as the quarterly moving totals in

Table 14.8
Deseasonalized expenditures for new plant and equipment by quarters

Year	Quarter	(1) Expenditures ($ billions) Y	(2) Seasonal indexes SI	(3) Deseasonalized expenditures ($ billions) $100Y/SI$
1973	I	$ 7.80	87.7	$ 8.89
	II	9.16	100.1	9.15
	III	9.62	99.6	9.66
	IV	11.43	112.5	10.13
1974	I	9.49	87.7	10.82
	II	11.27	100.1	11.26
	III	11.62	99.6	11.67
	IV	13.63	112.5	12.12
1975	I	10.84	87.7	12.36
	II	12.15	100.1	12.14
	III	11.67	99.6	11.72
	IV	13.30	112.5	11.82
1976	I	10.96	87.7	12.50
	II	12.66	100.1	12.65
	III	13.48	99.6	13.53
	IV	15.38	112.5	13.67
1977	I	12.52	87.7	14.28
	II	14.84	100.1	14.83
	III	15.60	99.6	15.66
	IV	17.19	112.5	15.28
1978	I	13.67	87.7	15.59
	II	16.76	100.1	16.74
	III	16.89	99.6	16.96
	IV	20.30	112.5	18.04
1979	I	15.88	87.7	18.11
	II	19.08	100.1	19.06
	III	20.11	99.6	20.19
	IV	23.84	112.5	21.19
1980	I	19.00	87.7	21.66
	II	22.14	100.1	22.12

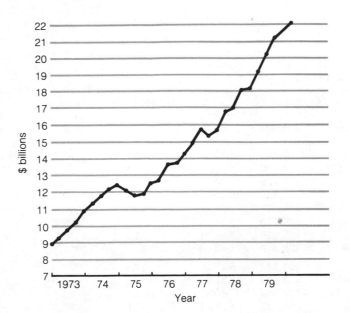

Figure 14.13
Deseasonalized figures for
new plant and equipment
expenditures

Deseasonalizing monthly sales of department stores

Table 14.9

Year	Month	(1) Sales Y	(2) 12-month	(3) 24-month	(4) Moving average MA	(5) Sales as percentage of moving average 100Y/MA	(6) Seasonal indexes SI	(7) Deseasonalized sales 100Y/SI ($ millions)
1974	July ...	$ 4,161					91.0	$4,572.5
	Aug ...	4,588					98.1	4,676.9
	Sept ...	4,242					95.2	4,455.9
	Oct ...	4,561					99.8	4,785.9
	Nov ...	5,208					116.7	4,462.7
	Dec ...	7,651					183.1	4,178.6
1975	Jan	3,257	$54,754	$109,745	$4,572.7	71.23	70.3	4,633.0
	Feb	3,159	54,991	110,227	4,592.8	68.78	69.0	4,578.3
	Mar ...	4,157	55,236	110,864	4,619.3	89.99	90.1	4,613.8
	Apr ...	4,242	55,628	111,487	4,645.3	91.32	93.8	4,522.4
	May ...	4,931	55,859	112,147	4,672.8	105.53	97.2	5,073.0
	June ...	4,597	56,288	113,730	4,738.8	97.01	95.7	4,803.6
	July ...	4,598	57,442	115,332	4,805.5	91.52	91.0	4,833.0
	Aug ...	4,833	57,890	116,204	4,841.8	99.82	98.1	4,926.6
	Sept ...	4,634	58,314	116,943	4,872.6	95.10	95.2	4,867.6
	Oct ...	4,792	58,629	118,025	4,917.7	97.44	99.8	4,801.6
	Nov ...	5,637	59,396	118,695	4,945.6	113.98	116.7	4,830.3
	Dec ...	8,805	59,299	119,001	4,958.4	177.58	183.1	4,808.8
1976	Jan	3,705	59,702	119,845	4,993.5	74.20	70.3	5,270.3
	Feb	3,583	60,143	120,477	5,019.9	71.38	69.0	5,192.8
	Mar ...	4,472	60,334	121,061	5,044.2	88.66	90.1	4,963.4
	Apr ...	5,009	60,727	122,022	5,084.2	98.52	93.8	5,340.1

Table 14.9 *(continued)*

		(1) $ millions Sales Y	(2) Moving total 12-month	(3) 24-month	(4) Moving average MA	(5) Sales as percentage of moving average 100 Y/MA	(6) Seasonal indexes SI	(7) Deseasonalized sales 100 Y/SI ($ millions)
Year	Month							
	May ...	4,834	61,295	123,195	5,133.1	96.12	97.2	4,973.3
	June ...	5,000	61,900	124,800	5,200.0	96.15	95.7	5,224.7
	July ...	4,839	62,900	125,935	5,247.3	92.22	91.0	5,317.6
	Aug ...	5,024	63,035	126,360	5,265.0	95.42	98.1	5,121.3
	Sept ...	5,027	63,325	127,222	5,300.9	94.83	95.2	5,280.5
	Oct ...	5,360	63,897	128,255	5,344.0	100.30	99.8	5,370.7
	Nov ...	6,242	64,358	129,339	5,389.1	115.83	116.7	5,348.7
	Dec ...	9,805	64,981	130,449	5,435.4	180.39	183.1	5,355.0
1977	Jan	3,840	65,468	131,589	5,482.9	70.03	70.3	5,462.3
	Feb	3,873	66,121	133,055	5,544.0	69.86	69.0	5,613.0
	Mar ...	5,044	66,934	134,638	5,609.9	89.91	90.1	5,598.2
	Apr ...	5,470	67,704	136,227	5,676.1	96.37	93.8	5,831.6
	May ...	5,457	68,523	138,094	5,753.9	94.84	97.2	5,614.2
	June ...	5,487	69,571	141,154	5,881.4	93.29	95.7	5,733.5
	July ...	5,492	71,583	143,651	5,985.5	91.76	91.0	6,035.2
	Aug ...	5,837	72,068	144,667	6,027.8	96.83	98.1	5,950.1
	Sept ...	5,797	72,599	146,021	6,084.2	95.28	95.2	6,089.3
	Oct ...	6,179	73,422	147,361	6,140.0	100.64	99.8	6,191.4
	Nov ...	7,290	73,939	148,822	6,200.9	117.56	116.7	6,246.8
	Dec ...	11,817	74,883	150,771	6,282.1	188.11	183.1	6,453.9
1978	Jan	4,325	75,888	152,221	6,342.5	68.19	70.8	6,108.8
	Feb	4,404	76,333	153,319	6,388.3	68.94	69.0	6,382.6
	Mar ...	5,867	76,986	154,613	6,442.2	91.07	90.1	6,511.7
	Apr ...	5,987	77,627	155,685	6,486.9	92.29	93.8	6,382.7
	May ...	6,401	78,058	156,734	6,530.6	98.02	97.2	6,585.4
	June ...	6,492	78,676	158,170	6,590.4	98.51	95.7	6,783.7
	July ...	5,937	79,494	159,410	6,642.1	89.38	91.0	6,524.2
	Aug ...	6,490	79,916	160,128	6,672.0	97.27	98.1	6,615.7
	Sept ...	6,438	80,212	160,861	6,702.5	96.05	95.2	6,762.6
	Oct ...	6,610	80,649	161,822	6,742.6	98.03	99.8	6,623.2
	Nov ...	7,908	81,173	162,891	6,787.1	116.52	116.7	6,776.3
	Dec ...	12,635	81,718	163,754	6,823.1	185.18	183.1	6,900.6
1979	Jan	4,747	82,036	164,531	6,855.5	69.24	70.8	6,704.8
	Feb	4,700	82,495	165,905	6,912.7	67.99	69.0	6,811.6
	Mar ...	6,304	83,410	167,487	6,978.6	90.33	90.1	6,996.7
	Apr ...	6,511	84,077	169,167	7,048.6	92.37	93.8	6,941.4
	May ...	6,946	85,090	171,639	7,151.6	97.13	97.2	7,146.1
	June ...	6,810	86,549	174,393	7,266.4	93.72	95.7	7,116.0
	July ...	6,396	87,844	176,429	7,351.2	87.01	91.0	7,028.6
	Aug ...	7,405	88,585	178,041	7,418.4	99.82	98.1	7,548.4
	Sept ...	7,105	89,456	179,378	7,474.1	95.06	95.2	7,463.2
	Oct ...	7,623	89,922	180,308	7,512.8	101.47	99.8	7,638.3
	Nov ...	9,367	90,386	181,562	7,565.1	123.82	116.7	8,075.0
	Dec ...	13,930	91,176	182,658	7,610.8	183.03	183.1	7,607.9

Table 14.9 *(continued)*

Year	Month	(1) Sales Y	(2) $ millions Moving total 12-month	(3) $ millions Moving total 24-month	(4) Moving average MA	(5) Sales as percentage of moving average 100 Y/MA	(6) Seasonal indexes SI	(7) Deseasonalized sales 100 Y/SI ($ millions)
1980	Jan	5,488	91,482				70.8	7,751.4
	Feb	5,571					69.0	8,073.9
	Mar . . .	6,770					90.1	7,513.9
	Apr . . .	6,975					93.8	7,436.0
	May . . .	7,736					97.2	7,958.8
	June . . .	7,116					95.7	7,435.7

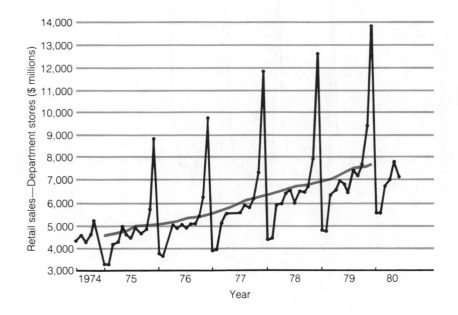

Figure 14.14

Monthly data and moving average for retail sales in department stores

Example D. After adding the first 12 months, we successively drop the first figure and add the next.

3. Column 3 is obtained by adding two 12-month totals and then successively dropping the first and adding the next.

4. The moving averages are obtained by dividing the corresponding 24-month moving total (Column 3) by 24.

Figure 14.14 presents the moving averages plotted against the actual monthly data.

5. To obtain the percent-of-moving-average figures, we divide the original data by the corresponding moving average and multiply by 100.

Figure 14.15 presents the percent-of-moving-average figures. Seasonal effects can be seen clearly.

6. The calculation of the seasonal indexes in Column 6 is shown in Table 14.10.

7. Finally, the deseasonalized values in Column 7 were obtained by dividing the original data by the corresponding seasonal index and multiplying by 100.

Figure 14.15

Percent-of-moving-average figures for retail sales in department stores

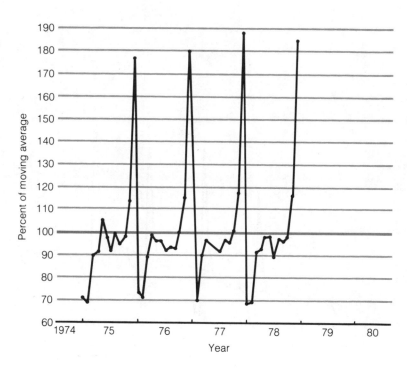

EXERCISES

14.7 What method is used to cancel out seasonal variations? Explain the rationale.

14.8 What components are contained in the moving average?

14.9 What components are contained in the percent-of-moving-average figures? What components are cancelled out?

14.10 In calculating seasonal indexes, why do we use modified means?

Calculation of monthly seasonal indexes for retail sales in department stores   Table 14.10

Year	Jan	Feb	Mar	Apr	May	June	July	Aug	Sept	Oct	Nov	Dec
1975	71.23	68.78	89.99	91.32	105.53	97.01	91.52	99.82	95.10	97.44	113.98	177.58
1976	74.20	71.38	88.66	98.52	96.12	96.15	92.22	95.42	94.83	100.30	115.83	180.39
1977	70.03	69.86	89.71	96.37	94.84	93.29	91.76	96.83	95.28	100.64	117.56	188.11
1978	68.19	68.94	91.07	92.29	98.02	98.51	89.38	97.27	96.05	98.03	116.52	185.18
1979	69.24	67.99	90.33	92.37	97.13	93.72	87.01	99.82	95.06	101.47	123.82	183.03
Modified means	70.2	68.9	90.0	93.7	97.1	95.6	90.9	98.0	95.1	99.7	116.6	182.9
Monthly seasonal indexes* ...	70.3	69.0	90.1	93.8	97.2	95.7	91.0	98.1	95.2	99.8	116.7	183.1

* Each obtained by multiplying the corresponding modified mean by an adjustment factor (1,200 divided by sum of modified means, i.e., 1200/1198.6 = 1.001).

14.11 What is meant by "deseasonalizing" data? How is this accomplished?

14.12 The following table presents the amount of consumer credit extended (in billions of dollars) monthly.

  a. Calculate the moving averages.
  b. Graph the actual data against the moving average.
  c. Calculate the percent-of-moving-average figures.
  d. Graph the percent-of-moving-average figures.
  e. Calculate the monthly seasonal indexes.
  f. Deseasonalize the data.

Year	Month	Credit extended ($ billions)	Year	Month	Credit extended ($ billions)
1974	July	$14.45	1977	July	$18.80
	Aug	14.30		Aug	21.31
	Sept	12.73		Sept	19.30
	Oct	12.83		Oct	18.78
	Nov	11.87		Nov	19.72
	Dec	13.62		Dec	21.43
1975	Jan	10.82	1978	Jan	16.72
	Feb	10.66		Feb	16.67
	Mar	11.96		Mar	21.98
	Apr	13.24		Apr	21.34
	May	13.38		May	24.00
	June	14.19		June	25.03
	July	15.06		July	22.42
	Aug	14.63		Aug	25.14
	Sept	14.40		Sept	21.89
	Oct	14.72		Oct	25.29
	Nov	13.98		Nov	25.71
	Dec	17.11		Dec	27.49

Year	Month	Credit extended ($ billions)	Year	Month	Credit extended ($ billions)
1976	Jan	12.89	1979	Jan	22.61
	Feb	13.12		Feb	22.00
	Mar	16.11		Mar	26.46
	Apr	15.91		Apr	27.02
	May	15.76		May	29.76
	June	17.77		June	28.02
	July	16.48		July	27.70
	Aug	17.25		Aug	30.51
	Sept	16.58		Sept	26.99
	Oct	15.06		Oct	28.09
	Nov	16.81		Nov	26.24
	Dec	19.59		Dec	27.16
1977	Jan	14.05	1980	Jan	23.41
	Feb	14.57		Feb	23.12
	Mar	18.90		Mar	25.48
	Apr	18.73		Apr	23.30
	May	19.28		May	22.15
	June	20.76		June	22.01
				July	25.87

14.6 EVALUATING FORECASTS

We have seen that there are basically three elements in a forecast:

1. Uncertainty: If it were not for the fact the future cannot be predicted with certainty, there would be no need for peering into the future. Broadly speaking, the greater the uncertainty, the greater the need for a workable forecasting tool.

2. Dependence on historical records: There is a familiar saying, "Those who do not look to the past are destined to repeat the errors of the past." This saying reminds us that forecasting the future is not done in a vacuum. Whether we like it or not, we are forced to make use of past and present experiences when we wish to "get a handle" on the future. This inescapable fact is at once a potential Achilles heel. Forecasting techniques assume stable conditions—that the future is a continuation of the past. When the economic or business behavior of interest is subject to many powerful and unpredictable influences (catastrophes, wars, embargoes on goods, etc.), the forecasts will not do much better than the crystal globes of carnival seers.

3. Time frame: The forecasts are always made within a time frame that may vary from hours, to days, to weeks, to months, and even to decades. In general, we would expect that the shorter the time frame, the more accurate the forecast. This is due to the fact that fewer unpredictable omitted variables are likely to intrude themselves over short periods of time. Thus, a TV weather

person can usually forecast with confidence what the weather will be during the next 12-hour period but be less assured about the weekend weather.

14.6.1   Sources of error

Recall that, in regression analysis, we referred to the scatter around the regression line as error. As long as we could assume the errors are independent, we were able to use this error to establish confidence and predictive intervals. However, when forecasting the future, we are understandably reluctant to project these intervals beyond the observed historical records. Moreover, in time series, the errors are often serially correlated. Consequently, regression analyses are not often used to evaluate forecasting models. Experience has shown that the actual errors tend to be larger than predicted.

Earlier we distinguished between two types of error—bias and random. Bias occurs when we introduce systematic errors so that the forecast is consistently "off target." These usually represent mistakes in the forecast model—failure to include the correct variables, using the wrong trend line, and failure to detect subtle but consistent changes in the historical record. Random errors, on the other hand, typically result from the interaction of many variables, some of which we have not identified and others that, for various reasons, we have been unable to include in the forecast model.

Figure 14.16 illustrates these two types of forecast errors.

Random and systematic (biased) errors in forecasting. The line represents the forecast over time and the dots represent the actual observations. Note that, in panel A, the errors appear to be distributed in a non-systematic fashion about the trend line (random error). In panel B, the errors are consistently above the trend line. Thus, the wrong trend line was used.   **Figure 14.16**

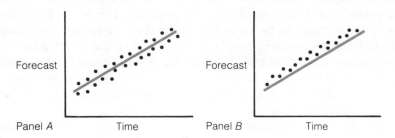

Panel A   Time   Panel B   Time

14.6.2   Assessing forecasting

In this section, we shall examine three commonly employed measures used in assessing forecasting—the mean absolute deviation (MAD); the mean error squared; and the running sum of forecast errors (RSFE), which is combined with MAD to provide "tracking signals" to detect possible bias in the forecast.

Table 14.11 shows a forecast for demand for automobiles during a six-month period, the actual number purchased, and the deviations needed in the calculation of MAD.

Month	Forecast	Actual purchases	Actual deviation	Absolute deviation
1	1,250	1,200	−50	50
2	1,000	950	−50	50
3	950	1,050	100	100
4	1,100	1,250	150	150
5	1,300	1,400	100	100
6	1,400	1,500	100	100
Sums			350	550

The mean absolute deviation is defined as

$$MAD = \frac{\Sigma(|A - F|)}{n}$$

where

$A =$ Actual demand for a given period
$F =$ The forecast for a given period

For the forecasts shown in Table 14.11,

$$MAD = \frac{550}{6} = 91.67$$

Note that the sum of the actual deviations is 350. If the trend line were better fitted to the data, the deviations above and below the forecast would tend to balance out. In fact, if the trend line had been perfectly fitted, the sums of the deviations would equal zero. The fact that this sum equals 350 suggests that the forecast model generally errs on the conservative side, i.e., it forecasts smaller demand than actually occurs. The MAD figure shows that, on the average, the forecasts were off by about 92 automobiles a month.

If the forecast errors are normally distributed, the standard deviation equals approximately 1.25 × MAD. Thus, in the present example, the standard deviation of MAD would be 1.25 × 91.67 = 114.59. However, since for expository purposes, the n was kept extremely small, the question of normality of errors is difficult to assess.

One feature of MAD is that errors of large magnitudes are treated arithmetically no differently from small errors. That is, there is no particular penalty for making a large forecast error. One large error may have no more effect on MAD than many small errors. Yet, from the point of view of management, a single large error in forecasting might have far more serious ramifications. Now if we were to square the deviations, sum, and divide by n, we would obtain the squared average deviation (SAD). Table 14.12 illustrates the effect on SAD of one large deviation in forecast compared to that of many smaller errors.

Table 14.12 shows that, although the mean absolute deviations of the forecasts were identical, the average of the squared deviations (SAD) was much higher in Forecast 2. This difference may be significant in management

Month	Forecast 1				Forecast 2			
	Fore-cast	Actual	Absolute deviation	Deviation squared	Fore-cast	Actual	Absolute deviation	Deviation squared
1	1,250	1,200	50	2,500	1,250	1,250	0	0
2	1,000	950	50	2,500	1,000	1,000	0	0
3	950	1,050	100	10,000	950	1,350	400	160,000
4	1,100	1,250	150	22,500	1,100	1,050	50	2,500
5	1,300	1,400	100	10,000	1,300	1,250	50	2,500
6	1,400	1,500	100	10,000	1,400	1,450	50	2,500
Sums ..			550	57,500			550	167,500

Table 14.12

Comparison of MAD and SAD measures of error in two different forecasts

$$MAD_1 = \frac{550}{6} = 91.67 \qquad MAD_2 = \frac{550}{6} = 91.67$$

$$SAD_1 = \frac{57,500}{6} = 9500 \qquad SAD_2 = \frac{167,500}{6} = 27916.67$$

decisions whenever emphasis is placed on minimizing large errors in forecasting.

The final measurement of error that we will be examining is called a *tracking signal,* which indicates the extent to which the forecast average is keeping pace with actual upward or downward changes in demand. Tracking signal is defined as

tracking signal

An indication of the extent to which the forecast average is keeping pace with actual changes in demand.

$$\text{Tracking signal} = \frac{\text{RSFE}}{\text{MAD}}$$

where

$$\text{RSFE} = \text{Running sum of forecast errors}$$

The running sum of forecast errors is obtained by cumulating the actual deviations and dividing by n. Table 14.13 illustrates the calculations of RSFE and tracking signal from the data in Table 14.11.

Month	Forecast	Actual purchases	Actual deviation	RSFE
1	1250	1200	−50	−50
2	1000	950	−50	−100
3	950	1050	100	0
4	1100	1250	150	150
5	1300	1400	100	250
6	1400	1500	100	350

Table 14.13

Calculation of RSFE and tracking signal for automobile data

$$MAD = \frac{550}{6} = 91.67$$

$$RSFE = \frac{350}{6} = 58.33$$

$$\text{Tracking signal} = \frac{58.33}{91.67} = 0.64$$

In an ideal forecasting model, we would expect MAD to be zero. In this case, tracking signal would also be zero. In actual practice, this will rarely be the case. In general, the larger RSFE relative to MAD, the greater the tracking signal. Large tracking signal errors indicate that the forecast is failing to keep pace with genuine trends in the data.

14.7 A FINAL WORD

We started this chapter out with the somewhat pessimistic view that the chances of correctly forecasting the future were somewhere between slim and none. That is not precisely the case. We have learned techniques that should enhance our predictions based on our knowledge of the past. However, we have merely scratched the surface. The procedures we have discussed are relatively simplistic. Thus predictions based on these procedures will be rough estimates, to say the least. However, all is not lost. For most of us, the estimates we obtain based on these rudimentary forecasting tools will be sufficient for our purposes. In many cases, we need only rough estimates of the future to plan intelligently.

For example, suppose you decide to go into a mail-order business that relies heavily on leads obtained through advertising. You now have the tools necessary to analyze ad-response and decide when to advertise heavily and when to cut back. Of course, we are assuming you would collect the appropriate data to make these decisions.

There are forecasting techniques available that are quite sophisticated and go considerably beyond the scope of what we have covered in this chapter. In many cases, these techniques use extremely complex equations and require computer analysis. Although we do not cover these procedures, be aware that they exist.

In this chapter we have attempted to introduce some of the fundamental concepts underlying more sophisticated prediction procedures. For example, we touched on the problem of business cycles and the inherent frustration in attempting to predict their duration, amplitude, etc. Economists have identified certain measures or indicators that, in general, tend to precede upswings or downswings in the economy. They call these *leading* indicators. Thus, if certain leading indicators go down this month, for example, we should expect a downturn in the economy. When? One economist has indicated that of 12 leading indicators, the median lead time (how soon they go up or down before the rest of the economy) averages about five months. Therefore, one is tempted to watch these indicators and make future predictions based on their behavior. However, the same economist reports that these same "leading" indicators have been known to turn up or down anywhere from two years *before* the rest of the economy to almost two years *after* the other indicators.

SUMMARY

1. In this chapter we focused our attention on a commonly used technique of forecasting—the time series model. Using this model, we analyzed

data over given time periods to learn historical patterns and estimate future ones.

2. Variations in time series consist of the following components:
    a. Trend—a long term upward or downward movement of a time series.
    b. Cyclical fluctuations—periodic ups and downs over periods of time that vary in length and size, but are greater than one year.
    c. Seasonal variations—cycles that complete themselves during a one-year period.
    d. Irregular fluctuation—variations that are random and erratic.

3. Trend movements in a time series may take many different forms. We concentrated on three different models and examined equations that describe the trends.
    a. A linear or straight line trend, which assumes that the values of the dependent variable increase (or decrease) by a constant absolute amount over the series.
    b. Exponential trend lines, which describe series that increase (or decrease) at a constant rate or by a constant percentage.
    c. Parabolic trend lines, which are useful for describing series that change direction (e.g., increase for a while and then decrease, or vice versa).

4. Cyclical fluctuations are the most difficult to analyze and predict because of their erratic nature. However, we identified four phases describing the upswings and downswings common to all business cycles.

    The most commonly used technique to separate cyclical variations from trend is to express the data as a percentage of its expected trend value. In other words, after removing the trend variations from the data, the resulting variations may be attributed to cyclical and irregular components.

5. Seasonal variations focus on movements in the data that occur within a calendar year and tend to recur in the same basic pattern over the years. If we are interested only in the pattern of seasonal variation, our goal is to obtain seasonal indexes that indicate the relative importance of each month (or quarter). However, if we wish to eliminate the seasonal variations from the data, we use these seasonal indexes to deseasonalize the data. A summary of the key steps:
    a. Remove the seasonal elements by obtaining a moving annual average. This moving average presumably contains the trend, cyclical, and irregular components.
    b. Express the data as a percentage of this moving average. In effect, this removes the trend and cyclical variations from the data, leaving only seasonal and irregular components.
    c. By combining the percentages for each month (or quarter) and removing the highest and lowest values, derive a modified mean that can be used to arrive at the seasonal indexes.
    d. Deseasonalize the data by dividing the original data by the corresponding seasonal index and multiplying it by 100. The resulting figures represent seasonally adjusted data.

time series **(441)**
trend **(441)**
cyclical fluctuations **(443)**
seasonal variations **(443)**
trough **(454)**

expansion **(454)**
peak **(454)**
recession **(454)**
tracking signal **(471)**

EXERCISES

14.13 The following table presents the total U.S. population (including armed forces overseas) rounded to the nearest million.
  a. Fit a linear trend line to these data.
  b. Graph the trend line against the actual data.

Year	Population (millions)	Year	Population (millions)
1960	181	1970	205
1961	184	1971	207
1962	187	1972	209
1963	189	1973	210
1964	192	1974	212
1965	194	1975	214
1966	197	1976	215
1967	199	1977	217
1968	201	1978	219
1969	203	1979	221

14.14 Many banks or lending institutions will not consider lending to a borrower who has been in business less than five years. Can you guess why?

14.15 Refer to Table 14.8, Example D.
  a. Using the deseasonalized values for new plant and equipment expenditures (Column 3), fit a parabolic trend line.
  b. Calculate each of the trend values.
  c. What do these trend values mean?
  d. Graph the trend line against the actual data.
  e. Divide each of the deseasonalized values by its corresponding trend value and multiply by 100. Graph the resulting figures and explain what is reflected by these figures.
  f. Predict the seasonally adjusted amount spent on new plant and equipment for the third and fourth quarters of 1980.

14.16 One real estate broker claimed that after President Reagan was elected, values of homes in the Pacific Palisades area surrounding his home were appreciating at the rate of 2 percent a month. What type of trend line would best describe this type of movement?

14.17 The following table presents quarterly sums of monthly averages of the prime rate from 1973 to 1980.
  a. Calculate the moving averages.
  b. Graph the actual data against the moving average.
  c. Calculate the percent-of-moving-average figures.

d. Graph the percent-of-moving-average figures.
e. Calculate the quarterly seasonal indexes.
f. Deseasonalize the data.
g. Graph the deseasonalized data.

Year	Quarter	Prime rate sums	Year	Quarter	Prime rate sums
1973	I	18.56	1977	I	14.44
	II	22.10		II	15.51
	III	29.56		III	17.47
	IV	26.74		IV	19.75
1974	I	24.98	1978	I	20.47
	II	31.08		II	21.99
	III	35.02		III	24.54
	IV	27.56		IV	30.40
1975	I	20.11	1979	I	30.24
	II	16.61		II	29.67
	III	19.97		III	32.31
	IV	17.79		IV	40.28
1976	I	15.25	1980	I	34.26
	II	16.33		II	33.54
	III	16.10		III	29.56
	IV	14.58		IV	39.91

14.18 Sally B. is in charge of allocating advertising expenditures within a specified annual budget of $100,000. Previously, the budget was equally divided by quarters. However, she suspects that certain times of the year yield greater value for the dollar in terms of higher response rate. She decides to analyze the number of responses received quarterly over a five-year period. Based on her analysis, how should she allocate her budget by quarters? Explain how you arrived at this decision.

The table below summarizes the number of responses received each quarter over a five-year period.

Year	Quarter	Number of responses (hundreds)	Year	Quarter	Number of responses (hundreds)
1	I	10.3	4	I	10.8
	II	11.2		II	12.2
	III	7.5		III	9.2
	IV	6.5		IV	7.8
2	I	10.2	5	I	10.9
	II	11.4		II	14.6
	III	8.0		III	10.6
	IV	6.8		IV	9.0
3	I	10.4			
	II	11.6			
	III	8.5			
	IV	7.2			

14.19 The following table shows the exports (in millions of dollars) of cigarettes from the United States.

a. Fit an exponential curve to these data.

b. Calculate the projected value for 1980.

Year	Exports ($ millions)
1973	$250
1974	301
1975	368
1976	510
1977	615
1978	749

Source: U.S. Bureau of the Census, *Statistical Abstract of the United States:* (Washington, D.C.: U.S. Government Printing Office, 1979), p. 867.

14.20 The following table presents the total population of persons 14 to 24 years old.

Year	Total number (000)
1930	123,077
1940	132,122
1950	152,271
1960	180,671
1970	204,878

Source: U.S. Bureau of the Census, *Statistical Abstract of the United States:* (Washington, D.C.: U.S. Printing Office, 1979), p. 32.

a. Fit a linear trend line to these data.

b. Calculate the projected values for 1975 and 1978.

c. Compare your projections with the actual figures (1975: 213,555; 1978: 218,548) and explain any disparities between the actual and predicted figures.

14.21 Following are two forecasts of the repair costs on capital equipment and the actual costs during the forecast period.

a. Calculate and compare the MAD, SAD, and tracking signals of the two forecasts.

b. Which forecast appears to be better?

c. Do the errors in Forecast 1 appear to be serially correlated?

Period	Forecast 1	Forecast 2	Actual
1	12,400	12,000	12,360
2	13,200	12,300	12,730
3	14,000	12,700	13,110
4	14,800	13,200	13,500
5	15,600	13,800	13,910
6	16,400	14,500	14,330
7	17,200	15,300	14,760
8	18,000	16,200	15,200
9	18,800	17,200	15,650
10	19,600	18,300	16,130

Index Numbers

15.1 THREE MAJOR TYPES OF INDEX NUMBERS

Lana T. was justifiably proud of herself. Several years before, she and her husband had pooled their life savings, refinanced the house, and in various other ways raised sufficient funds to open an appliance store at the east end of town. It was a calculated risk and they knew it since the population was relatively sparse on the east side. Friends had warned them that there might not be sufficient volume to maintain a profitable business. However, the town was in the sun belt, energy shortages and life style changes were causing increasing numbers of industries to locate in this region of the country, and the east side of town was an obvious locale for burgeoning growth. The gamble had paid off. The second gamble was the one in which Lana took special pride. She had convinced her husband, Ken, that they should invest in a small personalized computer. With it, she reasoned they could keep running accounts of sales, inventory, costs, profits, and the million-odd pieces of information demanded by Uncle Sam. The computer had been her responsibility, expanding with the business firm.

Today she was going to ask the computer to print out the sales of various classes of major appliances over the first 12 quarters of their business. She was then going to ask it to compile various types of *index numbers* so that she and her husband could evaluate the changes that had taken place since opening the appliance store.

The first print out dealt with sales of refrigerators. The summary of price, quantity, and total value appears in Table 15.1.

It was clear at a glance to Lana and her husband that the prices, quantity, and total value of refrigerators sold had all shown exuberant gains since the first quarter in 1979. What they wanted now was a quantitative expression of these changes. Lana requested that the computer execute the computation

index number

A ratio, usually expressed as a percentage, of prices, quantities or value that relates a given period with a comparison period.

Table 15.1

Summary of average price, quantity sold, and total value of refrigerators sold during the first 12 quarters of operation of an appliance store

Year	Quarter	Average price	Quantity	Total value
1979	1	$335	2	$ 670
	2	341	5	1,705
	3	347	8	2,776
	4	354	6	2,124
1980	5	363	4	1,452
	6	370	9	3,330
	7	377	11	4,147
	8	385	7	2,695
1981	9	394	6	2,364
	10	401	10	4,010
	11	412	15	6,180
	12	444	12	5,328

of index numbers on all three variables using the first quarter as a *base*. This is done by dividing each value of the variable by its value in the base quarter and multiplying by 100 to express the ratio as a percentage. This means that the index number for the base period will show up as 100. A value larger than 100 for a given period means a gain relative to the base period; a smaller value, a loss.

To illustrate, the index number for price is defined as

$$I_p = \frac{P_n}{P_0} \times 100 \qquad (15.1)$$

where

I_p = *Price index* or *price relative*
P_n = Price of a given product at a given time
P_0 = Price of the product in the base period

Thus, if the base period is the first quarter and P_n is the price in the seventh quarter, $I_p = \frac{377}{335} \times 100 = 113$.

Similarly, the *quantity index* or *quantity relative* is

$$I_q = \frac{Q_n}{Q_0} \times 100 \qquad (15.2)$$

where

I_q = Quantity index or quantity relative
Q_n = Quantity produced or sold in a given period
Q_0 = Quantity produced or sold in the base period

If the base period is the sixth quarter and Q_n is the quantity produced or sold in the 11th quarter, $I_q = \frac{15}{9} \times 100 = 167$.

Finally, the *value index* is the product of the price times quantity produced or sold in a given period divided by the product of the price times quantity produced or sold in the base period. Symbolically,

$$I_v = \frac{P_n Q_n}{P_0 Q_0} \times 100 \qquad (15.3)$$

where

I_v = Value index or value relative
$P_n Q_n$ = The product of the price times quantity produced or sold in a given period
$P_0 Q_0$ = The product of the price times quantity produced or sold in the base period

base

The price, quantity, or value of a period with which a given period is compared.

price index (price relative)

An index based on price.

quantity index (quantity relative)

An index based on quantity.

value index (value relative)

An index representing the product of price times quantity.

Table 15.2 summarizes the calculation of three major types of index numbers—price, quantity, and value.

Table 15.2
Calculation of index numbers for price, quantity, and value

		Price			Quantity			Value		
Quarter	Average price	Ratio	Index number	Number sold	Ratio	Index number	Total price	Ratio	Index number
1* ...	$335	$\frac{335}{335} = 1.00$	100	2	1.00	100	$ 670	1.00	100
2	341	$\frac{341}{335} = 1.02$	102	5	2.50	250	1,705	2.54	254
3	347	$\frac{347}{335} = 1.04$	104	8	4.00	400	2,776	4.14	414
4	354	$\frac{354}{335} = 1.06$	106	6	3.00	300	2,124	3.17	317
5	363	$\frac{363}{335} = 1.08$	108	4	2.00	200	1,452	2.17	217
6	370	$\frac{370}{335} = 1.10$	110	9	4.50	450	3,330	4.97	497
7	377	$\frac{377}{335} = 1.13$	113	11	5.50	550	4,147	6.18	619
8	385	$\frac{385}{335} = 1.15$	115	7	3.50	350	2,695	4.02	402
9	394	$\frac{394}{335} = 1.18$	118	6	3.00	300	2,364	3.53	353
10	401	$\frac{401}{335} = 1.20$	120	10	5.00	500	4,010	5.99	599
11	412	$\frac{412}{335} = 1.23$	123	15	7.50	750	6,180	9.22	922
12	444	$\frac{444}{335} = 1.33$	133	12	6.00	600	5,328	7.95	795

* Base period, for which all index numbers are 100.

Next, Lana charted all three index numbers on a single graph (see Figure 15.1). It was immediately clear to her that the percentage relatives for both quantity and value were far outstripping the increase in the average price for refrigerators. She was pleased that only a small proportion of the total gains could be attributed to inflationary factors. They were simply selling more refrigerators. She realized, however, that if there had been no increases in price, the index numbers for both quantity and price would have been identical for each quarter. Lana concluded that, for any given quarter, the difference between the percentage relative for value and the percentage relative for quantity represents the cumulative effect of the inflationary trend in the price of refrigerators. No surprises there, she thought. Everything is as it should be since value is a product of price and quantity. She generalized by noting that if she multiplied the ratio of price for any given quarter by the ratio of quantity for that quarter, she would obtain the ratio of value, allowing for small errors due to rounding. To confirm her generalization

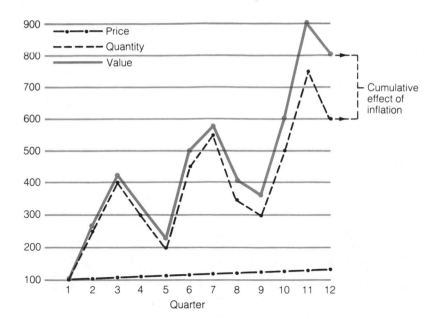

Figure 15.1
Index numbers for price, quantity, and value

she looked at the 10th quarter. She noted that the price ratio was 1.20 and the quantity ratio was 5.00. The product of the two is 6.00. The difference of 0.01 between this and the actual value of 5.99 she attributed to rounding error.

Satisfied that all was well in the land of refrigerators, Lana set her sights on some product lines about which she had grave misgivings. . . .

In this brief scenario, we have looked at simplified forms of three major types of index numbers or percentage relatives—price, quantity and value. In everyday life, virtually any variable that can be charted over time can also be converted into index numbers: crime statistics, economic measures, population changes, energy expenditures, resources discovered, knowledge accumulated, and so forth almost without limit. They may be used in business and economics as barometers of the economic climate or to assess short- and long-term trends (See Chapter 14). Most commonly they represent more than a single variable. When more than one variable makes up an index, it is referred to as a *composite* or *aggregate index*.

aggregate index (composite index)

An index that summarizes a group of items or components.

EXERCISES

15.1   Describe some of the reasons for calculating and using index numbers (percentage relatives).

15.2   List some of the index numbers with which you are already familiar.

15.3   Distinguish among the three major types of indexes.

15.4 Shown on the accompanying table is the monthly price per pound of butter and the quantities produced during a 10-month period in 1979 and 1980. Using December as the base, calculate unweighted index numbers for the period.

    a. Price.
    b. Quantity.
    c. Value.

Period	Price ($ per lb.)	Quantity (millions of lbs.)
September 1979	1.358	60.5
October 1979	1.342	78.0
November 1979	1.353	75.8
December 1979	1.366	84.0
January 1980	1.347	103.8
February 1980	1.357	99.1
March 1980	1.367	101.7
April 1980	1.396	111.1
May 1980	1.413	116.4
June 1980	1.424	93.8

15.2 UNWEIGHTED COMPOSITE OR AGGREGATE INDEXES

Suppose Ken and Lana wished to prepare a composite index of prices so that they could compare overall changes in prices during the various quarters that their store had been in operation. They had independently observed that four different items appeared to be good bellwethers of future economic activity. These were refrigerators, ranges, toasters, and TV sets. They requested the computer to print out a composite price index comparing the average prices of these four items during the 4th and 12th quarters. Note that the prices are simply added together to obtain the composite price for each quarter. This is the reason it is called an aggregate index. Since the sum does not take into account the quantities sold, it is referred to as an *unweighted aggregate index* or an unweighted aggregate price relative. The formula for calculating an unweighted aggregate price relative is

unweighted aggregate index

An index that treats each item or component as equal

$$\text{Unweighted aggregate price index} = \frac{\Sigma P_n}{\Sigma P_0} \times 100 \qquad (15.4)$$

where

ΣP_n = The sum of the prices for a given period
ΣP_0 = The sum of the prices for the base period

Table 15.3 shows the average prices of the four appliances during the 1st and 12th quarters and the calculation of the unweighted aggregate price index.

Compared to the base quarter, thus, the unweighted aggregate price increase of the four items was 28 percent.

	Average prices	
	Quarter 1 P_0	Quarter 12 P_n
Refrigerators	355	444
Ranges	425	588
Toasters	18	25
TV sets	390	440
	$\Sigma P_0 = 1168$	$\Sigma P_n = 1497$

Table 15.3
Average prices on four appliances during the 1st and 12th quarters of business

$$\text{Unweighted aggregate price index} = \frac{1{,}497}{1{,}168} \times 100 = 128$$

The analogous index for quantity—called the unweighted aggregate quantity index—involves summing the quantities sold or produced over a given period, dividing by the quantities sold or produced during the base period, and multiplying by 100. Thus,

$$\text{Unweighted aggregate quantity index} = \frac{\Sigma Q_n}{\Sigma Q_0} \times 100 \qquad (15.5)$$

where

$\Sigma Q_n =$ The sum of the quantity produced or sold during a given time period

$\Sigma Q_0 =$ The sum of the quantity produced or sold during the base time period

Table 15.4 shows the quantities of the four appliances sold over the two quarters and the calculation of the unweighted aggregate quantity index.

	Quantities	
	Quarter 1 Q_0	Quarter 12 Q_n
Refrigerators	2	12
Ranges	3	18
Toasters	22	58
TV sets	12	29
	$\Sigma Q_0 = 39$	$\Sigma Q_n = 117$

Table 15.4
Quantities of four appliances sold during the 1st and 12th quarters of operation of a retail store

$$\text{Unweighted aggregate quantity index} = \frac{117}{39} \times 100 = 300$$

Thus, the unweighted quantity index increased by 200 percent during the 12th quarter when compared to the base quarter.

Note that merely summing the prices without regard to quantities sold or the quantities sold without regard to price ignores the importance of quantity in the unweighted price index and price in the unweighted quantity index. From the point of view of the retailer, the sale of a major appliance is usually a more important event than the sale of a small appliance. A further difficulty with unweighted aggregate price indexes is that the value we obtain depends on the units of measure that we use. To illustrate, if we included the price of milk as an item in a grocery basket type of index, do we use the price of a pint, quart, half gallon, or gallon?

Suppose we obtained the simple aggregate price index of four commodities, using 1975 as the base year and 1978 as the given year. The results are shown in Table 15.5.

Table 15.5
Prices of four commodities in 1975 and 1978

	Price	
Commodity	1975	1978
Milk (pint)	19.6	21.4
Round steak (pound)	188.5	189.5
Flour, wheat (pound)	19.9	16.9
Eggs, Grade A large (dozen) ...	77.0	81.5

$$\text{Unweighted aggregate price index} = \frac{21.4 + 189.5 + 16.9 + 81.5}{19.6 + 188.5 + 19.9 + 77.0} \times 100$$

$$= \frac{309.3}{305} \times 100 = 101$$

Now, if we had used the price of a gallon of milk (156.8 in 1975 and 171.2 in 1978), our simple unweighted aggregate price index would have been

$$\frac{156.8 + 189.5 + 16.9 + 81.5}{171.2 + 188.5 + 19.9 + 77.0} \times 100$$

$$= \frac{444.7}{456.6} \times 100 = 97$$

One way out of this difficulty is to use an average of relatives. We obtain a ratio of the price in a given year to the price of a base year for each item separately. Then we obtain the mean of these four relatives (divide by n, the number of items) and multiply by 100. The calculation of the *average of relatives index* is shown in Table 15.6.

average of relatives index

An index in which we obtain separately the relative of each item or component and then find the mean of these relatives.

$$\text{Simple average of relatives index} = \frac{\sum \left(\frac{P_n}{P_0} \right)}{n} \times 100 \qquad (15.6)$$

$$= \frac{4.01}{4} \times 100 = 100$$

Note that, had we used a gallon instead of a pint of milk, the index would be unchanged since the price relative of a gallon of milk is 1.09, the same as for a pint.

Although this index overcomes the problem of the choice of units with which to express the commodity, none of the unweighted indexes meets the difficulty that *all prices in the index are considered of equal importance.* To overcome this problem, we must turn to weighted indexes.

Commodity	Price		P_n/P_0
	1975 (P_0)	1978 (P_n)	
Milk (pint)......................	19.6	21.4	1.09
Round steak (pound)	188.5	189.5	1.01
Flour, wheat (pound)	19.9	16.9	0.85
Eggs, Grade A large (dozen)........	77.0	81.5	1.06
		$\Sigma \left(\dfrac{P_n}{P_0}\right) = 4.01$	

Table 15.6
Calculation of the simple average of relatives index of four commodities in 1975 and 1978

EXERCISES

15.5 What is the value of aggregate indexes?

15.6 Discuss some of the problems associated with using unweighted indexes.

15.7 What is the purpose of an average of relatives index?

15.8 The following tables shows the producer prices in dollars per unit of four different lumber and wood products in 1973 and 1979. Calculate an unweighted aggregate price index for these two periods, using 1973 as the base.

	1973	1979
Softwood plywood, interior grade (1,000 square feet)	$112	$210
Redwood boards, clear (1,000 board feet)	382	931
Maple # 1, common (1,000 board feet)	220	322
Southern pine, finish (1,000 board feet)	275	535

15.9 Using the data in Exercise 15.8, calculate the simple average of relatives index, using 1973 as the base year.

15.10 The table below shows the quantities of four different mineral products produced in the United States in 1970 and 1977. Calculate an unweighted aggregate quantity index, using 1970 as the base year.

	1970	1978
Copper		
(thousands of short tons)	1,720	1,504
Petroleum		
(millions of barrels)	3,517	3,009
Gold		
(thousands of troy ounces)	1,743	1,110
Silver		
(millions of troy ounces)	45.0	38.2

15.11 Using the data in Exercise 15.10, calculate a simple average of relatives index, using 1970 as the base year.

15.3   WEIGHTED AGGREGATE INDEXES

When we treat price alone or quantity alone in the preparation of an aggregate index, we are actually giving each item the same weight. However, as we have just seen, this equal weighting can produce a distorted index when items vary widely in price or in quantity. However, when we weight the price by the quantities produced or sold in a given period, we eliminate much of the inequity. The weighting procedure consists of multiplying the price by the quantity sold or produced. Note that this is a variation of the value index discussed earlier (Section 15.1).

In practice, there are four different strategies for weighting.

weighted index

An index in which the price is weighted by the quantity produced or sold (weighted price index) or the quantity is weighted by the price (weighted quantity index).

1. The simple *weighted index* uses the base year for both price and quantity in the denominator and the given year for both price and quantity in the numerator. Thus, the simple weighted aggregate index is

$$\frac{\Sigma P_n Q_n}{\Sigma P_0 Q_0} \times 100 \tag{15.7}$$

2. A second index, called the Paasche index, uses the quantity in the given year as the weight in the denominator. Hence,

$$\text{Paasche index} = \frac{\Sigma P_n Q_n}{\Sigma P_0 Q_n} \times 100 \tag{15.8}$$

3. The Laspeyres index modifies only the weight for quantity in the numerator of the aggregate index, substituting Q_0 for Q_n. Thus,

$$\text{Laspeyres index} = \frac{\Sigma P_n Q_0}{\Sigma P_0 Q_0} \times 100 \tag{15.9}$$

4. The fixed weight index employs weights that are based on a period or periods that are considered representative. The weights and base prices may or may not come from the same period.

$$\text{Fixed weight aggregate index} = \frac{\Sigma P_n Q_f}{\Sigma P_0 Q_f} \times 100 \qquad (15.10)$$

where Q_f is the quantity produced or sold during the period selected as representative.

Let us look at each of these aggregate indexes in greater detail.

15.3.1 Simple weighted aggregate index

We previously obtained unweighted composite indexes on four appliances selected by Lana and Ken as bellwether items. We may exemplify the calculation of a simple weighted aggregate index by combining the information in Tables 15.3 and 15.4. The result is shown in Table 15.7.

	Quarter 1			Quarter 12		
	P_0	Q_0	P_0Q_0	P_n	Q_n	P_nQ_n
Refrigerators	335	2	670	444	12	5,328
Ranges	425	3	1,275	588	18	10,584
Toasters	18	22	396	25	58	1,450
TV sets	390	12	4,680	440	29	12,760
Sums			7,021			30,122

Table 15.7
Calculation of a simple weighted aggregate index for four appliances

$$\text{Simple weighted aggregate index} = \frac{\Sigma P_n Q_n}{\Sigma P_0 Q_0} \times 100$$

$$= \frac{30,122}{7,021} \times 100 = 429$$

The main advantage of this index is its simplicity. However, the base period need not be the first quarter of operation. Indeed, as most uses of the index are for assessing business changes over time, the first quarter would probably not be retained as the base period since it is likely to be unrepresentative. In all likelihood, a period would be selected during which sales had stabilized somewhat.

15.3.2 The Paasche index

Recall that one possible drawback of the simple weighted aggregate index is that the quantities sold (or produced) during the base period may not be considered representative of overall performance. If Lana and Ken wished to express changes, using the first quarter of business as the base, they might consider the Paasche index. Recall that this index uses the prices of the base period but the quantities of the period of interest. Table 15.8 shows the calculation of the Paasche index from the sales figures for the four appliances.

	Denominator			Numerator		
	P_0	Q_n	$P_0 Q_n$	P_n	Q_n	$P_n Q_n$
Refrigerators	335	12	4,020	444	12	5,328
Ranges	425	18	7,650	588	18	10,584
Toasters	18	58	1,044	25	58	1,450
TV sets	390	29	11,310	440	29	12,760
Sums			24,024			30,122

$$\text{Paasche index} = \frac{\Sigma P_n Q_n}{\Sigma P_0 Q_n} \times 100 = \frac{30,122}{24,024} \times 100 = 125$$

Unlike the simple weighted aggregate index, the Paasche index reflects changed patterns of production or consumption. A disadvantage is that a separate index must be computed for each period of interest, substituting the quantity values of that period. Thus, new inputs of quantity data would be required for each period to which we direct our attention. This can be both costly and time-consuming.

15.3.3   Laspeyres index

This method of weighting is more often used than the Paasche index since it uses the quantity weights only for the base period. Thus, it does not require fresh input of quantity data when calculating the price index of a given period. Moreover, the index numbers for different periods are directly comparable. Table 15.9 illustrates the calculation of the Laspeyres Index from the appliance data.

	Denominator			Numerator		
	P_0	Q_0	$P_0 Q_0$	P_n	Q_0	$P_n Q_0$
Refrigerators	335	2	670	444	2	888
Ranges	425	3	1,275	588	3	1,764
Toasters	18	22	396	25	22	550
TV sets	390	12	4,680	440	12	5,280
Sums			7,021			8,482

$$\text{Laspeyres index} = \frac{\Sigma P_n Q_0}{\Sigma P_0 Q_0} \times 100 = \frac{8,482}{7,021} \times 100 = 121$$

15.3.4   Fixed weight aggregate index

We have seen that the Paasche index uses the quantities for the period of interest as weights, whereas the Laspeyres index uses the quantities of the base period as weights. Theoretically, however, there is no restriction on what period may be used to obtain weights. Given this flexibility, we may feel free to avoid periods that are unusual or abnormal and focus on those we feel are more representative of usual quantities. To illustrate, during the spring and summer of 1980, a severe and prolonged drought wreaked havoc on the cotton crops of Texas, Arkansas, and other southwestern states. To use production figures for that year as a quantity weight would distort all comparisons based on these weights.

One further point: the fixed weights need not all be based on a single period. Some weights may be based in one period, others on another period, and some may be averages of several periods. Many governmental agencies use quantity weights that are averages of several periods (e.g., Department of Agriculture indexes of prices received by farmers). Moreover, even the base periods for *prices* may represent an average of two or more periods (e.g., Producer Price Index).

Table 15.10 demonstrates one technique of applying fixed weights. Using the appliance data, the weights for each appliance are obtained by finding the average number sold over two or more periods. In this example, periods 1 and 12 were used in calculating the index for the 12th quarter, using the first quarter as base.

Calculation of fixed weight aggregate index using mean quantities sold during two periods as weights   Table 15.10

	Q_0	Q_n	$\overline{Q}$	P_0	P_n	$\overline{Q}P_0$	$\overline{Q}P_n$
Refrigerators	2	12	7	335	444	2,345.0	3,108
Ranges	3	18	10.5	425	588	4,662.5	6,174
Toasters	22	58	40	18	25	720.0	1,000
TV sets	12	29	20.5	390	440	7,995.0	9,020
Sums						15,522.5	19,302

$$\text{Fixed weight aggregate index} = \frac{\Sigma(\overline{Q}P_n)}{\Sigma(\overline{Q}P_0)} \times 100 = \frac{19,302}{15,522.5} \times 100 = 124$$

EXERCISES

15.12 Discuss the reasons for calculating weighted as opposed to unweighted indexes.

15.13 Enumerate four different strategies for weighting.

15.14 Discuss the advantages and the disadvantages of a Paasche Index.

15.15 Discuss the advantages and the disadvantages of a Laspeyres Index.

15.16 Discuss the features of the fixed weight aggregate index.

15.17 The table below shows the prices *(P)* and total consumption *(Q)* of various foods in the United States during 1977 and 1979. Using 1977 as base year, calculate the

   a.   Simple weighted aggregate index.

   b.   Paasche index.

   c.   Laspeyres index.

	1977		1979	
	P	*Q*	*P*	*Q*
Poultry	$ 0.237*	11,916†	$ 0.260*	13,820†
Eggs	0.624‡	180§	0.662‡	192§
Cattle	38.74‖	38,717#	77.60‖	31,504#
Calves	48.19‖	4,696#	92.24‖	2,499#
Sheep and hogs	53.38‖	6,133#	67.12‖	4,833#

  * Per pound.
  † Millions of pounds.
  ‡ Per dozen.
  § Thousands of cases.
  ‖ Per 100 pounds.
  # Thousands of head.

15.4   QUANTITY INDEXES: WEIGHTED AVERAGE-OF-RELATIVES

In Section 15.2, we looked at the unweighted or simple average of relatives price index. Recall that it may be used to overcome the difficulty posed by the selection of the units of measurement for expressing a given item. A modified weighted form of this index may be used to compute quantity indexes. The procedures are essentially the same as with price indexes except that prices instead of quantities are used as weights.

The general formula for the weighted average-of-relatives quantity index is

$$\frac{\sum \left(\frac{Q_n}{Q_0} \times 100 \right) (Q_0 P_0)}{\sum Q_0 P_0} \qquad (15.11)$$

Table 15.11 illustrates the computation of the weighted average-of-relatives quantity index, using the quantities and prices of wheat, flour, poultry, and hogs in July 1979 and July 1980.

As with price indexes, the weights used in the computation may reflect the base period (as in the present example), the period of interest, a period when prices appear typical, or an average of several periods.

Table 15.11
Computation of weighted
average-of-relatives
quantity index

	Quantities		Price				
	July 1979 Q_0	July 1980 Q_n	July 1979 P_0	$\dfrac{Q_n}{Q_0} \times 100$	$Q_0 P_0$	$\dfrac{Q_n}{Q_0} \times 100 \times Q_0 P_0$	
Wheat flour ..	23,508*	23,144*	$10.51†	99	247,069.08	24,459,838.92	
Poultry	1,241‡	1,211‡	0.235§	98	291.635	28,580.23	
Hogs	6,734‖	6,910‖	38.58†	103	259,797.72	26,759,165.16	
Sums					507,158.435	51,247,584.31	

* Thousands of 100-pound sacks.
† Per 100 pounds.
‡ Millions of pounds.
‖ Thousands of head.
§ Per pound.

$$\text{Weighted average-of-relatives quantity index} = \frac{\sum \left(\dfrac{Q_n}{Q_0} \times 100 \right)(Q_0 P_0)}{\sum (Q_0 P_0)}$$

$$= \frac{51,247,584.31}{507,158.435} = 101$$

EXERCISES

15.18 Discuss the reasons for using weighted average-of-relatives indexes for quantity data.

15.19 Using the data in Exercise 15.17, compute weighted average-of-relatives quantity index, using the prices in 1977 as weights.

15.20 Using the data in Exercise 15.17, construct a weighted average-of-relatives quantity index, using prices in 1979 as weights.

15.5 SHIFTING BASES OF INDEX NUMBERS

There are times when we want to change the base of an index number. This may be done to update the index or to make it comparable to other indexes that are in use. This procedure is easily accomplished as long as we continue to use the same series of index numbers, complete with original base. We simply divide each index number by the index number for the new base year and multiply by 100.

Table 15.12 illustrates the procedures for converting the Consumer Price Index (all items) from a base of 1967 to a base of 1975.

Table 15.12
Procedures for shifting the
CPI from one base
period to another

Year	CPI (1967 = 100) (all items)	$\dfrac{CPI_n}{CPI_{1975}}$	CPI (1975 = 100) (all items)
1964	92.9	0.576	57.6
1965	94.5	0.586	58.6
1966	97.2	0.608	60.3
1967	100.0	0.620	62.0
1968	104.2	0.646	64.6
1969	109.8	0.681	68.1
1970	116.3	0.721	72.1
1971	121.3	0.752	75.2
1972	125.3	0.777	77.7
1973	133.1	0.826	82.6
1974	147.7	0.916	91.6
1975	161.2	1.000	100.0
1976	170.5	1.058	105.8
1977	181.5	1.126	112.6
1978	195.4	1.212	121.2
1979	217.4	1.349	134.9

15.6 WORDS OF CAUTION CONCERNING THE USE OF INDEX NUMBERS

Most of the indexes in common use employ selected items as the basis for compiling estimates of changes over time of the dimension of interest. For example, the Consumer Price Index is based on prices of food, clothing, shelter, fuels, transportation, drugs, dental and medical care, and other goods and services. With thousands of possible items to choose from, those actually selected are presumed to be representative of the total. Similarly, the Dow Jones averages are based on 65 common stocks. Needless to say, an index is as good as the representativeness of the sample components making up the index. Care must be taken when selecting items for inclusion in the index and revisions must be made from time to time to reflect changed patterns of production and consumption.

In addition, the information compiled on each of the items making up an index may be subject to sampling error. For example, prices for the CPI are collected in 85 urban areas across the country from 18,000 tenants, 18,000 housing units, and about 24,000 establishments, including grocery stores, department stores, hospitals, etc. Note also that the CPI applies to urban dwellers only. The CPI-U applies to all urban consumers, which includes about 80 percent of the noninstitutionalized population. The CPI-W for urban wage earners and clerical workers represents about 50 percent of the population included in CPI-U.

Many of the governmental indexes are published monthly and quarterly. An increase in an index from one period to another cannot be interpreted as a percentage increase from the preceding period. It represents a percentage increase only relative to the base year. To illustrate, if the CPI-U for July

1980 is 247.8, it means that the prices of items are 147.8 percent higher than they were in the base year (1967). If we wish to show the percentage change from June 1980 (247.6), we subtract the June index from the July index, divide by the June index, and multiply by 100. Thus, the percentage change in this illustration is

$$\frac{247.8 - 247.6}{247.6} \times 100 = 0.08$$

When multiplied by 12, this translates into an annual percentage rate of increase of just under 1 percent.

Finally, the selection of the base period can produce erroneous impressions and may be used by unscrupulous people or organizations to mislead. To illustrate, if you wish to convey the impression of large increases in sales, production, dollar volume, etc., you may purposely select a base period that is low on the variable of interest. Thus, if the base period is 85 and the period of interest is 280, the percentage relative is 329. However, had a base period of 200 been selected a different impression of change would be conveyed (i.e., 140). Actually, as we have seen (Section 15.6), the two index numbers provide identical information, although it might not appear so to an unsophisticated reader.

EXERCISES

15.21 Shown below is a manufacturing index showing output per hour in the United States in selected years since 1960. Shift the base, using 1975 as base year.

Year	Output per hour (1967 = 100)
1960	78.9
1965	98.3
1970	104.5
1973	118.8
1974	112.6
1975	118.2
1976	123.2
1977	126.1
1978	129.2

Source: Bureau of Labor Statistics, U.S. Department of Labor.

15.22 The table below shows index numbers of total individual production in the United States for selected years between 1963 and 1979. Shift the base to 1963.

Year	Total production (1967 = 100)
1963	76.5
1965	89.8
1970	107.8
1975	117.8
1976	130.5
1977	138.2
1978	146.1
1979	152.1

Source: Federal Reserve System; F.W. Dodge Division, McGraw-Hill; U.S. Department of Labor; U.S. Department of Commerce.

15.23 Looking at the index numbers obtained as answers to Exercises 15.21 and 15.22, what conclusion could erroneously be drawn by a careless person or by an unsophisticated reader?

15.24 Shown below are index numbers from CPI-W in June and July, 1980 for seven different categories of expenditures.
  a. Convert each to a percent change.
  b. Express each change as an annual percentage rate.

	Index	
Item	June 1980	July 1980
Food and beverages	246.4	249.1
Housing	266.9	265.1
Apparel and upkeep	176.0	175.4
Transportation	250.6	251.9
Medical care	265.9	267.8
Entertainment	204.0	204.4
Other goods and services	212.1	212.9

SUMMARY

In this chapter we looked at one of the most common devices used to monitor changes in variables over spans of time—index numbers, also referred to as percentage relatives.

1. The three major type of index numbers are price, quantity, and value.
2. Many widely used index numbers are comprised of a number of different elements. For example, prices of a number of different foods may be used in a food price index.
3. Unweighted composite or aggregate indexes treat each component with

equal importance. They also present difficulties concerning the choice of units of measurement.

4. The latter problem may be handled with an unweighted index by computing an average of relatives index.

5. Weighted aggregate price indexes use quantity measures as weights to assure that each component is weighted according to its importance.

6. Four different strategies for constructing weighted aggregate indexes were discussed. The simple weighted aggregate index uses the quantities during the base year as weights. The Paasche index uses the quantities for the given year in the denominator of the index, and the Laspeyres uses the quantities for the given year in the numerator of the index. Fixed weight techniques may use weights based on "normal" periods or averages of several periods. Many governmental indexes use fixed weights.

7. We showed the computation of the weighted average-of-relatives for quantity indexes. In quantity indexes, prices are used as weights.

8. It is frequently desirable to shift a base to a different, usually more recent, time period. This is easily done as long as we continue to use the same series of index numbers, complete with the original base.

9. We noted that most indexes are made up of items thought to be representative of the dimensions of interest. They are as good as the assumption is correct. Also, like any statistic, index numbers are subject to sampling error since the data base uses sampling techniques.

10. An index number shows the percentage of change during one period relative to the base period. The number of points of change from one period (not the base) to another is not a percentage. Procedures for converting to percentages were demonstrated.

11. Selecting the base period is an important consideration in the construction of index numbers. Selecting the wrong base period can convey erroneous impressions of change.

TERMS TO REMEMBER

index number (478)

base (479)

price index (price relative) (479)

quantity index (quantity relative) (479)

value index (value relative) (479)

aggregate index (composite index) (481)

unweighted aggregate index (482)

average of relatives index (484)

weighted index (486)

EXERCISES

15.25 The following table shows the U.S. energy consumption, by fuel type, in 1970 and 1978. Compute a simple unweighted aggregate quantity index for 1978, using 1970 as the base period.

	Quadrillion Btus	
	1970	1978
Coal	12.66	14.09
Natural gas	21.80	19.82
Petroleum	29.52	37.79
Hydropower	2.65	3.15
Nuclear	0.24	2.98
Geothermal and other	0.02	0.07

15.26 The table below shows the number of fatal injuries in U.S. coal mines for selected years between 1930 and 1975. Prepare a simple index of number of fatalities, using 1960 as the base year.

Year	Fatalities
1930	2,063
1935	1,242
1940	1,388
1945	1,068
1950	643
1955	448
1960	325
1965	259
1970	260
1975	155

15.27 The index number calculated in Exercise 15.26 does not take into account the number of miners employed in the mines. Shown below are the average numbers of miners working daily during each of these years.

    a. Devise a method for using the average number of men employed daily as weights for the fatality data.

    b. Prepare a weighted aggregate index for the years shown.

Year	Average number of miners (000)
1930	644
1935	565
1940	533
1945	438
1950	483
1955	260
1960	190
1965	149
1970	144
1975	218

15.28 Compare the index numbers obtained in Exercises 15.26 and 15.27. Account for the large disparities.

15.29 Many data available from governmental and business sources present value data (e.g., millions of dollars spent in new home construction). Thus, weighting has already taken place. Shown below are the number of dollars spent, on various types of construction, in July 1979 and 1980. Prepare an aggregate price index for the data using July 1979 as base.

Category	July 1979 ($ millions)	July 1980 ($ millions)
Residential	$9,448	$4,986
Industrial	1,382	1,134
Commercial	2,275	2,547
Public building	1,456	1,654
Military facilities	134	143

15.30 Shown in the table below are indexes of industrial output per hour of Japan and the United Kingdom. Shift the base of both indexes to 1975.

	Japan	U.K.
1960	52.6	77.0
1965	79.1	92.5
1970	146.5	108.8
1973	181.2	127.6
1974	181.7	127.7
1975	174.6	124.2
1976	188.7	127.9
1977	199.2	126.5
1978	215.7	128.6

Source: U.S. Department of Labor, 1967 = 100.

15.31 Refer to Exercise 15.30. Why is it that the index numbers in Exercise 15.30 appear to show smaller increases per hour than the index numbers using 1975 as the base?

15.32 The table below shows the production per ton and prices per ton of five metals in U.S. plants in 1966 and 1976. Calculate a simple weighted aggregate index.

	1966		1976	
	P	Q	P	Q
Iron	56.25	91,500	182.33	86,870
Aluminum	249.80	2,968.4	444.90	4,251.0
Copper	365.70	1,429.2	695.60	1,611.3
Zinc	145.00	1,025.1	370.10	498.9
Lead	231.00	327.4	151.20	609.5

15.33 Refer to Exercise 15.32. Calculate a Paasche Index from the data.

15.34 Refer to Exercise 15.32. Calculate a Laspeyres index.

15.35 Refer to Exercise 15.32. Find the mean quantities for 1966 and 1976 and use them as weights for finding a fixed weight aggregate index.

15.36 Refer to Exercise 15.32. Using the price in 1976 as the weight, find the weighted average of the quantity relatives.

Appendixes

Appendix A

Glossary of Symbols

The following are definitions of symbols that appear in the text, followed by the page number indicating the first reference to the symbol.

Some of the symbols have more than one definition. Therefore you must judge the meaning by the context.

Greek letters are listed separately in approximately alphabetical order.

Symbol	Definition	Page
	Greek letters	
α	Probability of a Type I error (probability of rejecting a true H_0)	227
β	Probability of a Type II error (probability of accepting a false H_0)	227
λ	Mean *rate* of occurrence of an event in a Poisson process	134
μ	Population mean	
	Mean occurrence of an event per interval of time, or mean proportion of space occupied by an event in the Poisson distribution	134
	Mean of a binomial variable	137
μ_d	Population mean of the difference between paired scores	242
μ_0	Value of the population mean under the null hypothesis	233
μ_R	Expected value of R (number of runs)	361
μ_U	Mean of the Mann-Whitney U-distribution	356
$\mu_{Y \cdot X}$	A specific individual value in the population model	384
$\mu_{\bar{X}}$	Mean of a sampling distribution of sample means	173
ρ	Population correlation coefficient	408
σ	Standard deviation of a population	68
σ_R	Standard deviation of R (number of runs)	361
σ_U	Population standard deviation of the Mann-Whitney U-distribution	356
$\sigma_{Y \cdot X}$	Population standard error of estimate	396
$\sigma_{\bar{X}}$	Standard error of the mean; standard deviation of sampling distribution of sample means	173

Symbol	*Definition*	*Page*
σ^2	Variance of a population	**67**
σ_a^2	Treatment effects in ANOVA	**300**
σ_e^2	Random error	**300**
χ^2	Chi-square	**331**

English letters

a	Y-intercept; value of Y when $X = 0$	**381**
	Estimated A	**385**
A	Population Y-intercept	**384**
$\overline{A}$	Complement of A; not A	**85**
AB	Number of treatment combinations in a two-way-analysis-of-variance factorial design	**313**
ANOVA	Analysis of variance	**298**
b	Number of blocks in a randomized block design	**320**
	Slope of a straight line	**381**
	Estimated B	**385**
B	Population slope of the regression line	**384**
c	Number of column classes in a contingency table	**342**
C	Cyclical component in a time series	**444**
cf	Cumulative frequency	**35**
d	Difference between paired scores	**241**
$\overline{d}$	Mean of the difference of the paired scores	**241**
df	Degrees of freedom	**203**
$df_{A \times B}$	Interaction degrees of freedom	**313**
df_{bet}	Between-group degrees of freedom	**300**
df_{blks}	Degrees of freedom for blocks in a randomized block design	**320**
df_{tot}	Total degrees of freedom	**300**
df_{treat}	Treatment degrees of freedom	**319**
df_w	Within-group degrees of freedom	**300**
e	Allowable error $(\overline{X} - \mu)$	**215**
	Dispersion or scatter in the population associated with omitted, unobserved, and unobservable variables	**384**
$e^{-\mu}$	Base of the natural logarithm system raised to the negative mean power	**134**
$E(X)$	Expected value of a variable	**118**
$E(\overline{X})$	Expected value of the sample mean	**185**
$E(Y\|X)$	Expected value of Y given X	**385**
f	Frequency of occurrence	**18**
Σf	Sum of the frequencies; n	**18**
f_e	Expected frequency	**331**
f_0	Observed or obtained frequency	**331**
F	Cumulative frequency of class immediately preceding the median class	**60**
fcf	Finite correction factor	**180**

Symbol	Definition	Page
H_1	Alternative hypothesis	225
H_0	Null hypothesis	225
i	Width of a class	60
I	Irregular component in a time-series	444
I_p	Price index or price relative	479
I_q	Quantity index or quantity relative	479
I_v	Value index or value relative	479
k	Number of classes	24
	In Chebyshev's theorem, the number of standard deviations from the mean	72
	Number of groups or conditions in one-way ANOVA	300
	Number of means in Tukey HSD test	307
	Number of mutually exclusive and exhaustive categories in a multinomial distribution	330
k^2	Coefficient of nondetermination	405
Lm	Lower limit of the median class	60
MA	Moving average	458
MAD	Mean absolute deviation (a measure used to assess forecasting)	469
n	Sample size	18
	Number of observations	25
n_i	Number of observations in the ith condition	303
N	Number in the population	52
$n - x$	Number of times an observation falls in the Q-category	124
p	Combined proportion estimated from sample proportions (pooled estimate of population proportion)	246
$\bar{p}$	Estimated proportion in the P-category	213
	Obtained sample proportion	244
$p(A)$	Probability of A	85
	Marginal probability of A	96
$p(A|B)$	Conditional probability; probability of A *given* that B has occurred (or is certain to occur)	97
$p(A$ and $B)$	Probability of both A and B occurring;	91
	Joint probability	95
$p(A$ or $B)$	Probability of A or B occurring	91
P	Probability that event of interest will occur	122
P_n	Price of given product at a given time	479
P_0	Known or hypothesized population proportion	244
	Price of the product in the base period	479
$\bar{q}$	Estimated proportion in the Q-category	213
	$1 - \bar{p}$	244
$q\alpha$	Tabled value for Tukey HSD test at a given α-level for df_w and number of means	307
Q	Probability that event of interest will not occur	122
Q_1	First quartile	66

Symbol	Definition	Page
Q_2	Second quartile (median)	66
Q_3	Third quartile	66
Q_f	Quantity produced or sold during the period selected as representative	487
Q_n	Quantity produced or sold in a given period	479
Q_0	Quantity produced or sold in the base period	479
r	Number of row classes in a contingency table	342
	Pearson r correlation coefficient	405
r^2	Coefficient of determination	403
R	Number of runs	360
R_1	Sum of the ranks assigned to the group with sample size n_1 (Mann-Whitney U-test)	355
R_2	Sum of the ranks assigned to the group with sample size n_2 (Mann-Whitney U-test)	355
R^2	Coefficient of multiple determination	429
RSFE	Running sum of forecast errors (a measure used to assess forecasting)	469
s	Standard deviation of a sample	68
s_b	Estimated standard error of b	410
s_d	Standard deviation of the difference scores	241
$s_{\bar{d}}$	Estimated standard error of the difference between dependent means	241
$s_{\bar{p}}$	Estimated standard error of the proportion	213
$s_{\bar{p}_1-\bar{p}_2}$	Pooled estimate of population standard error of the difference in proportions	247
s_r	Estimated standard error of r	409
$s_{\bar{X}}$	Estimated standard error of the mean (Standard deviation of the sampling distribution of sample means)	179
$s_{\bar{X}_1-\bar{X}_2}$	Estimated standard error of the difference between means	209
s_{Y_c}	Estimated standard error of the conditional mean	398
$s_{Y_{ind}}$	Estimated standard error of the individual scores	399
$s_{Y \cdot X}$	Estimated standard error of estimate	395
$s_{Y \cdot X_1 X_2}$	Estimated standard error of estimate with two independent variables	427
s^2	Variance of a sample	68
$s^2_{A \times B}$	Interaction variance estimate	313
s^2_{bet}	Between-group variance estimate	299
s^2_{blks}	Blocks variance estimate	322
s^2_{treat}	Treatment variance estimate	322
s^2_w	Within-group variance estimate	299
sk	Pearson's coefficient of skew	75
S	Seasonal component in a time-series	444
SAD	Squared average deviation (a measure used to assess forecasting)	470

Symbol	Definition	Page
SS	Sum of squares $[\Sigma(X - \overline{X})^2]$	299
SS_{bet}	Between-group sum of squares in ANOVA	301
SS_{blks}	Sum of squares for blocks in a randomized block design	319
SS_{tot}	Total sum of squares in ANOVA	301
SS_{treat}	Treatment sum of squares	319
SS_w	Within-group sum of squares in ANOVA	301
t	Number of units of space or the amount of time in a Poisson process	134
t-distributions	Theoretical sampling distributions with a mean of zero and a standard deviation that varies as a function of sample size	203
t_{df}	t-value for a given number of degrees of freedom	205
T	Trend component in a time series	444
U and/or (U_1, U_2)	Mann-Whitney U-test statistics	353,355
w	Weighting factor in weighted mean	56
x	Number of observed outcomes favoring an event	85
	Number of successes in n trials; number of times an observation falls in P-category	124
	Coded time value $(X - \overline{X})$	445
$x!$	Factorial of event of interest	126
$\overline{X}$	Sample mean	4
	Arithmetic mean	51
$\overline{X}_i$	Mean of the ith condition	303
ΣX^2	Sum of the squares of all measurements	69
$(\Sigma X)^2$	Square of the sum of the measurements	69
$\overline{X}_{tot}$	Grand mean; mean of all the measurements	302
$\overline{X}_w$	Weighted mean	56
Y_c	Predicted score on the Y-variable for a given X; point estimate for $E(Y|X)$	385
Y_t	Predicted trend value	445
z	Deviation of a score from the mean expressed in standard deviation units	139

Appendix B

List of Formulas

Listed below are the formulas appearing in the text, followed by a verbal description. The page number refers to the first time the formula appears.

Number	*Formula*	*Verbal description*	*Page*
(3.1)	$\bar{X} = \dfrac{\Sigma X}{n}$	Arithmetic mean	**51**
(3.2)	$\bar{X} = \dfrac{\Sigma fX}{n}$	Arithmetic mean for grouped frequency distribution	**53**
(3.3)	$\bar{X}_w = \dfrac{\Sigma wX}{\Sigma w}$	Weighted mean	**56**
(3.4)	$Mdn = \left(\dfrac{0.5n - F}{f}\right)i + Lm$	Median for grouped data	**60**
(3.5)	$s^2 = \dfrac{\Sigma (X - \bar{X})^2}{n - 1}$	Variance of a sample	**68**
(3.6)	$s = \sqrt{\dfrac{\Sigma (X - \bar{X})^2}{n - 1}}$	Standard deviation of a sample	**68**
(3.7)	$\Sigma (X - \bar{X})^2 = \Sigma X^2 - \dfrac{(\Sigma X)^2}{n}$	Raw score formula	**69**
(3.8)	$s^2 = \dfrac{\Sigma f(X - \bar{X})^2}{n - 1}$	Variance for grouped data	**70**
(3.9)	$s = \sqrt{\dfrac{\Sigma f(X - \bar{X})^2}{n - 1}}$	Standard deviation for grouped data	**70**
(3.10)	$\Sigma f(X - \bar{X})^2 = \Sigma fX^2 - \dfrac{(\Sigma fX)^2}{n}$	Raw score formula for grouped data	**70**
(3.11)	$sk = \dfrac{3(\bar{X} - \text{Median})}{s}$	Pearson's coefficient of skew	**75**
(4.1)	$p(A) = \dfrac{\text{Number of observed outcomes favoring } A}{\text{Total number of observed outcomes}}$	Empirical or relative frequency probability	**85**
(4.2)	$p(A \text{ or } B) = p(A) + p(B) - p(A \text{ and } B)$	General addition rule for events that are not mutually exclusive	**91**

505

506

Number	Formula	Verbal description	Page
(4.3)	$p(A \text{ or } B) = p(A) + p(B)$	Addition rule for mutually exclusive events	93
(4.4)	$p(A\|B) = \dfrac{p(A \text{ and } B)}{p(B)}$	Conditional probability: probability of A given that B has occurred (or is certain to occur)	97
(4.5)	$p(A \text{ and } B) = p(B)p(A\|B)$	Multiplication rule for dependent events	98
(4.6)	$p(A \text{ and } B) = p(A)p(B)$	Multiplication rule for independent events	98
(5.1)	$E(X) = \Sigma X p(X) = \text{Mean of } X \text{ (i.e., } \mu)$	Expected value of a variable	118
(5.2)	$p(x) = \dfrac{n!}{x!(n-x)!} P^x Q^{n-x}$	Probability of x successes in n trials	126
(5.3)	$p(x) = \dfrac{(\mu^x)(e^{-\mu})}{x!}$	Probability of event x using the Poisson distribution	134
(5.4)	$\mu = \lambda t$	Mean of a Poisson distribution	135
(5.5)	$\mu = nP$	Mean of a binomial distribution	137
(5.6)	$z = \dfrac{X - \mu}{\sigma}$	z-score transformation	139
(5.7)	$X = \mu + z\sigma$	Value of X given μ, σ, and z	144
(5.8)	$\sigma = \sqrt{nPQ}$	Standard deviation of a binomial distribution	147
(6.1)	$\sigma_{\bar{x}} = \dfrac{\sigma}{\sqrt{n}}$	Standard error of the mean (standard deviation of sampling distribution of sample means)	173
(6.2)	$\text{est. } \sigma_{\bar{X}} = s_{\bar{X}} = \dfrac{s}{\sqrt{n}}$	Estimated standard error of the mean	179
(6.3)	$\text{fcf} = \sqrt{\dfrac{N-n}{N-1}}$	Finite correction factor	180
(6.4)	$\sigma_{\bar{X}} = \left(\dfrac{\sigma}{\sqrt{n}}\right)\left(\sqrt{\dfrac{N-n}{N-1}}\right)$	Standard error of the mean when the population is small	180
(6.5)	$s_{\bar{X}} = \left(\dfrac{s}{\sqrt{n}}\right)\sqrt{\dfrac{N-n}{N-1}}$	Estimated standard error of the mean when the population is small	181
(7.1)	$\bar{X} \pm z\sigma_{\bar{X}}$	General formula for a confidence interval when σ is known and sample is large	200

Number	Formula	Verbal description	Page
(7.2)	$\bar{X} \pm zs_{\bar{X}}$	General formula for a confidence interval when σ is unknown and sample is large	200
(7.3)	$\bar{X} \pm t_{df}s_{\bar{X}}$	General formula for a confidence interval when σ is unknown and sample is small	205
(7.4)	$s_{\bar{X}_1 - \bar{X}_2} = \sqrt{\dfrac{n_1 s_1^2 + n_2 s_2^2}{n_1 n_2}}$	Estimate of the standard error of the difference between means	209
(7.5)	$s_{\bar{X}_1 - \bar{X}_2} = \sqrt{\dfrac{s_1^2}{n_1} + \dfrac{s_2^2}{n_2}}$	Estimate of the standard error of the difference between means when $n_1 = n_2$	209
(7.6)	$s_{\bar{p}} = \sqrt{\dfrac{\bar{p}\bar{q}}{n}}$	Estimated standard error of a proportion	213
(7.7)	$\bar{p} \pm zs_{\bar{p}}$	General formula for a confidence interval of proportion	213
(7.8)	$n = \dfrac{(z\sigma)^2}{e^2}$	Estimating sample size for interval estimation of the mean	215
(8.1)	$z = \dfrac{\bar{X} - \mu_0}{s_{\bar{X}}}$	Test statistic used to test null hypotheses involving a single mean (large sample)	233
(8.2)	$t = \dfrac{\bar{X} - \mu_0}{s_{\bar{X}}}$	Test statistic used to test null hypotheses involving a single mean (small sample)	234
(8.3)	$z = \dfrac{\bar{X}_1 - \bar{X}_2}{s_{\bar{X}_1 - \bar{X}_2}}$	Test statistic used to test hypotheses involving differences between means (large samples)	237
(8.4)	$t = \dfrac{\bar{X}_1 - \bar{X}_2}{s_{\bar{X}_1 - \bar{X}_2}}$	Test statistic used to test hypotheses involving differences between means (small samples)	239
(8.5)	$t = \dfrac{\bar{d}}{s_{\bar{d}}}$	Test statistic used to test hypotheses between *related* means	241
(8.6)	$s_{\bar{d}} = \dfrac{s_d}{\sqrt{n}}$	The standard error of the difference between related means	241

Number	Formula	Verbal description	Page
(8.7)	$s_d = \sqrt{\dfrac{\Sigma d^2 - \dfrac{(\Sigma d)^2}{n}}{n-1}}$	Standard deviation of the difference (related means)	241
(8.8)	$s_{\bar{d}} = \sqrt{\dfrac{\Sigma d^2 - \dfrac{(\Sigma d)^2}{n}}{n(n-1)}}$	Standard error of the difference, related means	241
(8.9)	$z = \dfrac{\bar{p} - P_0}{s_{\bar{p}}}$	Test statistic used to test hypotheses about a single proportion	244
(8.10)	$p = \dfrac{n_1 \bar{p}_1 + n_2 \bar{p}_2}{n_1 + n_2}$	Pooled estimate of the population proportion	246
(8.11)	$s_{\bar{p}1-\bar{p}2} = \sqrt{pq\left(\dfrac{1}{n_1} + \dfrac{1}{n_2}\right)}$	Pooled standard error of the difference in proportion	247
(8.12)	$z = \dfrac{\bar{p}_1 - \bar{p}_2}{s_{\bar{p}1-\bar{p}2}}$	Test statistic for testing difference between two proportions	247
(10.1)	$F = \dfrac{s^2_{\text{bet}}}{s^2_{\text{w}}}$	F-ratio in one-way analysis of variance	299
(10.2)	$df_{\text{bet}} = k - 1$	Degrees of freedom for between-groups in a one-way ANOVA	300
(10.3)	$df_{\text{w}} = n - k$	Degrees of freedom for within-groups in a one-way ANOVA	300
(10.4)	$SS_{\text{tot}} = \Sigma X^2_{\text{tot}} - \dfrac{(\Sigma X_{\text{tot}})^2}{n}$	Raw score formula for total sum of squares in ANOVA	304
(10.5)	$SS_{\text{bet}} = \Sigma \dfrac{(\Sigma X_i)^2}{n_i} - \dfrac{(\Sigma X_{\text{tot}})^2}{n}$	Raw score formula for between-group sum of squares for ANOVA	305
(10.6)	$HSD = q_\alpha \sqrt{\dfrac{s^2_{\text{w}}}{n}}$	Test statistic for Tukey's HSD pairwise comparison test for a significant F-ratio in ANOVA	307
(10.7)	$SS_{\text{bet}} = SS_A + SS_B + SS_{A \times B}$	Partitioning the between-group sum of squares in a two-way analysis of variance	312
(10.8)	$s^2_A = \dfrac{SS_A}{df_A} \quad df_A = A - 1$	Variance estimates for the two variables in a two-way analysis of variance	
(10.9)	$s^2_B = \dfrac{SS_B}{df_B} \quad df_B = B - 1$		312

Number	Formula	Verbal description	Page
(10.10)	$s_{A \times B}^2 = \dfrac{SS_{A \times B}}{df_{A \times B}}$ $df_{A \times B} = df_A \times df_B$	Variance estimate for interaction in two-way analysis of variance	313
(11.1)	$\chi^2 = \displaystyle\sum_{i=1}^{k} \dfrac{(f_0 - f_e)^2}{f_e}$	Test statistic for multicategory case	331
(11.2)	$\chi^2 = \displaystyle\sum_{r=1}^{k} \sum_{c=1}^{k} \dfrac{(f_0 - f_e)^2}{f_e}$	Test statistic for two-variable (contingency table) case	342
(11.3)	$U_1 = n_1 n_2 + \dfrac{n_1(n_1 + 1)}{2} - R_1$	Test statistic for Mann-Whitney U-test	355
(11.4)	$U_2 = n_1 n_2 + \dfrac{n_2(n_2 + 1)}{2} - R_2$		
(11.5)	$\mu_U = \dfrac{n_1 n_2}{2}$	Mean of the U-distribution	356
(11.6)	$\sigma_U = \sqrt{\dfrac{n_1 n_2 (n_1 + n_2 + 1)}{12}}$	Standard deviation of the U-distribution	356
(11.7)	$\mu_R = \dfrac{2 n_1 n_2}{n_1 + n_2} + 1$	Mean of the distribution of R in the runs test	361
(11.8)	$\sigma_R = \sqrt{\dfrac{2 n_1 n_2 (2 n_1 n_2 - n_1 - n_2)}{(n_1 + n_2)^2 (n_1 + n_2 - 1)}}$	Standard deviation of the distribution of R in the runs test	361
(12.1)	$Y_c = a + bX$	Estimating equation for linear regression	385
(12.2)	$b = \dfrac{n(\Sigma XY) - (\Sigma X)(\Sigma Y)}{n(\Sigma X^2) - (\Sigma X)^2}$	The slope of the regression line	388
(12.3)	$a = \dfrac{\Sigma Y - b\Sigma X}{n}$	Y-intercept	388
(12.4)	$s_{Y \cdot X} = \sqrt{\dfrac{\Sigma Y^2 - a\Sigma Y - b\Sigma XY}{n - 2}}$	Standard deviation around regression line; standard error of estimate	396
(12.5)	$Y_c \pm t s_{Yc}$	Confidence interval estimate for the conditional mean	398
(12.6)	$s_{Yc} = s_{Y \cdot X} \sqrt{\dfrac{1}{n} + \dfrac{(X - \bar{X})^2}{\Sigma X^2 - \dfrac{(\Sigma X)^2}{n}}}$	Estimated standard error of the conditional mean	398
(12.7)	$Y_c \pm t s_{\text{ind}}$	Predictive interval estimate for an individual score	399
(12.8)	$s_{Y_{\text{ind}}} = s_{Y \cdot X} \sqrt{1 + \dfrac{1}{n} + \dfrac{(X - \bar{X})^2}{\Sigma X^2 - \dfrac{(\Sigma X)^2}{n}}}$	Standard error of the individual score	399

510

Number	Formula	Verbal description	Page
(12.9)	$r^2 = \dfrac{a\Sigma Y + b\Sigma XY - n\bar{Y}^2}{\Sigma Y^2 - n\bar{Y}^2}$	Coefficient of determination	**405**
(12.10)	$r = \dfrac{\Sigma XY - \dfrac{(\Sigma X)(\Sigma Y)}{n}}{\sqrt{\left[\Sigma X^2 - \dfrac{(\Sigma X)^2}{n}\right]\left[\Sigma Y^2 - \dfrac{(\Sigma Y)^2}{n}\right]}}$	Coefficient of correlation	**407**
(12.11)	$s_r = \sqrt{\dfrac{(1-r)^2}{n-2}}$	Standard error of r	**409**
(12.12)	$t = \dfrac{r}{s_r} = \dfrac{r}{\sqrt{\dfrac{(1-r^2)}{n-2}}}$	Test statistic for H_0: $\rho = 0$	**409**
(12.13)	$s_b = s_{Y\cdot X}\sqrt{\dfrac{1}{\Sigma X^2 - \dfrac{(\Sigma X)^2}{n}}}$	Standard error of the slope of the regression line	**410**
(12.14)	$t = \dfrac{b}{s_b}$	Test statistic for H_0: $B = 0$	**410**
(13.1)	$Y_c = a + b_1 X_1 + b_2 X_2$	Regression equation for two independent variables	**421**
(13.2)	$\Sigma Y = na + b_1\Sigma X_1 + b_2\Sigma X_2$	Simultaneous equations for obtaining values of a, b_1, and b_2	
(13.3)	$\Sigma X_1 Y = a\Sigma X_1 + b_1\Sigma X_1^2 + b_2\Sigma X_1 X_2$		
(13.4)	$\Sigma X_2 Y = a\Sigma X_2 + b_1\Sigma X_1 X_2 + b_2\Sigma X_2^2$		**422**
(13.5)	$s_{Y\cdot X_1 X_2} = \sqrt{\dfrac{\Sigma Y^2 - a\Sigma Y - b_1\Sigma X_1 Y - b_2\Sigma X_2 Y}{n-3}}$	Standard error of estimate for multiple regression	**428**
(13.6)	$R^2 = \dfrac{a\Sigma Y + b_1\Sigma X_1 Y + b_2\Sigma X_2 Y - \dfrac{(\Sigma Y)^2}{n}}{\Sigma Y^2 - \dfrac{(\Sigma Y)^2}{n}}$	Coefficient of multiple determination	**429**
(14.1)	$Y = T \times C \times S \times I$	Time series model	**443**
(14.2)	$Y_t = a + bx$	Linear trend line	**445**
(14.3)	$x = X - \bar{X}$	Coded score for years	**445**
(14.4)	$a = \bar{Y}$	Coefficients for linear trend line	
(14.5)	$b = \dfrac{\Sigma xY}{\Sigma x^2}$		**446**
(14.6)	$Y_t = ab^x$	Exponential trend line	**447**
(14.7)	$\log a = \dfrac{\Sigma \log Y}{n}$	Coefficients for exponential trend line	**448**
(14.8)	$\log b = \dfrac{\Sigma x \log Y}{\Sigma x^2}$		

Number	Formula	Verbal description	Page
(14.9)	$Y_t = a + bx + cx^2$	Parabolic trend line	**449**
(14.10) (14.11)	$\left.\begin{array}{l} \Sigma Y = na + c\Sigma x^2 \\ \Sigma x^2 Y = a\Sigma x^2 + c\Sigma x^4 \end{array}\right\}$	Simultaneous equations for obtaining values of a and c for a parabolic trend line	**449**
(15.1)	$I_p = \dfrac{P_n}{P_0} \times 100$	Index number for price	**479**
(15.2)	$I_q = \dfrac{Q_n}{Q_0} \times 100$	Index number for quantity	**479**
(15.3)	$I_v = \dfrac{P_n Q_n}{P_0 Q_0} \times 100$	Value index	**479**
(15.4)	$\dfrac{\Sigma P_n}{\Sigma P_0} \times 100$	Unweighted aggregate price index	**482**
(15.5)	$\dfrac{\Sigma Q_n}{\Sigma Q_0} \times 100$	Unweighted aggregate quantity index	**483**
(15.6)	$\dfrac{\Sigma \left(\dfrac{P_n}{P_0}\right)}{n} \times 100$	Simple average of relatives index	**484**
(15.7)	$\dfrac{\Sigma P_n Q_n}{\Sigma P_0 Q_0} \times 100$	Simple weighted aggregate index	**486**
(15.8)	$\dfrac{\Sigma P_n Q_n}{\Sigma P_0 Q_n} \times 100$	Paasche index	**486**
(15.9)	$\dfrac{\Sigma P_n Q_0}{P_0 Q_0} \times 100$	Laspeyres index	**486**
(15.10)	$\dfrac{P_n Q_f}{P_0 Q_f} \times 100$	Fixed weight aggregate index	**487**
(15.11)	$\dfrac{\Sigma \left(\dfrac{Q_n}{Q_0} \times 100\right)(Q_0 P_0)}{\Sigma Q_0 P_0}$	Weighted average-of-relatives quantity index	**490**

Appendix C

Tables

THE USE OF TABLE I—BINOMIAL PROBABILITIES

Example: $n = 9$; $P = 0.25$; $O = 0.75$

1. Exact probability (x)

$$p(x = 2) = 0.3003$$

Exact probabilities are obtained *directly* from the table.

2. At least x, x or more

$$p(x \geq 2) = 0.3003 + 0.2336 + 0.1168 + 0.0389 + 0.0087$$
$$+ 0.0012 + 0.0001 = 0.6996$$

or $\quad p(x \geq 2) = 1 - p(x \geq 1) = 1 - 0.2253 - 0.0751$
$$= 0.6996$$

3. x or less, x or fewer

$$p(x \leq 2) = 0.3003 + 0.2253 + 0.0751$$
$$= 0.6007$$

or $\quad p(x \leq 2) = 1 - p(x \geq 3)$
$$= 1 - 0.2336 - 0.1168 - 0.0389 - 0.0087$$
$$- 0.0012 - 0.0001$$
$$= 1 - 0.3983 = 0.6007$$

4. Exact probability $(n - x)$

$$p[(n - x) = 2] = p(x = 7) = 0.0012$$

5. At least $(n - x)$, $(n - x)$ or more

$$p[(n - x) \geq 2] = p(x \leq 7) = 0.9997$$

6. $(n - x)$ or less, $(n - x)$ or fewer

$$p[(n - x) \leq 2] = p(x \geq 7) = 0.0012 + 0.0001 = 0.0013$$

Table I

Binomial probabilities [$p(x)$]

P

n	x	.01	.05	.10	.15	.20	.25	.30	.35	.40	.45	.50
1	0	.9900	.9500	.9000	.8500	.8000	.7500	.7000	.6500	.6000	.5500	.5000
	1	.0100	.0500	.1000	.1500	.2000	.2500	.3000	.3500	.4000	.4500	.5000
2	0	.9801	.9025	.8100	.7225	.6400	.5625	.4900	.4225	.3600	.3025	.2500
	1	.0198	.0950	.1800	.2550	.3200	.3750	.4200	.4550	.4800	.4950	.5000
	2	.0001	.0025	.0100	.0225	.0400	.0625	.0900	.1225	.1600	.2025	.2500
3	0	.9703	.8574	.7290	.6141	.5120	.4219	.3430	.2746	.2160	.1664	.1250
	1	.0294	.1354	.2430	.3251	.3840	.4219	.4410	.4436	.4320	.4084	.3750
	2	.0003	.0071	.0270	.0574	.0960	.1406	.1890	.2389	.2880	.3341	.3750
	3	.0000	.0001	.0010	.0034	.0080	.0156	.0270	.0429	.0640	.0911	.1250
4	0	.9606	.8145	.6561	.5220	.4096	.3164	.2401	.1785	.1296	.0915	.0625
	1	.0388	.1715	.2916	.3685	.4096	.4219	.4116	.3845	.3456	.2995	.2500
	2	.0006	.0135	.0486	.0975	.1536	.2109	.2646	.3105	.3456	.3675	.3750
	3	.0000	.0005	.0036	.0115	.0258	.0469	.0756	.1115	.1536	.2005	.2500
	4	.0000	.0000	.0001	.0005	.0016	.0039	.0081	.0150	.0256	.0410	.0625
5	0	.9510	.7738	.5905	.4437	.3277	.2373	.1681	.1160	.0778	.0503	.0312
	1	.0480	.2036	.3280	.3915	.4096	.3955	.3602	.3124	.2592	.2059	.1562
	2	.0010	.0214	.0729	.1382	.2048	.2637	.3087	.3364	.3456	.3369	.3125
	3	.0000	.0011	.0081	.0244	.0512	.0879	.1323	.1811	.2304	.2757	.3125
	4	.0000	.0000	.0004	.0022	.0064	.0146	.0284	.0488	.0768	.1128	.1562
	5	.0000	.0000	.0000	.0001	.0003	.0010	.0024	.0053	.0102	.0185	.0312
6	0	.9415	.7351	.5314	.3771	.2621	.1780	.1176	.0754	.0467	.0277	.0156
	1	.0571	.2321	.3543	.3993	.3932	.3560	.3025	.2437	.1866	.1359	.0938
	2	.0014	.0305	.0984	.1762	.2458	.2966	.3241	.3280	.3110	.2780	.2344
	3	.0000	.0021	.0146	.0415	.0819	.1318	.1852	.2355	.2765	.3032	.3125
	4	.0000	.0001	.0012	.0055	.0154	.0330	.0595	.0951	.1382	.1861	.2344
	5	.0000	.0000	.0001	.0004	.0015	.0044	.0102	.0205	.0369	.0609	.0938
	6	.0000	.0000	.0000	.0000	.0001	.0002	.0007	.0018	.0041	.0083	.0156
7	0	.9321	.6983	.4783	.3206	.2097	.1335	.0824	.0490	.0280	.0152	.0078
	1	.0659	.2573	.3720	.3960	.3670	.3115	.2471	.1848	.1306	.0872	.0547
	2	.0020	.0406	.1240	.2097	.2753	.3115	.3177	.2985	.2613	.2140	.1641
	3	.0000	.0036	.0230	.0617	.1147	.1730	.2269	.2679	.2903	.2918	.2734
	4	.0000	.0002	.0026	.0109	.0287	.0577	.0972	.1442	.1935	.2388	.2734
	5	.0000	.0000	.0002	.0012	.0043	.0115	.0250	.0466	.0774	.1172	.1641
	6	.0000	.0000	.0000	.0001	.0004	.0013	.0036	.0084	.0172	.0320	.0547
	7	.0000	.0000	.0000	.0000	.0000	.0001	.0002	.0006	.0016	.0037	.0078
8	0	.9227	.6634	.4305	.2725	.1678	.1002	.0576	.0319	.0168	.0084	.0039
	1	.0746	.2793	.3826	.3847	.3355	.2670	.1977	.1373	.0896	.0548	.0312
	2	.0026	.0515	.1488	.2376	.2936	.3115	.2065	.2587	.2090	.1569	.1094
	3	.0001	.0054	.0331	.0839	.1468	.2076	.2541	.2786	.2787	.2568	.2188
	4	.0000	.0004	.0046	.0185	.0459	.0865	.1361	.1875	.2322	.2627	.2734
	5	.0000	.0000	.0004	.0026	.0092	.0231	.0467	.0808	.1239	.1719	.2188
	6	.0000	.0000	.0004	.0026	.0011	.0038	.0100	.0217	.0413	.0403	.1094
	7	.0000	.0000	.0000	.0000	.0001	.0004	.0012	.0033	.0079	.0164	.0312
	8	.0000	.0000	.0000	.0000	.0000	.0000	.0001	.0002	.0007	.0017	.0039
9	0	.9135	.6302	.3874	.2316	.1342	.0751	.0404	.0207	.0101	.0046	.0020
	1	.0830	.2985	.3874	.3679	.3020	.2253	.1556	.1004	.0605	.0339	.0176
	2	.0034	.0629	.1722	.2597	.3020	.3003	.2668	.2162	.1612	.1110	.0703

Table I *(continued)*

P

n	x	.01	.05	.10	.15	.20	.25	.30	.35	.40	.45	.50
9	3	.0001	.0077	.0446	.1069	.1762	.2336	.2668	.2716	.2508	.2119	.1641
	4	.0000	.0006	.0074	.0283	.0661	.1168	.1715	.2194	.2508	.2600	.2461
	5	.0000	.0000	.0008	.0050	.0165	.0389	.0735	.1181	.1672	.2128	.2461
	6	.0000	.0000	.0001	.0006	.0028	.0087	.0210	.0424	.0743	.1160	.1641
	7	.0000	.0000	.0000	.0000	.0003	.0012	.0039	.0098	.0212	.0407	.0703
	8	.0000	.0000	.0000	.0000	.0000	.0001	.0004	.0013	.0035	.0083	.0176
	9	.0000	.0000	.0000	.0000	.0000	.0000	.0000	.0001	.0003	.0008	.0020
10	0	.9044	.5987	.3487	.1969	.1074	.0563	.0282	.0135	.0060	.0025	.0010
	1	.0914	.3151	.3874	.3474	.2684	.1877	.1211	.0725	.0403	.0207	.0098
	2	.0042	.0746	.1937	.2759	.3020	.2816	.2335	.1757	.1209	.0763	.0439
	3	.0001	.0105	.0574	.1298	.2013	.2503	.2668	.2522	.2150	.1665	.1172
	4	.0000	.0010	.0112	.0401	.0881	.1460	.2001	.2377	.2508	.2384	.2051
	5	.0000	.0001	.0015	.0085	.0264	.0584	.1029	.1536	.2007	.2340	.2461
	6	.0000	.0000	.0001	.0012	.0055	.0162	.0368	.0689	.1115	.1596	.2051
	7	.0000	.0000	.0000	.0001	.0008	.0031	.0090	.0212	.0425	.0746	.1172
	8	.0000	.0000	.0000	.0000	.0001	.0004	.0014	.0043	.0106	.0229	.0439
	9	.0000	.0000	.0000	.0000	.0000	.0000	.0001	.0005	.0016	.0042	.0098
	10	.0000	.0000	.0000	.0000	.0000	.0000	.0000	.0000	.0001	.0003	.0010
11	0	.8953	.5688	.3138	.1673	.0859	.0422	.0198	.0088	.0036	.0014	.0005
	1	.0995	.3293	.3835	.3248	.2362	.1549	.0932	.0518	.0266	.0125	.0054
	2	.0050	.0867	.2131	.2866	.2953	.2581	.1998	.1395	.0887	.0513	.0269
	3	.0002	.0137	.0710	.1517	.2215	.2581	.2568	.2254	.1774	.1259	.0806
	4	.0000	.0010	.0112	.0401	.0881	.1460	.2001	.2377	.2508	.2384	.2051
	5	.0000	.0001	.0025	.0132	.0388	.0803	.1321	.1830	.2207	.2360	.2256
	6	.0000	.0000	.0003	.0023	.0097	.0268	.0566	.0985	.1471	.1931	.2256
	7	.0000	.0000	.0000	.0003	.0017	.0064	.0173	.0379	.0701	.1128	.1611
	8	.0000	.0000	.0000	.0000	.0002	.0011	.0037	.0102	.0234	.0462	.0806
	9	.0000	.0000	.0000	.0000	.0000	.0001	.0005	.0018	.0052	.0126	.0269
	10	.0000	.0000	.0000	.0000	.0000	.0000	.0000	.0002	.0007	.0021	.0054
	11	.0000	.0000	.0000	.0000	.0000	.0000	.0000	.0000	.0000	.0002	.0005
12	0	.8864	.5404	.2824	.1422	.0687	.0317	.0138	.0057	.0022	.0008	.0002
	1	.1074	.3413	.3766	.3012	.2062	.1267	.0712	.0368	.0174	.0075	.0029
	2	.0060	.0988	.2301	.2924	.2835	.2323	.1678	.1088	.0639	.0339	.0161
	3	.0002	.0173	.0852	.1720	.2362	.2581	.2397	.1954	.1419	.0923	.0537
	4	.0000	.0021	.0213	.0683	.1329	.1936	.2311	.2367	.2128	.1700	.1204
	5	.0000	.0002	.0038	.0193	.0532	.1032	.1585	.2039	.2270	.2225	.1934
	6	.0000	.0000	.0005	.0040	.0155	.0401	.0792	.1281	.1766	.2124	.2256
	7	.0000	.0000	.0000	.0006	.0033	.0115	.0291	.0591	.1009	.1489	.1934
	8	.0000	.0000	.0000	.0001	.0005	.0024	.0078	.0199	.0420	.0762	.1208
	9	.0000	.0000	.0000	.0000	.0001	.0004	.0015	.0048	.0125	.0277	.0537
	10	.0000	.0000	.0000	.0000	.0000	.0000	.0002	.0008	.0025	.0068	.0161
	11	.0000	.0000	.0000	.0000	.0000	.0000	.0000	.0001	.0003	.0010	.0029
	12	.0000	.0000	.0000	.0000	.0000	.0000	.0000	.0000	.0000	.0001	.0002
13	0	.8775	.5133	.2542	.1209	.0550	.0238	.0097	.0037	.0013	.0004	.0001
	1	.1152	.3512	.3672	.2774	.1787	.1029	.0540	.0259	.0113	.0045	.0016
	2	.0070	.1109	.2448	.2937	.2680	.2059	.1388	.0836	.0453	.0220	.0095
	3	.0003	.0214	.0997	.1900	.2457	.2517	.2181	.1651	.1107	.0660	.0349
	4	.0000	.0028	.0277	.0838	.1535	.2097	.2337	.2222	.1845	.1350	.0873

Table I *(continued)*

P

n	x	.01	.05	.10	.15	.20	.25	.30	.35	.40	.45	.50
13	5	.0000	.0003	.0055	.0266	.0691	.1258	.1803	.2154	.2214	.1989	.1571
	6	.0000	.0000	.0008	.0063	.0230	.0559	.1030	.1546	.1968	.2169	.2095
	7	.0000	.0000	.0001	.0011	.0058	.0186	.0442	.0833	.1312	.1775	.2095
	8	.0000	.0000	.0001	.0001	.0011	.0047	.0142	.0336	.0656	.1089	.1571
	9	.0000	.0000	.0000	.0000	.0001	.0009	.0034	.0101	.0243	.0495	.0873
	10	.0000	.0000	.0000	.0000	.0000	.0001	.0006	.0022	.0065	.0162	.0349
	11	.0000	.0000	.0000	.0000	.0000	.0000	.0001	.0003	.0012	.0036	.0095
	12	.0000	.0000	.0000	.0000	.0000	.0000	.0000	.0000	.0001	.0005	.0016
	13	.0000	.0000	.0000	.0000	.0000	.0000	.0000	.0000	.0000	.0000	.0001
14	0	.8687	.4877	.2288	.1028	.0440	.0178	.0068	.0024	.0008	.0002	.0001
	1	.1229	.3593	.3559	.2539	.1539	.0832	.0407	.0181	.0073	.0027	.0009
	2	.0081	.1229	.2570	.2912	.2501	.1802	.1134	.0634	.0317	.0141	.0056
	3	.0003	.0259	.1142	.2056	.2501	.2402	.1943	.1366	.0845	.0462	.0222
	4	.0000	.0037	.0349	.0998	.1720	.2202	.2290	.2022	.1549	.1040	.0611
	5	.0000	.0004	.0078	.0352	.0860	.1468	.1963	.2178	.2066	.1701	.1222
	6	.0000	.0000	.0013	.0093	.0322	.0734	.1262	.1759	.2066	.2088	.1833
	7	.0000	.0000	.0002	.0019	.0092	.0280	.0618	.1082	.1574	.1952	.2095
	8	.0000	.0000	.0000	.0003	.0020	.0082	.0232	.0510	.0918	.1398	.1833
	9	.0000	.0000	.0000	.0000	.0003	.0018	.0066	.0183	.0408	.0762	.1222
	10	.0000	.0000	.0000	.0000	.0000	.0003	.0014	.0049	.0136	.0312	.0611
	11	.0000	.0000	.0000	.0000	.0000	.0000	.0002	.0010	.0033	.0093	.0222
	12	.0000	.0000	.0000	.0000	.0000	.0000	.0000	.0001	.0005	.0019	.0056
	13	.0000	.0000	.0000	.0000	.0000	.0000	.0000	.0000	.0001	.0002	.0009
	14	.0000	.0000	.0000	.0000	.0000	.0000	.0000	.0000	.0000	.0000	.0001
15	0	.8601	.4633	.2059	.0874	.0352	.0134	.0047	.0016	.0005	.0001	.0000
	1	.1303	.3658	.3432	.2312	.1319	.0668	.0305	.0126	.0047	.0016	.0005
	2	.0092	.1348	.2669	.2856	.2309	.1559	.0916	.0476	.0219	.0090	.0032
	3	.0004	.0307	.1285	.2184	.2501	.2252	.1700	.1110	.0634	.0318	.0139
	4	.0000	.0049	.0428	.1156	.1876	.2252	.2186	.1792	.1268	.0780	.0417
	5	.0000	.4633	.2059	.0449	.0352	.0134	.0047	.0016	.0005	(.000)	.0000
	6	.0000	.0000	.0019	.0132	.0430	.0917	.1472	.1906	.2066	.1914	.1527
	7	.0000	.0000	.0003	.0030	.0138	.0393	.0811	.1319	.1771	.2013	.1964
	8	.0000	.0000	.0000	.0005	.0035	.0131	.0348	.0710	.1181	.1647	.1964
	9	.0000	.0000	.0000	.0001	.0007	.0034	.0116	.0298	.0612	.1048	.1527
	10	.0000	.0000	.0000	.0000	.0001	.0007	.0030	.0096	.0245	.0515	.0916
	11	.0000	.0000	.0000	.0000	.0000	.0001	.0006	.0024	.0074	.0191	.0417
	12	.0000	.0000	.0000	.0000	.0000	.0000	.0001	.0004	.0016	.0052	.0139
	13	.0000	.0000	.0000	.0000	.0000	.0000	.0000	.0001	.0003	.0010	.0032
	14	.0000	.0000	.0000	.0000	.0000	.0000	.0000	.0000	.0000	.0001	.0005
	15	.0000	.0000	.0000	.0000	.0000	.0000	.0000	.0000	.0000	.0000	.0000
16	0	.8515	.4404	.1853	.0743	.0281	.0100	.0033	.0010	.0003	.0001	.0000
	1	.1376	.3706	.3294	.2097	.1126	.0535	.0228	.0087	.0030	.0009	.0002
	2	.0104	.1463	.2745	.2775	.2111	.1336	.0732	.0353	.0150	.0056	.0018
	3	.0005	.0359	.1423	.2285	.2463	.2079	.1465	.0888	.0468	.0215	.0085
	4	.0000	.0061	.0514	.1311	.2001	.2252	.2040	.1553	.1014	.0572	.0278
	5	.0000	.0008	.0137	.0555	.1201	.1802	.2099	.2008	.1623	.1123	.0667
	6	.0000	.0001	.0028	.0180	.0550	.1101	.1649	.1982	.1983	.1684	.1222

Table I *(continued)*

P

n	x	.01	.05	.10	.15	.20	.25	.30	.35	.40	.45	.50
16	7	.0000	.0000	.0004	.0045	.0197	.0524	.1010	.1524	.1889	.1969	.1746
	8	.0000	.0000	.0001	.0009	.0055	.0197	.0487	.0923	.1417	.1812	.1964
	9	.0000	.0000	.0000	.0001	.0012	.0058	.0185	.0442	.0840	.1318	.1746
	10	.0000	.0000	.0000	.0000	.0002	.0014	.0056	.0167	.0392	.0755	.1222
	11	.0000	.0000	.0000	.0000	.0000	.0002	.0013	.0049	.0142	.0337	.0667
	12	.0000	.0000	.0000	.0000	.0000	.0000	.0002	.0011	.0040	.0115	.0278
	13	.0000	.0000	.0000	.0000	.0000	.0000	.0000	.0002	.0008	.0029	.0085
	14	.0000	.0000	.0000	.0000	.0000	.0000	.0000	.0000	.0001	.0005	.0018
	15	.0000	.0000	.0000	.0000	.0000	.0000	.0000	.0000	.0000	.0001	.0002
	16	.0000	.0000	.0000	.0000	.0000	.0000	.0000	.0000	.0000	.0000	.0000
17	0	.8429	.4181	.1668	.0631	.0225	.0075	.0023	.0007	.0002	.0000	.0000
	1	.1447	.3741	.3150	.1893	.0957	.0426	.0169	.0060	.0019	.0005	.0001
	2	.0177	.1575	.2800	.2673	.1914	.1136	.0581	.0260	.0102	.0035	.0010
	3	.0006	.0415	.1556	.2359	.2393	.1893	.1245	.0701	.0341	.0144	.0052
	4	.0000	.0076	.0605	.1457	.2093	.2209	.1868	.1320	.0796	.0411	.0182
	5	.0000	.0010	.0175	.0668	.1361	.1914	.2081	.1849	.1379	.0875	.0472
	6	.0000	.0001	.0039	.0236	.0680	.1276	.1784	.1991	.1839	.1432	.0944
	7	.0000	.0000	.0007	.0065	.0267	.0668	.1201	.1685	.1927	.1841	.1484
	8	.0000	.0000	.0001	.0014	.0084	.0279	.0644	.1134	.1606	.1883	.1855
	9	.0000	.0000	.0000	.0003	.0021	.0093	.0276	.0611	.1070	.1540	.1855
	10	.0000	.0000	.0000	.0000	.0004	.0025	.0095	.0263	.0571	.1008	.1484
	11	.0000	.0000	.0000	.0000	.0001	.0005	.0026	.0090	.0242	.0525	.0944
	12	.0000	.0000	.0000	.0000	.0000	.0001	.0006	.0024	.0081	.0215	.0472
	13	.0000	.0000	.0000	.0000	.0000	.0000	.0001	.0005	.0021	.0068	.0182
	14	.0000	.0000	.0000	.0000	.0000	.0000	.0000	.0001	.0004	.0016	.0052
	15	.0000	.0000	.0000	.0000	.0000	.0000	.0000	.0000	.0001	.0003	.0010
	16	.0000	.0000	.0000	.0000	.0000	.0000	.0000	.0000	.0000	.0000	.0001
	17	.0000	.0000	.0000	.0000	.0000	.0000	.0000	.0000	.0000	.0000	.0000
18	0	.8345	.3972	.1501	.0536	.0180	.0056	.0016	.0004	.0001	.0003	.0010
	1	.1517	.3763	.3002	.1704	.0811	.0338	.0126	.0042	.0012	.0003	.0001
	2	.0130	.1683	.2835	.2556	.1723	.0958	.0458	.0190	.0069	.0022	.0006
	3	.0007	.0473	.1680	.2406	.2297	.1704	.1046	.0547	.0246	.0095	.0001
	4	.0000	.0093	.0700	.1592	.2153	.2130	.1681	.1104	.0614	.0291	.0117
	5	.0000	.0014	.0218	.0787	.1507	.1988	.2017	.1664	.1146	.0666	.0327
	6	.0000	.0002	.0052	.0301	.0816	.1436	.1873	.1941	.1655	.1181	.0708
	7	.0000	.0000	.0010	.0091	.0350	.0820	.1376	.1792	.1892	.1657	.1214
	8	.0000	.0000	.0002	.0022	.0120	.0376	.0811	.1327	.1734	.1864	.1669
	9	.0000	.0000	.0000	.0004	.0033	.0139	.0386	.0794	.1284	.1694	.1855
	10	.0000	.0000	.0000	.0001	.0008	.0042	.0149	.0385	.0771	.1248	.1669
	11	.0000	.0000	.0000	.0000	.0001	.0010	.0046	.0151	.0374	.0742	.1214
	12	.0000	.0000	.0000	.0000	.0000	.0002	.0012	.0047	.0145	.0354	.0708
	13	.0000	.0000	.0000	.0000	.0000	.0000	.0002	.0012	.0045	.0134	.0327
	14	.0000	.0000	.0000	.0000	.0000	.0000	.0000	.0002	.0011	.0039	.0117
	15	.0000	.0000	.0000	.0000	.0000	.0000	.0000	.0000	.0002	.0009	.0031
	16	.0000	.0000	.0000	.0000	.0000	.0000	.0000	.0000	.0000	.0001	.0006
	17	.0000	.0000	.0000	.0000	.0000	.0000	.0000	.0000	.0000	.0000	.0001
	18	.0000	.0000	.0000	.0000	.0000	.0000	.0000	.0000	.0000	.0000	.0000

Table I *(continued)*

P

n	x	.01	.05	.10	.15	.20	.25	.30	.35	.40	.45	.50
19	0	.8262	.3774	.1351	.0456	.0144	.0042	.0011	.0003	.0001	.0000	.0000
	1	.1586	.3774	.2852	.1529	.0685	.0268	.0093	.0029	.0008	.0002	.0000
	2	.0144	.1787	.2852	.2428	.1540	.0803	.0358	.0138	.0046	.0013	.0003
	3	.0008	.0533	.1796	.2428	.2182	.1517	.0869	.0422	.0175	.0062	.0018
	4	.0000	.0112	.0798	.1714	.2182	.2023	.1491	.0909	.0467	.0203	.0074
	5	.0000	.0018	.0266	.0907	.1636	.2023	.1916	.1468	.0933	.0497	.0222
	6	.0000	.0002	.0069	.0374	.0955	.1574	.1916	.1844	.1451	.0949	.0518
	7	.0000	.0000	.0014	.0122	.0443	.0974	.1525	.1844	.1797	.1443	.0961
	8	.0000	.0000	.0002	.0032	.0166	.0487	.0981	.1489	.1797	.1771	.1442
	9	.0000	.0000	.0000	.0007	.0051	.0198	.0514	.0980	.1464	.1771	.1762
	10	.0000	.0000	.0000	.0001	.0013	.0066	.0220	.0528	.0976	.1449	.1762
	11	.0000	.0000	.0000	.0000	.0003	.0018	.0077	.0233	.0532	.0970	.1442
	12	.0000	.0000	.0000	.0000	.0000	.0004	.0022	.0083	.0237	.0529	.0961
	13	.0000	.0000	.0000	.0000	.0000	.0001	.0005	.0024	.0085	.0233	.0518
	14	.0000	.0000	.0000	.0000	.0000	.0000	.0001	.0006	.0024	.0082	.0222
	15	.0000	.0000	.0000	.0000	.0000	.0000	.0000	.0001	.0005	.0022	.0074
	16	.0000	.0000	.0000	.0000	.0000	.0000	.0000	.0000	.0001	.0005	.0018
	17	.0000	.0000	.0000	.0000	.0000	.0000	.0000	.0000	.0000	.0001	.0003
	18	.0000	.0000	.0000	.0000	.0000	.0000	.0000	.0000	.0000	.0000	.0000
	19	.0000	.0000	.0000	.0000	.0000	.0000	.0000	.0000	.0000	.0000	.0000
20	0	.8179	.3585	.1216	.0388	.0115	.0032	.0008	.0002	.0000	.0000	.0000
	1	.1652	.3774	.2702	.1368	.0576	.0211	.0068	.0020	.0005	.0001	.0000
	2	.0159	.1887	.2852	.2293	.1369	.0669	.0278	.0100	.0031	.0008	.0002
	3	.0010	.0596	.1901	.2428	.2054	.1339	.0718	.0323	.0123	.0040	.0011
	4	.0000	.0133	.0898	.1821	.2182	.1897	.1304	.0738	.0350	.0139	.0046
	5	.0000	.0022	.0319	.1028	.1746	.2023	.1789	.1272	.0746	.0365	.0148
	6	.0000	.0003	.0089	.0454	.1091	.1686	.1916	.1712	.1244	.0746	.0370
	7	.0000	.0000	.0020	.0160	.0545	.1124	.1643	.1844	.1659	.1221	.0739
	8	.0000	.0000	.0004	.0046	.0222	.0609	.1144	.1614	.1797	.1623	.1201
	9	.0000	.0000	.0001	.0011	.0074	.0271	.0654	.1158	.1597	.1771	.1602
	10	.0000	.0000	.0000	.0002	.0020	.0099	.0308	.0686	.1171	.1593	.1762
	11	.0000	.0000	.0000	.0000	.0005	.0030	.0120	.0336	.0710	.1185	.1602
	12	.0000	.0000	.0000	.0000	.0001	.0008	.0039	.0136	.0355	.0727	.1201
	13	.0000	.0000	.0000	.0000	.0000	.0002	.0010	.0045	.0146	.0366	.0739
	14	.0000	.0000	.0000	.0000	.0000	.0000	.0002	.0012	.0049	.0150	.0370
	15	.0000	.0000	.0000	.0000	.0000	.0000	.0000	.0003	.0013	.0049	.0148
	16	.0000	.0000	.0000	.0000	.0000	.0000	.0000	.0000	.0003	.0013	.0046
	17	.0000	.0000	.0000	.0000	.0000	.0000	.0000	.0000	.0000	.0002	.0011
	18	.0000	.0000	.0000	.0000	.0000	.0000	.0000	.0000	.0000	.0000	.0002
	19	.0000	.0000	.0000	.0000	.0000	.0000	.0000	.0000	.0000	.0000	.0000
	20	.0000	.0000	.0000	.0000	.0000	.0000	.0000	.0000	.0000	.0000	.0000
25	0	.7778	.2774	.0718	.0172	.0038	.0008	.0001	.0000	.0000	.0000	.0000
	1	.1964	.3650	.1994	.0759	.0236	.0063	.0014	.0003	.0000	.0000	.0000
	2	.0238	.2305	.2659	.1607	.0708	.0251	.0074	.0018	.0004	.0001	.0000
	3	.0018	.0930	.2265	.2174	.1358	.0641	.0243	.0076	.0019	.0004	.0001
	4	.0001	.0269	.1384	.2110	.1867	.1175	.0572	.0224	.0071	.0018	.0004

Table I *(concluded)*

P

n	x	.01	.05	.10	.15	.20	.25	.30	.35	.40	.45	.50
25	5	.0000	.0060	.0646	.1564	.1960	.1645	.1030	.0506	.0199	.0063	.0016
	6	.0000	.0010	.0239	.0920	.1633	.1828	.1472	.0908	.0442	.0172	.0053
	7	.0000	.0001	.0072	.0441	.1108	.1654	.1712	.1327	.0800	.0381	.0143
	8	.0000	.0000	.0018	.0175	.0623	.1241	.1651	.1607	.1200	.0701	.0322
	9	.0000	.0000	.0004	.0058	.0294	.0781	.1336	.1635	.1511	.1084	.0609
	10	.0000	.0000	.0000	.0016	.0118	.0417	.0916	.1409	.1612	.1419	.0974
	11	.0000	.0000	.0000	.0004	.0040	.0189	.0536	.1034	.1465	.1583	.1328
	12	.0000	.0000	.0000	.0000	.0012	.0074	.0268	.0650	.1140	.1511	.1550
	13	.0000	.0000	.0000	.0000	.0003	.0025	.0115	.0350	.0760	.1236	.1550
	14	.0000	.0000	.0000	.0000	.0000	.0007	.0042	.0161	.0434	.0867	.1328
	15	.0000	.0000	.0000	.0000	.0000	.0002	.0013	.0064	.0212	.0520	.0974
	16	.0000	.0000	.0000	.0000	.0000	.0000	.0004	.0021	.0088	.0266	.0609
	17	.0000	.0000	.0000	.0000	.0000	.0000	.0001	.0006	.0031	.0115	.0322
	18	.0000	.0000	.0000	.0000	.0000	.0000	.0000	.0001	.0009	.0042	.0143
	19	.0000	.0000	.0000	.0000	.0000	.0000	.0000	.0000	.0002	.0013	.0053
	20	.0000	.0000	.0000	.0000	.0000	.0000	.0000	.0000	.0000	.0001	.0016
	21	.0000	.0000	.0000	.0000	.0000	.0000	.0000	.0000	.0000	.0000	.0004
	22	.0000	.0000	.0000	.0000	.0000	.0000	.0000	.0000	.0000	.0000	.0001
30	0	.7397	.2146	.0424	.0076	.0012	.0002	.0000	.0000	.0000	.0000	.0000
	1	.2242	.3389	.1413	.0404	.0093	.0018	.0003	.0000	.0000	.0000	.0000
	2	.0328	.2586	.2277	.1034	.0337	.0086	.0018	.0003	.0000	.0000	.0000
	3	.0031	.1270	.2361	.1703	.0785	.0269	.0072	.0015	.0003	.0000	.0000
	4	.0002	.0451	.1771	.2028	.1325	.0604	.0208	.0056	.0012	.0002	.0000
	5	.0000	.0124	.1023	.1861	.1723	.1047	.0464	.0157	.0041	.0008	.0001
	6	.0000	.0027	.0474	.1368	.1795	.1455	.0829	.0353	.0115	.0029	.0006
	7	.0000	.0005	.0180	.0828	.1538	.1662	.1219	.0652	.0263	.0081	.0019
	8	.0000	.0001	.0058	.0420	.1106	.1593	.1501	.1009	.0505	.0191	.0055
	9	.0000	.0000	.0016	.0181	.0676	.1298	.1573	.1328	.0823	.0382	.0133
	10	.0000	.0000	.0004	.0067	.0355	.0909	.1416	.1502	.1152	.0656	.0280
	11	.0000	.0000	.0001	.0022	.0161	.0551	.1103	.1471	.1396	.0976	.0509
	12	.0000	.0000	.0000	.0006	.0064	.0291	.0749	.1254	.1474	.1265	.0806
	13	.0000	.0000	.0000	.0001	.0020	.0134	.0444	.0935	.1360	.1433	.1115
	14	.0000	.0000	.0000	.0000	.0007	.0054	.0231	.0611	.1101	.1424	.1354
	15	.0000	.0000	.0000	.0000	.0002	.0019	.0106	.0351	.0783	.1242	.1445
	16	.0000	.0000	.0000	.0000	.0000	.0006	.0042	.0177	.0489	.0953	.1354
	17	.0000	.0000	.0000	.0000	.0000	.0002	.0015	.0079	.0269	.0642	.1115
	18	.0000	.0000	.0000	.0000	.0000	.0000	.0005	.0031	.0129	.0379	.0806
	19	.0000	.0000	.0000	.0000	.0000	.0000	.0001	.0010	.0054	.0196	.0509
	20	.0000	.0000	.0000	.0000	.0000	.0000	.0000	.0003	.0020	.0088	.0280
	21	.0000	.0000	.0000	.0000	.0000	.0000	.0000	.0001	.0006	.0034	.0133
	22	.0000	.0000	.0000	.0000	.0000	.0000	.0000	.0000	.0002	.0012	.0055
	23	.0000	.0000	.0000	.0000	.0000	.0000	.0000	.0000	.0000	.0003	.0019
	24	.0000	.0000	.0000	.0000	.0000	.0000	.0000	.0000	.0000	.0001	.0006
	25	.0000	.0000	.0000	.0000	.0000	.0000	.0000	.0000	.0000	.0000	.0001

Table II

Values of $e^{-\mu}$

μ	$e^{-\mu}$	μ	$e^{-\mu}$	μ	$e^{-\mu}$	μ	$e^{-\mu}$	μ	$e^{-\mu}$
.01	.990050	.28	.755784	.75	.472367	3.20	.0407622	5.90	.00273945
.02	.980199	.29	.748264	.80	.449329	3.30	.0368832	6.00	.00247875
.03	.970446	.30	.740818	.85	.427415	3.40	.0333733	6.10	.00224287
.04	.960789	.31	.733467	.90	.406570	3.50	.0301974	6.20	.00202943
.05	.951229	.32	.726149	.95	.386741	3.60	.0273237	6.30	.00183631
.06	.941765	.33	.718924	1.00	.367879	3.70	.0247235	6.40	.00166156
.07	.932394	.34	.711770	1.10	.332871	3.80	.0223708	6.50	.00150344
.08	.923116	.35	.704688	1.20	.301194	3.90	.0202419	6.60	.00136037
.09	.913931	.36	.697676	1.30	.272532	4.00	.0183156	6.70	.00123091
.10	.904837	.37	.690743	1.40	.246597	4.10	.0165727	6.80	.00111378
.11	.895834	.38	.683861	1.50	.223130	4.20	.0149956	6.90	.00100779
.12	.886920	.39	.677057	.1.60	.201897	4.30	.0135686	7.00	.00091188
.13	.878095	.40	.670320	1.70	.182684	4.40	.0122773	7.50	.00055308
.14	.869358	.41	.663650	1.80	.165299	4.50	.0111090	8.00	.00033546
.15	.860708	.42	.657047	1.90	.149569	4.60	.0100518	8.50	.00020347
.16	.852144	.43	.650509	2.00	.135335	4.70	.00909528	9.00	.00012341
.17	.843665	.44	.644036	2.10	.122456	4.80	.00822975	9.50	.00007485
.18	.835270	.45	.637628	2.20	.110803	4.90	.00744658	10.00	.00004540
.19	.826959	.46	.631284	2.30	.100259	5.00	.00673795	10.50	.00002754
.20	.818731	.47	.625002	2.40	.0907180	5.10	.00609675	11.00	.00001670
.21	.810584	.48	.618783	2.50	.0820850	5.20	.00551656	11.50	.00001013
.22	.802519	.49	.612626	2.60	.0742736	5.30	.00499159	12.00	.00000614
.23	.794534	.50	.606531	2.70	.0672055	5.40	.00451658	12.50	.00000373
.24	.786628	.55	.576950	2.80	.0608101	5.50	.00408677	13.00	.00000226
.25	.778801	.60	.548812	2.90	.0550232	5.60	.00369786		
.26	.771052	.65	.522046	3.00	.0497871	5.70	.00334597		
.27	.763379	.70	.496585	3.10	.0450492	5.80	.00302756		

Table III
Factorials of numbers 0 to 20

x	x!
0	1
1	1
2	2
3	6
4	24
5	120
6	720
7	5 040
8	40 320
9	362 880
10	3 628 800
11	39 916 800
12	479 001 600
13	6 227 020 800
14	87 178 291 200
15	1 307 674 368 000
16	20 922 789 888 000
17	355 687 428 096 000
18	6 402 373 705 728 000
19	121 645 100 408 832 000
20	2 432 902 008 176 640 000

Table IV
Poisson probability distribution [$p(x)$]

Example:
$$\mu = 0.40$$
$$p(x = 2) = 0.0536$$
$$p(x < 2) = p(x \le 1) = 0.2681 + 0.6703 = 0.9384$$
$$p(x \ge 4) = 0.0007 + 0.0001 = 0.0008$$

x	μ									
	0.10	0.20	0.30	0.40	0.50	0.60	0.70	0.80	0.90	1.00
0	.9048	.8187	.7408	.6703	.6065	.5488	.4966	.4493	.4066	.3679
1	.0905	.1637	.2222	.2681	.3033	.3293	.3476	.3595	.3659	.3679
2	.0045	.0164	.0333	.0536	.0758	.0988	.1217	.1438	.1647	.1839
3	.0002	.0011	.0033	.0072	.0126	.0198	.0284	.0383	.0494	.0613
4	.0000	.0001	.0003	.0007	.0016	.0030	.0050	.0077	.0111	.0153
5	.0000	.0000	.0000	.0001	.0002	.0004	.0007	.0012	.0020	.0031
6	.0000	.0000	.0000	.0000	.0000	.0000	.0001	.0002	.0003	.0005
7	.0000	.0000	.0000	.0000	.0000	.0000	.0000	.0000	.0000	.0001

Table IV *(continued)*

	μ									
x	1.10	1.20	1.30	1.40	1.50	1.60	1.70	1.80	1.90	2.00
0	.3329	.3012	.2725	.2466	.2231	.2019	.1827	.1653	.1496	.1353
1	.3662	.3614	.3543	.3452	.3347	.3230	.3106	.2975	.2842	.2707
2	.2014	.2169	.2303	.2417	.2510	.2584	.2640	.2678	.2700	.2707
3	.0738	.0867	.0998	.1128	.1255	.1378	.1496	.1607	.1710	.1804
4	.0203	.0260	.0324	.0395	.0471	.0551	.0636	.0723	.0812	.0902
5	.0045	.0062	.0084	.0111	.0141	.0176	.0216	.0260	.0309	.0361
6	.0008	.0012	.0018	.0026	.0035	.0047	.0061	.0078	.0098	.0120
7	.0001	.0002	.0003	.0005	.0008	.0011	.0015	.0020	.0027	.0034
8	.0000	.0000	.0001	.0001	.0001	.0002	.0003	.0005	.0006	.0009
9	.0000	.0000	.0000	.0000	.0000	.0000	.0001	.0001	.0000	.0002

	μ									
x	2.10	2.20	2.30	2.40	2.50	2.60	2.70	2.80	2.90	3.00
0	.1225	.1108	.1003	.0907	.0821	.0743	.0672	.0608	.0550	.0498
1	.2572	.2438	.2306	.2177	.2052	.1931	.1815	.1703	.1596	.1494
2	.2700	.2681	.2652	.2613	.2565	.2510	.2450	.2384	.2314	.2240
3	.1890	.1966	.2033	.2090	.2138	.2176	.2205	.2225	.2237	.2240
4	.0992	.1082	.1169	.1254	.1336	.1414	.1488	.1557	.1622	.1680
5	.0417	.0476	.0538	.0602	.0668	.0735	.0804	.0872	.0940	.1008
6	.0146	.0174	.0206	.0241	.0278	.0319	.0362	.0407	.0455	.0504
7	.0044	.0055	.0068	.0083	.0099	.0118	.0139	.0163	.0188	.0216
8	.0011	.0015	.0019	.0025	.0031	.0038	.0047	.0057	.0068	.0081
9	.0003	.0004	.0005	.0007	.0009	.0011	.0014	.0018	.0022	.0027
10	.0001	.0001	.0001	.0002	.0002	.0003	.0004	.0005	.0006	.0008
11	.0000	.0000	.0000	.0000	.0000	.0001	.0001	.0001	.0002	.0002
12	.0000	.0000	.0000	.0000	.0000	.0000	.0000	.0000	.0000	.0001

	μ									
x	3.10	3.20	3.30	3.40	3.50	3.60	3.70	3.80	3.90	4.00
0	.0450	.0408	.0369	.0034	.0302	.0273	.0247	.0224	.0202	.0183
1	.1397	.1304	.1217	.1135	.1057	.0984	.0915	.0850	.0789	.0733
2	.2165	.2087	.2008	.1929	.1850	.1771	.1692	.1615	.1539	.1465
3	.2237	.2226	.2209	.2186	.2158	.2125	.2087	.2046	.2001	.1954
4	.1733	.1781	.1823	.1858	.1888	.1912	.1931	.1944	.1951	.1954

Table IV *(continued)*

	μ									
x	3.10	3.20	3.30	3.40	3.50	3.60	3.70	3.80	3.90	4.00
5	.1075	.1140	.1203	.1264	.1322	.1377	.1429	.1477	.1522	.1563
6	.0555	.0608	.0662	.0716	.0771	.0826	.0881	.0936	.0989	.1042
7	.0246	.0278	.0312	.0348	.0385	.0425	.0466	.0508	.0551	.0595
8	.0095	.0111	.0129	.0148	.0169	.0191	.0215	.0241	.0269	.0298
9	.0033	.0040	.0047	.0056	.0066	.0076	.0089	.0102	.0116	.0132
10	.0010	.0013	.0016	.0019	.0023	.0028	.0033	.0039	.0045	.0053
11	.0003	.0004	.0005	.0006	.0007	.0009	.0011	.0013	.0016	.0019
12	.0001	.0001	.0001	.0002	.0002	.0003	.0003	.0004	.0005	.0006
13	.0000	.0000	.0000	.0000	.0001	.0001	.0001	.0001	.0002	.0002
14	.0000	.0000	.0000	.0000	.0000	.0000	.0000	.0000	.0000	.0001

	μ									
x	4.10	4.20	4.30	4.40	4.50	4.60	4.70	4.80	4.90	5.00
0	.0166	.0150	.0136	.0123	.0111	.0101	.0091	.0082	.0074	.0067
1	.0679	.0630	.0583	.0540	.0500	.0462	.0427	.0395	.0365	.0337
2	.1393	.1323	.1254	.1188	.1125	.1063	.1005	.0948	.0894	.0842
3	.1904	.1852	.1798	.1743	.1687	.1631	.1574	.1517	.1460	.1404
4	.1951	.1944	.1933	.1917	.1898	.1875	.1849	.1820	.1789	.1755
5	.1600	.1633	.1662	.1687	.1708	.1725	.1738	.1747	.1753	.1755
6	.1093	.1143	.1191	.1237	.1281	.1323	.1362	.1398	.1432	.1462
7	.0640	.0686	.0732	.0778	.0824	.0869	.0914	.0959	.1002	.1044
8	.0328	.0360	.0393	.0428	.0463	.0500	.0537	.0575	.0614	.0653
9	.0150	.0168	.0188	.0209	.0232	.0255	.0281	.0307	.0334	.0363
10	.0061	.0071	.0081	.0092	.0104	.0118	.0132	.0147	.0164	.0181
11	.0023	.0027	.0032	.0037	.0043	.0049	.0056	.0064	.0073	.0082
12	.0008	.0009	.0011	.0013	.0016	.0019	.0022	.0026	.0030	.0034
13	.0002	.0003	.0004	.0005	.0006	.0007	.0008	.0009	.0011	.0013
14	.0001	.0001	.0001	.0001	.0002	.0002	.0003	.0003	.0004	.0005
15	.0000	.0000	.0000	.0000	.0001	.0001	.0001	.0001	.0001	.0002

Table IV *(continued)*

					μ					
x	5.10	5.20	5.30	5.40	5.50	5.60	5.70	5.80	5.90	6.00
0	.0061	.0055	.0050	.0045	.0041	.0037	.0033	.0030	.0027	.0025
1	.0311	.0287	.0265	.0244	.0225	.0207	.0191	.0176	.0162	.0149
2	.0793	.0746	.0701	.0659	.0618	.0580	.0544	.0509	.0477	.0446
3	.1348	.1293	.1239	.1185	.1133	.1082	.1033	.0985	.0938	.0892
4	.1719	.1681	.1641	.1600	.1558	.1515	.1472	.1428	.1383	.1339
5	.1753	.1748	.1740	.1728	.1714	.1697	.1678	.1656	.1632	.1606
6	.1490	.1515	.1537	.1555	.1571	.1584	.1594	.1601	.1605	.1606
7	.1086	.1125	.1163	.1200	.1234	.1267	.1298	.1326	.1353	.1377
8	.0692	.0731	.0771	.0810	.0849	.0887	.0925	.0962	.0998	.1033
9	.0392	.0423	.0454	.0486	.0519	.0552	.0586	.0620	.0654	.0688
10	.0200	.0220	.0241	.0262	.0285	.0309	.0334	.0359	.0386	.0413
11	.0093	.0104	.0116	.0129	.0143	.0157	.0173	.0190	.0207	.0225
12	.0039	.0045	.0051	.0058	.0065	.0073	.0082	.0092	.0102	.0113
13	.0015	.0018	.0021	.0024	.0028	.0032	.0036	.0041	.0046	.0052
14	.0006	.0007	.0008	.0009	.0011	.0013	.0015	.0017	.0019	.0022
15	.0002	.0002	.0003	.0003	.0004	.0005	.0006	.0007	.0008	.0009
16	.0001	.0001	.0001	.0001	.0001	.0002	.0002	.0002	.0003	.0003
17	.0000	.0000	.0000	.0000	.0000	.0001	.0001	.0001	.0001	.0001

					μ					
x	6.10	6.20	6.30	6.40	6.50	6.60	6.70	6.80	6.90	7.00
0	.0022	.0020	.0018	.0017	.0015	.0014	.0012	.0011	.0010	.0009
1	.0137	.0126	.0116	.0106	.0098	.0090	.0082	.0076	.0070	.0064
2	.0417	.0390	.0364	.0340	.0318	.0296	.0276	.0258	.0240	.0223
3	.0848	.0806	.0765	.0726	.0688	.0652	.0617	.0584	.0552	.0521
4	.1294	.1249	.1205	.1161	.1118	.1076	.1034	.0992	.0952	.0912
5	.1579	.1549	.1519	.1487	.1454	.1420	.1385	.1349	.1314	.1277
6	.1605	.1601	.1595	.1586	.1575	.1562	.1546	.1529	.1511	.1490
7	.1399	.1418	.1435	.1450	.1462	.1472	.1480	.1486	.1489	.1490
8	.1066	.1099	.1130	.1160	.1188	.1215	.1240	.1263	.1284	.1304
9	.0723	.0757	.0791	.0825	.0858	.0891	.0923	.0954	.0985	.1014
10	.0441	.0469	.0498	.0528	.0558	.0588	.0618	.0649	.0679	.0710
11	.0244	.0265	.0285	.0307	.0330	.0353	.0377	.0401	.0426	.0452
12	.0124	.0137	.0150	.0164	.0179	.0194	.0210	.0227	.0245	.0263
13	.0058	.0065	.0073	.0081	.0089	.0099	.0108	.0119	.0130	.0142
14	.0025	.0029	.0033	.0037	.0041	.0046	.0052	.0058	.0064	.0071

					μ					
x	6.10	6.20	6.30	6.40	6.50	6.60	6.70	6.80	6.90	7.00
15	.0010	.0012	.0014	.0016	.0018	.0020	.0023	.0026	.0029	.0033
16	.0004	.0005	.0005	.0006	.0007	.0008	.0010	.0011	.0013	.0014
17	.0001	.0002	.0002	.0002	.0003	.0003	.0004	.0004	.0005	.0006
18	.0000	.0001	.0001	.0001	.0001	.0001	.0001	.0002	.0002	.0002
19	.0000	.0000	.0000	.0000	.0000	.0000	.0001	.0001	.0001	.0001

Table IV (continued)

					μ					
x	7.10	7.20	7.30	7.40	7.50	7.60	7.70	7.80	7.90	8.00
0	.0008	.0007	.0007	.0006	.0006	.0005	.0005	.0004	.0004	.0003
1	.0059	.0054	.0049	.0045	.0041	.0038	.0035	.0032	.0029	.0027
2	.0208	.0194	.0180	.0167	.0156	.0145	.0134	.0125	.0116	.0107
3	.0492	.0464	.0438	.0413	.0389	.0366	.0345	.0324	.0305	.0286
4	.0874	.0836	.0799	.0764	.0729	.0696	.0663	.0632	.0602	.0573
5	.1241	.1204	.1167	.1130	.1094	.1057	.1021	.0986	.0951	.0916
6	.1468	.1445	.1420	.1394	.1367	.1339	.1311	.1282	.1252	.1221
7	.1489	.1486	.1481	.1474	.1465	.1454	.1442	.1428	.1413	.1396
8	.1321	.1337	.1351	.1363	.1373	.1381	.1388	.1392	.1395	.1396
9	.1042	.1070	.1096	.1121	.1144	.1167	.1187	.1207	.1224	.1241
10	.0740	.0770	.0800	.0829	.0858	.0887	.0914	.0941	.0967	.0993
11	.0478	.0504	.0531	.0558	.0585	.0613	.0640	.0667	.0695	.0722
12	.0283	.0303	.0323	.0344	.0366	.0388	.0411	.0434	.0457	.0481
13	.0154	.0168	.0181	.0196	.0211	.0227	.0243	.0260	.0278	.0296
14	.0078	.0086	.0095	.0104	.0113	.0123	.0134	.0145	.0157	.0169
15	.0037	.0041	.0046	.0051	.0057	.0062	.0069	.0075	.0083	.0090
16	.0016	.0019	.0021	.0024	.0026	.0030	.0033	.0037	.0041	.0045
17	.0007	.0008	.0009	.0010	.0012	.0013	.0015	.0017	.0019	.0021
18	.0003	.0003	.0004	.0004	.0005	.0006	.0006	.0007	.0008	.0009
19	.0001	.0001	.0001	.0002	.0002	.0002	.0003	.0003	.0003	.0004
20	.0000	.0000	.0000	.0001	.0001	.0001	.0001	.0001	.0001	.0002
21	.0000	.0000	.0000	.0000	.0000	.0000	.0000	.0000	.0001	.0001

					μ					
x	8.10	8.20	8.30	8.40	8.50	8.60	8.70	8.80	8.90	9.00
0	.0003	.0003	.0002	.0002	.0002	.0002	.0002	.0002	.0001	.0001
1	.0025	.0023	.0021	.0019	.0017	.0016	.0014	.0013	.0012	.0011
2	.0100	.0092	.0086	.0079	.0074	.0068	.0063	.0058	.0054	.0050
3	.0269	.0252	.0237	.0222	.0208	.0195	.0183	.0171	.0160	.0150
4	.0544	.0517	.0491	.0466	.0443	.0420	.0398	.0377	.0357	.0337
5	.0882	.0849	.0816	.0784	.0752	.0722	.0692	.0663	.0635	.0607
6	.1191	.1160	.1128	.1097	.1066	.1034	.1003	.0972	.0941	.0911
7	.1378	.1358	.1338	.1317	.1294	.1271	.1247	.1222	.1197	.1171
8	.1395	.1392	.1388	.1382	.1375	.1366	.1356	.1344	.1332	.1318
9	.1256	.1269	.1280	.1290	.1299	.1306	.1311	.1315	.1317	.1318
10	.1017	.1040	.1063	.1084	.1104	.1123	.1140	.1157	.1172	.1186
11	.0749	.0776	.0802	.0828	.0853	.0878	.0902	.0925	.0948	.0970
12	.0505	.0530	.0555	.0579	.0604	.0629	.0654	.0679	.0703	.0728
13	.0315	.0334	.0354	.0374	.0395	.0416	.0438	.0459	.0481	.0504
14	.0182	.0196	.0210	.0225	.0240	.0256	.0272	.0289	.0306	.0324
15	.0098	.0107	.0116	.0126	.0136	.0147	.0158	.0169	.0182	.0194
16	.0050	.0055	.0060	.0066	.0072	.0079	.0086	.0093	.0101	.0109
17	.0024	.0026	.0029	.0033	.0036	.0040	.0044	.0048	.0053	.0058
18	.0011	.0012	.0014	.0015	.0017	.0019	.0021	.0024	.0026	.0029
19	.0005	.0005	.0006	.0007	.0008	.0009	.0010	.0011	.0012	.0014
20	.0002	.0002	.0002	.0003	.0003	.0004	.0004	.0005	.0005	.0006
21	.0001	.0001	.0001	.0001	.0001	.0002	.0002	.0002	.0002	.0003
22	.0000	.0000	.0000	.0000	.0001	.0001	.0001	.0001	.0001	.0001

Table IV *(continued)*

					μ					
x	9.10	9.20	9.30	9.40	9.50	9.60	9.70	9.80	9.90	10.00
0	.0001	.0001	.0001	.0001	.0001	.0001	.0001	.0001	.0001	.0000
1	.0010	.0009	.0009	.0008	.0007	.0007	.0006	.0005	.0005	.0005
2	.0046	.0043	.0040	.0037	.0034	.0031	.0029	.0027	.0025	.0023
3	.0140	.0131	.0123	.0115	.0107	.0100	.0093	.0087	.0081	.0076
4	.0319	.0302	.0285	.0269	.0254	.0240	.0226	.0213	.0201	.0189
5	.0581	.0555	.0530	.0506	.0483	.0460	.0439	.0418	.0398	.0378
6	.0881	.0851	.0822	.0793	.0764	.0736	.0709	.0682	.0656	.0631
7	.1145	.1118	.1091	.1064	.1037	.1010	.0982	.0955	.0928	.0901
8	.1302	.1286	.1269	.1251	.1232	.1212	.1191	.1170	.1148	.1126
9	.1317	.1315	.1311	.1306	.1300	.1293	.1284	.1274	.1263	.1251

					μ					
x	9.10	9.20	9.30	9.40	9.50	9.60	9.70	9.80	9.90	10.00
10	.1198	.1210	.1219	.1228	.1235	.1241	.1245	.1249	.1250	.1251
11	.0991	.1012	.1031	.1049	.1067	.1083	.1098	.1112	.1125	.1137
12	.0752	.0776	.0799	.0822	.0844	.0866	.0888	.0908	.0928	.0948
13	.0526	.0549	.0572	.0594	.0617	.0640	.0662	.0685	.0707	.0729
14	.0342	.0361	.0380	.0399	.0419	.0439	.0459	.0479	.0500	.0521
15	.0208	.0221	.0235	.0250	.0265	.0281	.0297	.0313	.0330	.0347
16	.0118	.0127	.0137	.0147	.0157	.0168	.0180	.0192	.0204	.0217
17	.0063	.0069	.0075	.0081	.0088	.0095	.0103	.0111	.0119	.0128
18	.0032	.0035	.0039	.0042	.0046	.0051	.0055	.0060	.0065	.0071
19	.0015	.0017	.0019	.0021	.0023	.0026	.0028	.0031	.0034	.0037
20	.0007	.0008	.0009	.0010	.0011	.0012	.0014	.0015	.0017	.0019
21	.0003	.0003	.0004	.0004	.0005	.0006	.0006	.0007	.0008	.0009
22	.0001	.0001	.0002	.0002	.0002	.0002	.0003	.0003	.0004	.0004
23	.0000	.0001	.0001	.0001	.0001	.0001	.0001	.0001	.0002	.0002
24	.0000	.0000	.0000	.0000	.0000	.0000	.0000	.0001	.0001	.0001

Table IV *(concluded)*

					μ					
x	11.	12.	13.	14.	15.	16.	17.	18.	19.	20.
0	.0000	.0000	.0000	.0000	.0000	.0000	.0000	.0000	.0000	.0000
1	.0002	.0001	.0000	.0000	.0000	.0000	.0000	.0000	.0000	.0000
2	.0010	.0004	.0002	.0001	.0000	.0000	.0000	.0000	.0000	.0000
3	.0037	.0018	.0008	.0004	.0002	.0001	.0000	.0000	.0000	.0000
4	.0102	.0053	.0027	.0013	.0006	.0003	.0001	.0001	.0000	.0000
5	.0224	.0127	.0070	.0037	.0019	.0010	.0005	.0002	.0001	.0001
6	.0411	.0255	.0152	.0087	.0048	.0026	.0014	.0007	.0004	.0002
7	.0646	.0437	.0281	.0174	.0104	.0060	.0034	.0019	.0010	.0005
8	.0888	.0655	.0457	.0304	.0194	.0120	.0072	.0042	.0024	.0013
9	.1085	.0874	.0661	.0473	.0324	.0213	.0135	.0083	.0050	.0029
10	.1194	.1048	.0859	.0663	.0486	.0341	.0230	.0150	.0095	.0058
11	.1194	.1144	.1015	.0844	.0663	.0496	.0355	.0245	.0164	.0106
12	.1094	.1144	.1099	.0984	.0829	.0661	.0504	.0368	.0259	.0176
13	.0926	.1056	.1099	.1060	.0956	.0814	.0658	.0509	.0378	.0271
14	.0728	.0905	.1021	.1060	.1024	.0930	.0800	.0655	.0514	.0387
15	.0534	.0724	.0885	.0989	.1024	.0992	.0906	.0786	.0650	.0516
16	.0367	.0543	.0719	.0866	.0960	.0992	.0963	.0884	.0772	.0646
17	.0237	.0383	.0550	.0713	.0847	.0934	.0963	.0936	.0863	.0760
18	.0145	.0256	.0397	.0554	.0706	.0830	.0909	.0936	.0911	.0844
19	.0084	.0161	.0272	.0409	.0557	.0699	.0814	.0887	.0911	.0888
20	.0046	.0097	.0177	.0286	.0418	.0559	.0692	.0798	.0866	.0888
21	.0024	.0055	.0109	.0191	.0299	.0426	.0560	.0684	.0783	.0846
22	.0012	.0030	.0065	.0121	.0204	.0310	.0433	.0560	.0676	.0769
23	.0006	.0016	.0037	.0074	.0133	.0216	.0320	.0438	.0559	.0669
24	.0003	.0008	.0020	.0043	.0083	.0144	.0226	.0329	.0442	.0557
25	.0001	.0004	.0010	.0024	.0050	.0092	.0154	.0237	.0336	.0446
26	.0000	.0002	.0005	.0013	.0029	.0057	.0101	.0164	.0246	.0343
27	.0000	.0001	.0002	.0007	.0016	.0034	.0063	.0109	.0173	.0254
28	.0000	.0000	.0001	.0003	.0009	.0019	.0038	.0070	.0117	.0181
29	.0000	.0000	.0001	.0002	.0004	.0011	.0023	.0044	.0077	.0125
30	.0000	.0000	.0000	.0001	.0002	.0006	.0013	.0026	.0049	.0083
31	.0000	.0000	.0000	.0000	.0001	.0003	.0007	.0015	.0030	.0054
32	.0000	.0000	.0000	.0000	.0001	.0001	.0004	.0009	.0018	.0034
33	.0000	.0000	.0000	.0000	.0000	.0001	.0002	.0005	.0010	.0020
34	.0000	.0000	.0000	.0000	.0000	.0000	.0001	.0002	.0006	.0012
35	.0000	.0000	.0000	.0000	.0000	.0000	.0000	.0001	.0003	.0007
36	.0000	.0000	.0000	.0000	.0000	.0000	.0000	.0001	.0002	.0004
37	.0000	.0000	.0000	.0000	.0000	.0000	.0000	.0000	.0001	.0002
38	.0000	.0000	.0000	.0000	.0000	.0000	.0000	.0000	.0000	.0001
39	.0000	.0000	.0000	.0000	.0000	.0000	.0000	.0000	.0000	.0001

THE USE OF TABLE V—PROPORTIONS OF AREA UNDER THE NORMAL CURVE

The use of Table V requires that the raw score be transformed into a z-score and that the values be normally distributed.

The values in Table V represent the proportion of area in the standard normal curve, which has a mean of 0, a standard deviation of 1.00, and a total area equal to 1.00.

Since the normal curve is symmetrical, only the areas corresponding to positive z-values are presented. Negative z-values will have exactly the same proportions of area as their positive counterparts.

Column B presents the proportion of area between the mean and a given z.

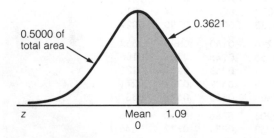
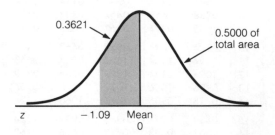

To find the proportion of area between two z-scores:

Area between $z = -1.09$ and $z = 1.09$: $0.3621 + 0.3621 = 0.7242$

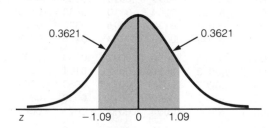

Area between $z = 1.09$ and $z = 1.30$ or $z = -1.09$ and $z = -1.30$: $0.4032 - 0.3621 = 0.0411$

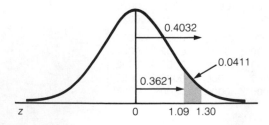
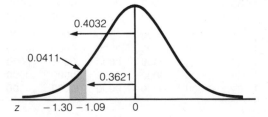

Column C presents the proportion of area *beyond* a given z-score.

Area corresponding to $z \geq 1.09$

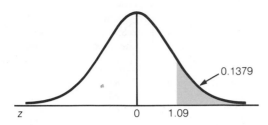

Area corresponding to $z \leq -1.09$

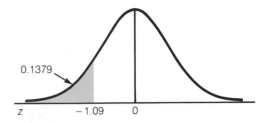

Area corresponding to $z \geq |1.09|$ or $z \geq 1.09$ and $z \leq -1.09 = 2(0.1379) = 0.3758$

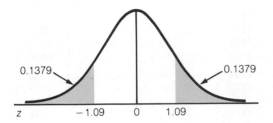

Table V
Proportions of area under the normal curve

(A) z	(B) Area between mean and z	(C) Area beyond z	(A) z	(B) Area between mean and z	(C) Area beyond z	(A) z	(B) Area between mean and z	(C) Area beyond z
0.00	.0000	.5000	0.20	.0793	.4207	0.40	.1554	.3446
0.01	.0040	.4960	0.21	.0832	.4168	0.41	.1591	.3409
0.02	.0080	.4920	0.22	.0871	.4129	0.42	.1628	.3372
0.03	.0120	.4880	0.23	.0910	.4090	0.43	.1664	.3336
0.04	.0160	.4840	0.24	.0948	.4052	0.44	.1700	.3300
0.05	.0199	.4801	0.25	.0987	.4013	0.45	.1736	.3264
0.06	.0239	.4761	0.26	.1026	.3974	0.46	.1772	.3228
0.07	.0279	.4721	0.27	.1064	.3936	0.47	.1808	.3192
0.08	.0319	.4681	0.28	.1103	.3897	0.48	.1844	.3156
0.09	.0359	.4641	0.29	.1141	.3859	0.49	.1879	.3121
0.10	.0398	.4602	0.30	.1179	.3821	0.50	.1915	.3085
0.11	.0438	.4562	0.31	.1217	.3783	0.51	.1950	.3050
0.12	.0478	.4522	0.32	.1255	.3745	0.52	.1985	.3015
0.13	.0517	.4483	0.33	.1293	.3707	0.53	.2019	.2981
0.14	.0557	.4443	0.34	.1331	.3669	0.54	.2054	.2946
0.15	.0596	.4404	0.35	.1368	.3632	0.55	.2088	.2912
0.16	.0636	.4364	0.36	.1406	.3594	0.56	.2123	.2877
0.17	.0675	.4325	0.37	.1443	.3557	0.57	.2157	.2843
0.18	.0714	.4286	0.38	.1480	.3520	0.58	.2190	.2810
0.19	.0753	.4247	0.39	.1517	.3483	0.59	.2224	.2776

Table V (continued)

(A) z	(B) Area between mean and z	(C) Area beyond z	(A) z	(B) Area between mean and z	(C) Area beyond z	(A) z	(B) Area between mean and z	(C) Area beyond z
0.60	.2257	.2743	1.05	.3531	.1469	1.50	.4332	.0668
0.61	.2291	.2709	1.06	.3554	.1446	1.51	.4345	.0655
0.62	.2324	.2676	1.07	.3577	.1423	1.52	.4357	.0643
0.63	.2357	.2643	1.08	.3599	.1401	1.53	.4370	.0630
0.64	.2389	.2611	1.09	.3621	.1379	1.54	.4382	.0618
0.65	.2422	.2578	1.10	.3643	.1357	1.55	.4394	.0606
0.66	.2454	.2546	1.11	.3665	.1335	1.56	.4406	.0594
0.67	.2486	.2514	1.12	.3686	.1314	1.57	.4418	.0582
0.68	.2517	.2483	1.13	.3708	.1292	1.58	.4429	.0571
0.69	.2549	.2451	1.14	.3729	.1271	1.59	.4441	.0559
0.70	.2580	.2420	1.15	.3749	.1251	1.60	.4452	.0548
0.71	.2611	.2389	1.16	.3770	.1230	1.61	.4463	.0537
0.72	.2642	.2358	1.17	.3790	.1210	1.62	.4474	.0526
0.73	.2673	.2327	1.18	.3810	.1190	1.63	.4484	.0516
0.74	.2704	.2296	1.19	.3830	.1170	1.64	.4495	.0505
0.75	.2734	.2266	1.20	.3849	.1151	1.65	.4505	.0495
0.76	.2764	.2236	1.21	.3869	.1131	1.66	.4515	.0485
0.77	.2794	.2206	1.22	.3888	.1112	1.67	.4525	.0475
0.78	.2823	.2177	1.23	.3907	.1093	1.68	.4535	.0465
0.79	.2852	.2148	1.24	.3925	.1075	1.69	.4545	.0455
0.80	.2881	.2119	1.25	.3944	.1056	1.70	.4554	.0446
0.81	.2910	.2090	1.26	.3962	.1038	1.71	.4564	.0436
0.82	.2939	.2061	1.27	.3980	.1020	1.72	.4573	.0427
0.83	.2967	.2033	1.28	.3997	.1003	1.73	.4582	.0418
0.84	.2995	.2005	1.29	.4015	.0985	1.74	.4591	.0409
0.85	.3023	.1977	1.30	.4032	.0968	1.75	.4599	.0401
0.86	.3051	.1949	1.31	.4049	.0951	1.76	.4608	.0392
0.87	.3078	.1922	1.32	.4066	.0934	1.77	.4616	.0384
0.88	.3106	.1894	1.33	.4082	.0918	1.78	.4625	.0375
0.89	.3133	.1867	1.34	.4099	.0901	1.79	.4633	.0367
0.90	.3159	.1841	1.35	.4115	.0885	1.80	.4641	.0359
0.91	.3186	.1814	1.36	.4131	.0869	1.81	.4649	.0351
0.92	.3212	.1788	1.37	.4147	.0853	1.82	.4656	.0344
0.93	.3238	.1762	1.38	.4162	.0838	1.83	.4664	.0336
0.94	.3264	.1736	1.39	.4177	.0823	1.84	.4671	.0329
0.95	.3289	.1711	1.40	.4192	.0808	1.85	.4678	.0322
0.96	.3315	.1685	1.41	.4207	.0793	1.86	.4686	.0314
0.97	.3340	.1660	1.42	.4222	.0778	1.87	.4693	.0307
0.98	.3365	.1635	1.43	.4236	.0764	1.88	.4699	.0301
0.99	.3389	.1611	1.44	.4251	.0749	1.89	.4706	.0294
1.00	.3413	.1587	1.45	.4265	.0735	1.90	.4713	.0287
1.01	.3438	.1562	1.46	.4279	.0721	1.91	.4719	.0281
1.02	.3461	.1539	1.47	.4292	.0708	1.92	.4726	.0274
1.03	.3485	.1515	1.48	.4306	.0694	1.93	.4732	.0268
1.04	.3508	.1492	1.49	.4319	.0681	1.94	.4738	.0262

Table V *(concluded)*

(A) z	(B) Area between mean and z	(C) Area beyond z	(A) z	(B) Area between mean and z	(C) Area beyond z	(A) z	(B) Area between mean and z	(C) Area beyond z
1.95	.4744	.0256	2.42	.4922	.0078	2.89	.4981	.0019
1.96	.4750	.0250	2.43	.4925	.0075	2.90	.4981	.0019
1.97	.4756	.0244	2.44	.4927	.0073	2.91	.4982	.0018
1.98	.4761	.0239	2.45	.4929	.0071	2.92	.4982	.0018
1.99	.4767	.0233	2.46	.4931	.0069	2.93	.4983	.0017
2.00	.4772	.0228	2.47	.4932	.0068	2.94	.4984	.0016
2.01	.4778	.0222	2.48	.4934	.0066	2.95	.4984	.0016
2.02	.4783	.0217	2.49	.4936	.0064	2.96	.4985	.0015
2.03	.4788	.0212	2.50	.4938	.0062	2.97	.4985	.0015
2.04	.4793	.0207	2.51	.4940	.0060	2.98	.4986	.0014
2.05	.4798	.0202	2.52	.4941	.0059	2.99	.4986	.0014
2.06	.4803	.0197	2.53	.4943	.0057	3.00	.4987	.0013
2.07	.4808	.0192	2.54	.4945	.0055	3.01	.4987	.0013
2.08	.4812	.0188	2.55	.4946	.0054	3.02	.4987	.0013
2.09	.4817	.0183	2.56	.4948	.0052	3.03	.4988	.0012
2.10	.4821	.0179	2.57	.4949	.0051	3.04	.4988	.0012
2.11	.4826	.0174	2.58	.4951	.0049	3.05	.4989	.0011
2.12	.4830	.0170	2.59	.4952	.0048	3.06	.4989	.0011
2.13	.4834	.0166	2.60	.4953	.0047	3.07	.4989	.0011
2.14	.4838	.0162	2.61	.4955	.0045	3.08	.4990	.0010
2.15	.4842	.0158	2.62	.4956	.0044	3.09	.4990	.0010
2.16	.4846	.0154	2.63	.4957	.0043	3.10	.4990	.0010
2.17	.4850	.0150	2.64	.4959	.0041	3.11	.4991	.0009
2.18	.4854	.0146	2.65	.4960	.0040	3.12	.4991	.0009
2.19	.4857	.0143	2.66	.4961	.0039	3.13	.4991	.0009
2.20	.4861	.0139	2.67	.4962	.0038	3.14	.4992	.0008
2.21	.4864	.0136	2.68	.4963	.0037	3.15	.4992	.0008
2.22	.4868	.0132	2.69	.4964	.0036	3.16	.4992	.0008
2.23	.4871	.0129	2.70	.4965	.0035	3.17	.4992	.0008
2.24	.4875	.0125	2.71	.4966	.0034	3.18	.4993	.0007
2.25	.4878	.0122	2.72	.4967	.0033	3.19	.4993	.0007
2.26	.4881	.0119	2.73	.4968	.0032	3.20	.4993	.0007
2.27	.4884	.0116	2.74	.4969	.0031	3.21	.4993	.0007
2.28	.4887	.0113	2.75	.4970	.0030	3.22	.4994	.0006
2.29	.4890	.0110	2.76	.4971	.0029	3.23	.4994	.0006
2.30	.4893	.0107	2.77	.4972	.0028	3.24	.4994	.0006
2.31	.4896	.0104	2.78	.4973	.0027	3.25	.4994	.0006
2.32	.4898	.0102	2.79	.4974	.0026	3.30	.4995	.0005
2.33	.4901	.0099	2.80	.4974	.0026	3.35	.4996	.0004
2.34	.4904	.0096	2.81	.4975	.0025	3.40	.4997	.0003
2.35	.4906	.0094	2.82	.4976	.0024	3.45	.4997	.0003
2.36	.4909	.0091	2.83	.4977	.0023	3.50	.4998	.0002
2.37	.4911	.0089	2.84	.4977	.0023	3.60	.4998	.0002
2.38	.4913	.0087	2.85	.4978	.0022	3.70	.4999	.0001
2.39	.4916	.0084	2.86	.4979	.0021	3.80	.4999	.0001
2.40	.4918	.0082	2.87	.4979	.0021	3.90	.49995	.00005
2.41	.4920	.0080	2.88	.4980	.0020	4.00	.49997	.00003

Table VI
Random digits

Row number										
00000	10097	32533	76520	13586	34673	54876	80959	09117	39292	74945
00001	37542	04805	64894	74296	24805	24037	20636	10402	00822	91665
00002	08422	68953	19645	09303	23209	02560	15953	34764	35080	33606
00003	99019	02529	09376	70715	38311	31165	88676	74397	04436	27659
00004	12807	99970	80157	36147	64032	36653	98951	16877	12171	76833
00005	66065	74717	34072	76850	36697	36170	65813	39885	11199	29170
00006	31060	10805	45571	82406	35303	42614	86799	07439	23403	09732
00007	85269	77602	02051	65692	68665	74818	73053	85247	18623	88579
00008	63573	32135	05325	47048	90553	57548	28468	28709	83491	25624
00009	73796	45753	03529	64778	35808	34282	60935	20344	35273	88435
00010	98520	17767	14905	68607	22109	40558	60970	93433	50500	73998
00011	11805	05431	39808	27732	50725	68248	29405	24201	52775	67851
00012	83452	99634	06288	98033	13746	70078	18475	40610	68711	77817
00013	88685	40200	86507	58401	36766	67951	90364	76493	29609	11062
00014	99594	67348	87517	64969	91826	08928	93785	61368	23478	34113
00015	65481	17674	17468	50950	58047	76974	73039	57186	40218	16544
00016	80124	35635	17727	08015	45318	22374	21115	78253	14385	53763
00017	74350	99817	77402	77214	43236	00210	45521	64237	96286	02655
00018	69916	26803	66252	29148	36936	87203	76621	13990	94400	56418
00019	09893	20505	14225	68514	46427	56788	96297	78822	54382	14598
00020	91499	14523	68479	27686	46162	83554	94750	89923	37089	20048
00021	80336	94598	26940	36858	70297	34135	53140	33340	42050	82341
00022	44104	81949	85157	47954	32979	26575	57600	40881	22222	06413
00023	12550	73742	11100	02040	12860	74697	96644	89439	28707	25815
00024	63606	49329	16505	34484	40219	52563	43651	77082	07207	31790
00025	61196	90446	26457	47774	51924	33729	65394	59593	42582	60527
00026	15474	45266	95270	79953	59367	83848	82396	10118	33211	59466
00027	94557	28573	67897	54387	54622	44431	91190	42592	92927	45973
00028	42481	16213	97344	08721	16868	48767	03071	12059	25701	46670
00029	23523	78317	73208	89837	68935	91416	26252	29663	05522	82562
00030	04493	52494	75246	33824	45862	51025	61962	79335	65337	12472
00031	00549	97654	64051	88159	96119	63896	54692	82391	23287	29529
00032	35963	15307	26898	09354	33351	35462	77974	50024	90103	39333
00033	59808	08391	45427	26842	83609	49700	13021	24892	78565	20106
00034	46058	85236	01390	92286	77281	44077	93910	83647	70617	42941
00035	32179	00597	87379	25241	05567	07007	86743	17157	85394	11838
00036	69234	61406	20117	45204	15956	60000	18743	92423	97118	96338
00037	19565	41430	01758	75379	40419	21585	66674	36806	84962	85207
00038	45155	14938	19476	07246	43667	94543	59047	90033	20826	69541
00039	94864	31994	36168	10851	34888	81553	01540	35456	05014	51176
00040	98086	24826	45240	28404	44999	08896	39094	73407	35441	31880
00041	33185	16232	41941	50949	89435	48581	88695	41994	37548	73043
00042	80951	00406	96382	70774	20151	23387	25016	25298	94624	61171
00043	79752	49140	71961	28296	69861	02591	74852	20539	00387	59579
00044	18633	32537	98145	06571	31010	24674	05455	61427	77938	91936

Table VI *(continued)*

Row number										
00045	74029	43902	77557	32270	97790	17119	52527	58021	80814	51748
00046	54178	45611	80993	37143	05335	12969	56127	19255	36040	90324
00047	11664	49883	52079	84827	59381	71539	09973	33440	88461	23356
00048	48324	77928	31249	64710	02295	36870	32307	57546	15020	09994
00049	69074	94138	87637	91976	35584	04401	10518	21615	01848	76938
00050	09188	20097	32825	39527	04220	86304	83389	87374	64278	58044
00051	90045	85497	51981	50654	94938	81997	91870	76150	68476	64659
00052	73189	50207	47677	26269	62290	64464	27124	67018	41361	82760
00053	75768	76490	20971	87749	90429	12272	95375	05871	93823	43178
00054	54016	44056	66281	31003	00682	27398	20714	53295	07706	17813
00055	08358	69910	78542	42785	13661	58873	04618	97553	31223	08420
00056	28306	03264	81333	10591	40510	07893	32604	60475	94119	01840
00057	53840	86233	81594	13628	51215	90290	28466	68795	77762	20791
00058	91757	53741	61613	62669	50263	90212	55781	76514	83483	47055
00059	89415	92694	00397	58391	12607	17646	48949	72306	94541	37408
00060	77513	03820	86864	29901	68414	82774	51908	13980	72893	55507
00061	19502	37174	69979	20288	55210	29773	74287	75251	65344	67415
00062	21818	59313	93278	81757	05686	73156	07082	85046	31853	38452
00063	51474	66499	68107	23621	94049	91345	42836	09191	08007	45449
00064	99559	68331	62535	24170	69777	12830	74819	78142	43860	72834
00065	33713	48007	93584	72869	51926	64721	58303	29822	93174	93972
00066	85274	86893	11303	22970	28834	34137	73515	90400	71148	43643
00067	84133	89640	44035	52166	73852	70091	61222	60561	62327	18423
00068	56732	16234	17395	96131	10123	91622	85496	57560	81604	18880
00069	65138	56806	87648	85261	34313	65861	45875	21069	85644	47277
00070	38001	02176	81719	11711	71602	92937	74219	64049	65584	49698
00071	37402	96397	01304	77586	56271	10086	47324	62605	40030	37438
00072	97125	40348	87083	31417	21815	39250	75237	62047	15501	29578
00073	21826	41134	47143	34072	64638	85902	49139	06441	03856	54552
00074	73135	42742	95719	09035	85794	74296	08789	88156	64691	19202
00075	07638	77929	03061	18072	96207	44156	23821	99538	04713	66994
00076	60528	83441	07954	19814	59175	20695	05533	52139	61212	06455
00077	83596	35655	06958	92983	05128	09719	77433	53783	92301	50498
00078	10850	62746	99599	10507	13499	06319	53075	71839	06410	19362
00079	39820	98952	43622	63147	64421	80814	43800	09351	31024	73167
00080	59580	06478	75569	78800	88835	54486	23768	06156	04111	08408
00081	38508	07341	23793	48763	90822	97022	17719	04207	95954	49953
00082	30692	70668	94688	16127	56196	80091	82067	63400	05462	69200
00083	65443	95659	18238	27437	49632	24041	08337	65676	96299	90836
00084	27267	50264	13192	72294	07477	44606	17985	48911	97341	30358
00085	91307	06991	19072	24210	36699	53728	28825	35793	28976	66252
00086	68434	94688	84473	13622	62126	98408	12843	82590	09815	93146
00087	48908	15877	54745	24591	35700	04754	83824	52692	54130	55160
00088	06913	45197	42672	78601	11883	09528	63011	98901	14974	40344
00089	10455	16019	14210	33712	91342	37821	88325	80851	43667	70883

Table VI *(continued)*

Row number										
00090	12883	97343	65027	61184	04285	01392	17974	15077	90712	26769
00091	21778	30976	38807	36961	31649	42096	63281	02023	08816	47449
00092	19523	59515	65122	59659	86283	68258	69572	13798	16435	91529
00093	67245	52670	35583	16563	79246	86686	76463	34222	26655	90802
00094	60584	47377	07500	37992	45134	26529	26760	83637	41326	44344
00095	53853	41377	36066	94850	58838	73859	49364	73331	96240	43642
00096	24637	38736	74384	89342	52623	07992	12369	18601	03742	83873
00097	83080	12451	38992	22815	07759	51777	97377	27585	51972	37867
00098	16444	24334	36151	99073	27493	70939	85130	32552	54846	54759
00099	60790	18157	57178	65762	11161	78576	45819	52979	65130	04860
00100	03991	10461	93716	16894	66083	24653	84609	58232	88618	19161
00101	38555	95554	32886	59780	08355	60860	29735	47762	71299	23853
00102	17546	73704	92052	46215	55121	29281	59076	07936	27954	58909
00103	32643	52861	95819	06831	00911	98936	76355	93779	80863	00514
00104	69572	68777	39510	35905	14060	40619	29549	69616	33564	60780
00105	24122	66591	27699	06494	14845	46672	61958	77100	90899	75754
00106	61196	30231	92962	61773	41839	55382	17267	70943	78038	70267
00107	30532	21704	10274	12202	39685	23309	10061	68829	55986	66485
00108	03788	97599	75867	20717	74416	53166	35208	33374	87539	08823
00109	48228	63379	85783	47619	53152	67433	35663	52972	16818	60311
00110	60365	94653	35075	33949	42614	29297	01918	28316	98953	73231
00111	83799	42402	56623	34442	34994	41374	70071	14736	09958	18065
00112	32960	07405	36409	83232	99385	41600	11133	07586	15917	06253
00113	19322	53845	57620	52606	66497	68646	78138	66559	19640	99413
00114	11220	94747	07399	37408	48509	23929	27482	45476	85244	35159
00115	31751	57260	68980	05339	15470	48355	88651	22596	03152	19121
00116	88492	99382	14454	04504	20094	98977	74843	93413	22109	78508
00117	30934	47744	07481	83828	73788	06533	28597	20405	94205	20380
00118	22888	48893	27499	98748	60530	45128	74022	84617	82037	10268
00119	78212	16993	35902	91386	44372	15486	65741	14014	87481	37220
00120	41849	84547	46850	52326	34677	58300	74910	64345	19325	81549
00121	46352	33049	69248	93460	45305	07521	61318	31855	14413	70951
00122	11087	96294	14013	31792	59747	67277	76503	34513	39663	77544
00123	52701	08337	56303	87315	16520	69676	11654	99893	02181	68161
00124	57275	36898	81304	48585	68652	27376	92852	55866	88448	03584
00125	20857	73156	70284	24326	79375	95220	01159	63267	10622	48391
00126	15633	84924	90415	93614	33521	26665	55823	47641	86225	31704
00127	92694	48297	39904	02115	59589	49067	66821	41575	49767	04037
00128	77613	19019	88152	00080	20554	91409	96277	48257	50816	97616
00129	38688	32486	45134	63545	59404	72059	43947	51680	43852	59693
00130	25163	01889	70014	15021	41290	67312	71857	15957	68971	11403
00131	65251	07629	37239	33295	05870	01119	92784	26340	18477	65622
00132	36815	43625	18637	37509	82444	99005	04921	73701	14707	93997
00133	64397	11692	05327	82162	20247	81759	45197	25332	83745	22567
00134	04515	25624	95096	67946	48460	85558	15191	18782	16930	33361

Table VI (continued)

Row number										
00135	83761	60873	43253	84145	60833	25983	01291	41349	20368	07126
00136	14387	06345	80854	09279	43529	06318	38384	74761	41196	37480
00137	51321	92246	80088	77074	88722	56736	66164	49431	66919	31678
00138	72472	00008	80890	18002	94813	31900	54155	83436	35352	54131
00139	05466	55306	93128	18464	74457	90561	72848	11834	79982	68416
00140	39528	72484	82474	25593	48545	35247	18619	13674	18611	19241
00141	81616	18711	53342	44276	75122	11724	74627	73707	58319	15997
00142	07586	16120	82641	22820	92904	13141	32392	19763	61199	67940
00143	90767	04235	13574	17200	69902	63742	78464	22501	18627	90872
00144	40188	28193	29593	88627	94972	11598	62095	36787	00441	58997
00145	34414	82157	86887	55087	19152	00023	12302	80783	32624	68691
00146	63439	75363	44989	16822	36024	00867	76378	41605	65961	73488
00147	67049	09070	93399	45547	94458	74284	05041	49807	20288	34060
00148	79495	04146	52162	90286	54158	34243	46978	35482	59362	95938
00149	91704	30552	04737	21031	75051	93029	47665	64382	99782	93478
00150	94015	46874	32444	48277	59820	96163	64654	25843	41145	42820
00151	74108	88222	88570	74015	25704	91035	01755	14750	48968	38603
00152	62880	87873	95160	59221	22304	90314	72877	17334	39283	04149
00153	11748	12102	80580	41867	17710	59621	06554	07850	73950	79552
00154	17944	05600	60478	03343	25852	58905	57216	39618	49856	99326
00155	66067	42792	95043	52680	46780	56487	09971	59481	37006	22186
00156	54244	91030	45547	70818	59849	96169	61459	21647	87417	17198
00157	30945	57589	31732	57260	47670	07654	46376	25366	94746	49580
00158	69170	37403	86995	90307	94304	71803	26825	05511	12459	91314
00159	08345	88975	35841	85771	08105	59987	87112	21476	14713	71181
00160	27767	43584	85301	88977	29490	69714	73035	41207	74699	09310
00161	13025	14338	54066	15243	47724	66733	47431	43905	31048	56699
00162	80217	36292	98525	24335	24432	24896	43277	58874	11466	16082
00163	10875	62004	90391	61105	57411	06368	53856	30743	08670	84741
00164	54127	57326	26629	19087	24472	88779	30540	27886	61732	75454
00165	60311	42824	37301	42678	45990	43242	17374	52003	70707	70214
00166	49739	71484	92003	98086	76668	73209	59202	11973	02902	33250
00167	78626	51594	16453	94614	39014	97066	83012	09832	25571	77628
00168	66692	13986	99837	00582	81232	44987	09504	96412	90193	79568
00169	44071	28091	07362	97703	76447	42537	98524	97831	65704	09514
00170	41468	85149	49554	17994	14924	39650	95294	00556	70481	06905
00171	94559	37559	49678	53119	70312	05682	66986	34099	74474	20740
00172	41615	70360	64114	58660	90850	64618	80620	51790	11436	38072
00173	50273	93113	41794	86861	24781	89683	55411	85667	77535	99892
00174	41396	80504	90670	08289	40902	05069	95083	06783	28102	57816
00175	25807	24260	71529	78920	72682	07385	90726	57166	98884	08583
00176	06170	97965	88302	98041	21443	41808	68984	83620	89747	98882
00177	60808	54444	74412	81105	01176	28838	36421	16489	18059	51061
00178	80940	44893	10408	36222	80582	71944	92638	40333	67054	16067
00179	19516	90120	46759	71643	13177	55292	21036	82808	77501	97427

Table VI *(concluded)*

Row number										
00180	49386	54480	23604	23554	21785	41104	91178	10174	29420	90438
00181	06312	88940	15995	69321	47458	64809	98189	81851	29651	84215
00182	60942	00307	11897	92674	40405	68032	96717	54244	10701	41393
00183	92329	98932	78284	46347	71209	92061	39448	93136	25722	08564
00184	77936	63574	31384	51924	85561	29671	58137	17820	22751	36518
00185	38101	77756	11657	13897	95889	57067	47648	13885	70669	93406
00186	39641	69457	91339	22502	92613	89719	11947	56203	19324	20504
00187	84054	40455	99396	63680	67667	60631	69181	96845	38525	11600
00188	47468	03577	57649	63266	24700	71594	14004	23153	69249	05747
00189	43321	31370	28977	23896	76479	68562	62342	07589	08899	05985
00190	64281	61826	18555	64937	13173	33365	78851	16499	87064	13075
00191	66847	70495	32350	02985	86716	38746	26313	77463	55387	72681
00192	72461	33230	21529	53424	92581	02262	78438	66276	18396	73538
00193	21032	91050	13058	16218	12470	56500	15292	76139	59526	52113
00194	95362	67011	06651	16136	01016	00857	55018	56374	35824	71708
00195	49712	97380	10404	55452	34030	60726	75211	10271	36633	68424
00196	58275	61764	97586	54716	50259	46345	87195	46092	26787	60939
00197	89514	11788	68224	23417	73959	76145	30342	40277	11049	72049
00198	15472	50669	48139	36732	46874	37088	63465	09819	58869	35220
00199	12120	86124	51247	44302	60883	52109	21437	36786	49226	77837

Source: Adapted from RAND Corporation, *A Million Random Variables* (Glencoe, Ill.: Free Press of Glencoe, 1955).

THE USE OF TABLE VII—CRITICAL VALUES OF *t*

For any given df, the table shows the values of *t* corresponding to various levels of probability. Obtained *t* is significant at a given level if it is equal to or *greater than* the value shown in the table.

The confidence coefficients corresponding to the various levels of significance are shown in the bottom row.

Table VII

Critical values of t

	Level of significance for one-tailed test					
	.10	.05	.025	.01	.005	.0005
	Level of significance for two-tailed test					
df	.20	.10	.05	.02	.01	.001
1	3.078	6.314	12.706	31.821	63.657	636.619
2	1.886	2.920	4.303	6.965	9.925	31.598
3	1.638	2.353	3.182	4.541	5.841	12.941
4	1.533	2.132	2.776	3.747	4.604	8.610
5	1.476	2.015	2.571	3.365	4.032	6.859
6	1.440	1.943	2.447	3.143	3.707	5.959
7	1.415	1.895	2.365	2.998	3.499	5.405
8	1.397	1.860	2.306	2.896	3.355	5.041
9	1.383	1.833	2.262	2.821	3.250	4.781
10	1.372	1.812	2.228	2.764	3.169	4.587
11	1.363	1.796	2.201	2.718	3.106	4.437
12	1.356	1.782	2.179	2.681	3.055	4.318
13	1.350	1.771	2.160	2.650	3.012	4.221
14	1.345	1.761	2.145	2.624	2.977	4.140
15	1.341	1.753	2.131	2.602	2.947	4.073
16	1.337	1.746	2.120	2.583	2.921	4.015
17	1.333	1.740	2.110	2.567	2.898	3.965
18	1.330	1.734	2.101	2.552	2.878	3.922
19	1.328	1.729	2.093	2.539	2.861	3.883
20	1.325	1.725	2.086	2.528	2.845	3.850
21	1.323	1.721	2.080	2.518	2.831	3.819
22	1.321	1.717	2.074	2.508	2.819	3.792
23	1.319	1.714	2.069	2.500	2.807	3.767
24	1.318	1.711	2.064	2.492	2.797	3.745
25	1.316	1.708	2.060	2.485	2.787	3.725
26	1.315	1.706	2.056	2.479	2.779	3.707
27	1.314	1.703	2.052	2.473	2.771	3.690
28	1.313	1.701	2.048	2.467	2.763	3.674
29	1.311	1.699	2.045	2.462	2.756	3.659
30	1.310	1.697	2.042	2.457	2.750	3.646
40	1.303	1.684	2.021	2.423	2.704	3.551
60	1.296	1.671	2.000	2.390	2.660	3.460
120	1.289	1.658	1.980	2.358	2.617	3.373
∞	1.282	1.645	1.960	2.326	2.576	3.291
Confidence coefficient	0.80	0.90	0.95	0.98	0.99	0.999

Source: Adapted from R. A. Fisher and F. Yates, *Statistical Tables for Biological, Agricultural, and Medical Research*, 2d ed. (Edinburgh: Oliver and Boyd, Ltd., 1974), Table III.

Table VIII

Critical values of F

The obtained F is significant at a given level if it is equal to or *greater than* the value shown in the table. [0.05 (light row) and 0.01 **(dark row)** points for the distribution of F.] The values shown are the right tail of the distribution obtained by dividing the appropriate variance estimate by the error variance estimate.

Degrees of freedom for numerator

df (denom.)	1	2	3	4	5	6	7	8	9	10	11	12	14	16	20	24	30	40	50	75	100	200	500	∞
1	161 / **4052**	200 / **4999**	216 / **5403**	225 / **5625**	230 / **5764**	234 / **5859**	237 / **5928**	239 / **5981**	241 / **6022**	242 / **6056**	243 / **6082**	244 / **6106**	245 / **6142**	246 / **6169**	248 / **6208**	249 / **6234**	250 / **6258**	251 / **6286**	252 / **6302**	253 / **6323**	253 / **6334**	254 / **6352**	254 / **6361**	254 / **6366**
2	18.51 / **98.49**	19.00 / **99.01**	19.16 / **99.17**	19.25 / **99.25**	19.30 / **99.30**	19.33 / **99.33**	19.36 / **99.34**	19.37 / **99.36**	19.38 / **99.38**	19.39 / **99.40**	19.40 / **99.41**	19.41 / **99.42**	19.42 / **99.43**	19.43 / **99.44**	19.44 / **99.45**	19.45 / **99.46**	19.46 / **99.47**	19.47 / **99.48**	19.47 / **99.48**	19.48 / **99.49**	19.49 / **99.49**	19.49 / **99.49**	19.50 / **99.50**	19.50 / **99.50**
3	10.13 / **34.12**	9.55 / **30.81**	9.28 / **29.46**	9.12 / **28.71**	9.01 / **28.24**	8.94 / **27.91**	8.88 / **27.67**	8.84 / **27.49**	8.81 / **27.34**	8.78 / **27.23**	8.76 / **27.13**	8.74 / **27.05**	8.71 / **26.92**	8.69 / **26.83**	8.66 / **26.69**	8.64 / **26.60**	8.62 / **26.50**	8.60 / **26.41**	8.58 / **26.30**	8.57 / **26.27**	8.56 / **26.23**	8.54 / **26.18**	8.54 / **26.14**	8.53 / **26.12**
4	7.71 / **21.20**	6.94 / **18.00**	6.59 / **16.69**	6.39 / **15.98**	6.26 / **15.52**	6.16 / **15.21**	6.09 / **14.98**	6.04 / **14.80**	6.00 / **14.66**	5.96 / **14.54**	5.93 / **14.45**	5.91 / **14.37**	5.87 / **14.24**	5.84 / **14.15**	5.80 / **14.02**	5.77 / **13.93**	5.74 / **13.83**	5.71 / **13.74**	5.70 / **13.69**	5.68 / **13.61**	5.66 / **13.57**	5.65 / **13.52**	5.64 / **13.48**	5.63 / **13.46**
5	6.61 / **16.26**	5.79 / **13.27**	5.41 / **12.06**	5.19 / **11.39**	5.05 / **10.97**	4.95 / **10.67**	4.88 / **10.45**	4.82 / **10.27**	4.78 / **10.15**	4.74 / **10.05**	4.70 / **9.96**	4.68 / **9.89**	4.64 / **9.77**	4.60 / **9.68**	4.56 / **9.55**	4.53 / **9.47**	4.50 / **9.38**	4.46 / **9.29**	4.44 / **9.24**	4.42 / **9.17**	4.40 / **9.13**	4.38 / **9.07**	4.37 / **9.04**	4.36 / **9.02**
6	5.99 / **13.74**	5.14 / **10.92**	4.76 / **9.78**	4.53 / **9.15**	4.39 / **8.75**	4.28 / **8.47**	4.21 / **8.26**	4.15 / **8.10**	4.10 / **7.98**	4.06 / **7.87**	4.03 / **7.79**	4.00 / **7.72**	3.96 / **7.60**	3.92 / **7.52**	3.87 / **7.39**	3.84 / **7.31**	3.81 / **7.23**	3.77 / **7.14**	3.75 / **7.09**	3.72 / **7.02**	3.71 / **6.99**	3.69 / **6.94**	3.68 / **6.90**	3.67 / **6.88**
7	5.59 / **12.25**	4.74 / **9.55**	4.35 / **8.45**	4.12 / **7.85**	3.97 / **7.46**	3.87 / **7.19**	3.79 / **7.00**	3.73 / **6.84**	3.68 / **6.71**	3.63 / **6.62**	3.60 / **6.54**	3.57 / **6.47**	3.52 / **6.35**	3.49 / **6.27**	3.44 / **6.15**	3.41 / **6.07**	3.38 / **5.98**	3.34 / **5.90**	3.32 / **5.85**	3.29 / **5.78**	3.28 / **5.75**	3.25 / **5.70**	3.24 / **5.67**	3.23 / **5.65**
8	5.32 / **11.26**	4.46 / **8.65**	4.07 / **7.59**	3.84 / **7.01**	3.69 / **6.63**	3.58 / **6.37**	3.50 / **6.19**	3.44 / **6.03**	3.39 / **5.91**	3.34 / **5.82**	3.31 / **5.74**	3.28 / **5.67**	3.23 / **5.56**	3.20 / **5.48**	3.15 / **5.36**	3.12 / **5.28**	3.08 / **5.20**	3.05 / **5.11**	3.03 / **5.06**	3.00 / **5.00**	2.98 / **4.96**	2.96 / **4.91**	2.94 / **4.88**	2.93 / **4.86**
9	5.12 / **10.56**	4.26 / **8.02**	3.86 / **6.99**	3.63 / **6.42**	3.48 / **6.06**	3.37 / **5.80**	3.29 / **5.62**	3.23 / **5.47**	3.18 / **5.35**	3.13 / **5.26**	3.10 / **5.18**	3.07 / **5.11**	3.02 / **5.00**	2.98 / **4.92**	2.93 / **4.80**	2.90 / **4.73**	2.86 / **4.64**	2.82 / **4.56**	2.80 / **4.51**	2.77 / **4.45**	2.76 / **4.41**	2.73 / **4.36**	2.72 / **4.33**	2.71 / **4.31**
10	4.96 / **10.04**	4.10 / **7.56**	3.71 / **6.55**	3.48 / **5.99**	3.33 / **5.64**	3.22 / **5.39**	3.14 / **5.21**	3.07 / **5.06**	3.02 / **4.95**	2.97 / **4.85**	2.94 / **4.78**	2.91 / **4.71**	2.86 / **4.60**	2.82 / **4.52**	2.77 / **4.41**	2.74 / **4.33**	2.70 / **4.25**	2.67 / **4.17**	2.64 / **4.12**	2.61 / **4.05**	2.59 / **4.01**	2.56 / **3.96**	2.55 / **3.93**	2.54 / **3.91**
11	4.84 / **9.65**	3.98 / **7.20**	3.59 / **6.22**	3.36 / **5.67**	3.20 / **5.32**	3.09 / **5.07**	3.01 / **4.88**	2.95 / **4.74**	2.90 / **4.63**	2.86 / **4.54**	2.82 / **4.46**	2.79 / **4.40**	2.74 / **4.29**	2.70 / **4.21**	2.65 / **4.10**	2.61 / **4.02**	2.57 / **3.94**	2.53 / **3.86**	2.50 / **3.80**	2.47 / **3.74**	2.45 / **3.70**	2.42 / **3.66**	2.41 / **3.62**	2.40 / **3.60**
12	4.75 / **9.33**	3.88 / **6.93**	3.49 / **5.95**	3.26 / **5.41**	3.11 / **5.06**	3.00 / **4.82**	2.92 / **4.65**	2.85 / **4.50**	2.80 / **4.39**	2.76 / **4.30**	2.72 / **4.22**	2.69 / **4.16**	2.64 / **4.05**	2.60 / **3.98**	2.54 / **3.86**	2.50 / **3.78**	2.46 / **3.70**	2.42 / **3.61**	2.40 / **3.56**	2.36 / **3.49**	2.35 / **3.46**	2.32 / **3.41**	2.31 / **3.38**	2.30 / **3.36**
13	4.67 / **9.07**	3.80 / **6.70**	3.41 / **5.74**	3.18 / **5.20**	3.02 / **4.86**	2.92 / **4.62**	2.84 / **4.44**	2.77 / **4.30**	2.72 / **4.19**	2.67 / **4.10**	2.63 / **4.02**	2.60 / **3.96**	2.55 / **3.85**	2.51 / **3.78**	2.46 / **3.67**	2.42 / **3.59**	2.38 / **3.51**	2.34 / **3.42**	2.32 / **3.37**	2.28 / **3.30**	2.26 / **3.27**	2.24 / **3.21**	2.22 / **3.18**	2.21 / **3.16**
14	4.60 / **8.86**	3.74 / **6.51**	3.34 / **5.56**	3.11 / **5.03**	2.96 / **4.69**	2.85 / **4.46**	2.77 / **4.28**	2.70 / **4.14**	2.65 / **4.03**	2.60 / **3.94**	2.56 / **3.86**	2.53 / **3.80**	2.48 / **3.70**	2.44 / **3.62**	2.39 / **3.51**	2.35 / **3.43**	2.31 / **3.34**	2.27 / **3.26**	2.24 / **3.21**	2.21 / **3.14**	2.19 / **3.11**	2.16 / **3.06**	2.14 / **3.02**	2.13 / **3.00**
15	4.54 / **8.68**	3.68 / **6.36**	3.29 / **5.42**	3.06 / **4.89**	2.90 / **4.56**	2.79 / **4.32**	2.70 / **4.14**	2.64 / **4.00**	2.59 / **3.89**	2.55 / **3.80**	2.51 / **3.73**	2.48 / **3.67**	2.43 / **3.56**	2.39 / **3.48**	2.33 / **3.36**	2.29 / **3.29**	2.25 / **3.20**	2.21 / **3.12**	2.18 / **3.07**	2.15 / **3.00**	2.12 / **2.97**	2.10 / **2.92**	2.08 / **2.89**	2.07 / **2.87**

Degrees of freedom for denominator

df																								
16	2.01	2.02	2.04	2.07	2.09	2.13	2.16	2.20	2.24	2.28	2.33	2.37	2.42	2.45	2.49	2.54	2.59	2.66	2.74	2.85	3.01	3.24	3.63	4.49
	2.75	**2.77**	**2.80**	**2.86**	**2.89**	**2.96**	**3.01**	**3.10**	**3.18**	**3.25**	**3.37**	**3.45**	**3.55**	**3.61**	**3.69**	**3.78**	**3.89**	**4.03**	**4.20**	**4.44**	**4.77**	**5.29**	**6.23**	**8.53**
17	1.96	1.97	1.99	2.02	2.04	2.08	2.11	2.15	2.19	2.23	2.29	2.33	2.38	2.41	2.45	2.50	2.55	2.62	2.70	2.81	2.96	3.20	3.59	4.45
	2.65	**2.67**	**2.70**	**2.76**	**2.79**	**2.86**	**2.92**	**3.00**	**3.08**	**3.16**	**3.27**	**3.35**	**3.45**	**3.52**	**3.59**	**3.68**	**3.79**	**3.93**	**4.10**	**4.34**	**4.67**	**5.18**	**6.11**	**8.40**
18	1.92	1.93	1.95	1.98	2.00	2.04	2.07	2.11	2.15	2.19	2.25	2.29	2.34	2.37	2.41	2.46	2.51	2.58	2.66	2.77	2.93	3.16	3.55	4.41
	2.57	**2.59**	**2.62**	**2.68**	**2.71**	**2.78**	**2.83**	**2.91**	**3.00**	**3.07**	**3.19**	**3.27**	**3.37**	**3.44**	**3.51**	**3.60**	**3.71**	**3.85**	**4.01**	**4.25**	**4.58**	**5.09**	**6.01**	**8.28**
19	1.88	1.90	1.91	1.94	1.96	2.00	2.02	2.07	2.11	2.15	2.21	2.26	2.31	2.34	2.38	2.43	2.48	2.55	2.63	2.74	2.90	3.13	3.52	4.38
	2.49	**2.51**	**2.54**	**2.60**	**2.63**	**2.70**	**2.76**	**2.84**	**2.92**	**3.00**	**3.12**	**3.19**	**3.30**	**3.36**	**3.43**	**3.52**	**3.63**	**3.77**	**3.94**	**4.17**	**4.50**	**5.01**	**5.93**	**8.18**
20	1.84	1.85	1.87	1.90	1.92	1.96	1.99	2.04	2.08	2.12	2.18	2.23	2.28	2.31	2.35	2.40	2.45	2.52	2.60	2.71	2.87	3.10	3.49	4.35
	2.42	**2.44**	**2.47**	**2.53**	**2.56**	**2.63**	**2.69**	**2.77**	**2.86**	**2.94**	**3.05**	**3.13**	**3.23**	**3.30**	**3.37**	**3.45**	**3.56**	**3.71**	**3.87**	**4.10**	**4.43**	**4.94**	**5.85**	**8.10**
21	1.81	1.82	1.84	1.87	1.90	1.93	1.96	2.00	2.05	2.09	2.15	2.20	2.25	2.28	2.32	2.37	2.42	2.49	2.57	2.68	2.84	3.07	3.47	4.32
	2.36	**2.38**	**2.42**	**2.47**	**2.51**	**2.58**	**2.63**	**2.72**	**2.80**	**2.88**	**2.99**	**3.07**	**3.17**	**3.24**	**3.31**	**3.40**	**3.51**	**3.65**	**3.81**	**4.04**	**4.37**	**4.87**	**5.78**	**8.02**
22	1.78	1.80	1.81	1.84	1.87	1.91	1.93	1.98	2.03	2.07	2.13	2.18	2.23	2.26	2.30	2.35	2.40	2.47	2.55	2.66	2.82	3.05	3.44	4.30
	2.31	**2.33**	**2.37**	**2.42**	**2.46**	**2.53**	**2.58**	**2.67**	**2.75**	**2.83**	**2.94**	**3.02**	**3.12**	**3.18**	**3.26**	**3.35**	**3.45**	**3.59**	**3.76**	**3.99**	**4.31**	**4.82**	**5.72**	**7.94**
23	1.76	1.77	1.79	1.82	1.84	1.88	1.91	1.96	2.00	2.04	2.10	2.14	2.20	2.24	2.28	2.32	2.38	2.45	2.53	2.64	2.80	3.03	3.42	4.28
	2.26	**2.28**	**2.32**	**2.37**	**2.41**	**2.48**	**2.53**	**2.62**	**2.70**	**2.78**	**2.89**	**2.97**	**3.07**	**3.14**	**3.21**	**3.30**	**3.41**	**3.54**	**3.71**	**3.94**	**4.26**	**4.76**	**5.66**	**7.88**
24	1.73	1.74	1.76	1.80	1.82	1.86	1.89	1.94	1.98	2.02	2.09	2.13	2.18	2.22	2.26	2.30	2.36	2.43	2.51	2.62	2.78	3.01	3.40	4.26
	2.21	**2.23**	**2.27**	**2.33**	**2.36**	**2.44**	**2.49**	**2.58**	**2.66**	**2.74**	**2.85**	**2.93**	**3.03**	**3.09**	**3.17**	**3.25**	**3.36**	**3.50**	**3.67**	**3.90**	**4.22**	**4.72**	**5.61**	**7.82**
25	1.71	1.72	1.74	1.77	1.80	1.84	1.87	1.92	1.96	2.00	2.06	2.11	2.16	2.20	2.24	2.28	2.34	2.41	2.49	2.60	2.76	2.99	3.38	4.24
	2.17	**2.19**	**2.23**	**2.29**	**2.32**	**2.40**	**2.45**	**2.54**	**2.62**	**2.70**	**2.81**	**2.89**	**2.99**	**3.05**	**3.13**	**3.21**	**3.32**	**3.46**	**3.63**	**3.86**	**4.18**	**4.68**	**5.57**	**7.77**
26	1.69	1.70	1.72	1.76	1.78	1.82	1.85	1.90	1.95	1.99	2.05	2.10	2.15	2.18	2.22	2.27	2.32	2.39	2.47	2.59	2.74	2.97	3.37	4.22
	2.13	**2.15**	**2.19**	**2.25**	**2.28**	**2.36**	**2.41**	**2.50**	**2.58**	**2.66**	**2.77**	**2.86**	**2.96**	**3.02**	**3.09**	**3.17**	**3.29**	**3.42**	**3.59**	**3.82**	**4.14**	**4.64**	**5.53**	**7.72**
27	1.67	1.68	1.71	1.74	1.76	1.80	1.84	1.88	1.93	1.97	2.03	2.08	2.13	2.16	2.20	2.25	2.30	2.37	2.46	2.57	2.73	2.96	3.35	4.21
	2.10	**2.12**	**2.16**	**2.21**	**2.25**	**2.33**	**2.38**	**2.47**	**2.55**	**2.63**	**2.74**	**2.83**	**2.93**	**2.98**	**3.06**	**3.14**	**3.26**	**3.39**	**3.56**	**3.79**	**4.11**	**4.60**	**5.49**	**7.68**
28	1.65	1.67	1.69	1.72	1.75	1.78	1.81	1.87	1.91	1.96	2.02	2.06	2.12	2.15	2.19	2.24	2.29	2.36	2.44	2.56	2.71	2.95	3.34	4.20
	2.06	**2.09**	**2.13**	**2.18**	**2.22**	**2.30**	**2.35**	**2.44**	**2.52**	**2.60**	**2.71**	**2.80**	**2.90**	**2.95**	**3.03**	**3.11**	**3.23**	**3.36**	**3.53**	**3.76**	**4.07**	**4.57**	**5.45**	**7.64**
29	1.64	1.65	1.68	1.71	1.73	1.77	1.80	1.85	1.90	1.94	2.00	2.05	2.10	2.14	2.18	2.22	2.28	2.35	2.43	2.54	2.70	2.93	3.33	4.18
	2.03	**2.06**	**2.10**	**2.15**	**2.19**	**2.27**	**2.32**	**2.41**	**2.49**	**2.57**	**2.68**	**2.77**	**2.87**	**2.92**	**3.00**	**3.08**	**3.20**	**3.33**	**3.50**	**3.73**	**4.04**	**4.54**	**5.42**	**7.60**
30	1.62	1.64	1.66	1.69	1.72	1.76	1.79	1.84	1.89	1.93	1.99	2.04	2.09	2.12	2.16	2.21	2.27	2.34	2.42	2.53	2.69	2.92	3.32	4.17
	2.01	**2.03**	**2.07**	**2.13**	**2.16**	**2.24**	**2.29**	**2.38**	**2.47**	**2.55**	**2.66**	**2.74**	**2.84**	**2.90**	**2.98**	**3.06**	**3.17**	**3.30**	**3.47**	**3.70**	**4.02**	**4.51**	**5.39**	**7.56**
32	1.59	1.61	1.64	1.67	1.69	1.74	1.76	1.82	1.86	1.91	1.97	2.02	2.07	2.10	2.14	2.19	2.25	2.32	2.40	2.51	2.67	2.90	3.30	4.15
	1.96	**1.98**	**2.02**	**2.08**	**2.12**	**2.20**	**2.25**	**2.34**	**2.42**	**2.51**	**2.62**	**2.70**	**2.80**	**2.86**	**2.94**	**3.01**	**3.12**	**3.25**	**3.42**	**3.66**	**3.97**	**4.46**	**5.34**	**7.50**
34	1.57	1.59	1.61	1.64	1.67	1.71	1.74	1.80	1.84	1.89	1.95	2.00	2.05	2.08	2.12	2.17	2.23	2.30	2.38	2.49	2.65	2.88	3.28	4.13
	1.91	**1.94**	**1.98**	**2.04**	**2.08**	**2.15**	**2.21**	**2.30**	**2.38**	**2.47**	**2.58**	**2.66**	**2.76**	**2.82**	**2.89**	**2.97**	**3.08**	**3.21**	**3.38**	**3.61**	**3.93**	**4.42**	**5.29**	**7.44**
36	1.55	1.56	1.59	1.62	1.65	1.69	1.72	1.78	1.82	1.87	1.93	1.98	2.03	2.06	2.10	2.15	2.21	2.28	2.36	2.48	2.63	2.86	3.26	4.11
	1.87	**1.90**	**1.94**	**2.00**	**2.04**	**2.12**	**2.17**	**2.26**	**2.35**	**2.43**	**2.54**	**2.62**	**2.72**	**2.78**	**2.86**	**2.94**	**3.04**	**3.18**	**3.35**	**3.58**	**3.89**	**4.38**	**5.25**	**7.39**
38	1.53	1.54	1.57	1.60	1.63	1.67	1.71	1.76	1.80	1.85	1.92	1.96	2.02	2.05	2.09	2.14	2.19	2.26	2.35	2.46	2.62	2.85	3.25	4.10
	1.84	**1.86**	**1.90**	**1.97**	**2.00**	**2.08**	**2.14**	**2.22**	**2.32**	**2.40**	**2.51**	**2.59**	**2.69**	**2.75**	**2.82**	**2.91**	**3.02**	**3.15**	**3.32**	**3.54**	**3.86**	**4.34**	**5.21**	**7.35**
40	1.51	1.53	1.55	1.59	1.61	1.66	1.69	1.74	1.79	1.84	1.90	1.95	2.00	2.04	2.07	2.12	2.18	2.25	2.34	2.45	2.61	2.84	3.23	4.08
	1.81	**1.84**	**1.88**	**1.94**	**1.97**	**2.05**	**2.11**	**2.20**	**2.29**	**2.37**	**2.49**	**2.56**	**2.66**	**2.73**	**2.80**	**2.88**	**2.99**	**3.12**	**3.29**	**3.51**	**3.83**	**4.31**	**5.18**	**7.31**

Degrees of freedom for denominator

Table VIII *(concluded)*
0.05 (light row) and 0.01 **(dark row)** points for the distribution of F

Degrees of freedom for numerator

df₂	p	1	2	3	4	5	6	7	8	9	10	11	12	14	16	20	24	30	40	50	75	100	200	500	∞
42	.05	4.07	3.22	2.83	2.59	2.44	2.32	2.24	2.17	2.11	2.06	2.02	1.99	1.94	1.89	1.82	1.78	1.73	1.68	1.64	1.60	1.57	1.54	1.51	1.49
	.01	**7.27**	**5.15**	**4.29**	**3.80**	**3.49**	**3.26**	**3.10**	**2.96**	**2.86**	**2.77**	**2.70**	**2.64**	**2.54**	**2.46**	**2.35**	**2.26**	**2.17**	**2.08**	**2.02**	**1.94**	**1.91**	**1.85**	**1.80**	**1.78**
44	.05	4.06	3.21	2.82	2.58	2.43	2.31	2.23	2.16	2.10	2.05	2.01	1.98	1.92	1.88	1.81	1.76	1.72	1.66	1.63	1.58	1.56	1.52	1.50	1.48
	.01	**7.24**	**5.12**	**4.26**	**3.78**	**3.46**	**3.24**	**3.07**	**2.94**	**2.84**	**2.75**	**2.68**	**2.62**	**2.52**	**2.44**	**2.32**	**2.24**	**2.15**	**2.06**	**2.09**	**1.92**	**1.88**	**1.82**	**1.78**	**1.75**
46	.05	4.05	3.20	2.81	2.57	2.42	2.30	2.22	2.14	2.09	2.04	2.00	1.97	1.91	1.87	1.80	1.75	1.71	1.65	1.62	1.57	1.54	1.51	1.48	1.46
	.01	**7.21**	**5.10**	**4.24**	**3.76**	**3.44**	**3.22**	**3.05**	**2.92**	**2.82**	**2.73**	**2.66**	**2.60**	**2.50**	**2.42**	**2.30**	**2.22**	**2.13**	**2.04**	**1.98**	**1.90**	**1.86**	**1.80**	**1.76**	**1.72**
48	.05	4.04	3.19	2.80	2.56	2.41	2.30	2.21	2.14	2.08	2.03	1.99	1.96	1.90	1.86	1.79	1.74	1.70	1.64	1.61	1.56	1.53	1.50	1.47	1.45
	.01	**7.19**	**5.08**	**4.22**	**3.74**	**3.42**	**3.20**	**3.04**	**2.90**	**2.80**	**2.71**	**2.64**	**2.58**	**2.48**	**2.40**	**2.28**	**2.20**	**2.11**	**2.02**	**1.96**	**1.88**	**1.84**	**1.78**	**1.73**	**1.70**
50	.05	4.03	3.18	2.79	2.56	2.40	2.29	2.20	2.13	2.07	2.02	1.98	1.95	1.90	1.85	1.78	1.74	1.69	1.63	1.60	1.55	1.52	1.48	1.46	1.44
	.01	**7.17**	**5.06**	**4.20**	**3.72**	**3.41**	**3.18**	**3.02**	**2.88**	**2.78**	**2.70**	**2.62**	**2.56**	**2.46**	**2.39**	**2.26**	**2.18**	**2.10**	**2.00**	**1.94**	**1.86**	**1.82**	**1.76**	**1.71**	**1.68**
55	.05	4.02	3.17	2.78	2.54	2.38	2.27	2.18	2.11	2.05	2.00	1.97	1.93	1.88	1.83	1.76	1.72	1.67	1.61	1.58	1.52	1.50	1.46	1.43	1.41
	.01	**7.12**	**5.01**	**4.16**	**3.68**	**3.37**	**3.15**	**2.98**	**2.85**	**2.75**	**2.66**	**2.59**	**2.53**	**2.43**	**2.35**	**2.23**	**2.15**	**2.06**	**1.96**	**1.90**	**1.82**	**1.78**	**1.71**	**1.66**	**1.64**
60	.05	4.00	3.15	2.76	2.52	2.37	2.25	2.17	2.10	2.04	1.99	1.95	1.92	1.86	1.81	1.75	1.70	1.65	1.59	1.56	1.50	1.48	1.44	1.41	1.39
	.01	**7.08**	**4.98**	**4.13**	**3.65**	**3.34**	**3.12**	**2.95**	**2.82**	**2.72**	**2.63**	**2.56**	**2.50**	**2.40**	**2.32**	**2.20**	**2.12**	**2.03**	**1.93**	**1.87**	**1.79**	**1.74**	**1.68**	**1.63**	**1.60**
65	.05	3.99	3.14	2.75	2.51	2.36	2.24	2.15	2.08	2.02	1.98	1.94	1.90	1.85	1.80	1.73	1.68	1.63	1.57	1.54	1.49	1.46	1.42	1.39	1.37
	.01	**7.04**	**4.95**	**4.10**	**3.62**	**3.31**	**3.09**	**2.93**	**2.79**	**2.70**	**2.61**	**2.54**	**2.47**	**2.37**	**2.30**	**2.18**	**2.09**	**2.00**	**1.90**	**1.84**	**1.76**	**1.71**	**1.64**	**1.60**	**1.56**
70	.05	3.98	3.13	2.74	2.50	2.35	2.23	2.14	2.07	2.01	1.97	1.93	1.89	1.84	1.79	1.72	1.67	1.62	1.56	1.53	1.47	1.45	1.40	1.37	1.35
	.01	**7.01**	**4.92**	**4.08**	**3.60**	**3.29**	**3.07**	**2.91**	**2.77**	**2.67**	**2.59**	**2.51**	**2.45**	**2.35**	**2.28**	**2.15**	**2.07**	**1.98**	**1.88**	**1.82**	**1.74**	**1.69**	**1.62**	**1.56**	**1.53**
80	.05	3.96	3.11	2.72	2.48	2.33	2.21	2.12	2.05	1.99	1.95	1.91	1.88	1.82	1.77	1.70	1.65	1.60	1.54	1.51	1.45	1.42	1.38	1.35	1.32
	.01	**6.96**	**4.88**	**4.04**	**3.56**	**3.25**	**3.04**	**2.87**	**2.74**	**2.64**	**2.55**	**2.48**	**2.41**	**2.32**	**2.24**	**2.11**	**2.03**	**1.94**	**1.84**	**1.78**	**1.70**	**1.65**	**1.57**	**1.52**	**1.49**
100	.05	3.94	3.09	2.70	2.46	2.30	2.19	2.10	2.03	1.97	1.92	1.88	1.85	1.79	1.75	1.68	1.63	1.57	1.51	1.48	1.42	1.39	1.34	1.30	1.28
	.01	**6.90**	**4.82**	**3.98**	**3.51**	**3.20**	**2.99**	**2.82**	**2.69**	**2.59**	**2.51**	**2.43**	**2.36**	**2.26**	**2.19**	**2.06**	**1.98**	**1.89**	**1.79**	**1.73**	**1.64**	**1.59**	**1.51**	**1.46**	**1.43**
125	.05	3.92	3.07	2.68	2.44	2.29	2.17	2.08	2.01	1.95	1.90	1.86	1.83	1.77	1.72	1.65	1.60	1.55	1.49	1.45	1.39	1.36	1.31	1.27	1.25
	.01	**6.84**	**4.78**	**3.94**	**3.47**	**3.17**	**2.95**	**2.79**	**2.65**	**2.56**	**2.47**	**2.40**	**2.33**	**2.23**	**2.15**	**2.03**	**1.94**	**1.85**	**1.75**	**1.68**	**1.59**	**1.54**	**1.46**	**1.40**	**1.37**
150	.05	3.91	3.06	2.67	2.43	2.27	2.16	2.07	2.00	1.94	1.89	1.85	1.82	1.76	1.71	1.64	1.59	1.54	1.47	1.44	1.37	1.34	1.29	1.25	1.22
	.01	**6.81**	**4.75**	**3.91**	**3.44**	**3.13**	**2.92**	**2.76**	**2.62**	**2.53**	**2.44**	**2.37**	**2.30**	**2.20**	**2.12**	**2.00**	**1.91**	**1.83**	**1.72**	**1.66**	**1.56**	**1.51**	**1.43**	**1.37**	**1.33**
200	.05	3.89	3.04	2.65	2.41	2.26	2.14	2.05	1.98	1.92	1.87	1.83	1.80	1.74	1.69	1.62	1.57	1.52	1.45	1.42	1.35	1.32	1.26	1.22	1.19
	.01	**6.76**	**4.71**	**3.88**	**3.41**	**3.11**	**2.90**	**2.73**	**2.60**	**2.50**	**2.41**	**2.34**	**2.28**	**2.17**	**2.09**	**1.97**	**1.88**	**1.79**	**1.69**	**1.62**	**1.53**	**1.48**	**1.39**	**1.33**	**1.28**
400	.05	3.86	3.02	2.62	2.39	2.23	2.12	2.03	1.96	1.90	1.85	1.81	1.78	1.72	1.67	1.60	1.54	1.49	1.42	1.38	1.32	1.28	1.22	1.16	1.13
	.01	**6.70**	**4.66**	**3.83**	**3.36**	**3.06**	**2.85**	**2.69**	**2.55**	**2.46**	**2.37**	**2.29**	**2.23**	**2.12**	**2.04**	**1.92**	**1.84**	**1.74**	**1.64**	**1.57**	**1.47**	**1.42**	**1.32**	**1.24**	**1.19**
1000	.05	3.85	3.00	2.61	2.38	2.22	2.10	2.02	1.95	1.89	1.84	1.80	1.76	1.70	1.65	1.58	1.53	1.47	1.41	1.36	1.30	1.26	1.19	1.13	1.08
	.01	**6.66**	**4.62**	**3.80**	**3.34**	**3.04**	**2.82**	**2.66**	**2.53**	**2.43**	**2.34**	**2.26**	**2.20**	**2.09**	**2.01**	**1.89**	**1.81**	**1.71**	**1.61**	**1.54**	**1.44**	**1.38**	**1.28**	**1.19**	**1.11**
∞	.05	3.84	2.99	2.60	2.37	2.21	2.09	2.01	1.94	1.88	1.83	1.79	1.75	1.69	1.64	1.57	1.52	1.46	1.40	1.35	1.28	1.24	1.17	1.11	1.00
	.01	**6.64**	**4.60**	**3.78**	**3.32**	**3.02**	**2.80**	**2.64**	**2.51**	**2.41**	**2.32**	**2.24**	**2.18**	**2.07**	**1.99**	**1.87**	**1.79**	**1.69**	**1.59**	**1.52**	**1.41**	**1.36**	**1.25**	**1.15**	**1.00**

Degrees of freedom for denominator

Source: Adapted from G. W. Snedecor and W. G. Cochran, *Statistical Methods*, 6th ed. (Ames, Iowa: Iowa State University, 1967). © 1967 by Iowa State University.

Table IX
Percentage points of the Studentized range

The values shown in the body of the table correspond to q_α in Tukey's HSD test.

Error df	α	k = Number of means or number of steps between ordered means									
		2	3	4	5	6	7	8	9	10	11
5	.05	3.64	4.60	5.22	5.67	6.03	6.33	6.58	6.80	6.99	7.17
	.01	5.70	6.98	7.80	8.42	8.91	9.32	9.67	9.97	10.24	10.48
6	.05	3.46	4.34	4.90	5.30	5.63	5.90	6.12	6.32	6.49	6.65
	.01	5.24	6.33	7.03	7.56	7.97	8.32	8.61	8.87	9.10	9.30
7	.05	3.34	4.16	4.68	5.06	5.36	5.61	5.82	6.00	6.16	6.30
	.01	4.95	5.92	6.54	7.01	7.37	7.68	7.94	8.17	8.37	8.55
8	.05	3.26	4.04	4.53	4.89	5.17	5.40	5.60	5.77	5.92	6.05
	.01	4.75	5.64	6.20	6.62	6.96	7.24	7.47	7.68	7.86	8.03
9	.05	3.20	3.95	4.41	4.76	5.02	5.24	5.43	5.59	5.74	5.87
	.01	4.60	5.43	5.96	6.35	6.66	6.91	7.13	7.33	7.49	7.65
10	.05	3.15	3.88	4.33	4.65	4.91	5.12	5.30	5.46	5.60	5.72
	.01	4.48	5.27	5.77	6.14	6.43	6.67	6.87	7.05	7.21	7.36
11	.05	3.11	3.82	4.26	4.57	4.82	5.03	5.20	5.35	5.49	5.61
	.01	4.39	5.15	5.62	5.97	6.25	6.48	6.67	6.84	6.99	7.13
12	.05	3.08	3.77	4.20	4.51	4.75	4.95	5.12	5.27	5.39	5.51
	.01	4.32	5.05	5.50	5.84	6.10	6.32	6.51	6.67	6.81	6.94
13	.05	3.06	3.73	4.15	4.45	4.69	4.88	5.05	5.19	5.32	5.43
	.01	4.26	4.96	5.40	5.73	5.98	6.19	6.37	6.53	6.67	6.79
14	.05	3.03	3.70	4.11	4.41	4.64	4.83	4.99	5.13	5.25	5.36
	.01	4.21	4.89	5.32	5.63	5.88	6.08	6.26	6.41	6.54	6.66
15	.05	3.01	3.67	4.08	4.37	4.59	4.78	4.94	5.08	5.20	5.31
	.01	4.17	4.84	5.25	5.56	5.80	5.99	6.16	6.31	6.44	6.55
16	.05	3.00	3.65	4.05	4.33	4.56	4.74	4.90	5.03	5.15	5.26
	.01	4.13	4.79	5.19	5.49	5.72	5.92	6.08	6.22	6.35	6.46
17	.05	2.98	3.63	4.02	4.30	4.52	4.70	4.86	4.99	5.11	5.21
	.01	4.10	4.74	5.14	5.43	5.66	5.85	6.01	6.15	6.27	6.38
18	.05	2.97	3.61	4.00	4.28	4.49	4.67	4.82	4.96	5.07	5.17
	.01	4.07	4.70	5.09	5.38	5.60	5.79	5.94	6.08	6.20	6.31
19	.05	2.96	3.59	3.98	4.25	4.47	4.65	4.79	4.92	5.04	5.14
	.01	4.05	4.67	5.05	5.33	5.55	5.73	5.89	6.02	6.14	6.25
20	.05	2.95	3.58	3.96	4.23	4.45	4.62	4.77	4.90	5.01	5.11
	.01	4.02	4.64	5.02	5.29	5.51	5.69	5.84	5.97	6.09	6.19
24	.05	2.92	3.53	3.90	4.17	4.37	4.54	4.68	4.81	4.92	5.01
	.01	3.96	4.55	4.91	5.17	5.37	5.54	5.69	5.81	5.92	6.02
30	.05	2.89	3.49	3.85	4.10	4.30	4.46	4.60	4.72	4.82	4.92
	.01	3.89	4.45	4.80	5.05	5.24	5.40	5.54	5.65	5.76	5.85
40	.05	2.86	3.44	3.79	4.04	4.23	4.39	4.52	4.63	4.73	4.82
	.01	3.82	4.37	4.70	4.93	5.11	5.26	5.39	5.50	5.60	5.69
60	.05	2.83	3.40	3.74	3.98	4.16	4.31	4.44	4.55	4.65	4.73
	.01	3.76	4.28	4.59	4.82	4.99	5.13	5.25	5.36	5.45	5.53
120	.05	2.80	3.36	3.68	3.92	4.10	4.24	4.36	4.47	4.56	4.64
	.01	3.70	4.20	4.50	4.71	4.87	5.01	5.12	5.21	5.30	5.37
∞	.05	2.77	3.31	3.63	3.86	4.03	4.17	4.29	4.39	4.47	4.55
	.01	3.64	4.12	4.40	4.60	4.76	4.88	4.99	5.08	5.16	5.23

Source: Adapted from E. S. Pearson and H. O. Hartley, *Biometrika Tables for Statisticians*, vol. 1, 2d ed. (New York: Cambridge University Press, 1958).

Table X
Table of χ^2

Tabled values are two-tailed. To be significant, obtained χ^2 must equal or exceed the tabled value at the specified α-level.

Degrees of freedom df	.10	.05	.025	.01
1	2.706	3.841	5.024	6.635
2	4.605	5.991	7.378	9.210
3	6.251	7.815	9.348	11.345
4	7.779	9.488	11.143	13.277
5	9.236	11.070	12.832	15.086
6	10.645	12.592	14.449	16.812
7	12.017	14.067	16.013	18.475
8	13.362	15.507	17.535	20.090
9	14.684	16.919	19.023	21.666
10	15.987	18.307	20.483	23.209
11	17.275	19.675	21.920	24.725
12	18.549	21.026	23.337	26.217
13	19.812	22.362	24.736	27.688
14	21.064	23.685	26.119	29.141
15	22.307	24.996	27.488	30.578
16	23.542	26.296	28.845	32.000
17	24.769	27.587	30.191	33.409
18	25.989	28.869	31.526	34.805
19	27.204	30.144	32.852	36.191
20	28.412	31.410	34.170	37.566
21	29.615	32.671	35.479	38.932
22	30.813	33.924	36.781	40.289
23	32.007	35.172	38.076	41.638
24	33.196	36.415	39.364	42.980
25	34.382	37.652	40.646	44.314
26	35.563	38.885	41.923	45.642
27	36.741	40.113	43.194	46.963
28	37.916	41.337	44.461	48.278
29	39.087	42.557	45.722	49.588
30	40.256	43.773	46.979	50.892
40	51.805	55.758	59.342	63.691
50	63.167	67.505	71.420	76.154
60	74.397	79.082	83.298	88.379
70	85.527	90.531	95.023	100.425
80	96.578	101.879	106.629	112.329
90	107.565	113.145	118.136	124.116
100	118.498	124.342	129.561	135.807

Source: Adapted from R. A. Fisher, *Statistical Methods for Research Workers,* 14th ed. (New York: Macmillan, 1970). © 1970 by the University of Adelaide.

Table XI (a)

Critical values of U_1 and U_2 for a one-tailed test at $\alpha = 0.005$ or a two-tailed test at $\alpha = 0.01$

To be significant for any given n_1 and n_2, a given U_1 or U_2 must be equal to or less than the smaller value, or equal to or greater than the larger value, shown in the body of the table.

Example: If $\alpha = 0.01$ (two-tailed test), $n_1 = 9$, $n_2 = 11$, and obtained $U_1 = 67$, we fail to reject H_0 since the obtained value is within the lower (16) and upper (83) critical values.

n_2 \ n_1	1	2	3	4	5	6	7	8	9	10	11	12	13	14	15	16	17	18	19	20
1	—	—	—	—	—	—	—	—	—	—	—	—	—	—	—	—	—	—	—	—
2	—	—	—	—	—	—	—	—	—	—	—	—	—	—	—	—	—	—	—	—
3	—	—	—	—	—	—	—	—	0/27	0/30	0/33	1/35	1/38	1/41	2/43	2/46	2/49	2/52	3/54	3/57
4	—	—	—	—	—	—	0/28	1/31	1/35	2/38	2/42	3/45	3/49	4/52	5/55	5/59	6/62	6/66	7/69	8/72
5	—	—	—	—	0/25	1/29	1/34	2/38	3/42	4/46	5/50	6/54	7/58	7/63	8/67	9/71	10/75	11/79	12/83	13/87
6	—	—	—	0/24	1/29	2/34	3/39	4/44	5/49	6/54	7/59	9/63	10/68	11/73	12/78	13/83	15/87	16/92	17/97	18/102
7	—	—	—	0/28	1/34	3/39	4/45	6/50	7/56	9/61	10/67	12/72	13/78	15/83	16/89	18/94	19/100	21/105	22/111	24/116
8	—	—	—	1/31	2/38	4/44	6/50	7/57	9/63	11/69	13/75	15/81	17/87	18/94	20/100	22/106	24/112	26/118	28/124	30/130

9	—	—	0 27	1 35	3 42	5 49	7 56	9 63	11 70	13 77	16 83	18 90	20 97	22 104	24 111	27 117	29 124	31 131	33 138	36 144
10	—	—	0 30	2 38	4 46	6 54	9 61	11 69	13 77	16 84	18 92	21 99	24 106	26 114	29 121	31 129	34 136	37 143	39 151	42 158
11	—	—	0 33	2 42	5 50	7 59	10 67	13 75	16 83	18 92	21 100	24 108	27 116	30 124	33 132	36 140	39 148	42 156	45 164	48 172
12	—	—	1 35	3 45	6 54	9 63	12 72	15 81	18 90	21 99	24 108	27 117	31 125	34 134	37 143	41 151	44 160	47 169	51 177	54 186
13	—	—	1 38	3 49	7 58	10 68	13 78	17 87	20 97	24 106	27 116	31 125	34 125	38 144	42 153	45 163	49 172	53 181	56 191	60 200
14	—	—	1 41	4 52	7 63	11 73	15 83	18 94	22 104	26 114	30 124	34 134	38 144	42 154	46 164	50 174	54 184	58 194	63 203	67 213
15	—	—	2 43	5 55	8 67	12 78	16 89	20 100	24 111	29 121	33 132	37 143	42 153	46 164	51 174	55 185	60 195	64 206	69 216	73 227
16	—	—	2 46	5 59	9 71	13 83	18 94	22 106	27 117	31 129	36 140	41 151	45 163	50 174	55 185	60 196	65 207	70 218	74 230	79 241
17	—	—	2 49	6 62	10 75	15 87	19 100	24 112	29 124	34 148	39 148	44 160	49 172	54 184	60 195	65 207	70 219	75 231	81 242	86 254
18	—	—	2 52	6 66	11 79	16 92	21 105	26 118	31 131	37 143	42 156	47 169	53 181	58 194	64 206	70 218	75 231	81 243	87 255	92 268
19	—	0 38	3 54	7 69	12 83	17 97	22 111	28 124	33 138	39 151	45 164	51 177	56 191	63 203	69 216	74 230	81 242	87 255	93 268	99 281
20	—	0 40	3 57	8 72	13 87	18 102	24 116	30 130	36 144	42 158	48 172	54 186	60 200	67 213	73 227	79 241	86 254	92 268	99 281	105 295

Note: Dashes in the body of the table indicate that no decision is possible at the stated level of significance.

Table XI (b)

Critical values of U_1 and U_2 for a one-tailed test at $\alpha = 0.01$ or a two-tailed test at $\alpha = 0.02$

To be significant for any given n_1 and n_2, a given U_1 or U_2 must be equal to or less than the smaller value, or equal to or greater than the larger value shown in the body of the table.

n_1 / n_2	1	2	3	4	5	6	7	8	9	10	11	12	13	14	15	16	17	18	19	20
1	—	—	—	—	—	—	—	—	—	—	—	—	—	—	—	—	—	—	—	—
2	—	—	—	—	—	—	—	—	—	—	—	—	0/26	0/28	0/30	0/32	0/34	0/36	1/37	1/39
3	—	—	—	—	—	—	0/21	0/24	1/26	1/29	1/32	2/34	2/37	2/40	3/42	3/45	4/47	4/50	4/52	5/55
4	—	—	—	—	0/20	1/23	1/27	2/30	3/33	3/37	4/40	5/43	5/47	6/50	7/53	7/57	8/60	9/63	9/67	10/70
5	—	—	—	0/20	1/24	2/28	3/32	4/36	5/40	6/44	7/48	8/52	9/56	10/60	11/64	12/68	13/72	14/76	15/80	16/84
6	—	—	—	1/23	2/28	3/33	4/38	6/42	7/47	8/52	9/57	11/61	12/66	13/71	15/75	16/80	18/84	19/89	20/94	22/98
7	—	—	0/21	1/27	3/32	4/38	6/43	7/49	9/54	11/59	12/65	14/70	16/75	17/81	19/86	21/91	23/96	24/102	26/107	28/112
8	—	—	0/24	2/30	4/36	6/42	7/49	9/55	11/61	13/67	15/73	17/79	20/84	22/90	24/96	26/102	28/108	30/114	32/120	34/126
9	—	—	1/26	3/33	5/40	7/47	9/54	11/61	14/67	16/74	18/81	21/87	23/94	26/100	28/107	31/113	33/120	36/126	38/133	40/140

	1	2	3	4	5	6	7	8	9	10	11	12	13	14	15	16	17	18	19	20
10	—	—	1/29	3/37	6/44	8/52	11/59	13/67	16/74	19/81	22/88	24/96	27/103	30/110	33/117	36/124	38/132	41/139	44/146	47/153
11	—	—	1/32	4/40	7/48	9/57	12/65	15/73	18/81	22/88	25/96	28/104	31/112	34/120	37/128	41/135	44/143	47/151	50/159	53/167
12	—	—	2/34	5/43	8/52	11/61	14/70	17/79	21/87	24/96	28/104	31/113	35/121	38/130	42/138	46/146	49/155	53/163	56/172	60/180
13	—	0/26	2/37	5/47	9/56	12/66	16/75	20/84	23/94	27/103	31/112	35/121	39/130	43/139	47/148	51/157	55/166	59/175	63/184	67/193
14	—	0/28	2/40	6/50	10/60	13/71	17/81	22/90	26/100	30/110	34/120	38/130	43/139	47/149	51/159	56/168	60/178	65/187	69/197	73/207
15	—	0/30	3/42	7/53	11/64	15/75	19/86	24/96	28/107	33/117	37/128	42/138	47/148	51/159	56/169	61/179	66/189	70/200	75/210	80/220
16	—	0/32	3/45	7/57	12/68	16/80	21/91	26/102	31/113	36/124	41/135	46/146	51/157	56/168	61/179	66/190	71/201	76/212	82/222	87/233
17	—	0/34	4/47	8/60	13/72	18/84	23/96	28/108	33/120	38/132	44/143	49/155	55/166	60/178	66/189	71/201	77/212	82/224	88/234	93/247
18	—	0/36	4/50	9/63	14/76	19/89	24/102	30/114	36/126	41/139	47/151	53/163	59/175	65/187	70/200	76/212	82/224	88/236	94/248	100/260
19	—	1/37	4/53	10/67	15/80	20/94	26/107	32/120	38/133	44/146	50/159	56/172	63/184	69/197	75/210	82/222	88/235	94/248	101/260	107/273
20	—	1/39	5/55	10/70	16/84	22/98	28/112	34/126	40/140	47/153	53/167	60/180	67/193	73/207	80/220	87/233	93/247	100/260	107/273	114/286

Note: Dashes in the body of the table indicate that no decision is possible at the stated level of significance.

Table XI (c)

Critical values of U_1 and U_2 for a one-tailed test at $\alpha = 0.025$ or a two-tailed test at $\alpha = 0.05$

To be significant for any given n_1 and n_2, a given U_1 or U_2 must be equal to or less than the smaller value, or equal to or greater than the larger value, shown in the body of the table.

In each cell the smaller value is given first and the larger (underlined) value second.

$n_2 \backslash n_1$	1	2	3	4	5	6	7	8	9	10	11	12	13	14	15	16	17	18	19	20
1	—	—	—	—	—	—	—	—	—	—	—	—	—	—	—	—	—	—	—	—
2	—	—	—	—	—	—	—	0/16	0/18	0/20	0/22	1/23	1/25	1/27	1/29	1/31	2/32	2/34	2/36	2/38
3	—	—	—	—	0/15	1/17	1/20	2/22	2/25	3/27	3/30	4/32	4/35	5/37	5/40	6/42	6/45	7/47	7/50	8/52
4	—	—	—	0/16	1/19	2/22	3/25	4/28	4/32	5/35	6/38	7/41	8/44	9/47	10/50	11/53	11/57	12/60	13/63	13/67
5	—	—	0/15	1/19	2/23	3/27	5/30	6/34	7/38	8/42	9/46	11/49	12/53	13/57	14/61	15/65	17/68	18/72	19/76	20/80
6	—	—	1/17	2/22	3/27	5/31	6/36	8/40	10/44	11/49	13/53	14/58	16/62	17/67	19/71	21/75	22/80	24/84	25/89	27/93
7	—	—	1/20	3/25	5/30	6/36	8/41	10/46	12/51	14/56	16/61	18/66	20/71	22/76	24/81	26/86	28/91	30/96	32/101	34/106
8	—	0/16	2/22	4/28	6/34	8/40	10/46	13/51	15/57	17/63	19/69	22/74	24/80	26/86	29/91	31/97	34/102	36/108	38/114	41/119
9	—	0/18	2/25	4/32	7/38	10/44	12/51	15/57	17/64	20/70	23/76	26/82	28/89	31/95	34/101	37/107	39/114	42/120	45/126	48/132

10	—	0 / 20	3 / 27	5 / 35	8 / 42	11 / 49	14 / 56	17 / 63	20 / 70	23 / 77	26 / 84	29 / 91	33 / 97	36 / 104	39 / 111	42 / 118	45 / 125	48 / 132	52 / 138
																			55 / 145
11	—	0 / 22	3 / 30	6 / 38	9 / 46	13 / 53	16 / 61	19 / 69	23 / 76	26 / 84	30 / 91	33 / 99	37 / 106	40 / 114	44 / 121	47 / 129	51 / 136	55 / 143	58 / 151
																			62 / 158
12	—	1 / 23	4 / 32	7 / 41	11 / 49	14 / 58	18 / 66	22 / 74	26 / 82	29 / 91	33 / 99	37 / 107	41 / 115	45 / 123	49 / 131	53 / 139	57 / 147	61 / 155	65 / 163
																			69 / 171
13	—	1 / 25	4 / 35	8 / 44	12 / 53	16 / 62	20 / 71	24 / 80	28 / 89	33 / 97	37 / 106	41 / 115	45 / 124	50 / 132	54 / 141	59 / 149	63 / 158	67 / 167	72 / 175
																			76 / 184
14	—	1 / 27	5 / 37	9 / 47	13 / 57	17 / 67	22 / 76	26 / 86	31 / 95	36 / 104	40 / 114	45 / 123	50 / 132	55 / 141	59 / 151	64 / 160	67 / 171	74 / 178	78 / 188
																			83 / 197
15	—	1 / 29	5 / 40	10 / 50	14 / 61	19 / 71	24 / 81	29 / 91	34 / 101	39 / 111	44 / 121	49 / 131	54 / 141	59 / 151	64 / 161	70 / 170	75 / 180	80 / 190	85 / 200
																			90 / 210
16	—	1 / 31	6 / 42	11 / 53	15 / 65	21 / 75	26 / 86	31 / 97	37 / 107	42 / 118	47 / 129	53 / 139	59 / 149	64 / 160	70 / 170	75 / 181	81 / 191	86 / 202	92 / 212
																			98 / 222
17	—	2 / 32	6 / 45	11 / 57	17 / 68	22 / 80	28 / 91	34 / 102	39 / 114	45 / 125	51 / 136	57 / 147	63 / 158	67 / 171	75 / 180	81 / 191	87 / 202	93 / 213	99 / 224
																			105 / 235
18	—	2 / 34	7 / 47	12 / 60	18 / 72	24 / 84	30 / 96	36 / 108	42 / 120	48 / 132	55 / 143	61 / 155	67 / 167	74 / 178	80 / 190	86 / 202	93 / 213	99 / 225	106 / 236
																			112 / 248
19	—	2 / 36	7 / 50	13 / 63	19 / 76	25 / 89	32 / 101	38 / 114	45 / 126	52 / 138	58 / 151	65 / 163	72 / 175	78 / 188	85 / 200	92 / 212	99 / 224	106 / 236	113 / 248
																			119 / 261
20	—	2 / 38	8 / 52	13 / 67	20 / 80	27 / 93	34 / 106	41 / 119	48 / 132	55 / 145	62 / 158	69 / 171	76 / 184	83 / 197	90 / 210	98 / 222	105 / 235	112 / 248	119 / 261
																			127 / 273

Note: Dashes in the body of the table indicate that no decision is possible at the stated level of significance.

Table XI (d)

Critical values of U_1 and U_2 for a one-tailed test at $\alpha = 0.05$ or a two-tailed test at $\alpha = 0.10$

To be significant for any given n_1 and n_2, a given U_1 or U_2 must be equal to or less than the smaller value, or equal to or greater than the larger value, shown in the body of the table.

n_1 \ n_2	1	2	3	4	5	6	7	8	9	10	11	12	13	14	15	16	17	18	19	20
1	—	—	—	—	—	—	—	—	—	—	—	—	—	—	—	—	—	—	—	0/20
2	—	—	—	—	0/10	0/12	0/14	1/15	1/17	1/19	1/21	2/22	2/24	2/26	3/27	3/29	3/31	4/32	4/34	4/36
3	—	—	0/9	0/12	1/14	2/16	2/19	3/21	3/24	4/26	5/28	5/31	6/33	7/35	7/38	8/40	9/42	9/45	10/47	11/49
4	—	—	0/12	1/15	2/18	3/21	4/24	5/27	6/30	7/33	8/36	9/39	10/42	11/45	12/48	14/50	15/53	16/56	17/59	18/62
5	—	0/10	1/14	2/18	4/21	5/25	6/29	8/32	9/36	11/39	12/43	13/47	15/50	16/54	18/57	19/61	20/65	22/68	23/72	25/75
6	—	0/12	2/16	3/21	5/25	7/29	8/34	10/38	12/42	14/46	16/50	17/55	19/59	21/63	23/67	25/71	26/76	28/80	30/84	32/88
7	—	0/14	2/19	4/24	6/29	8/34	11/38	13/43	15/48	17/53	19/58	21/63	24/67	26/72	28/77	30/82	33/86	35/91	37/96	39/101
8	—	1/15	3/21	5/27	8/32	10/38	13/43	15/49	18/54	20/60	23/65	26/70	28/76	31/81	33/87	36/92	39/97	41/103	44/108	47/113
9	—	1/17	3/24	6/30	9/36	12/42	15/48	18/54	21/60	24/66	27/72	30/78	33/84	36/90	39/96	42/102	45/108	48/114	51/120	54/126

n_1	1	2	3	4	5	6	7	8	9	10	11	12	13	14	15	16	17	18	19	20
10	—	1/19	4/26	7/33	11/39	14/46	17/53	20/60	24/66	27/73	31/79	34/86	37/93	41/99	44/106	48/112	51/119	55/125	58/132	62/138
11	—	1/21	5/28	8/36	12/43	16/50	19/58	23/65	27/72	31/79	34/87	38/94	42/101	46/108	50/115	54/122	57/130	61/137	65/144	69/151
12	—	2/22	5/31	9/39	13/47	17/55	21/63	26/70	30/78	34/86	38/94	42/102	47/109	51/117	55/125	60/132	64/140	68/148	72/156	77/163
13	—	2/24	6/33	10/42	15/50	19/59	24/67	28/76	33/84	37/93	42/101	47/109	51/118	56/126	61/134	65/143	70/151	75/159	80/167	84/176
14	—	2/26	7/35	11/45	16/54	21/63	26/72	31/81	36/90	41/99	46/108	51/117	56/126	61/135	66/144	71/153	77/161	82/170	87/179	92/188
15	—	3/27	7/38	12/48	18/57	23/67	28/77	33/87	39/96	44/106	50/115	55/125	61/134	66/144	72/153	77/163	83/172	88/182	94/191	100/200
16	—	3/29	8/40	14/50	19/61	25/71	30/82	36/92	42/102	48/112	54/122	60/132	65/143	71/153	77/163	83/173	89/183	95/193	101/203	107/213
17	—	3/31	9/42	15/53	20/65	26/76	33/86	39/97	45/108	51/119	57/130	64/140	70/151	77/161	83/172	89/183	96/193	102/204	109/214	115/225
18	—	4/32	9/45	16/56	22/68	28/80	35/91	41/103	48/114	55/123	61/137	68/148	75/159	82/170	88/182	95/193	102/204	109/215	116/226	123/237
19	0/19	4/34	10/47	17/59	23/72	30/84	37/96	44/108	51/120	58/132	65/144	72/156	80/167	87/179	94/191	101/203	109/215	116/226	123/238	130/250
20	0/20	4/36	11/49	18/62	25/75	32/88	39/101	47/113	54/126	62/138	69/151	77/163	84/176	92/188	100/200	107/213	115/225	123/237	130/250	138/262

Note: Dashes in the body of the table indicate that no decision is possible at the stated level of significance.

Source: Adapted from H. B. Mann and D. R. Whitney, "On a test of whether one of two variables is stochastically larger than the other," *Annals of Mathematical Statistics*, 18 (1947):52–54.

D. Auble, "Extended Tables for the Mann-Whitney Statistic," *Bulletin of the Institute of Educational Research at Indiana University*, 1, No. 2 (1953).

Table XII
Critical values of R in the runs test

Each cell shows two values for n_1 and n_2. If R is equal to or less than the upper value, or equal to or greater than the lower value, in each cell, the sequence is nonrandom at $\alpha = 0.05$ (two-tailed test) or 0.025 (one-tailed test). Dashes indicate that no decision is possible for the indicated values of n_1 and n_2.

n_2 \ n_1	2	3	4	5	6	7	8	9	10	11	12	13	14	15	16	17	18	19	20
2	—	—	—	—	—	—	—	—	—	—	2	2	2	2	2	2	2	2	2
	—	—	—	—	—	—	—	—	—	—	—	—	—	—	—	—	—	—	—
3	—	—	—	—	2	2	2	2	2	2	2	2	2	3	3	3	3	3	3
	—	—	—	—	—	—	—	—	—	—	—	—	—	—	—	—	—	—	—
4	—	—	—	2	2	2	3	3	3	3	3	3	3	3	4	4	4	4	4
	—	—	—	9	9	—	—	—	—	—	—	—	—	—	—	—	—	—	—
5	—	—	2	2	3	3	3	3	3	4	4	4	4	4	4	4	5	5	5
	—	—	9	10	10	11	11	—	—	—	—	—	—	—	—	—	—	—	—
6	—	2	2	3	3	3	3	4	4	4	4	5	5	5	5	5	5	6	6
	—	—	9	10	11	12	12	13	13	13	13	—	—	—	—	—	—	—	—
7	—	2	2	3	3	3	4	4	5	5	5	5	5	6	6	6	6	6	6
	—	—	11	12	13	13	14	14	14	14	15	15	15	—	—	—	—	—	—
8	—	2	3	3	3	4	4	5	5	5	6	6	6	6	6	7	7	7	7
	—	—	—	11	12	13	14	14	15	15	16	16	16	16	17	17	17	17	17
9	—	2	3	3	4	4	5	5	6	6	6	6	7	7	7	7	8	8	8
	—	—	—	13	14	14	15	16	16	16	17	17	18	18	18	18	18	18	18
10	—	2	3	4	4	5	5	6	6	7	7	7	8	8	8	9	9	9	9
	—	—	—	13	14	15	16	16	17	17	18	18	18	19	19	19	20	20	20
11	—	2	3	4	4	5	5	6	6	7	7	7	8	8	8	9	9	9	9
	—	—	—	13	14	15	16	17	17	18	19	19	19	20	20	20	21	21	21
12	2	2	3	4	4	5	6	6	7	7	7	8	8	8	9	9	9	10	10
	—	—	—	—	13	15	16	17	18	19	19	19	20	20	21	21	21	22	22
13	2	2	3	4	5	5	6	6	7	7	8	8	9	9	9	10	10	10	10
	—	—	—	—	—	15	16	17	18	19	19	20	20	21	21	22	22	23	23
14	2	2	3	4	5	5	6	7	7	8	8	9	9	9	10	10	10	11	11
	—	—	—	—	—	15	16	18	18	19	20	20	21	22	22	23	23	23	24
15	2	3	3	4	5	6	6	7	7	8	8	9	9	10	10	11	11	11	12
	—	—	—	—	—	—	16	18	19	20	20	21	22	22	23	23	23	23	23
16	2	3	4	4	5	6	6	7	8	8	9	9	10	10	11	11	11	12	12
	—	—	—	—	—	—	17	18	19	20	21	21	22	23	23	24	25	25	25
17	2	3	4	4	5	6	7	7	8	9	9	10	10	11	11	11	12	12	13
	—	—	—	—	—	—	17	18	19	20	21	22	23	23	24	25	25	26	26
18	2	3	4	5	5	6	7	8	8	9	9	10	10	11	11	12	12	13	13
	—	—	—	—	—	—	17	18	20	21	21	22	23	23	25	25	26	26	27
19	2	3	4	5	6	6	7	8	8	9	10	10	11	11	12	12	13	13	13
	—	—	—	—	—	—	17	18	20	21	22	23	23	23	25	26	26	27	27
20	2	3	4	5	6	6	7	8	9	9	10	10	11	12	12	13	13	13	14
	—	—	—	—	—	—	17	18	20	21	22	23	24	23	25	26	27	27	28

Source: F. Swed and C. Eisenhart, "Tables for Testing Randomness of Grouping in a Sequence of Alternatives," *Annals of Mathematical Statistics* 14 (1943).

THE USE OF TABLE XIII—FOUR-PLACE COMMON LOGARITHMS

A common logarithm of any positive number is the power to which the base 10 must be raised to give that number. The number itself is called the *antilogarithm*.

The logarithm of a number consists of two parts: the *mantissa* and the *characteristic*. For example,

$$\log 6233 = 3.7947$$

characteristic mantissa

1. To find the *characteristic,* we use the following:

Range of number	Characteristic
≥1 but <10	0
≥10 but <100	1
≥100 but <1,000	2
≥1,000 but <10,000	3
≥10,000 but <100,000	4

Thus, the characteristic of the number 6233 is 3;
the characteristic of the number 623.3 is 2;
the characteristic of the number 62.33 is 1;
the characteristic of the number 6.233 is 0.

2. To find the *mantissa* of 6233, we first locate 62 in the left-hand column. We locate the column headed by 3 and find 7945 in the body of the table. This gives us the mantissa of 623. Now we must add to this the proportional part corresponding to the fourth number, i.e., 3. We find the value 2, which is added to 7945. Thus, the mantissa of 6233 (or 623.3 or 62.33 or 6.233) is 7947.

3. Putting steps 1 and 2 together, we find

$$\log 6233 = 3.7947$$
$$\log 623.3 = 2.7947$$
$$\log 62.33 = 1.7947$$
$$\log 6.233 = 0.7947$$

4. To find the antilogarithm of a number, we simply reverse the procedure. Thus, the antilogarithm of 1.7947 is 62.33.

Table XIII
Four-place common logarithms

	0	1	2	3	4	5	6	7	8	9		Proportional parts								
												1	2	3	4	5	6	7	8	9
10	0000	0043	0086	0128	0170	0212	0253	0294	0334	0374		4	8	12	17	21	25	29	33	37
11	0414	0453	0492	0531	0569	0607	0645	0682	0719	0755		4	8	11	15	19	23	26	30	34
12	0792	0828	0864	0899	0934	0969	1004	1038	1072	1106		3	7	10	14	17	21	24	28	31
13	1139	1173	1206	1239	1271	1303	1335	1367	1399	1430		3	6	10	13	16	19	23	26	29
14	1461	1492	1523	1553	1584	1614	1644	1673	1703	1732		3	6	9	12	15	18	21	24	27
15	1761	1790	1818	1847	1875	1903	1931	1959	1987	2014		3	6	8	11	14	17	20	22	25
16	2041	2068	2095	2122	2148	2175	2201	2227	2253	2279		3	5	8	11	13	16	18	21	24
17	2304	2330	2355	2380	2405	2430	2455	2480	2504	2529		2	5	7	10	12	15	17	20	22
18	2553	2577	2601	2625	2648	2672	2695	2718	2742	2765		2	5	7	9	12	14	16	19	21
19	2788	2810	2833	2856	2878	2900	2923	2945	2967	2989		2	4	7	9	11	13	16	18	20
20	3010	3032	3054	3075	3096	3118	3139	3160	3181	3201		2	4	6	8	11	13	15	17	19
21	3222	3243	3263	3284	3304	3324	3345	3365	3385	3404		2	4	6	8	10	12	14	16	18
22	3424	3444	3464	3483	3502	3522	3541	3560	3579	3598		2	4	6	8	10	12	14	15	17
23	3617	3636	3655	3674	3692	3711	3729	3747	3766	3784		2	4	6	7	9	11	13	15	17
24	3802	3820	3838	3856	3874	3892	3909	3927	3945	3962		2	4	5	7	9	11	12	14	16
25	3979	3997	4014	4031	4048	4065	4082	4099	4116	4133		2	3	5	7	9	10	12	14	15
26	4150	4166	4183	4200	4216	4232	4249	4265	4281	4298		2	3	5	7	8	10	11	13	15
27	4314	4330	4346	4362	4378	4393	4409	4425	4440	4456		2	3	5	6	8	9	11	13	14
28	4472	4487	4502	4518	4533	4548	4564	4579	4594	4609		2	3	5	6	8	9	11	12	14
29	4624	4639	4654	4669	4683	4698	4713	4728	4742	4757		1	3	4	6	7	9	10	12	13

N	0	1	2	3	4	5	6	7	8	9	1	2	3	4	5	6	7	8	9
30	4771	4786	4800	4814	4829	4843	4857	4871	4886	4900	1	3	4	6	7	9	10	11	13
31	4914	4928	4942	4955	4969	4983	4997	5011	5024	5038	1	3	4	6	7	8	10	11	12
32	5051	5065	5079	5092	5105	5119	5132	5145	5159	5172	1	3	4	5	7	8	9	11	12
33	5185	5198	5211	5224	5237	5250	5263	5276	5289	5302	1	3	4	5	6	8	9	10	12
34	5315	5328	5340	5353	5366	5378	5391	5403	5416	5428	1	3	4	5	6	8	9	10	11
35	5441	5453	5465	5478	5490	5502	5514	5527	5539	5551	1	2	4	5	6	7	9	10	11
36	5563	5575	5587	5599	5611	5623	5635	5647	5658	5670	1	2	4	5	6	7	8	10	11
37	5682	5694	5705	5717	5729	5740	5752	5763	5775	5786	1	2	3	5	6	7	8	9	10
38	5798	5809	5821	5832	5843	5855	5866	5877	5888	5899	1	2	3	5	6	7	8	9	10
39	5911	5922	5933	5944	5955	5966	5977	5988	5999	6010	1	2	3	4	5	7	8	9	10
40	6021	6031	6042	6053	6064	6075	6085	6096	6107	6117	1	2	3	4	5	6	8	9	10
41	6128	6138	6149	6160	6170	6180	6191	6201	6212	6222	1	2	3	4	5	6	7	8	9
42	6232	6243	6253	6263	6274	6284	6294	6304	6314	6325	1	2	3	4	5	6	7	8	9
43	6335	6345	6355	6365	6375	6385	6395	6405	6415	6425	1	2	3	4	5	6	7	8	9
44	6435	6444	6454	6464	6474	6484	6493	6503	6513	6522	1	2	3	4	5	6	7	8	9
45	6532	6542	6551	6561	6571	6580	6590	6599	6609	6618	1	2	3	4	5	6	7	8	9
46	6628	6637	6646	6656	6665	6675	6684	6693	6702	6712	1	2	3	4	5	6	7	7	8
47	6721	6730	6739	6749	6758	6767	6776	6785	6794	6803	1	2	3	4	5	5	6	7	8
48	6812	6821	6830	6839	6848	6857	6866	6875	6884	6893	1	2	3	4	4	5	6	7	8
49	6902	6911	6920	6928	6937	6946	6955	6964	6972	6981	1	2	3	4	4	5	6	7	8
50	6990	6998	7007	7016	7024	7033	7042	7050	7059	7067	1	2	3	3	4	5	6	7	8
51	7076	7084	7093	7101	7110	7118	7126	7135	7143.	7152	1	2	3	3	4	5	6	7	8
52	7160	7168	7177	7185	7193	7202	7210	7218	7226	7235	1	2	2	3	4	5	6	7	7
53	7243	7251	7259	7267	7275	7284	7292	7300	7308	7316	1	2	2	3	4	5	6	6	7
54	7324	7332	7340	7348	7356	7364	7372	7380	7388	7396	1	2	2	3	4	5	6	6	7

Table XIII *(concluded)*

	0	1	2	3	4	5	6	7	8	9	|	1	2	3	4	5	6	7	8	9
												|	Proportional parts								
55	7404	7412	7419	7427	7435	7443	7451	7459	7466	7474	|	1	2	2	3	4	5	5	6	7
56	7482	7490	7497	7505	7513	7520	7528	7536	7543	7551	|	1	2	2	3	4	5	5	6	7
57	7559	7566	7574	7582	7589	7597	7604	7612	7619	7627	|	1	2	2	3	4	5	5	6	7
58	7634	7642	7649	7657	7664	7672	7679	7686	7694	7701	|	1	2	2	3	4	4	5	6	7
59	7709	7716	7723	7731	7738	7745	7752	7760	7767	7774	|	1	1	2	3	4	4	5	6	7
60	7782	7789	7796	7803	7810	7818	7825	7832	7839	7846	|	1	1	2	3	4	4	5	6	6
61	7853	7860	7868	7875	7882	7889	7896	7903	7910	7917	|	1	1	2	3	4	4	5	6	6
62	7924	7931	7938	7945	7952	7959	7966	7973	7980	7987	|	1	1	2	3	4	4	5	6	6
63	7993	8000	8007	8014	8021	8028	8035	8041	8048	8055	|	1	1	2	3	3	4	5	5	6
64	8062	8069	8075	8082	8089	8096	8102	8109	8116	8122	|	1	1	2	3	3	4	5	5	6
65	8129	8136	8142	8149	8156	8162	8169	8176	8182	8189	|	1	1	2	3	3	4	5	5	6
66	8195	8202	8209	8215	8222	8228	8235	8241	8248	8254	|	1	1	2	3	3	4	5	5	6
67	8261	8267	8274	8280	8287	8293	8299	8306	8312	8319	|	1	1	2	3	3	4	5	5	6
68	8325	8331	8338	8344	8351	8357	8363	8370	8376	8382	|	1	1	2	3	3	4	4	5	6
69	8388	8395	8401	8407	8414	8420	8426	8432	8439	8445	|	1	1	2	2	3	4	4	5	6
70	8451	8457	8463	8470	8476	8482	8488	8494	8500	8506	|	1	1	2	2	3	4	4	5	6
71	8513	8519	8525	8531	8537	8543	8549	8555	8561	8567	|	1	1	2	2	3	4	4	5	5
72	8573	8579	8585	8591	8597	8603	8609	8615	8621	8627	|	1	1	2	2	3	4	4	5	5
73	8633	8639	8645	8651	8657	8663	8669	8675	8681	8686	|	1	1	2	2	3	4	4	5	5
74	8692	8698	8704	8710	8716	8722	8727	8733	8739	8745	|	1	1	2	2	3	4	4	5	5
75	8751	8756	8762	8768	8774	8779	8785	8791	8797	8802	|	1	1	2	2	3	4	4	5	5
76	8808	8814	8820	8825	8831	8837	8842	8848	8854	8859	|	1	1	2	2	3	3	4	5	5
77	8865	8871	8876	8882	8887	8893	8899	8904	8910	8915	|	1	1	2	2	3	3	4	4	5
78	8921	8927	8932	8938	8943	8949	8954	8960	8965	8971	|	1	1	2	2	3	3	4	4	5
79	8976	8982	8987	8993	8998	9004	9009	9015	9020	9025	|	1	1	2	2	3	3	4	4	5

	0	1	2	3	4	5	6	7	8	9	1	2	3	4	5	6	7	8	9
80	9031	9036	9042	9047	9053	9058	9063	9069	9074	9079	1	1	2	2	3	3	4	4	5
81	9085	9090	9096	9101	9106	9112	9117	9122	9128	9133	1	1	2	2	3	3	4	4	5
82	9138	9143	9149	9154	9159	9165	9170	9175	9180	9186	1	1	2	2	3	3	4	4	5
83	9191	9196	9201	9206	9212	9217	9222	9227	9232	9238	1	1	2	2	3	3	4	4	5
84	9243	9248	9253	9258	9263	9269	9274	9279	9284	9289	1	1	2	2	3	3	4	4	5
85	9294	9299	9304	9309	9315	9320	9325	9330	9335	9340	1	1	2	2	3	3	4	4	5
86	9345	9350	9355	9360	9365	9370	9375	9380	9385	9390	1	1	2	2	3	3	4	4	5
87	9395	9400	9405	9410	9415	9420	9425	9430	9435	9440	0	1	1	2	2	3	3	4	4
88	9445	9450	9455	9460	9465	9469	9474	9479	9484	9489	0	1	1	2	2	3	3	4	4
89	9494	9499	9504	9509	9513	9518	9523	9528	9533	9538	0	1	1	2	2	3	3	4	4
90	9542	9547	9552	9557	9562	9566	9571	9576	9581	9586	0	1	1	2	2	3	3	4	4
91	9590	9595	9600	9605	9609	9614	9619	9624	9628	9633	0	1	1	2	2	3	3	4	4
92	9638	9643	9647	9652	9657	9661	9666	9671	9675	9680	0	1	1	2	2	3	3	4	4
93	9685	9689	9694	9699	9703	9708	9713	9717	9722	9727	0	1	1	2	2	3	3	4	4
94	9731	9736	9741	9745	9750	9754	9759	9763	9768	9773	0	1	1	2	2	3	3	4	4
95	9777	9782	9786	9791	9795	9800	9805	9809	9814	9818	0	1	1	2	2	3	3	4	4
96	9823	9827	9832	9836	9841	9845	9850	9854	9859	9863	0	1	1	2	2	3	3	4	4
97	9868	9872	9877	9881	9886	9890	9894	9899	9903	9908	0	1	1	2	2	3	3	4	4
98	9912	9917	9921	9926	9930	9934	9939	9943	9948	9952	0	1	1	2	2	3	3	4	4
99	9956	9961	9965	9969	9974	9978	9983	9987	9991	9996	0	1	1	2	2	3	3	3	4

Appendix D

Answers to Odd-Numbered Exercises[1]

Chapter 1

1.1 b.

1.3 a.

1.5 b.

1.7 The demographics of the readership (age, income, race, gender, etc.) determines where advertisers will place their advertising dollars. For example, a manufacturer of accessory items for motorcycles would be most likely to place an ad in a magazine that appeals to a younger age group.

1.9 a) Gender. b) Employment status. c) Color of soap packaging. d) Brand of beer. e) Type of milk cow.

1.11 Individuals and groups with strong feelings against or in favor of an issue may consciously or unconsciously bias the results of the study. For example, they may select their sample from among friends and relatives who will probably not mirror the general population.

1.13 a) Discrete. b) Continuous. c) Discrete. d) Continuous.

1.15 Based on the survey results, you might confine your job hunting to smaller firms if salary is of primary concern.

1.17 a) Population. b) Data. c) Sample.

1.19 a) The population consists of all possible laboratory rats of the same strain used in the study.

    b) The sample is the 88 rats actually administered daily dosages of FD&C Red No. 2.

    c) The variable was dosage level (high versus low) of the red dye.

    d) The data were the status of each animal with respect to the presence of cancer.

1.21 a) Population: all of the bank's checking accounts. This sample is unlikely to mirror the population since the chances are they differ in many ways from accounts that show a great deal of monthly activity.

    b) Population: all people in the labor force of this particular industry. Since the study is being conducted by minority members, we may assume they hold strong feelings about this issue. Although they may attempt to conduct an objective study, they may unwittingly introduce their personal biases both in the collection and analysis of the data.

    c) Population: all members of this particular union. The question is worded in such a way that union members will probably vote *against* the proposal.

    d) Population: all employees of this particular firm. It is reasonable to assume that employee reactions will be highly influenced by the recent accident. Thus, the survey results may not reflect the true feelings in the population.

---

[1] Answers to many of these exercises (starting with Chapter 3) appear in the Study Guide. In addition, fully worked-out solutions to selected odd-numbered exercises are shown.

Chapter 2

2.1 a)

Industry	(Proportions)		
	Proprietorship	Partnerships	Corporations
Agriculture, etc.	0.31	0.11	0.03
Mining	0.01	0.02	0.01
Construction	0.09	0.05	0.09
Manufacturing	0.02	0.03	0.10
Transportation, etc.	0.03	0.02	0.04
Trade	0.20	0.18	0.31
Finance, etc.	0.07	0.41	0.20
Service	0.28	0.19	0.22
	1.01*	1.01*	1.00

* Slight discrepancy from 1.00 due to intermediate rounding.

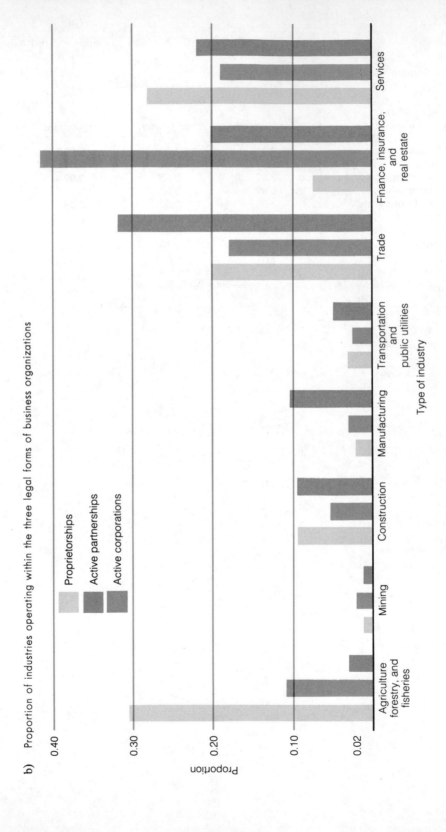

b) Proportion of industries operating within the three legal forms of business organizations

2.3 a) Approximate width $= \dfrac{741,272 - 194,846}{6}$

$= 91,071$

Using 90,000 as the width:

Class limits	f
190,000–279,999	11
280,000–369,999	1
370,000–459,999	4
460,000–549,999	4
550,000–639,999	3
640,000–729,999	1
	24

b) Histogram of frequency distribution of total takeoffs and landings at busiest airports in United States and Canada

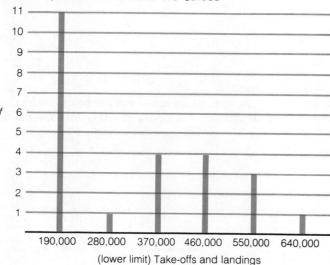

(lower limit) Take-offs and landings

2.5 a)

Manual	
Class limits	f
13–under 14	1
14–under 15	0
15–under 16	4
16–under 17	12
17–under 18	3
18–under 19	4
19–under 20	12
20–under 21	3
21–under 22	5
22–under 23	4
23–under 24	2
24–under 25	5
25–under 26	0
26–under 27	1
27–under 28	2
28–under 29	0
29–under 30	1
	59

Automatic	
Classes	f
10–under 11	8
11–under 12	19
12–under 13	16
13–under 14	10
14–under 15	0
15–under 16	10
16–under 17	2
	65

b) Frequency polygon (see accompanying chart).

It seems clear that automobiles with manual transmissions achieve better mileage ratings than those with automatic transmissions. Your chances of achieving 25 MPG with manual transmission are 4/59 or approximately 7 percent. You have *no* chance with automatic transmission.

Frequency polygons of gasoline mileage for automobiles with *manual* and *automatic* transmissions

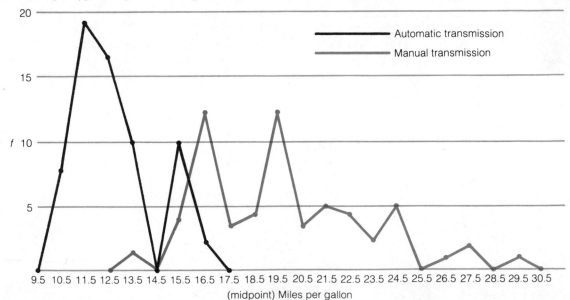

(midpoint) Miles per gallon

c)

Combined distribution	
Class limits	f
10–under 11	8
11–under 12	19
12–under 13	16
13–under 14	11
14–under 15	0
15–under 16	14
16–under 17	14
17–under 18	3
18–under 19	4
19–under 20	12
20–under 21	3
21–under 22	5
22–under 23	4
23–under 24	2
24–under 25	5
25–under 26	0
26–under 27	1
27–under 28	2
28–under 29	0
29–under 30	1
	124

d) Frequency polygon (see accompanying chart).

When we combine the data in this fashion, we no longer can see the superior miles-per-gallon ratings of the automobiles with manual transmissions.

Your chances of achieving at least 25 MPG are 4/124 or approximately 3 percent.

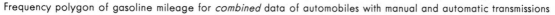

Frequency polygon of gasoline mileage for *combined* data of automobiles with manual and automatic transmissions

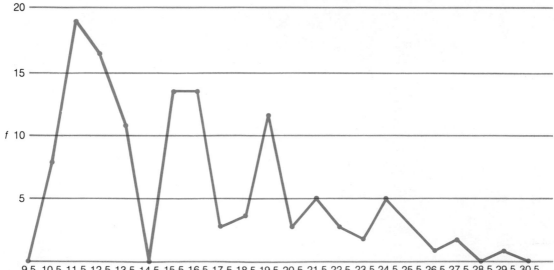

(midpoint) Miles per gallon

2.7 a)

Hours prior to replacement	f	cf
0—under 15	4	4
15—under 30	8	12
30—under 45	15	27
45—under 60	32	59
60—under 75	36	95
75—under 90	14	109
90—under 105	11	120
105—under 120	6	126

b)

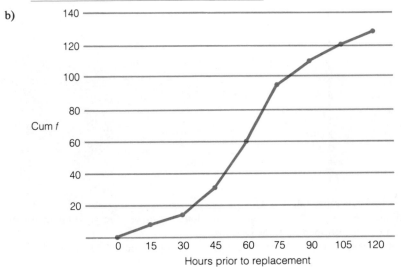

2.9 a)

Holder	Proportion	Degrees
Commercial banks	0.49	176.4
Finance companies	0.20	72.0
Credit unions	0.18	64.8
Retailers and others	0.13	46.8
	1.00	360.0

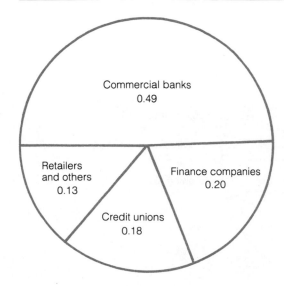

b)

Type	Proportion	Degrees
Automobiles...................	0.38	136.8
Mobile homes	0.07	25.2
Home improvement	0.06	21.6
Revolving	0.08	28.8
All others	0.41	147.6
	1.00	360.0

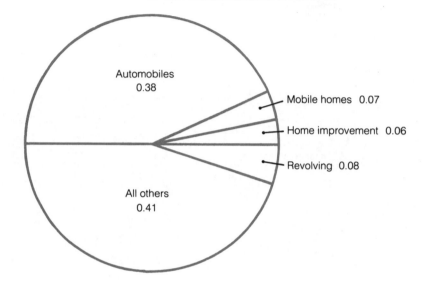

2.11

a)

Percent of total	Age group		
Year	16–17	18–20	24
1968–69	2.81	7.76	0.83
1969–70	4.29	6.44	1.16
1970–71	4.13	7.92	1.49
1971–72	5.45	21.95	2.31
1972–73	3.80	25.25	4.46

b)

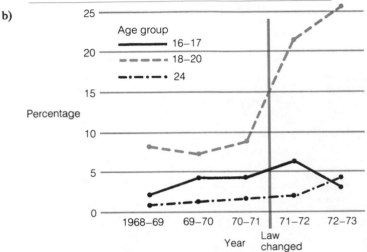

c) There seems to have been a precipitous rise in the proportion of alcohol-related collisions in the age group (18–20) affected by the ruling.

2.13 a) and b)

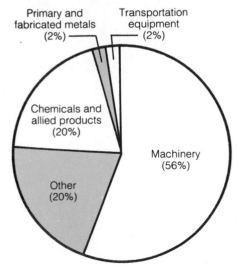

2.15

Percent distribution of all companies classified in
each enterprise industry category: 1972

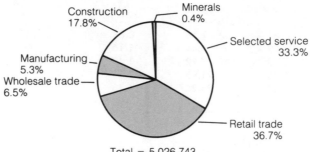

Construction
17.8%

Minerals
0.4%

Manufacturing
5.3%

Wholesale trade
6.5%

Selected service
33.3%

Retail trade
36.7%

Total = 5,026,743

Percent distribution of sales receipts of all
companies classified in each enterprise
industry category: 1972

Minerals
1.2%

Selected
service
5.7%

Construction
8.2%

Retail
trade
22.9%

Manufacturing
40.9%

Wholesale
trade
21.1%

Total = $1,992,451.9

Percent distribution of employment of all
companies classified in each enterprise
industry category: 1972

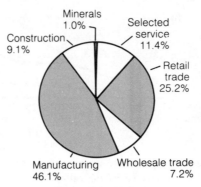

Minerals
1.0%

Selected
service
11.4%

Construction
9.1%

Retail
trade
25.2%

Manufacturing
46.1%

Wholesale trade
7.2%

Total = 45,810,804

2.17 a)

Class limits	f
360–under 390	13
390–under 420	14
420–under 450	4
450–under 480	0
480–under 510	3
510–under 540	2
540–under 570	3
	39

b)

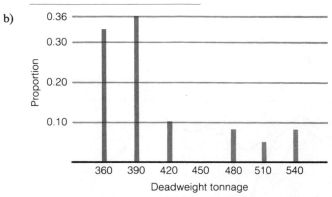

c) The proportion of tankers equipped to handle over 480,000 long tons is 8/39; therefore, you have approximately a 20 percent chance.

2.19 a)

Kind of transportation	Deaths per billion passenger miles
Passenger autos and taxis	12.99
Buses	1.73
Railroad passenger trains	1.27
Scheduled air transport planes	0.07

b)

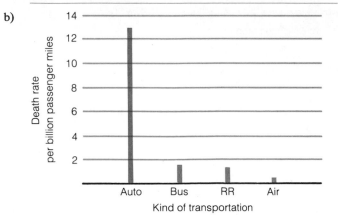

2.21 a)

	1973	1974	1975	1976
Jan	0.08	0.08	0.08	0.07
Feb	0.08	0.09	0.08	0.08
Mar	0.09	0.08	0.09	0.09
Apr	0.08	0.08	0.08	0.10
May	0.09	0.09	0.09	0.08
June	0.08	0.09	0.07	0.09
July	0.06	0.08	0.07	0.06
Aug	0.09	0.09	0.09	0.09
Sept	0.09	0.08	0.08	0.10
Oct	0.09	0.09	0.09	0.09
Nov	0.09	0.08	0.09	0.08
Dec	0.07	0.07	0.09	0.09
	0.99*	1.00	1.00	1.02*

* Slight disparity due to rounding

b)

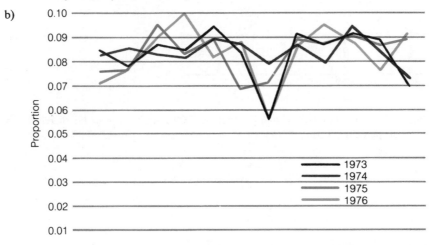

Athletic shoe production

2.23 a)

Class limits	f
600–under 700	5
700–under 800	6
800–under 900	6
900–under 1000	10
1000–under 1100	7
1100–under 1200	5
1200–under 1300	1
1300–under 1400	1
	41

b) Histogram of frequency distribution of gallons of water used daily in a four-unit building

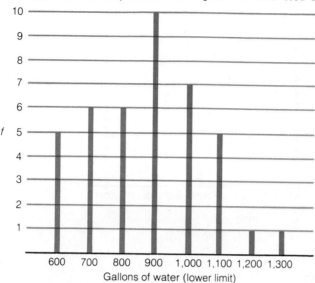

c) Since 14/41 exceed 1,000 gallons per day: 9.76 percent.

2.25 a)

	Proportion
Falls	0.31
Firearms	0.05
Fires, burns	0.22
Poison (solid, liquid)	0.12
Poison (gas)	0.04
Suffocation	0.11
Other	0.14

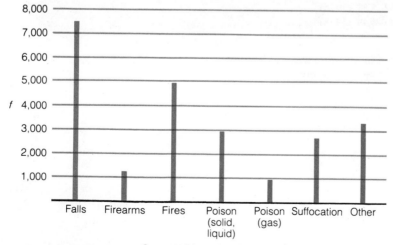

Causes of home accidental deaths

b)

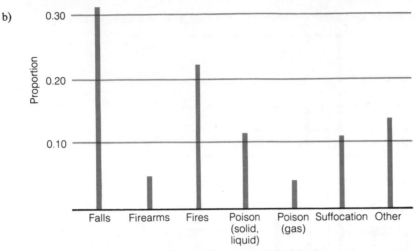

Causes of home accidental deaths

Chapter 3

3.1 a) $\bar{X} = 5$; $(-2, -1, 0, +1, +2 = 0)$.

 b) $\Sigma (X - \bar{X})^2 = 10$.

3.3 $\bar{X} = 16.37$.

3.5 13,075.

3.7 No, since he did not inquire about the quantities of each grade of carpeting. It may have turned out to be a good deal but he cannot know without this information. A decision in the absence of this information cannot be considered a good decision, even if it turned out to be "right."

3.9 a) Positively skewed. b) Negatively skewed. c) Symmetrical, bell-shaped.

3.11 a) Negatively skewed. b) Positively skewed. c) Symmetrical.

3.13 Food, range = 1912; Housing, range = 3372; Education, range = 411; Alcohol, range = 152.

3.15 Food, $s = 4.207$.

3.17 $13.50; $13.50

3.19 a) $4.43 \pm (2)(1.10) = 2.23–6.63$ (1973).

   $3.55 \pm (2)(0.94) = 1.67–5.43$ (1978).

 b) $4.43 \pm 1.10 = 3.33–5.53$ (1976).

   $3.55 \pm 0.94 = 2.61–4.49$ (1978).

Chapter 4

4.1 a) Empirical. b) Empirical. c) Subjective. d) Classical.

4.3 Zero, because it never happened before.

4.5 a) 0.0278.

 b) 0.0556.

 c) 0.0833.

 d) 0.1111.

 e) 0.1389.

 f) 0.1667.

 g) 0.1389.

 h) 0.1111.

 i) 0.0833.

 j) 0.0556.

 k) 0.0278.

The sum of the separate probabilities is 1.001; thus, allowing for rounding, the sum is 1.

4.7 $p(1) = 0.4929.$
$p(2) = 0.2400.$
$p(3) = 0.1200.$
$p(4) = 0.0588.$
$p(5) = 0.0294.$
$p(6) = 0.0143.$
$p(7) = 0.0071.$
$p(8) = 0.0035.$
$p(9) = 0.0017.$

4.9 $p(A \text{ or } B) = 35 \text{ percent} + 25 \text{ percent} = 60 \text{ percent}.$

4.11 a) 0.880.
b) 0.052.
c) 0.068.
d) 0.120.
e) 0.948.

4.13 A company granted an increase in 1978 has a slightly better chance of approval in 1980 ($p = 0.215$).

4.15 a) 0.95. b) 0.03. c) 0.08. d) 0.92. e) 0.05. f) 0.05. g) 0. h) 0.92.

4.17 a) 0.29.
b) 0.67.

4.19 a) The subjective approach. However, the commentator was a gifted retired tennis professional with a great deal of experience with shot-making and probabilities of success. His estimate should not be scoffed at.
b) Pat Summerall's rejoinder implies a common misconception that chance will necessarily even things out for each person.

4.21 $p(A \text{ and } B \text{ and } C \text{ and } D \text{ and } E) = (0.80)^5 = 0.33.$

4.23 a) No. b) No.

4.25 a) No. b) Yes.

4.27 0.80. The two events are independent.

4.29 a) (1,2) (1,3) (1,4) (1,5) (2,3) (2,4) (2,5) (3,4) (3,5) (4,5)
b) *(A)* = best worker chosen: (1,2) (1,3) (1,4) (1,5).
(B) = worst worker chosen: (1,5) (2,5) 3,5) (4,5).
c) 0.40.
d) 0.40.
e) 0.10.
f) No.
g) No.
h) 0.70.

4.31 94.12 percent.

4.33 a) 0.6141.
b) 0.7225.
c) 0.0034.

4.35 a) 0.1789.
b) 0.02.
c) 0.0036.

4.37 a) No.
b) Yes.
c) No.
d) No.
Gosden is using the subjective approach.

4.39 0.70.

4.41 0.000021.

Chapter 5

5.1 a) 0.3750. b) 0.1875. c) 0.1875.

5.3 No.

5.5 Since the expected return is positive, it will explore the possibility in greater depth.

5.7 Yes. $P = 0.05$, $Q = 0.95$, $n = 40$, $x = 5, 4, 3, 2, 1$, and 0; $(n - x) = 35, 36, 37, 38, 39$.

5.9 a) 0.4746.
b) 0.9829.

5.11 $P = 0.20$, $n = 10$.
a) $p(x \geq 5) = 0.0264 + 0.0055 + 0.0008 + 0.0001 = 0.0328$.
b) $p(x = 5) = 0.0264$.
c) $p(x = 0) = 0.1074$.

5.13 a) 0.9831.
b) 0.9999.

5.15 0.0657.

5.17 0.1558.

5.19 a) 0.2514.
b) 0.2514.
c) 0.5028.
d) 0.2528.
e) 0.4972.
f) 0.2472.

5.21 0.0749.

5.23 0.5000.

5.25 $498.50.

5.27 a) 0.1736.
b) 0.0011.
c) 519.

5.29 a) *(i)* 0.5987. *(ii)* 0.9999.
b) 1.0000.
c) 0.0016.

5.31 0.0228.

5.33 a) 0.5488.
b) 0.0034.

5.35 0.8815.

5.37 Money-market certificate.

5.39 $\mu = (8)(0.5) = 4.0$.
a) $p(x = 0) = 0.0183$.
b) $p(x = 1) = 0.0733$.
c) $p(x = 2) = 0.1465$.
d) $p(x = 3) = 0.1954$.
e) $p(x \geq 4) = 0.5665$.
Note: Since the events are mutually exclusive and exhaustive, $p(x \geq 4)$ can be obtained by $1 - [p(x = 0) + p(x = 1) + p(x = 2) + p(x = 3)]$.

5.41 $z = 13.42$.

5.43 $p = 0.0042$.

5.45 $z = 7.75$.

5.47 a) $2.50.
b) $-0.50.

5.49 a) 0.0392.
  b) 0.5267.
  c) 0.4607.
  d) 0.5392.

Chapter 6

6.1 Individuals and groups with strong feelings against or in favor of an issue may consciously or unconsciously bias the results of the study. For example, they may select their sample from among friends and relatives who will probably not be representative of the general population. They may phrase the questions in such a way as to bias the answer. Thus, the question, "Do you favor the continued planning and construction of nuclear-power plants?" may receive quite a different response from, "Do you favor the continued planning and construction of nuclear power plants even though the problem of the disposal of radioactive wastes has not been solved?"

6.3 This is an example of sampling error. In this case, the economist's data would reflect variations caused by an uncontrolled factor—government increases in a particular month.

6.5 The population of interest would be all NYSE stock market prices. A sample would probably be more appropriate since the time and cost involved to conduct a complete census would be prohibitive.

Stock market prices are extremely volatile—they seem to go up, down, and sideways as a result of any number of uncontrolled events. On any given day, one might find stock prices plummeting in the morning and then showing a sharp recovery by the end of the day, and vice versa. Hints of war, changes in OPEC oil prices, federal banking regulations, election results are just a small handful of variables that have been known to have sizable impact on the stock market.

6.7 This is convenience sampling. The population has not been clearly defined. What is a "typical" citizen? The interviewees are not likely to be representative of the entire population of the community (if that is the population of interest). The n is not sufficiently large, even if it were representative.

6.9 a) 0.9968. b) 0.9278. c) 0.8584. d) 0.5408.

6.11 a) 0.3174. b) 0.0013. c) $z = -5.00$.

6.13 a) 0.9974.
  b) 0.8806.

6.15 a) The population: all luggage in the new line. A sample would be more appropriate since the test is likely to destroy the luggage; a census would put the manufacturer out of business.
  b) Population: all pregnant women and all infants born. Since we are dealing with a population that is simply too large (and, in a theoretical sense, infinite), it would not be feasible to attempt a census. A sample from this population should be carefully selected and studied.
  c) Population: all registered voters who will vote. In this case, timeliness is perhaps the key factor influencing data collection via a sample.
  d) Population: all registered owners of automobiles. A sample would certainly be preferable to a census for several reasons. For one, many members of this population will either not respond or simply be inaccessible. Further, it is doubtful that the high costs involved in attempting a complete census could possibly be justified. A carefully selected sample will certainly suffice in this instance.

6.17 a) 0.7776.
  b) 0.9198.
  c) 0.9996.

6.19 a) 0.9030.
  b) 0.8882.

6.21 a) 0.0032.
  b) 0.1416.

6.23 Answers will vary.

6.25 Since the Los Angeles County unemployment rate is not seasonally adjusted, we should be cautious about conclusions drawn from this kind of information. This reported "surge" may simply be a reflection of sampling error.

6.27 No. A piece of limburger cheese is a homogeneous population. One part is very much like every other part. Thus, a single sample is likely to be representative of the whole. Books, by good as well as by poor authors, are not uniform. A good book may include many poor pages. Conversely, even a poor book may possess some good pages.

6.29 In all likelihood, many children will destroy, lose, or forget to deliver the questionnaire. Even if all the questionnaires were delivered, the survey leaves out voters who do not have children in public schools.

6.31 a) 0.9292; 0.0001. b) 0.9901; 0.00003.

6.35 a) Answers will vary.

    b) (1) 0.1335. (2) $z = \pm 5.56$. (3) 0.1203. (4) 0.00003.

    c) (1) 0.00003. (2) 0.9525. (3) 0.0467. (4) 0.0475.

Chapter 7

7.1 a) Estimator. b) Estimate. c) Estimate. d) Estimator. e) Estimate. f) Estimator.

7.3 a) Lack of bias. b) Efficiency. c) Consistency.

7.5 a) $213.75 \pm (1.96)(3.75) = 206.40$ to 221.10.

    b) $213.75 \pm (2.58)(3.75) = 204.07$ to 223.43.

    c) The more confident, the wider the interval.

7.7 a) $n = 16$, df $= 15$, $t = 2.131$, $s_{\bar{x}} = 27.5$.

       $450 \pm (2.131)(27.5) = 391.40$ to 508.60.

    b) $n = 25$, df $= 24$, $t = 2.064$, $s_{\bar{x}} = 24$.

       $450 \pm (2.064)(24) = 400.46$ to 499.54.

    Increased sample size narrows the interval.

7.9 a) 21.01 to 21.99. b) 63,303 to 63,597.

7.11 a) 10.64. b) 6.24 to 15.04.

7.13 a) 0.0476 to 0.0723. b) 48 to 72.

7.15 a) 2.392. b) 383.

7.17 $200 \pm (1.96) \left(\dfrac{30}{\sqrt{50}} \right)$: 191.68 to 208.32.

7.19 ± 1.08.

7.21 0.97 to 7.03.

7.23 222.

7.25 a) 13.38 to 18.06. b) 12.42 to 18.92.

7.27 a) 0.0163 to 0.0237. b) 12.225 to 17,775.

7.29 431.12 to 438.88.

7.31 a) 0.1233 to 0.1789. b) 4,533,000; 3,699,000 to 5,367,000.

7.33 a) 0.197 to 0.345.

    b) 49.25 to 86.25.

7.35 a) 0.7941 to 0.8259.

    b) 0.7891 to 0.8309.

Chapter 8

8.1 a) No error since a false H_0 was rejected.

    b) Type I error since a true H_0 was rejected.

    c) Type II error since a false H_0 was not rejected.

    d) No error since a true H_0 was not rejected.

8.3 a) One-tailed. b) One-tailed. c) Two-tailed. d) Two-tailed. e) One-tailed.

8.5 He has satisfied a rigorous criterion for rejecting H_0 and asserting that the program significantly reduced time lost due to accidents. However, in the real world, nothing is proved "beyond a shadow of a doubt." Occasionally, a statistically significant finding is due to the operation of chance factors. By stating a rigorous criterion for rejecting H_0, we will reduce the incidence of Type I errors but we will not eliminate them altogether.

8.7 $t = -5.817$ (two-tailed).

8.9 $t = 5.837$ (one-tailed).

8.11 $z = 17.42$ (two-tailed).

8.13 a) $z = 5.71$.

   b) $z = 9.54$.

   c) $z = 2.57$.

8.15 c) $z = 15.17$.

8.17 $z = 2.42$.

8.19 $t = -3.75$ (one-tailed).

8.21 $t = 14.545$ (one-tailed).

8.23 $z = -1.72$ (one-tailed).

8.25 $z = 2.97$ (one-tailed).

8.27 a) $z = -2.42$ (two-tailed).

   b) $z = 1.35$ (two-tailed).

   c) $z = -11.81$ (two-tailed).

8.29 a) $z = 2.34$.

   b) $z = 4.79$.

   c) $z = -1.73$.

8.31 $z = 15.08$.

Chapter 9

9.1 The answers given below are a few possible states of nature and payoff possibilities:

	Possible states of nature	Possible payoffs
a)	Weather conditions	Time spent on travel Number of people to see candidate along the route Nonrenewable energy resources expended.
b)	Economic conditions (improved, same, worse)	Profit
	Quality of prepublication review (favorable, neutral, unfavorable)	Profit
c)	Type of remuneration offered (salary, commission, piecework)	Income, personal satisfaction
	Advancement opportunities (high, medium, low)	Income, personal satisfaction
d)	Anticipated demand (or any factor affecting demand, such as oil cutoffs, costs of crude oil, etc.)	Profit
e)	Availability of mortgage money (high, medium, low)	Cost
	Possible changes in mortgage rates	Cost

9.3 a) Maximin. b) Maximax.

9.5 a) Taking a plane.

   b) Taking a plane.

   c) 22.95.

9.7 a) Building 3.

   b) $412.50.

   d) 5 percent: building 3; 10 percent: building 3; 15 percent: building 2; 20 percent: building 1.

   e) 264.48.

9.9 b) EVSI = $412.50.

9.11 Accept.

9.13 Call a tug.

9.15 a) Posterior probabilities (based on a prediction of 11 million): 0.0294; 0.2206; 0.7500.

b) 7,212.50.

9.17 a) Gross income ($000)

Number of bales raised	Number of bales contracted for		
	1,000	1,500	2,000
1,000	$450	$425	$400
1,500	640	675	650
2,000	830	865	900

b) Conditional profit ($000)

Number of bales raised	Number of bales contracted for		
	1,000	1,500	2,000
1,000	$110	$ 85	$ 60
1,500	130	165	140
2,000	150	185	220

c) Maximin: contract for 1,000 bales.

d) Maximax: contract for 2,000 bales.

9.19 a) Posterior probabilities: 0.1538; 0.7692; 0.0769.

b) Expected profit from 1,000-bale contract is $128,840; from 1,500-bale contract, $154,220; from 2,000-bale contract, $133,830.

c) Contract for 1,500 bales.

9.21 If $x \leq 2$: accept; if $x > 2$: reject.

9.23 b) Maximin: Buy 1,000.

Maximax: Buy 3,000.

c) Buy 2,000.

d) Buy 2,000.

e) $8,300.

9.25 b) 0.35; 0.385; 0.265.

c) Buy 1,000; buy 2,000; buy 3,000.

d) $1349.48. Firm 1.

9.27   a) and b)

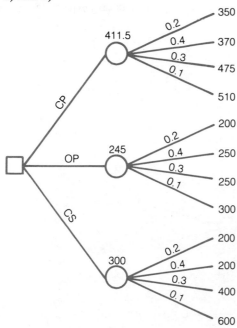

c)   The newspaper should select the offset press (OP) since the expected cost is less than with the other alternatives.

9.29   a)   Accommodate all customers in one restaurant versus opening the second restaurant.

b)   The size of the crowd (number of people waiting).
The length of time customers will have to wait for service.
The capability of opening the second restaurant, for example, in terms of adequate help.

c)   The economic consequences of opening the second restaurant, i.e., profit realized in a given period of time.
The expected monetary gain (or expected opportunity loss) of the two actions.
The loss at the gambling tables and slot machines if the customers spend too much time getting breakfast.
Utility in terms of satisfying the customers.

d)   Monetary gain: expected profit less the cost of opening the second restaurant compared to expected profit realized from operating only one restaurant.
Utility: satisfaction or dissatisfaction of the customers.

9.31 a) The revised probabilities are shown in parentheses on the final branches.
   b) The probabilities of the various reviews are shown on the branches relating to favorable, mediocre, and unfavorable.
   c)

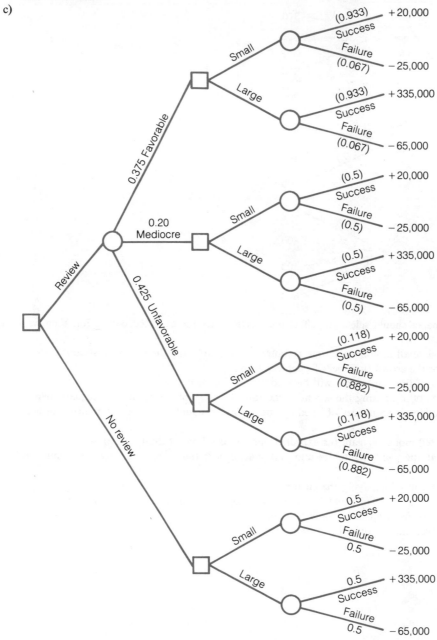

Chapter 10

10.1 $F = 1.05$.

10.3 $F = 19.45$.

HSD $= 3.10$.

10.5 a) Inspection of the graph suggests the possible effects of both the A and B variables but no interaction

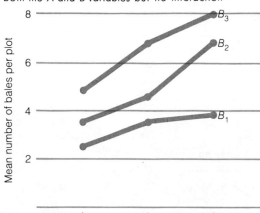

b) $(A \times B)$: $F = 1.42$; (A): $F = 15.96$; (B): $F = 38.74$.

c) HSD $= 1.17$.

10.7 (Blocks): $F = 2.83$; (treatments): $F = 3.75$.

10.9 a)

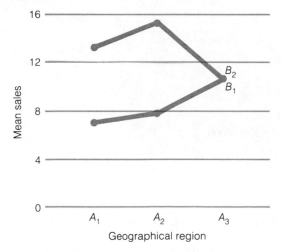

b) $(A \times B)$: $F = 1.22$; (A): $F = 0.08$; (B): $F = 5.16$.

10.11 (Blocks): $F = 1.81$; (treatments): $F = 2.19$.

10.13 a)

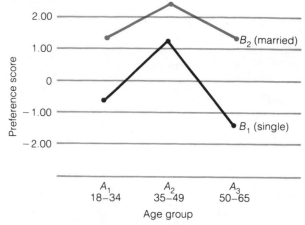

b) $(A \times B): F = 9.34; (A): F = 18.73: (B): F = 28.79.$

10.15 a)

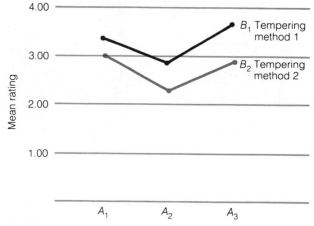

b) $(A \times B): F = 2.67; (A): F = 35.17; (B): F = 69.00.$

Chapter 11

11.1 $\chi^2 = 4.0888$, df $= 3$.

11.3 $\chi^2 = 3.768$, df $= 3$.

11.5 $\chi^2 = 11.0917$.

11.7 a) Ph.D.s in business and industry have statistically significant higher median salaries than those in educational institutions.

b) The median income of Ph.D.s in business and industry is statistically significantly higher than the total population of Ph.D.s.

c) $n = 11$, $p(x \geq 10)$ or $p[(n - x) \leq 1] = 0.0118$. Thus, the median salary of those working in educational institutions is statistically significantly below the total population of Ph.D.s.

11.9 $n_1 = 16$, $R_1 = 214$ (ranking from low—i.e., -17.55 rank of 1—to high—i.e., $+20.90$, rank of 26).

$n_2 = 10$ $R_2 = 137$.

$U_1 = 82$, $U_2 = 78$ Fail to reject H_0.

11.11 Runs test, $n_1 = 16$, $n_2 = 16$, $R = 14$.

Since obtained R does not fall in the critical region, we cannot reject H_0. There is no statistical basis for asserting that this sequence is not random.

11.13 $\chi^2 = 13.2383$.

11.15 Sign test. $n = 14$, $x = 10$, $p(x \geq 10) = 0.0839$. Accept H_0.

11.17 $\chi^2 = 33.196$.

11.19 Runs test. $n_1 = 7$, $n_2 = 7$. $R = 2$.
Reject H_0 and assert that these data do not follow a random sequence.

11.21 a) $\chi^2 = 84.656$.
b) $\chi^2 = 45.029$.

11.23 $z = 2.61$.

11.25 Runs test. $n_1 = 5$, $n_2 = 5$. $R = 2$.
Reject H_0. We may assert that the sequence does not appear to be random.

11.27 Runs test. $n_1 = 10$, $n_2 = 10$. $R = 4$. Reject H_0.
The behavior of the prime rate following the recession does not follow a random sequence.

11.29 $z = 0.91$.

11.31 $\chi^2 = 27.528$.

11.33 $\chi^2 = 23{,}854.263$.

11.35 a) Sign test. $n = 13$ (one-tie), $x = 13$, $p = 0.0002$. Reject H_0. The percentage of mining contracts containing these safety provisions is significantly higher than in manufacturing contracts.
b) Sign test. $n = 12$ (two ties), $x = 10$, $p = 0.0384$. Accept H_0.
c) Sign test. $n = 13$, $x = 12$, $p = 0.0034$. Reject H_0.

11.37 $z = 1.57$.

11.39 a) $z = 2.11$.
b) $z = -1.61$.
c) $z = -2.61$.

11.41 Sign test. $n = 22$. $x = 12$. Accept H_0. No statistical basis for asserting a difference between these two measures.

Chapter 12

12.1   **a)**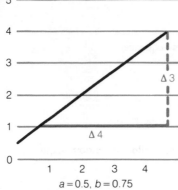

$a = 0.5, b = 0.75$

b)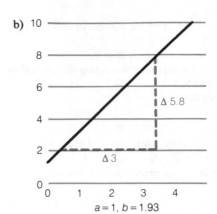

$a = 1, b = 1.93$

c)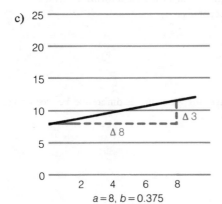

$a = 8, b = 0.375$

d)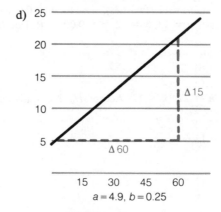

$a = 4.9, b = 0.25$

Note: These are estimated from line graphs and are only approximate. Your answers may differ slightly from the above.

12.3   Visual inspection of the scatter diagram suggests that a linear model is not appropriate.

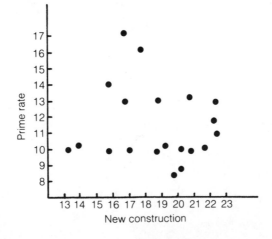

12.5   a) and b)

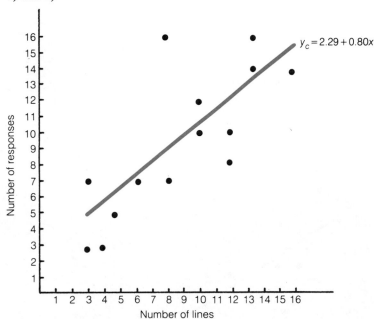

$y_c = 2.29 + 0.80x$

c) $X = 7,\ Y_c = 7.89.$
$X = 11,\ Y_c = 11.09.$
$X = 15,\ Y_c = 14.29.$

12.7   a)   8.02 to 13.60.
b)   2.59 to 19.03.

12.9 $r = 0.87.$

12.11   a) $r = 0.35.$
b) $r^2 = 0.12.$

12.13   For Exercise 12.10: $t = 1.324.$ For Exercise 12.11: $t = 2.303.$

12.15   a) and b)

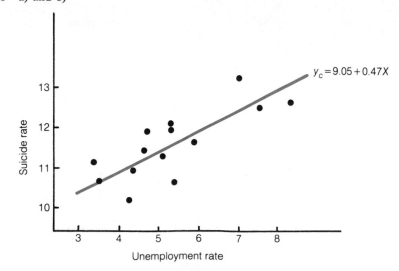

$y_c = 9.05 + 0.47X$

c) $r^2 = 0.6345$.

d) $r = 0.80$.

e) $t = 4.619$.

f) $s_{Y \cdot X} = 0.53$.

g) 12.18 to 13.44.

h) $t = 4.608$.

12.17 a) The data appear consistent with a linear model.

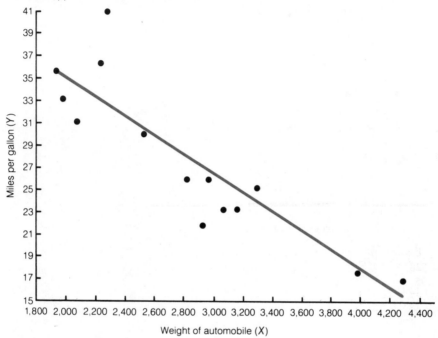

b) $Y_c = 51.74 - 0.009X$.

c) $s_{Y \cdot X} = 7.25$.

d) $t = -3.00$.

12.19 a) 24.07 to 34.41.

b) 12.28 to 46.20.

12.21 a) and b)   The data appear to conform to a linear model.

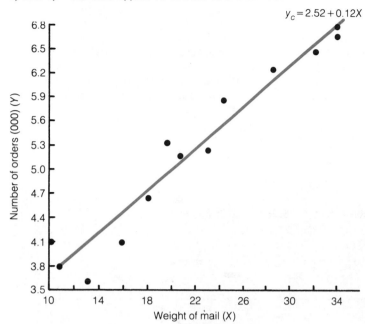

$y_c = 2.52 + 0.12X$

Number of orders (000) (Y) — vertical axis
Weight of mail (X) — horizontal axis

c) $X = 0$ lies outside the range of sample values.
d) $t = 4.528$.

12.23 a) $r = 0.95$.
b) $t = 9.127$, reject H_0.
c) $r^2 = 0.90$.

12.25 a) $Y_c = 8.43 + 0.35X$.
b) $s_{Y \cdot X} = 2.74$.
c) $r^2 = 0.0204$.
d) $r = \sqrt{0.0204} = 0.14$.
e) $s_r = 0.2221$.

$$t = \frac{0.14}{0.2221} = 0.630.$$

f)   Accept H_0

Chapter 13

13.1   No. The results do not show that sales of nondrug items *cause* losses. It is also possible that they contribute to profits in an indirect way. Customers may be attracted to the pharmacies because they offer nondrug items.

13.3   Since Y_c increases by 1.44 for every unit of increase in X_2 when X_1 is held constant and only by 0.307 with each unit of increase in X_1 when X_2 is held constant, work conditions appear to be a better predictor of productivity levels.

13.5   a) 4.02. b) 2.99. c) 2.78. d) 3.56.

13.7   a) $\Sigma Y^2 = 91.03$; $a = 2.726$; $\Sigma Y = 22.1$; $b_1 = 0.118$; $b_2 = 1.066$; $\Sigma X_1 Y = 1363.34$; $\Sigma X_2 Y = 160.56$; df = 5; $s_{y \cdot x_1 x_2} = 0.354$.
b) $3.10 \pm (4.604)(0.354) = 1.53$ to 4.79.

13.9   a)   Multiple $R^2 = 0.779$.
b)   The percentage of variation in Y accounted for by the regression plane. 79.9 percent.
c)   20.1 percent.

13.11 a) 0.84.

b) 84 percent.

13.13 a, b, and d. Possibly c.

13.15 All are likely to involve serial correlation.

13.17 Yes, multicollinearity. The two independent variables (number of employees and total wages) are likely to be highly correlated.

13.19 The three measures are serially correlated. Moreover, EPA also regulates these emissions. Effective enforcement would be expected to produce correlated decreases in all three measures of pollution.

13.21 a) 2.73.

b) 3.43.

c) 2.13.

d) 3.23.

e) 2.61.

Chapter 14

14.1 a) $\Sigma Y = 6088$; $a = \bar{Y} = 608.8$; $\Sigma xY = 8812$; $b = 26.7$; $\Sigma x^2 = 330$.

$x = 0$ for 1973½

$Y_t = 608.8 + 26.7x$.

b)

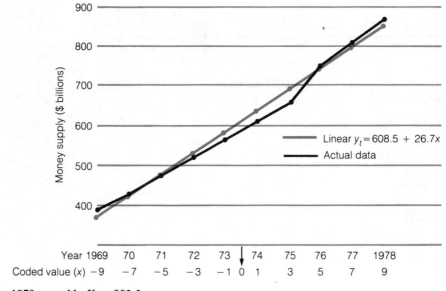

c) 1979; $x = 11$; $Y_t = 902.5$.

14.3 a) and b)

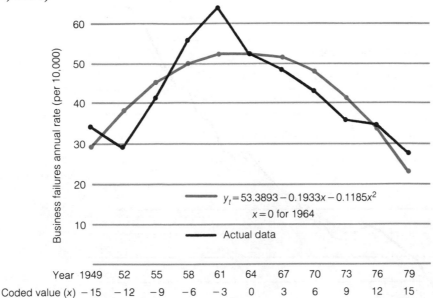

$$y_t = 53.3893 - 0.1933x - 0.1185x^2$$
$$x = 0 \text{ for } 1964$$

Actual data

Year	1949	52	55	58	61	64	67	70	73	76	79
Coded value (x)	-15	-12	-9	-6	-3	0	3	6	9	12	15

c) 1982; $x = 18$: $Y_t = 11.5$.

14.5

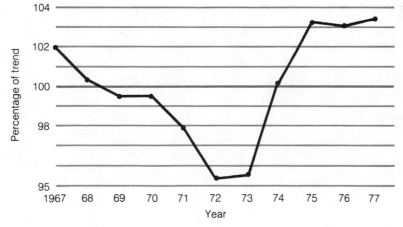

14.7 We use the moving average method. Since these yield annual averages, seasonal effects are canceled out since they occur only within a year.

14.9 The percent-of-moving-average contains the seasonal components and possibly some cyclical and irregular components.

14.11 We deseasonalize data by seasonally adjusting the data. This is accomplished by dividing the original data by the seasonal index and multiplying by 100.

14.13 a) $x = 0$ for 1969½.
$\Sigma Y = 4056$, $\overline{Y} = 202.8$.
$\Sigma xY = 2708$.
$\Sigma x^2 = 2660$.
$Y_t = 202.8 + 1.018x$.
b) See Figure 14.1.

14.15 d)

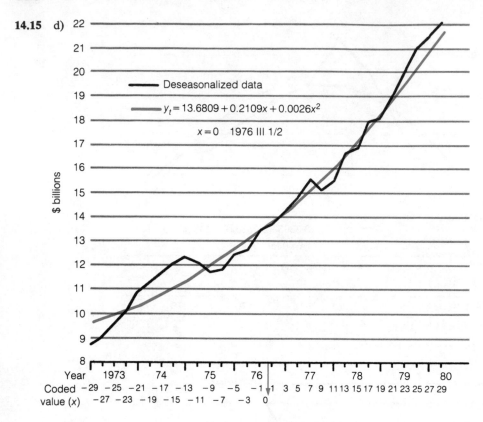

e)

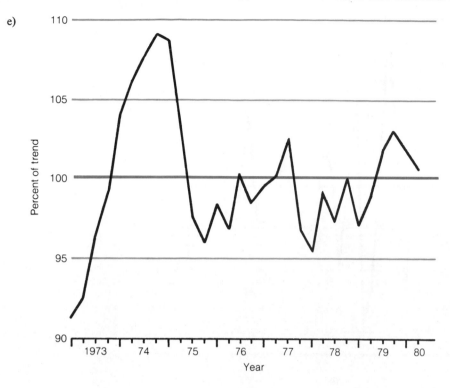

These figures reflect cyclical fluctuations primarily since trend and seasonal variation have been removed.

f)   1980 III: $x = 31$; $Y_t = 22.72$; seasonal index = 99.6; seasonally adjusted amount = 22.63.
     1980 IV: $x = 33$; $Y_t = 23.47$; seasonal index = 112.5; seasonally adjusted amount = 26.40.

14.17   b)

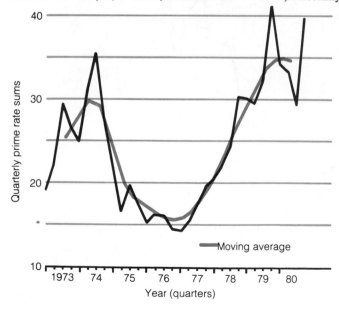

d)

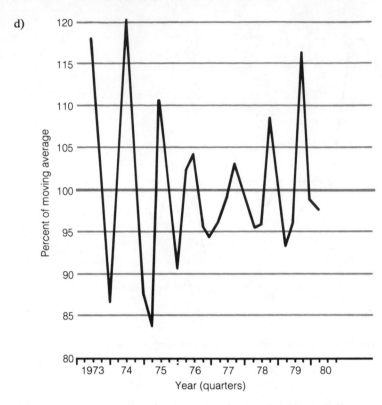

g)

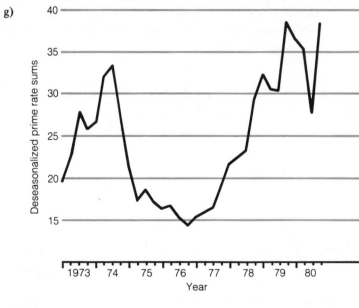

14.19 a) $Y_t = (432.1)(1.12)^x$; $x = 0$ in 1975½.
   b) 1980, $x = 9$, $Y_t = 1198.2$.

14.21 a)

Forecast 1	Forecast 2
MAD $= 1,551$	MAD $= 704$
SAD $= 4,553,330$	SAD $= 901,660$
Tracking signal $= -11.81$	Tracking signal $= -8.41$

  b) Forecast 2.

  c) The errors in Forecast 1 do not appear to be random. They appear to increase in magnitude with time. Thus, they are serially correlated.

Chapter 15

15.1 To keep records of changes over time of almost any variable that interests us including those in business, economics, education, sociology, psychology, etc; to act as barometers of the business climate; to assess long and short term trends; to influence legislation, etc.

15.3 Price and quantity indexes (or percentage relatives) show the changes that take place over time in the price goods and services and the quantities produced or sold. The value index takes into account both price and quantity.

15.5 Aggregate indexes permit us to track over time the behavior of broad classes of events—consumer prices, crime rates, producer prices, quantities manufactured, stocks traded, etc. If the items making up the index can be considered representative of the broader class of items, we can feel confident that the percentage relatives reflect the behavior of the entire class. Aggregate indexes may be used to forecast trends in finances, business, economics, etc.

15.7 An average-of-relatives index is independent of the units in terms of which the components or items are measured.

15.9 193.

15.11 Simple average-of-relatives index $= \dfrac{3.21}{4} \times 100 = 80$.

15.13 Base year is used for both price and quantity; the quantity in the given year is used as the weight in the denominator (Paasche index); the quantity in the base year is used as weight in the numerator; fixed weights based on averages or different base periods are used in the index.

15.15 The Laspeyres Index does not require fresh inputs of quantity weights for each period calculated. It permits direct comparison of index numbers for different periods. However, it may not be responsive to changed patterns of production or consumption.

15.17 a) Simple weighted aggregate index $= \dfrac{3003329.4}{2056512.8} = 146$.

  b) Paasche index $= \dfrac{3003329.4}{1602272.5} = 187$.

  c) Laspeyres index $= \dfrac{3852462.5}{2056512.8} = 187$.

15.19 78.

15.21

Year	Output per hour (1975 = 100)
1960	50.5
1965	83.2
1970	88.4
1973	100.5
1974	95.3
1975	100.0
1976	104.2
1977	106.7
1978	109.3

15.23 That output per hour is lagging far behind total production. The use of two different time bases can lead to this sort of confusion. The year selected for output per hour is one during which the output was high. Conversely, the index for total production uses a year when production was lowest.

15.25 $\dfrac{77.90}{66.89} \times 100 = 116.5.$

15.27

Year	Index number
1930	187.2
1935	128.5
1940	152.2
1945	142.5
1950	77.8
1955	101.2
1960	100.0
1965	101.6
1970	105.6
1975	41.6

15.29 $\dfrac{10464}{14695} \times 100 = 71.2.$

15.31 In 1975, the index numbers for Japan and the United Kingdom were relatively high. Thus, using these years as base periods, the hourly output for other years would appear to be relatively less. Note, however, that the ratios of the index numbers remained unchanged. For example, in Japan, when 1967 was used as the base year, the ratio of hourly output in 1978 to that of 1960 was 215.7/52.6 = 4.10. Similarly, in the revised index using 1975 as the base period, the ratio is 123.5/30.1 = 4.10.

15.33 $\dfrac{19,127,896.56}{6,751,149.81} \times 100 = 283.3.$

15.35 288.0

Index

This book has been set VideoComp in 10 and 9 point Times Roman, leaded 2 points. Chapter titles are 27 point Futura Demibold. The size of the overall type page is 36 by 49 picas.